$53.50

2007 NATIONAL REPAIR & REMODELING ESTIMATOR

by Albert S. Paxton
Edited by J. A. O'Grady

Includes inside the back cover:

- An estimating CD with all the costs in this book, plus,
- An estimating program that makes it easy to use these costs,
- An interactive video guide to the National Estimator program,
- A program that converts your estimates into invoices,
- A program that exports your estimates to QuickBooks Pro.

Monthly price updates on the Web are free and automatic all during 2007. You'll be prompted when it's time to collect the next update. A connection to the Web is required.

Download all of Craftsman's most popular costbooks for one low price with the Craftsman Site License. http://www.craftsmansitelicense.com

Craftsman Book Company
6058 Corte del Cedro / P.O. Box 6500 / Carlsbad, CA 92018

Preface

The author has corresponded with manufacturers and wholesalers of building material supplies and surveyed retail pricing services. From these sources, he has developed Average Material Unit Costs which should apply in most parts of the country.

Wherever possible, the author has listed Average Labor Unit Costs which are derived from the Average Manhours per Unit, the Crew Size, and the Wage Rates used in this book. Please read How to Use This Book for a more in-depth explanation of the arithmetic.

If you prefer, you can develop your own local labor unit costs. You can do this by simply multiplying the Average Manhours per Unit by your local crew wage rates per hour. Using your actual local labor wage rates for the trades will make your estimate more accurate.

What is a realistic labor unit cost to one reader may well be low or high to another reader, because of variations in labor efficiency. The Average Manhours per Unit figures were developed by time studies at job sites around the country. To determine the daily production rate for the crew, divide the total crew manhours per day by the Average Manhours per Unit.

The subject topics in this book are arranged in alphabetical order, A to Z. To help you find specific construction items, there is a complete alphabetical index at the end of the book, and a main subject index at the beginning of the book.

This manual shows crew, manhours, material, labor and equipment cost estimates based on Small Volume work, then a total cost and a total including overhead and profit. Only total cost and total including overhead and profit are shown for Large Volume. For those readers who need it, the breakdown for Large Volume is included on the National Estimator CD in the back of the book. No single price fits all repair and remodeling jobs. Generally, work done on smaller jobs costs more per unit installed and work on larger jobs costs less. The estimates in this book reflect that simple fact. The two estimates you find for each work item show the author's opinion of the likely range of costs for most contractors and for most jobs. So, which cost do you use, Low Volume or High Volume?

The only right price is the one that gets the job and earns a reasonable profit. Finding that price always requires estimating judgment. Use Small Volume cost estimates when some or most of the following conditions are likely:

- The crews won't work more than a few days on site.

- Better quality work is required.

- Productivity will probably be below average.

- Volume discounts on materials aren't available.

- Bidding is less competitive.

- Your overhead is higher than most contractors.

When few or none of those conditions apply, use Large Volume cost estimates.

Credits and Acknowledgments

This book has over 12,000 cost estimates for 2007. To develop these estimates, the author and editors relied on information supplied by hundreds of construction cost authorities. We offer our sincere thanks to the contractors, engineers, design professionals, construction estimators, material suppliers and manufacturers who, in the spirit of cooperation, have assisted in the preparation of this 30th edition of the *National Repair & Remodeling Estimator*. Our appreciation is extended to those listed below.

American Standard Products
DAP Products
Outwater Plastic Industries
Con-Rock Concrete
Georgia Pacific Products

Kohler Products
Wood Mode Cabinets
Transit Mixed Concrete
U.S. Gypsum Products
Henry Roofing Products

Special thanks to: Dal-Tile Corporation, 1713 Stewart, Santa Monica, California

About the Author

Albert Paxton is president of Professional Construction Analysts, Inc., located at 5823 Filaree Heights Ave., Malibu, CA 90265-3944; alpaxton@pca-group.net; fax (310) 589-0400. Mr. Paxton is a California licensed General Contractor (B1-425946). PCA's staff is comprised of estimators, engineers and project managers who are also expert witnesses, building appraisers and arbitrators operating throughout the United States.

PCA clients include property insurance carriers, financial institutions, self-insureds, and private individuals. The expertise of **PCA** is in both new and repair-remodel construction, both commercial and residential structures. In addition to individual structures, **PCA** assignments have included natural disasters such as Hurricanes Hugo, Andrew and Iniki, the Northridge earthquake in California, Hurricanes Charley, Frances, Ivan and Jeanne striking Florida and the Southeastern states, and the catastrophic Hurricane Katrina, whose destruction in the Gulf Coast will be felt in the building and repair industry for years to come.

Cover design by Bill Grote

Cover photos by Ron South Photography & Ed Kessler Studios

Main Subject Index

How to Use This Book

					Costs Based on Small Volume					Large Volume	
1	2	3	4	5	6	7	8	9	10	11	12
Description	Oper	Unit	Crew Size	Man-Hours Per Unit	Avg Mat'l Unit Cost	Avg Labor Unit Cost	Avg Equip Unit Cost	Avg Total Unit Cost	Avg Price Incl O&P	Avg Total Unit Cost	Avg Price Incl O&P

The descriptions and cost data in this book are arranged in a series of columns, which are described below. The cost data is divided into two categories: Costs Based On Large Volume and Costs Based On Small Volume. These two categories provide the estimator with a pricing range for each construction topic.

The Description column (1) contains the pertinent, specific information necessary to make the pricing information relevant and accurate.

The Operation column (2) contains a description of the construction repair or remodeling operation being performed. Generally the operations are Demolition, Install, and Reset.

The Unit column (3) contains the unit of measurement or quantity which applies to the item described.

The Crew Size column (4) contains a description of the trade that usually installs or labors on the specified item. It includes information on the labor trade that installs the material and the typical crew size. Letters and numbers are used in the abbreviations in the crew size column. Full descriptions of these abbreviations are in the Crew Compositions and Wage Rates table, beginning on page 15.

The Manhours Per Unit column (5) is for the listed operation and listed crew.

The units per day in this book don't take into consideration unusually large or small quantities. But items such as travel, accessibility to work, experience of workers, and protection of undamaged property, which can favorably or adversely affect productivity, have been considered in developing Average Manhours per Unit. For further information about labor, see "Notes — Labor" in the Notes Section of some specific items.

The Average Material Unit Cost column (6) contains an average material cost for products (including, in many cases, the by-products used in installing the products) for small volume. It doesn't include an allowance for sales tax, delivery charges, overhead and profit. Percentages for waste, shrinkage, or coverage have been taken into consideration unless indicated. For other information, see "Dimensions" or "Installation" in the Notes Section.

If the item described has many or very unusual by-products which are essential to determining the Average Material Unit Cost, the author has provided examples of material pricing. These examples are placed throughout the book in the Notes Section.

Average Daily Production and Average Material Unit Cost should assist the estimator in:

1. Checking prices quoted by others.

2. Developing local prices.

You should verify labor rates and material prices locally. Though the prices in this book are average material prices, prices vary from locality to locality. A local hourly wage rate should normally include taxes, benefits, and insurance. Some contractors may also include overhead and profit in the hourly rate.

The Average Labor Unit Cost column (7) contains an average labor cost for small volume based on the Average Manhours per Unit and the Crew Compositions and Wage Rates table. The average labor unit cost equals the Average Manhours per Unit multiplied by the Average Crew Rate per hour. The rates include fringe benefits, taxes, and insurance. Examples that show how to determine the average labor unit cost are provided in the Notes Section.

The Average Equipment Unit Cost column (8) contains an average equipment cost for small volume based on both the average daily rental and the cost per day if owned and depreciated. The costs of daily maintenance and the operator are included.

The Average Total Unit Cost column for Small Volume (9) includes the sum of the Material, Equipment, and Labor Cost columns. It doesn't include an allowance for overhead and profit. The Average Total Unit Cost column for Large Volume (11) has the same information, although you don't see those columns in the book. The National Estimator CD in the back of the book does include the Material, Equipment, and Labor Cost columns for Large Volume. Check it if you want to see how we arrived at the Total Unit Cost and Price Including Overhead & Profit for Large Volume.

The Average Total Price Including Overhead and Profit columns (10 and 12) result from adding an overhead and profit allowance to Total Cost. This allowance reflects the author's interpretation of average fixed and variable overhead expenses and the labor intensiveness of the operation vs. the costs of materials for the operation. This allowance factor varies throughout the book, depending on the operation. Each contractor interprets O&P differently. The range can be from 15% to 80% of the Average Total Unit Cost.

Estimating Techniques

Estimating Repair/Remodeling Jobs: The unforeseen, unpredictable, or unexpected can ruin you.

Each year, the residential repair and remodeling industry grows. It's currently outpacing residential new construction due to increases in land costs, labor wage rates, interest rates, material costs, and economic uncertainty. When people can't afford a new home, they tend to remodel their old one. And there are always houses that need repair, from natural disasters or accidents like fire. The professional repair and remodeling contractor is moving to the forefront of the industry.

Repair and remodeling spawns three occupations: the contractor and his workers, the insurance company property claims adjuster, and the property damage appraiser. Each of these professionals shares common functions, including estimating the cost of the repair or remodeling work.

Estimating isn't an exact science. Yet the estimate determines the profit or loss for the contractor, the fairness of the claim payout by the adjuster, and the amount of grant or loan by the appraiser. Quality estimating must be uppermost in the mind of each of these professionals. And accurate estimates are possible only when you know exactly what materials are needed and the number of manhours required for demolition, removal, and installation. Remember, profits follow the professional. To be profitable you must control costs — and cost control is directly related to accurate, professional estimates.

There are four general types of estimates, each with a different purpose and a corresponding degree of accuracy:

- The guess method: "All bathrooms cost $5,000." or "It looks like an $8,000 job to me."

- The per measure method: (I like to call it the surprise package.) "Remodeling costs $60 per SF, the job is 500 SF, so the price is $30,000."

These two methods are the least accurate and accomplish little for the adjuster or the appraiser. The contractor might use the methods for qualifying customers (e.g., "I thought a bathroom would only cost $2,000."), but never as the basis for bidding or negotiating a price.

- The *piece estimate* or *stick-by-stick* method.

- The *unit cost estimate* method.

These two methods yield a detailed estimate itemizing all of the material quantities and costs, the labor manhours and wage rates, the subcontract costs, and the allowance for overhead and profit.

Though time-consuming, the detailed estimate is the most accurate and predictable. It's a very satisfactory tool for negotiating either the contract price or the adjustment of a building loss. The piece estimate and the unit cost estimate rely on historical data, such as manhours per specific job operation and recent material costs. The successful repair and remodeling contractor, or insurance/appraisal company, maintains records of previous jobs detailing allocation of crew manhours per day and materials expended.

While new estimators don't have historical data records, they can rely on reference books, magazines, and newsletters to estimate manhours and material costs. It is important to remember that **the reference must pertain to repair and remodeling**. This book is designed *specifically* to meet this requirement.

The reference material must specialize in repair and remodeling work because there's a large cost difference between new construction and repair and remodeling. Material and labor construction costs vary radically with the size of the job or project. Economies of scale come into play. The larger the quantity of materials, the better the purchase price should be. The larger the number of units to be installed, the greater the labor efficiency.

Repair and remodeling work, compared to new construction, is more expensive due to a normally smaller volume of work. Typical repair work involves only two or three rooms of a house, or one roof. In new construction, the job size may be three to five complete homes or an entire development. And there's another factor: a lot of repair and remodeling is done with the house occupied, forcing the crew to work around the normal, daily activities of the occupants. In new construction, the approach is systematic and logical — work proceeds from the ground up to the roof and to the inside of the structure.

Since the jobs are small, the repair and remodeling contractor doesn't employ trade specialists. Repairers employ the "jack-of-all-trades" who is less specialized and therefore less efficient. This isn't to say the repairer is less professional than the trade specialist. On the contrary, the repairer must know about many more facets of construction: not just framing, but painting, finish carpentry, roofing, and electrical as well. But because the repairer has to spread his expertise over a greater area, he will be less efficient than the specialist who repeats the same operation all day long.

Another factor reducing worker efficiency is poor access to the work area. With new construction, where building is an orderly "from the ground up" approach, workers have easy access to the work area for any given operation. The workers can spread out as much as needed, which facilitates efficiency and minimizes the manhours required to perform a given operation.

The opposite situation exists with repair and remodeling construction. Consider an example where the work area involves fire damage on the second floor. Materials either go up through the interior stairs or through a second story window. Neither is easy when the exterior and interior walls have a finished covering such as siding and drywall. That results in greater labor costs with repair and remodeling because it takes more manhours to perform many of the same tasks.

If, as a professional estimator, you want to start collecting historical data, the place to begin is with daily worker time sheets that detail:

1. total hours worked by each worker per day

2. what specific operations each worker performed that day

3. how many hours (to the nearest tenth) each worker used in each operation performed that day.

Second, you must catalog all material invoices daily, being sure that quantities and unit costs per item are clearly indicated.

Third, maintain a record of overhead expenses attributable to the particular project. Then, after a number of jobs, you'll be able to calculate an average percentage of the job's gross amount that's attributable to overhead. Many contractors add 45% for overhead and profit to their total direct costs (direct labor, direct material, and direct subcontract costs). But that figure may not be right for your jobs.

Finally, each week you should reconcile in a job summary file the actual costs versus the estimated costs, and determine why there is any difference. This information can't immediately help you on this job since the contract has been signed, but it will be invaluable to you on your next job.

Up to now I've been talking about general estimating theory. Now let's be more specific. On page 8 is a Building Repair Estimate form. Each line is keyed to an explanation. A filled-out copy of the form is also provided, and on page 10, a blank, full-size copy that you can reproduce for your own use.

You can adapt the Building Repair Estimate form, whether you're a contractor, adjuster, or appraiser. Use of the form will yield a detailed estimate that will identify:

- The room or area involved, including sizes, dimensions and measurements.

- The kind and quality of material to be used.

- The quantities of materials to be used and verification of their prices.

- The type of work to be performed (demolish, remove, install, remove and reset) by what type of crew.

- The crew manhours per job operation and verification of the hourly wage scale.

- All arithmetical calculations that can be verified.

- Areas of difference between your estimate and others.

- Areas that will be a basis for negotiation and discussion of details.

Each job estimate begins with a visual inspection of the work site. If it's a repair job, you've got to see the damage. Without a visual inspection, you can't select a method of repair and you can't properly evaluate the opinions of others regarding repair or replacement. With either repair or remodeling work, the visual inspection is essential to uncover the "hiders" — the unpredictable, unforeseen, and unexpected problems that can turn profit into loss, or simplicity into nightmare. You're looking for the many variables and unknowns that exist behind an exterior or interior wall covering.

Along with the Building Repair Estimate form, use this checklist to make sure you're not forgetting anything.

Checklist

- Site accessibility: Will you store materials and tools in the garage? Is it secure? You can save a half hour to an hour each day by storing tools in the garage. Will the landscaping prevent trucks from reaching the work site? Are wheelbarrows or concrete pumpers going to be required?

- Soil: What type and how much water content? Will the soil change your excavation estimate?

- Utility lines: What's under the soil and where? Should you schedule the utilities to stake their lines?

- Soundness of the structure: If you're going to remodel, repair or add on, how sound is that portion of the house that you're going to have to work around? Where are the load-bearing walls? Are you going to remove and reset any walls? Do the floor joists sag?

- Roof strength: Can the roof support the weight of another layer of shingles. (Is four layers of composition shingles already too much?)

- Electrical: Is another breaker box required for the additional load?

This checklist is by no means complete, but it is a start. Take pictures! A digital camera will quickly pay for itself. When you're back at the office, the picture helps reconstruct the scene. Before and after pictures are also a sales tool representing your professional expertise.

During the visual inspection always be asking yourself "what if" this or that happened. Be looking for potential problem areas that would be extremely labor intensive or expensive in material to repair or replace.

Also spend some time getting to know your clients and their attitudes. Most of repair and remodeling work occurs while the house is occupied. If the work will be messy, let the homeowners know in advance. Their satisfaction is your ultimate goal — and their satisfaction will provide you a pleasant working atmosphere. You're there to communicate with them. At the end of an estimate and visual inspection, the homeowner should have a clear idea of what you can or can't do, how it will be done, and approximately how long it will take. Don't discuss costs now! Save the estimating for your quiet office with a print-out calculator and your cost files or reference books.

What you create on your estimate form during a visual inspection is a set of rough notes and diagrams that make the estimate speak. To avoid duplications and omissions, estimate in a *systematic sequence of inspection*. There are two questions to consider. First, where do you start the estimate? Second, in what order will you list the damaged or replaced items? It's customary to start in the room having either the most damage or requiring the most extensive remodeling. The sequence of listing is important. Start with either the floor or the ceiling. When starting with the floor, you might list items in the following sequence: Joists, subfloor, finish floor, base — listing from bottom to top. When starting with the ceiling, you reverse, and list from top to bottom. The important thing is to be *consistent* as you go from room to room! It's a good idea to figure the roof and foundation separately, instead of by the room.

After completing your visual inspection, go back to your office to cost out the items. Talk to your material supply houses and get unit costs for the quantity involved. Consult your job files or reference books and assign crew manhours to the different job operations.

There's one more reason for creating detailed estimates. Besides an estimate, what else have your notes given you? A material take-off sheet, a lumber list, a plan and specification sheet — the basis for writing a job summary for comparing estimated costs and profit versus actual costs and profit — and a project schedule that minimizes down time.

Here's the last step: Enter an amount for overhead and profit. No matter how small or large your work volume is, be realistic — everyone has overhead. An office,

even in your home, costs money to operate. If family members help out, pay them. Everyone's time is valuable!

Don't forget to charge for performing your estimate. A professional expects to be paid. You'll render a better product if you know you're being paid for your time. If you want to soften the blow to the owner, say the first hour is free or that the cost of the estimate will be deducted from the job price if you get the job.

In conclusion, whether you're a contractor, adjuster, or appraiser, you're selling your personal service, your ideas, and your reputation. To be successful you must:

● Know yourself and your capabilities.

● Know what the job will require by ferreting out the "hiders."

● Know your products and your work crew.

● Know your productivity and be able to deliver in a reasonable manner and within a reasonable time frame.

● Know your client and make it clear that all change orders, no matter how large or small, will cost money.

National Estimator '07 Inside the back cover of this book you'll find an envelope with a compact disk. The disk has National Estimator, an easy-to-use estimating program with all the cost estimates in this book. Insert the CD in your computer and wait a few seconds. Installation should begin automatically. (If not, click Start, Settings, Control Panel, double-click Add/Remove Programs and Install.) Select ShowMe from the installation menu and Dan will show you how to use National Estimator. When ShowMe is complete, select Install Program. When the National Estimator program has been installed, click Help on the menu bar, click Contents, click Print all Topics, click File and click Print Topic to print a 28-page instruction manual for National Estimator.

Building Repair Estimate

Insured	John Q. Smith		Date		
Loss Address	123 A. Main St.		Claim or Policy No. DP 0029	Page 1 of 2	
City	Anywhere, Anystate 00010		Home Ph. 555-1241	Cause of Loss Fire	
Bldg. R.C.V. 100,000	Bldg. A.C.V. 80,000		Bus. Ph. 555-1438	Other Ins. Y (N)	
			Insurance Amount $100,000		
Insurance Required R.C.V. (80%) A.C.V.(80%)					

Description of Item	Unit Cost or Material Price Only			Labor Price Only		
	Unit	Unit Price	Total (Col. A)	Hours	Rate	Total Col. B
Install 1/2" sheetrock (standard,) on walls, including tape and finish	400 (*page 104*)	0.43	172.00	9.6	48.14	462.14
Paint walls, roller, smooth finish						
1 coat sealer 600 (*page 205*)	600	0.07	42.00	4.2	49.33	207.19
2 coats latex flat 600 (*page 207*)	600	0.19	114.00	7.2	47.92	345.02
Totals			328.00			1014.35

Total Column A		328.00
		1342.35
6% Tax		80.54
		1422.89
10% Overhead		142.29
10% Profit		142.29
Grand Total		1707.47

This is not an order to Repair
The undersigned agrees to complete and guarantee repairs at a total of $
Repairer ABC Construction
Street 316 E. 2nd Street
City Anywhere   Phone
By Jack Williams
Adjuster Stan Jones   Date of A/P N/A
Adj. License No. (If Any) 561-84
Service Office Name Phoenix

Note: This form does not replace the need for field notes, sketches and measurements.

Building Repair Estimate

Insured (1)			Date (33)		
Loss Address (4)			Claim or Policy No. (2)	Page of (3)	
City			Home Ph. (5)	Cause of Loss (5)	
Bldg. R.C.V. (7)	Bldg. A.C.V. (8)		Bus. Ph.	Other Ins. Y N (6)	
			Insurance Amount (9)		
Insurance Required R.C.V. (10 %) A.C.V.(11 %) (12)					

Description of Item	Unit Cost or Material Price Only			Labor Price Only		
	Unit	Unit Price	Total (Col. A)	Hours	Rate	Total Col. B
(20)	(13)	(14)	(15)	(17)	(18)	(19)
Totals (21)			(22)			

Total Column A		(23)
		(24)
		(25)
		(26)
		(27)
Grand Total		(28)

This is not an order to Repair
The undersigned agrees to complete and guarantee repairs at a total of $ (29)
Repairer (30)
Street (31)
City (32)   Phone
By
Adjuster (34)   Date of A/P (35)
Adj. License No. (If Any)
Service Office Name (36)

Note: This form does not replace the need for field notes, sketches and measurements.

Keyed Explanations of the Building Repair Estimate Form

1. Insert name of insured(s).

2. Insert claim number or, if claim number is not available, insert policy number or binder number.

3. Insert the page number and the total number of pages.

4. Insert street address, city and state where loss or damage occurred.

5. Insert type of loss (wind, hail, fire, water, etc.)

6. Check YES if there is other insurance, whether collectible or not. Check NO if there's only one insurer.

7. Insert the present replacement cost of the building. What would it cost to build the structure today?

8. Insert present actual cash value of the building.

9. Insert the amount of insurance applicable. If there is more than one insurer, insert the total amount of applicable insurance provided by all insurers.

10. If the amount of insurance required is based on replacement cost value, circle RCV and insert the percent required by the policy, if any.

11. If the amount of insurance required is based on actual cash value, circle ACV and insert the percent required by the policy, if any.

   Note: (regarding 10 and 11) if there is a non-concurrency, i.e., one insurer requires insurance to 90% of value while another requires insurance to 80% of value, make a note here. Comment on the non-concurrency in the settlement report.

12. The installed price and/or material price only, as expressed in columns 13 through 15, may include any of the following (expressed in units and unit prices):

   Material only (no labor)

   Material and labor to replace

   Material and labor to remove and replace

   Unit Cost is determined by dividing dollar cost by quantity. The term cost, as used in unit cost, is not intended to include any allowance, percentage or otherwise, for overhead or profit. Usually, overhead and profit are expressed as a percentage of cost. Cost must be determined first. Insert a line or dash in a space(s) in columns 13, 14, 15, 17, 18 or 19 if the space is not to be used.

13. The *units* column includes both the quantity and the unit of measure, i.e., 100 SF, 100 BF, 200 CF, 100 CY, 20 ea., etc.

14. The *unit price* may be expressed in dollars, cents or both. If the units column has 100 SF and if the unit price column has $.10, this would indicate the price to be $.10 per SF.

15. The *total* column is merely the dollar product of the quantity (in column 13) times the price per unit measure (in column 14).

16-19. These columns are normally used to express labor as follows: hours times rate per hour. However, it is possible to express labor as a unit price, i.e., 100 SF in column 13, a dash in column 17, $.05 in column 18 and $5.00 in column 19.

20. Under *description of item*, the following may be included:

   Description of item to be repaired or replaced (studs 2" x 4" 8'0" #2 Fir, Sheetrock 1/2", etc.)

   Quantities or dimensions (20 pcs., 8'0" x 14'0", etc.)

   Location within a room or area (north wall, ceiling, etc.)

   Method of correcting damage (paint - 1 coat; sand, fill and finish; R&R; remove only; replace; resize; etc.)

21-22. Dollar totals of columns A and B respectively.

23-27. Spaces provided for items not included in the body of the estimate (subtotals, overhead, profit, sales tax, etc.)

28. Total cost of repair.

29. Insert the agreed amount here. The agreement may be between the claim representative and the insured or between the claim rep and the repairer. If the agreed price is different from the grand total, the reason(s) for the difference should be itemized on the estimate sheet. If there is no room, attach an additional estimate sheet.

30. PRINT the name of the insured or the repairer so that it is legible.

31. PRINT the address of the insured or repairer legibly. Include phone number.

32. The insured or a representative of the repairer should sign here indicating agreement with the claim rep's estimate.

33. Insured or representative of the repairer should insert date here.

34. Claim rep should sign here.

35. Claim rep should insert date here.

36. Insert name of service office here.

Building Repair Estimate

Date

Insured	Claim or Policy No.		Page of	
Loss Address	Home Ph.		Cause of Loss	
City	Bus. Ph.		Other Ins. Y N	
Building. R.C.V. Bldg. A.C.V.	Insurance Amount			

Insurance Required R.C.V.(%) A.C.V.(%)	Unit Cost or Material Price Only			Labor Price Only		
Description of Item	Unit	Unit Price	Total (Col. A)	Hours	Rate	Total Col. B)

THIS IS NOT AN ORDER TO REPAIR **TOTALS**						
The undersigned agrees to complete and guarantee repairs at a total of $	**Total Column A**		
Repairer			
Street			
City Phone			
By			
Adjuster Date of A/P			
Adj. License No. (If Any)	**Grand Total**		
Service Office Name			

Note: This form does not replace the need for field notes, sketches and measurements.

Wage Rates Used in This Book

Wage rates listed here and used in this book were compiled in the fall of 2006 and projected to mid-2007. Wage rates are in dollars per hour.

"Base Wage Per Hour" (Col. 1) includes items such as vacation pay and sick leave which are normally taxed as wages. Nationally, these benefits average 5.15% of the Base Wage Per Hour. This amount is paid by the Employer in addition to the Base Wage Per Hour.

"Liability Insurance and Employer Taxes" (Cols. 3 & 4) include national averages for state unemployment insurance (4.00%), federal unemployment insurance (0.80%), Social Security and Medicare tax (7.65%), lia-bility insurance (2.29%), and Workers' Compensation Insurance which varies by trade. This total percentage (Col. 3) is applied to the sum of Base Wage Per Hour and Taxable Fringe Benefits (Col. 1 + Col. 2) and is list-ed in Dollars (Col. 4). This amount is paid by the Employer in addition to the Base Wage Per Hour and the Taxable Fringe Benefits.

"Non-Taxable Fringe Benefits" (Col. 5) include employer-sponsored medical insurance and other bene-fits, which nationally average 4.55% of the Base Wage Per Hour.

"Total Hourly Cost Used In This Book" is the sum of Columns 1, 2, 4, & 5.

	1	2	3	4	5	6
Trade	Base Wage Per Hour	Taxable Fringe Benefits (5.15% of Base Wage)	Liability Insurance & Employer Taxes %	Liability Insurance & Employer Taxes $	Non-Taxable Fringe Benefits (4.55% of Base Wage)	Total Hourly Cost Used in This Book
Air Tool Operator	$21.23	$1.09	32.93%	$7.35	$0.97	$30.64
Bricklayer or Stone Mason	$23.84	$1.23	25.54%	$6.40	$1.08	$32.55
Bricktender	$18.26	$0.94	25.54%	$4.90	$0.83	$24.93
Carpenter	$22.42	$1.15	31.83%	$7.50	$1.02	$32.09
Cement Mason	$22.67	$1.17	23.35%	$5.57	$1.03	$30.44
Electrician, Journeyman Wireman	$26.30	$1.35	20.04%	$5.54	$1.20	$34.39
Equipment Operator	$26.37	$1.36	25.42%	$7.05	$1.20	$35.98
Fence Erector	$23.90	$1.23	26.21%	$6.59	$1.09	$32.81
Floorlayer: Carpet, Linoleum, Soft Tile	$21.88	$1.13	24.02%	$5.53	$1.00	$29.54
Floorlayer: Hardwood	$22.97	$1.18	24.02%	$5.80	$1.05	$31.00
Glazier	$21.73	$1.12	25.98%	$5.94	$0.99	$29.78
Laborer, General Construction	$18.46	$0.95	32.93%	$6.39	$0.84	$26.64
Lather	$23.57	$1.21	21.48%	$5.32	$1.07	$31.17
Marble and Terrazzo Setter	$21.06	$1.08	21.56%	$4.77	$0.96	$27.87
Painter, Brush	$23.79	$1.23	25.08%	$6.28	$1.08	$32.38
Painter, Spray-Gun	$24.50	$1.26	25.08%	$6.46	$1.11	$33.33
Paperhanger	$24.98	$1.29	25.08%	$6.59	$1.14	$34.00
Plasterer	$23.24	$1.20	28.78%	$7.03	$1.06	$32.53
Plumber	$26.84	$1.38	24.47%	$6.91	$1.22	$36.35
Reinforcing Ironworker	$22.07	$1.14	28.81%	$6.69	$1.00	$30.90
Roofer, Foreman	$25.17	$1.30	44.34%	$11.74	$1.15	$39.36
Roofer, Journeyman	$22.88	$1.18	44.34%	$10.67	$1.04	$35.77
Roofer, Hot Mop Pitch	$23.57	$1.21	44.34%	$10.99	$1.07	$36.84
Roofer, Wood Shingles	$24.02	$1.24	44.34%	$11.20	$1.09	$37.55
Sheet Metal Worker	$25.98	$1.34	26.21%	$7.16	$1.18	$35.66
Tile Setter	$23.26	$1.20	21.56%	$5.27	$1.06	$30.79
Tile Setter Helper	$17.91	$0.92	32.93%	$6.20	$0.81	$25.84
Truck Driver	$19.29	$0.99	26.42%	$5.36	$0.88	$26.52

Area Modification Factors

Construction costs are higher in some areas than in other areas. Add or deduct the percentages shown on the following pages to adapt the costs in this book to your job site. Adjust your cost estimate by the appropriate percentages in this table to find the estimated cost for the site selected. Where 0% is shown, it means no modification is required.

Modification factors are listed alphabetically by state and province. Areas within each state are listed by the first three digits of the postal zip code. For convenience, one representative city is identified in each three-digit zip or range of zips. Percentages are based on the average of all data points in the table.

Factors listed for each state and province are the average of all data points in that state or province. Figures for three-digit zips are the average of all five-digit zips in that area, and are the weighted average of factors for labor, material and equipment.

The National Estimator program will apply an area modification factor for any five-digit zip you select. Click Utilities. Click Options. Then select the Area Modification Factors tab.

These percentages are composites of many costs and will not necessarily be accurate when estimating the cost of any particular part of a building. But when used to modify costs for an entire structure, they should improve the accuracy of your estimates.

Alabama	-7%
368 Auburn	-10%
369 Bellamy	-9%
350-352 Birmingham	1%
363 Dothan	-10%
364 Evergreen	-15%
359 Gadsden	-11%
358 Huntsville	-4%
355 Jasper	-12%
365-366 Mobile	1%
360-361 Montgomery	-5%
357 Scottsboro	-7%
367 Selma	-11%
356 Sheffield	-4%
354 Tuscaloosa	-7%

Alaska	30%
995 Anchorage	40%
997 Fairbanks	36%
998 Juneau	29%
999 Ketchikan	12%
996 King Salmon	32%

Arizona	-6%
865 Chambers	-17%
855 Douglas	-10%
860 Flagstaff	-10%
864 Kingman	-1%
852-853 Mesa	3%
850 Phoenix	4%
863 Prescott	-7%
859 Show Low	-9%
856-857 Tucson	-6%

Arkansas	-10%
725 Batesville	-16%
717 Camden	-5%
727 Fayetteville	-5%
729 Fort Smith	-7%
726 Harrison	-18%
718 Hope	-12%
719 Hot Springs	-14%
724 Jonesboro	-12%
720-722 Little Rock	-4%
716 Pine Bluff	-7%
728 Russellville	-10%
723 West Memphis	-10%

California	10%
917-918 Alhambra	11%
932-933 Bakersfield	1%
922 El Centro	3%
955 Eureka	-1%
936-938 Fresno	0%
961 Herlong	4%

902-905 Inglewood	12%
926-927 Irvine	16%
934 Lompoc	5%
907-908 Long Beach	13%
900-901 Los Angeles	11%
959 Marysville	1%
953 Modesto	1%
935 Mojave	8%
949 Novato	16%
945-947 Oakland	21%
928 Orange	15%
930 Oxnard	6%
910-912 Pasadena	12%
956-957 Rancho Cordova	10%
960 Redding	1%
948 Richmond	20%
925 Riverside	5%
958 Sacramento	10%
939 Salinas	7%
923-924 San Bernardino	6%
919-921 San Diego	11%
941 San Francisco	28%
950-951 San Jose	20%
943-944 San Mateo	22%
931 Santa Barbara	11%
954 Santa Rosa	11%
952 Stockton	7%
940 Sunnyvale	22%
913-916 Van Nuys	11%
906 Whittier	11%

Colorado	1%
800-801 Aurora	9%
803-804 Boulder	6%
808-809 Colorado Springs	2%
802 Denver	9%
813 Durango	-5%
807 Fort Morgan	-2%
816 Glenwood Springs	6%
814-815 Grand Junction	-1%
806 Greeley	5%
805 Longmont	3%
811 Pagosa Springs	-7%
810 Pueblo	-7%
812 Salida	-8%

Connecticut	13%
066 Bridgeport	14%
060 Bristol	16%
064 Fairfield	16%
061 Hartford	15%
065 New Haven	13%
063 Norwich	10%

068-069 Stamford	18%
067 Waterbury	13%
062 West Hartford	6%

Delaware	6%
199 Dover	-2%
197 Newark	10%
198 Wilmington	9%

District of Columbia	12%
200-205 Washington	12%

Florida	-3%
327 Altamonte Springs	0%
342 Bradenton	-1%
346 Brooksville	-6%
321 Daytona Beach	-9%
333 Fort Lauderdale	6%
339 Fort Myers	0%
349 Fort Pierce	-7%
326 Gainesville	-9%
322 Jacksonville	1%
338 Lakeland	-7%
329 Melbourne	-3%
330-332 Miami	5%
341 Naples	5%
344 Ocala	-10%
328 Orlando	4%
325 Panama City	-11%
325 Pensacola	-8%
320 Saint Augustine	-3%
347 Saint Cloud	-1%
337 St Petersburg	-3%
323 Tallahassee	-6%
335-336 Tampa	0%
334 West Palm Beach	2%

Georgia	-5%
317 Albany	-12%
306 Athens	-5%
303 Atlanta	19%
308-309 Augusta	-7%
305 Buford	0%
307 Calhoun	-3%
318-319 Columbus	-11%
310 Dublin/Fort Valley	-10%
313 Hinesville	-5%
315 Kings Bay	-11%
312 Macon	-3%
300-302 Marietta	7%
314 Savannah	-1%
304 Statesboro	-14%
316 Valdosta	-12%

Hawaii	31%
968 Aliamanu	33%

967 Ewa	30%
967 Halawa Heights	30%
967 Hilo	30%
968 Honolulu	33%
968 Kailua	33%
967 Lualualei	30%
967 Mililani Town	30%
967 Pearl City	30%
967 Wahiawa	30%
967 Waianae	30%
967 Wailuku (Maui)	30%

Idaho	-7%
837 Boise	1%
838 Coeur d'Alene	-7%
834 Idaho Falls	-11%
835 Lewiston	-9%
836 Meridian	-7%
832 Pocatello	-11%
833 Sun Valley	-7%

Illinois	6%
600 Arlington Heights	22%
605 Aurora	22%
622 Belleville	-2%
617 Bloomington	3%
629 Carbondale	-6%
601 Carol Stream	22%
628 Centralia	-8%
618 Champaign	2%
606-608 Chicago	22%
623 Decatur	-6%
614 Galesburg	-7%
620 Granite City	-2%
612 Green River	0%
604 Joliet	19%
609 Kankakee	3%
624 Lawrenceville	-8%
603 Oak Park	24%
615-616 Peoria	7%
613 Peru	3%
602 Quincy	23%
610-611 Rockford	9%
625-627 Springfield	1%
619 Urbana	-3%

Indiana	-2%
470 Aurora	-8%
474 Bloomington	-2%
472 Columbus	-2%
465 Elkhart	0%
476-477 Evansville	2%
467-468 Fort Wayne	0%
463-464 Gary	8%

460-462 Indianapolis 8%
475 Jasper . . . .-8%
471 Jeffersonville . . . . . .-4%
469 Kokomo . . . .-5%
479 Lafayette . . . .-5%
473 Muncie . . . .-7%
466 South Bend 1%
478 Terre Haute . . . .-6%

Iowa   **-4%**
526 Burlington . . . .-4%
514 Carroll . . . .-12%
506 Cedar Falls . . . .-4%
522-524 Cedar Rapids . .3%
510 Cherokee . . . .-4%
515 Council Bluffs 4%
508 Creston . . . .-8%
527-528 Davenport 1%
521 Decorah 1%
500-503 Des Moines 4%
520 Dubuque . . . .-1%
505 Fort Dodge . . . .-5%
504 Mason City . . . .-3%
525 Ottumwa . . . .-7%
512 Sheldon . . . .-10%
516 Shenandoah . . . . .-16%
511 Sioux City 1%
513 Spencer . . . .-11%
507 Waterloo . . . .-7%

Kansas   **-10%**
677 Colby . . . .-14%
669 Concordia . . . .-15%
678 Dodge City . . . .-7%
668 Emporia . . . .-11%
667 Fort Scott . . . .-8%
676 Hays . . . .-14%
675 Hutchinson . . . .-8%
673 Independence . . . . .-13%
660-662 Kansas City 9%
679 Liberal . . . .-27%
674 Salina . . . .-8%
664-666 Topeka . . . .-4%
670-672 Wichita . . . .-5%

Kentucky   **-5%**
411-412 Ashland . . . . . .-7%
421 Bowling Green . . . . .-6%
413-414 Campton . . . . . .-1%
410 Covington 2%
427 Elizabethtown . . . . .-9%
406 Frankfort . . . .-6%
417-418 Hazard . . . .-2%
422 Hopkinsville . . . .-9%
403-405 Lexington 0%
407-409 London . . . .-7%
400-402 Louisville 1%
423 Owensboro . . . .-9%
420 Paducah . . . .-7%
415-416 Pikeville . . . . . .-8%
425-426 Somerset . . . .-12%
424 White Plains 0%

Louisiana   **0%**
713-714 Alexandria . . .-8%
707-708 Baton Rouge . . .4%
703 Houma 10%
705 Lafayette 3%
706 Lake Charles . . . . .-2%
704 Mandeville 3%
710 Minden . . . .-8%
712 Monroe . . . .-10%
700-701 New Orleans . .10%
711 Shreveport . . . .-6%

Maine   **-6%**
042 Auburn . . . .-6%
043 Augusta . . . .-7%
044 Bangor . . . .-6%
045 Bath . . . .-7%
039-040 Brunswick 0%
048 Camden . . . .-8%
046 Cutler . . . .-9%
049 Dexter . . . .-6%
047 Northern Area . . . .-13%
041 Portland 4%

Maryland   **3%**
214 Annapolis 11%
210-212 Baltimore 8%
208-209 Bethesda 17%
216 Church Hill . . . .-1%
215 Cumberland . . . . . .-11%
219 Elkton . . . .-3%
217 Frederick 6%
206-207 Laurel 11%
218 Salisbury . . . .-6%

Massachusetts   **15%**
015-016 Ayer 10%
017 Bedford 21%
021-022 Boston 37%
023-024 Brockton 24%
026 Cape Cod 8%
010 Chicopee 10%
019 Dedham 19%
014 Fitchburg 15%
020 Hingham 25%
018 Lawrence 19%
025 Nantucket 13%
027 New Bedford 8%
013 Northfield 2%
012 Pittsfield 2%
011 Springfield 12%

Michigan   **3%**
490-491 Battle Creek . . .1%
481-482 Detroit 13%
484-485 Flint 0%
493-495 Grand Rapids . .3%
497 Grayling . . . .-5%
492 Jackson 0%
488-489 Lansing 4%
498-499 Marquette . . . . .-2%
483 Pontiac 16%
480 Royal Oak 13%
486-487 Saginaw . . . . . .-4%
496 Traverse City . . . . . .-1%

Minnesota   **2%**
566 Bemidji . . . .-1%
564 Brainerd . . . .-1%
556-558 Duluth 2%
565 Fergus Falls . . . . . .-6%
561 Magnolia . . . .-9%
560 Mankato 4%
553-555 Minneapolis . . .15%
559 Rochester 0%
563 St Cloud 3%
550-551 St Paul 14%
567 Thief River Falls . . . .-2%
562 Willmar 1%

Mississippi   **-11%**
386 Clarksdale . . . .-8%
397 Columbus . . . .-11%
387 Greenville . . . .-17%
389 Greenwood . . . . . .-13%
395 Gulfport . . . .-2%
390-392 Jackson . . . . . .-7%
394 Laurel . . . .-11%

396 McComb . . . .-16%
393 Meridian . . . .-11%
388 Tupelo . . . .-12%

Missouri   **-4%**
637 Cape Girardeau . . . .-7%
638 Caruthersville . . . .-11%
646 Chillicothe . . . .-11%
652 Columbia . . . .-2%
647 East Lynne . . . .-5%
636 Farmington . . . .-12%
634 Hannibal 3%
640 Independence 7%
650-651 Jefferson City . .-3%
648 Joplin . . . .-12%
641 Kansas City 9%
635 Kirksville . . . .-17%
653 Knob Noster . . . . . .-7%
654-655 Lebanon . . . . .-14%
639 Poplar Bluff . . . . . .-10%
633 Saint Charles 5%
644-645 Saint Joseph . .-1%
656-658 Springfield . . . .-9%
630-631 St Louis 12%

Montana   **-5%**
590-591 Billings . . . .-1%
597 Butte . . . .-5%
592 Fairview . . . .-8%
594 Great Falls . . . .-1%
595 Havre . . . .-12%
596 Helena . . . .-4%
599 Kalispell . . . .-6%
593 Miles City . . . .-3%
598 Missoula . . . .-6%
Nebraska . . . .-9%
693 Alliance . . . .-12%
686 Columbus . . . .-7%
688 Grand Island . . . . . .-9%
689 Hastings . . . .-11%
683-685 Lincoln . . . .-3%
690 McCook . . . .-14%
687 Norfolk . . . .-13%
691 North Platte . . . . . .-12%
680-681 Omaha 2%
692 Valentine . . . .-12%

Nevada   **5%**
897 Carson City 0%
898 Elko 14%
893 Ely . . . .-8%
894 Fallon 6%
889-891 Las Vegas 10%
895 Reno 6%

New Hampshire   **2%**
036 Charlestown . . . . . .-2%
034 Concord 4%
038 Dover 5%
037 Lebanon . . . .-1%
035 Littleton . . . .-6%
032-033 Manchester 5%
030-031 New Boston . . .8%

New Jersey   **17%**
080-084 Atlantic City . . .10%
087 Brick 11%
078 Dover 18%
088-089 Edison 19%
076 Hackensack 16%
077 Monmouth 21%
071-073 Newark 17%
070 Passaic 18%
074-075 Paterson 15%
085 Princeton 16%
079 Summit 24%
086 Trenton 14%

New Mexico   **-10%**
883 Alamogordo . . . . . .-11%
870-871 Albuquerque . . .-1%
881 Clovis . . . .-13%
874 Farmington . . . .-6%
882 Fort Sumner . . . . . .-8%
873 Gallup . . . .-6%
877 Holman . . . .-10%
880 Las Cruces . . . . . .-13%
875 Santa Fe . . . .-7%
878 Socorro . . . .-14%
879 Truth or Consequences
. . . .-15%
884 Tucumcari . . . .-14%

New York   **10%**
120-123 Albany 8%
117 Amityville 15%
140 Batavia 2%
137-139 Binghamton 0%
104 Bronx 17%
112 Brooklyn 11%
142 Buffalo 3%
149 Elmira . . . .-2%
113 Flushing 25%
115 Garden City 20%
118 Hicksville 18%
148 Ithaca . . . .-2%
114 Jamaica 24%
147 Jamestown . . . .-6%
124 Kingston . . . .-2%
111 Long Island 37%
119 Montauk 14%
100-102 New York
(Manhattan) 37%
100-102 New York City .37%
128 Newcomb 0%
143 Niagara Falls . . . . . .-4%
129 Plattsburgh . . . .-2%
125-126 Poughkeepsie . .4%
110 Queens 24%
144-146 Rochester 4%
116 Rockaway 17%
133-134 Rome . . . .-4%
103 Staten Island 19%
127 Stewart 1%
130-132 Syracuse 3%
141 Tonawanda 1%
135 Utica . . . .-4%
136 Watertown . . . .-4%
109 West Point 13%
105-108 White Plains . .20%

North Carolina   **-4%**
287-289 Asheville . . . . . .-6%
280-282 Charlotte 5%
277 Durham 2%
279 Elizabeth City . . . . .-5%
283 Fayetteville . . . .-9%
275 Goldsboro 0%
274 Greensboro 0%
286 Hickory . . . .-8%
285 Kinston . . . .-11%
276 Raleigh 4%
278 Rocky Mount . . . . . .-7%
284 Wilmington . . . .-7%
270-273 Winston-Salem .-2%

North Dakota   **-4%**
585 Bismarck . . . .-3%
586 Dickinson . . . .-4%
580-581 Fargo 0%
582 Grand Forks 0%
584 Jamestown . . . .-6%

587 Minot ....-1%
583 Nekoma ....-9%
588 Williston ....-5%

Ohio **0%**
442-443 Akron ....-2%
446-447 Canton ....-2%
456 Chillicothe ....-5%
450-452 Cincinnati 5%
440-441 Cleveland 6%
432 Columbus 8%
453-455 Dayton 0%
458 Lima ....-5%
457 Marietta ....-5%
433 Marion ....-7%
430-431 Newark 6%
448-449 Sandusky 0%
439 Steubenville 6%
434-436 Toledo 5%
444 Warren ....-5%
445 Youngstown ......-2%
437-438 Zanesville ....-4%

Oklahoma **-10%**
739 Adams ....-14%
734 Ardmore ....-11%
736 Clinton ....-12%
747 Durant ....-11%
737 Enid ....-8%
735 Lawton ....-13%
745 McAlester ....-17%
744 Muskogee ....-12%
730 Norman ....-7%
731 Oklahoma City ....-5%
746 Ponca City ....-9%
749 Poteau ....-10%
743 Pryor ....-15%
748 Shawnee ....-16%
740-741 Tulsa ....-2%
738 Woodward ....-3%

Oregon **-4%**
979 Adrian ....-16%
977 Bend ....-2%
974 Eugene 1%
975 Grants Pass ....-5%
976 Klamath Falls .....-11%
978 Pendleton ....-6%
970-972 Portland 10%
973 Salem 0%

Pennsylvania **-1%**
181 Allentown 8%
166 Altoona ....-9%
178 Beaver Springs ....-2%
180 Bethlehem 9%
167 Bradford ....-9%
160 Butler ....-3%
172 Chambersburg .....-5%
168 Clearfield ....-1%
158 DuBois ....-12%
183 East Stroudsburg ...0%
164-165 Erie ....-9%
169 Genesee ....-13%
156 Greensburg ....-3%
170-171 Harrisburg 4%
182 Hazleton ....-4%
159 Johnstown ....-11%
162 Kittanning ....-8%
175-176 Lancaster 9%
163 Meadville ....-12%
188 Montrose ....-9%
161 New Castle ....-3%
190-191 Philadelphia ...15%
152 Pittsburgh 5%

179 Pottsville ....-7%
157 Punxsutawney ....-11%
195-196 Reading 4%
184-185 Scranton ......-1%
155 Somerset ....-9%
193 Southeastern 11%
154 Uniontown ....-8%
194 Valley Forge 17%
189 Warminster 15%
150-151 Warrendale ..4%
153 Washington 4%
186-187 Wilkes Barre ...-2%
177 Williamsport ......-3%
173-174 York 4%

Rhode Island **8%**
028 Bristol 7%
028 Coventry 7%
029 Cranston 9%
028 Davisville 7%
028 Narragansett 7%
028 Newport 7%
029 Providence 9%
028 Warwick 7%

South Carolina **-3%**
298 Aiken 4%
299 Beaufort ....-2%
294 Charleston ....-2%
290-292 Columbia .....-5%
296 Greenville ....-4%
295 Myrtle Beach ......-8%
297 Rock Hill ....-6%
293 Spartanburg ......-5%

South Dakota **-10%**
574 Aberdeen ....-7%
573 Mitchell ....-10%
576 Mobridge ....-20%
575 Pierre ....-12%
577 Rapid City ....-8%
570-571 Sioux Falls ....-5%
572 Watertown ....-9%

Tennessee **-3%**
374 Chattanooga 0%
370 Clarksville 3%
373 Cleveland ....-3%
384 Columbia ....-8%
385 Cookeville ....-8%
383 Jackson ....-6%
376 Kingsport ....-7%
377-379 Knoxville ......-3%
382 McKenzie ....-9%
380-381 Memphis 1%
371-372 Nashville 7%

Texas **-5%**
795-796 Abilene ......-11%
790-791 Amarillo .....-7%
760 Arlington 3%
786-787 Austin 5%
774 Bay City 13%
776-777 Beaumont ....-2%
768 Brownwood ......-12%
778 Bryan ....-9%
792 Childress ....-21%
783-784 Corpus Christi .-5%
751-753 Dallas 6%
788 Del Rio ....-19%
798-799 El Paso ......-13%
761-762 Fort Worth ...3%
775 Galveston 5%
789 Giddings ....-6%
754 Greenville 2%
770-772 Houston 7%

773 Huntsville 6%
756 Longview ....-5%
793-794 Lubbock .....-12%
759 Lufkin ....-12%
785 McAllen ....-17%
797 Midland ....-4%
758 Palestine ....-10%
750 Plano 6%
769 San Angelo ....-9%
780-782 San Antonio ...-3%
755 Texarkana ....-9%
757 Tyler ....-8%
779 Victoria ....-2%
765-767 Waco ....-7%
763 Wichita Falls ......-10%
764 Woodson ....-9%

Utah **-6%**
840 Clearfield ....-3%
845 Green River ....-4%
843-844 Ogden ......-12%
846-847 Provo ....-10%
841 Salt Lake City .....-1%

Vermont **-4%**
58 Albany ....-6%
53 Battleboro ....-3%
59 Beecher Falls ......-8%
52 Bennington ....-5%
54 Burlington 3%
56 Montpelier ....-4%
57 Rutland ....-7%
51 Springfield ....-5%
50 White River Junction .-5%

Virginia **-5%**
242 Abingdon ....-13%
220-223 Alexandria 16%
229 Charlottesville .....-6%
233 Chesapeake ......-4%
227 Culpeper 0%
239 Farmville ....-15%
224-225 Fredericksburg .-2%
243 Galax ....-14%
228 Harrisonburg ......-7%
245 Lynchburg ....-11%
235-237 Norfolk ......-1%
238 Petersburg ....-5%
241 Radford ....-12%
232 Richmond 5%
240 Roanoke ....-10%
244 Staunton ....-9%
246 Tazewell ....-12%
234 Virginia Beach .....-3%
230-231 Williamsburg ...-1%
226 Winchester ....-3%

Washington **1%**
994 Clarkston ....-1%
982 Everett 4%
985 Olympia 1%
993 Pasco ....-3%
980-981 Seattle 14%
990-992 Spokane ......-3%
983-984 Tacoma 5%
986 Vancouver 3%
988 Wenatchee ....-7%
989 Yakima ....-6%

West Virginia **-9%**
258-259 Beckley ....-6%
247-248 Bluefield .....-10%
250-253 Charleston 0%
263-264 Clarksburg ....-9%
266 Fairmont ....-10%
255-257 Huntington ....-6%

249 Lewisburg ....-11%
254 Martinsburg ....-7%
265 Morgantown ......-12%
262 New Martinsville ..-13%
261 Parkersburg ....-4%
267 Romney ....-12%
268 Sugar Grove .....-15%
260 Wheeling ....-3%

Wisconsin Average **3%**
540 Amery 4%
535 Beloit 8%
545 Clam Lake ....-3%
547 Eau Claire 1%
541-543 Green Bay 3%
546 La Crosse ....-2%
548 Ladysmith ....-5%
537 Madison 12%
530-534 Milwaukee 10%
549 Oshkosh 4%
539 Portage 4%
538 Prairie du Chien ...-2%
544 Wausau 2%

Wyoming **-5%**
826 Casper ....-2%
820 Cheyenne/Laramie .-6%
827 Gillette 1%
824 Powell ....-8%
823 Rawlins ....-3%
825 Riverton ....-8%
829-831 Rock Springs ...1%
828 Sheridan ....-8%
822 Wheatland ....-11%

CANADA

Alberta **39.80%**
Calgary 37.00%
Edmonton 40.60%
Ft. McMurray 41.80%
British Columbia **39.60%**
Fraser Valley 38.10%
Okanagan 34.20%
Vancouver 46.60%
Manitoba **43.00%**
North Manitoba 50.90%
South Manitoba 42.80%
Selkirk 37.30%
Winnipeg 41.00%
New Brunswick **36.90%**
Moncton 36.90%
Nova Scotia **43.50%**
Amherst 42.80%
Nova Scotia 39.60%
Sydney 48.10%
Newfoundland &
Labrador **37.40%**
Ontario **42.00%**
London 44.50%
Thunder Bay 36.30%
Toronto 45.30%
Quebec **44.70%**
Montreal 47.80%
Quebec 41.70%
Saskatchewan **44.10%**
La Ronge 46.90%
Prince Albert 43.40%
Saskatoon 41.90%

14

Crew Compositions & Wage Rates

Crew Code	Manhours per day	Total costs $/Hr.	$/Day	Average Crew Rate (ACR) per Hour	Average Crew Rate (ACROP) per Hour w/O&P
AB					
1 Air tool operator	8.00	$30.64	$245.11		
1 Laborer	8.00	$26.64	$213.12		
TOTAL	**16.00**		**$458.23**	**$28.64**	**$42.39**
AD					
2 Air tool operators	16.00	$30.64	$490.22		
1 Laborer	8.00	$26.64	$213.12		
TOTAL	**24.00**		**$703.34**	**$29.31**	**$43.37**
BD					
3 Bricklayers	24.00	$32.55	$781.20		
2 Bricktenders	16.00	$24.93	$398.88		
TOTAL	**40.00**		**$1,180.08**	**$29.50**	**$43.96**
BK					
1 Bricklayer	8.00	$32.55	$260.40		
1 Bricktender	8.00	$24.93	$199.44		
TOTAL	**16.00**		**$459.84**	**$28.74**	**$42.82**
BO					
2 Bricklayers	16.00	$32.55	$520.80		
2 Bricktenders	16.00	$24.93	$398.88		
TOTAL	**32.00**		**$919.68**	**$28.74**	**$42.82**
2C					
2 Carpenters	16.00	$32.09	$513.44	$32.09	$48.14
CA					
1 Carpenter	8.00	$32.09	$256.72	$32.09	$48.14
CH					
1 Carpenter	8.00	$32.09	$256.72		
½ Laborer	4.00	$26.64	$106.56		
TOTAL	**12.00**		**$363.28**	**$30.27**	**$45.41**
CJ					
1 Carpenter	8.00	$32.09	$256.72		
1 Laborer	8.00	$26.64	$213.12		
TOTAL	**16.00**		**$469.84**	**$29.37**	**$44.05**
CN					
2 Carpenters	16.00	$32.09	$513.44		
½ Laborer	4.00	$26.64	$106.56		
TOTAL	**20.00**		**$620.00**	**$31.00**	**$46.50**
CS					
2 Carpenters	16.00	$32.09	$513.44		
1 Laborer	8.00	$26.64	$213.12		
TOTAL	**24.00**		**$726.56**	**$30.27**	**$45.41**
CU					
4 Carpenters	32.00	$32.09	$1,026.88		
1 Laborer	8.00	$26.64	$213.12		
TOTAL	**40.00**		**$1,240.00**	**$31.00**	**$46.50**
CW					
2 Carpenters	16.00	$32.09	$513.44		
2 Laborers	16.00	$26.64	$426.24		
TOTAL	**32.00**		**$939.68**	**$29.37**	**$44.05**

Crew Code	Manhours per day	Total costs $/Hr.	$/Day	Average Crew Rate (ACR) per Hour	Average Crew Rate (ACROP) per Hour w/O&P
CX					
4 Carpenters	32.00	$32.09	$1,026.88	$32.09	$48.14
CY					
3 Carpenters	24.00	$32.09	$770.16		
2 Laborers	16.00	$26.64	$426.24		
1 Equipment operator	8.00	$35.98	$287.84		
1 Laborer	8.00	$26.64	$213.12		
TOTAL	**56.00**		**$1,697.36**	**$30.31**	**$45.47**
CZ					
4 Carpenters	32.00	$32.09	$1,026.88		
3 Laborers	24.00	$26.64	$639.36		
1 Equipment operator	8.00	$35.98	$287.84		
1 Laborer	8.00	$26.64	$213.12		
TOTAL	**72.00**		**$2,167.20**	**$30.10**	**$45.15**
DD					
2 Cement masons	16.00	$30.44	$487.04		
1 Laborer	8.00	$26.64	$213.12		
TOTAL	**24.00**		**$700.16**	**$29.17**	**$43.18**
DF					
3 Cement masons	24.00	$30.44	$730.56		
5 Laborers	40.00	$26.64	$1,065.60		
TOTAL	**64.00**		**$1,796.16**	**$28.07**	**$41.54**
EA					
1 Electrician	8.00	$34.39	$275.12	$34.39	$50.21
EB					
2 Electricians	16.00	$34.39	$550.24	$34.39	$50.21
ED					
1 Electrician	8.00	$34.39	$275.12		
1 Carpenter	8.00	$32.09	$256.72		
TOTAL	**16.00**		**$531.84**	**$33.24**	**$48.53**
FA					
1 Floorlayer	8.00	$29.54	$236.32	$29.54	$43.42
FB					
2 Floorlayers	16.00	$29.54	$472.64		
1/4 Laborer	2.00	$26.64	$53.28		
TOTAL	**18.00**		**$525.92**	**$29.22**	**$42.95**
FC					
1 Floorlayer (hardwood)	8.00	$31.00	$248.03	$31.00	$45.58
FD					
2 Floorlayers (hardwood)	16.00	$31.00	$496.06		
1/4 Laborer	2.00	$26.64	$53.28		
TOTAL	**18.00**		**$549.34**	**$30.52**	**$44.86**
GA					
1 Glazier	8.00	$29.78	$238.24	$29.78	$44.07
GB					
2 Glaziers	16.00	$29.78	$476.48	$29.78	$44.07
GC					
3 Glaziers	24.00	$29.78	$714.72	$29.78	$44.07
HA					
1 Fence erector	8.00	$32.81	$262.49	$32.81	$48.89
HB					
2 Fence erectors	16.00	$32.81	$524.99		
1 Laborer	8.00	$26.64	$213.12		
TOTAL	**24.00**		**$738.11**	**$30.75**	**$45.82**

Crew Code	Manhours per day	Total costs $/Hr.	$/Day	Average Crew Rate (ACR) per Hour	Average Crew Rate (ACROP) per Hour w/O&P
1L					
1 Laborer	8.00	$26.64	$213.12	$26.64	$39.69
LB					
2 Laborers	16.00	$26.64	$426.24	$26.64	$39.69
LC					
2 Laborers	16.00	$26.64	$426.24		
1 Carpenter	8.00	$32.09	$256.72		
TOTAL	**24.00**		**$682.96**	**$28.46**	**$42.40**
LD					
2 Laborers	16.00	$26.64	$426.24		
2 Carpenters	16.00	$32.09	$513.44		
TOTAL	**32.00**		**$939.68**	**$29.37**	**$43.75**
LG					
5 Laborers	40.00	$26.64	$1,065.60		
1 Carpenter	8.00	$32.09	$256.72		
TOTAL	**48.00**		**$1,322.32**	**$27.55**	**$41.05**
LH					
3 Laborers	24.00	$26.64	$639.36	$26.64	$39.69
LJ					
4 Laborers	32.00	$26.64	$852.48	$26.64	$39.69
LK					
2 Laborers	16.00	$26.64	$426.24		
2 Carpenters	16.00	$32.09	$513.44		
1 Equipment operator	8.00	$35.98	$287.84		
1 Laborer	8.00	$26.64	$213.12		
TOTAL	**48.00**		**$1,440.64**	**$30.01**	**$44.72**
LL					
3 Laborers	24.00	$26.64	$639.36		
1 Carpenter	8.00	$32.09	$256.72		
1 Equipment operator	8.00	$35.98	$287.84		
1 Laborer	8.00	$26.64	$213.12		
TOTAL	**48.00**		**$1,397.04**	**$29.11**	**$43.37**
LR					
1 Lather	8.00	$31.17	$249.36	$31.17	$46.44
ML					
2 Bricklayers	16.00	$32.55	$520.80		
1 Bricktender	8.00	$24.93	$199.44		
TOTAL	**24.00**		**$720.24**	**$30.01**	**$44.71**
NA					
1 Painter (brush)	8.00	$32.38	$259.04	$32.38	$47.92
NC					
1 Painter (spray)	8.00	$33.33	$266.67	$33.33	$49.33
P3					
2 Plasterers	16.00	$32.53	$520.48		
1 Laborer	8.00	$26.64	$213.12		
TOTAL	**24.00**		**$733.60**	**$30.57**	**$45.54**
PE					
3 Plasterers	24.00	$32.53	$780.72		
2 Laborers	16.00	$26.64	$426.24		
TOTAL	**40.00**		**$1,206.96**	**$30.17**	**$44.96**

Crew Code	Manhours per day	Total costs $/Hr.	$/Day	Average Crew Rate (ACR) per Hour	Average Crew Rate (ACROP) per Hour w/O&P
QA					
1 Paperhanger	8.00	$34.00	$272.00	$34.00	$50.32
2R					
2 Roofers (composition)	16.00	$35.77	$572.32	$35.77	$55.44
RG					
2 Roofers (composition)	16.00	$35.77	$572.32		
1 Laborer	8.00	$26.64	$213.12		
TOTAL	**24.00**		**$785.44**	**$32.73**	**$50.73**
RJ					
2 Roofers (wood shingles)	16.00	$37.55	$600.86	$37.55	$58.21
RL					
2 Roofers (composition)	16.00	$35.77	$572.32		
1/2 Laborer	4.00	$26.64	$106.56		
TOTAL	**20.00**		**$678.88**	**$33.94**	**$52.61**
RM					
2 Roofers (wood shingles)	16.00	$37.55	$600.86		
1/2 Laborer	4.00	$26.64	$106.56		
TOTAL	**20.00**		**$707.42**	**$35.37**	**$54.83**
RQ					
2 Roofers (wood shingles)	16.00	$37.55	$600.86		
7/8 Laborer	7.00	$26.64	$186.48		
TOTAL	**23.00**		**$787.34**	**$34.23**	**$53.06**
RS					
1 Roofer (foreman)	8.00	$39.36	$314.86		
3 Roofers (pitch)	24.00	$36.84	$884.07		
2 Laborers	16.00	$26.64	$426.24		
TOTAL	**48.00**		**$1,625.18**	**$33.86**	**$52.48**
RT					
2 Roofers (pitch)	16.00	$36.84	$589.38		
1 Laborer	8.00	$26.64	$213.12		
TOTAL	**24.00**		**$802.50**	**$33.44**	**$51.83**
SA					
1 Plumber	8.00	$36.35	$290.80	$36.35	$53.43
SB					
1 Plumber	8.00	$36.35	$290.80		
1 Laborer	8.00	$26.64	$213.12		
TOTAL	**16.00**		**$503.92**	**$31.50**	**$46.30**
SC					
1 Plumber	8.00	$36.35	$290.80		
1 Electrician	8.00	$34.39	$275.12		
TOTAL	**16.00**		**$565.92**	**$35.37**	**$51.99**
SD					
1 Plumber	8.00	$36.35	$290.80		
1 Laborer	8.00	$26.64	$213.12		
1 Electrician	8.00	$34.39	$275.12		
TOTAL	**24.00**		**$779.04**	**$32.46**	**$47.72**
SE					
2 Plumbers	16.00	$36.35	$581.60		
1 Laborer	8.00	$26.64	$213.12		
1 Electrician	8.00	$34.39	$275.12		
TOTAL	**32.00**		**$1,069.84**	**$33.43**	**$49.15**

Crew Code	Manhours per day	Total costs $/Hr.	$/Day	Average Crew Rate (ACR) per Hour	Average Crew Rate (ACROP) per Hour w/O&P
SF					
2 Plumbers	16.00	$36.35	$581.60		
1 Laborer	8.00	$26.64	$213.12		
TOTAL	**24.00**		**$794.72**	**$33.11**	**$48.68**
SG					
3 Plumbers	24.00	$36.35	$872.40		
1 Laborer	8.00	$26.64	$213.12		
TOTAL	**32.00**		**$1,085.52**	**$33.92**	**$49.87**
TB					
1 Tile setter (ceramic)	8.00	$30.79	$246.32		
1 Tile setter's helper (ceramic)	8.00	$25.84	$206.69		
TOTAL	**16.00**		**$453.01**	**$28.31**	**$41.62**
UA					
1 Sheet metal worker	8.00	$35.66	$285.28	$35.66	$52.78
UB					
2 Sheet metal workers	16.00	$35.66	$570.56	$35.66	$52.78
UC					
2 Sheet metal workers	16.00	$35.66	$570.56		
2 Laborers	16.00	$26.64	$426.24		
TOTAL	**32.00**		**$996.80**	**$31.15**	**$46.10**
UD					
1 Sheet metal worker	8.00	$35.66	$285.28		
1 Laborer	8.00	$26.64	$213.12		
TOTAL	**16.00**		**$498.40**	**$31.15**	**$46.10**
UE					
1 Sheet metal worker	8.00	$35.66	$285.28		
1 Laborer	8.00	$26.64	$213.12		
½ Electrician	4.00	$34.39	$137.56		
TOTAL	**20.00**		**$635.96**	**$31.80**	**$47.06**
UF					
2 Sheet metal workers	16.00	$35.66	$570.56		
1 Laborer	8.00	$26.64	$213.12		
TOTAL	**24.00**		**$783.68**	**$32.65**	**$48.33**
VB					
1 Equipment operator	8.00	$35.98	$287.84		
1 Laborer	8.00	$26.64	$213.12		
TOTAL	**16.00**		**$500.96**	**$31.31**	**$46.34**

Abbreviations Used in This Book

ABS	acrylonitrile butadiene styrene	**F.H.A.**	Federal Housing Administration	**PSI**	per square inch
ACR	average crew rate	**fl. oz.**	fluid ounce	**PVC**	polyvinyl chloride
AGA	American Gas Association	**flt**	flight	**Qt.**	quart
AMP	ampere	**ft.**	foot	**R.E.**	rounded edge
Approx.	approximately	**ga.**	gauge	**R/L**	random length
ASME	American Society of Mechanical Engineers	**gal**	gallon	**RS**	rapid start (lamps)
auto.	automatic	**galv.**	galvanized	**R/W/L**	random width and length
Avg.	average	**GFI**	ground fault interrupter	**S4S**	surfaced-four-sides
Bdle.	bundle	**GPH**	gallons per hour	**SF**	square foot
BTU	British thermal unit	**GPM**	gallons per minute	**SL**	slimline (lamps)
BTUH	British thermal unit per hour	**H**	height or high	**Sq.**	1 square or 100 square feet
C	100	**HP, hp**	horsepower	**S.S.B.**	single strength, B quality
cc	center to center or cubic centimeter	**Hr.**	hour	**std.**	standard
CF	cubic foot	**HVAC**	heating, ventilating, air conditioning	**SY**	square yard
CFM	cubic foot per minute	**i.d.**	inside diameter	**T**	thick
CLF	100 linear feet	**i.e.**	that is	**T&G**	tongue and groove
Const.	construction	**Inst**	install	**U**	thermal conductivity
Corr.	corrugated	**I.P.S.**	iron pipe size	**U.I.**	united inches
CSF	100 square feet	**KD**	knocked down	**UL**	Underwriters Laboratories
CSY	100 square yards	**KW, kw**	kilowatts	**U.S.G.**	United States Gypsum
Ctn	carton	**L**	length or long	**VLF**	vertical linear feet
CWT	100 pounds	**lb, lbs.**	pound(s)	**W**	width or wide
CY	cubic yard	**LF**	linear feet	**yr.**	year
Cu.	cubic	**LS**	lump sum		
d	penny	**M**	1000		
D	deep or depth	**Mat'l**	material		
Demo	demolish	**Max.**	maximum		
dia.	diameter	**MBF**	1000 board feet		
D.S.A.	double strength, A grade	**MBHP**	1000 boiler horsepower		
D.S.B.	double strength, B grade	**Mi**	miles		
Ea	each	**Min.**	minimum		
e.g.	for example	**MSF**	1000 square feet		
etc.	et cetera	**O.B.**	opposed blade		
exp.	exposure	**oc**	on center		
FAS	First and Select grade	**o.d.**	outside dimension		
		oz.	ounce		
		pcs.	pieces		
		pkg.	package		

Symbols

/	per
%	percent
"	inches
'	foot or feet
x	by
o	degree
#	number or pounds
$	dollar
+/-	plus or minus

For crew abbreviations, please see Crew Compositions & Wage Rates chart, pages 15 to 19.

Acoustical and insulating tile

1. **Dimensions**

   a. Acoustical tile. $\frac{1}{2}$" thick x 12" x 12", 24".

   b. Insulating tile, decorative. $\frac{1}{2}$" thick x 12" x 12", 24"; $\frac{1}{2}$" thick x 16" x 16", 32".

2. **Installation.** Tile may be applied to existing plaster (if joist spacing is suitable) or to wood furring strips. Tile may have a square edge or flange. Depending on the type and shape of the tile, you can use adhesive, staples, nails or clips to attach the tile.

3. **Estimating Technique.** Determine area and add 5% to 10% for waste.

4. **Notes on Material Pricing.** A material price of $19.00 a gallon for adhesive was used to compile the Average Material Cost/Unit on the following pages. Here are the coverage rates:

12" x 12"	1.25 Gal/CSF
12" x 24"	0.95 Gal/CSF
16" x 16"	0.75 Gal/CSF
16" x 32"	0.55 Gal/CSF

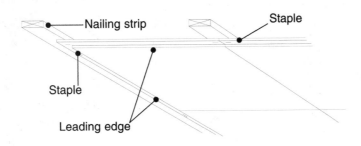

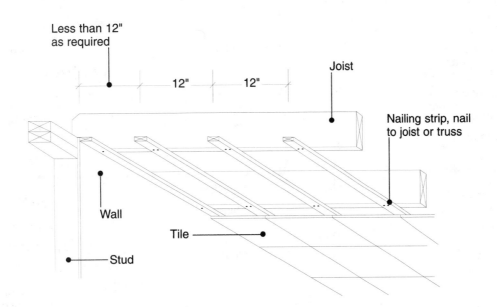

					Costs Based On Small Volume					Large Volume	
Description	Oper	Unit	Crew Size	Man-Hours Per Unit	Avg Mat'l Unit Cost	Avg Labor Unit Cost	Avg Equip Unit Cost	Avg Total Unit Cost	Avg Price Incl O&P	Avg Total Unit Cost	Avg Price Incl O&P

Acoustical treatment

See also Suspended ceiling systems, page 271

Ceiling and wall tile

Adhesive set

Description	Oper	Unit	Crew Size	Man-Hrs	Mat'l	Labor	Equip	Total	Price O&P	LV Total	LV Price
Tile only, no grid system	Demo	SF	LB	.018	---	.48	---	.48	.71	.32	.48
Tile on furring strips	Demo	SF	LB	.013	---	.35	---	.35	.52	.24	.36

Mineral fiber, vinyl coated, tile only

Applied in square pattern by adhesive to solid backing; 5% tile waste

1/2" thick, 12" x 12" or 12" x 24"

Description	Oper	Unit	Crew Size	Man-Hrs	Mat'l	Labor	Equip	Total	Price O&P	LV Total	LV Price
Economy, mini perforated	Inst	SF	2C	.026	1.10	.83	---	1.93	2.46	1.59	1.98
Standard, random perforated	Inst	SF	2C	.026	1.54	.83	---	2.37	2.95	2.00	2.43
Designer, swirl perforation	Inst	SF	2C	.026	2.47	.83	---	3.30	3.97	2.86	3.37
Deluxe, sculptured face	Inst	SF	2C	.026	2.82	.83	---	3.65	4.35	3.18	3.73

5/8" thick, 12" x 12" or 12" x 24"

Description	Oper	Unit	Crew Size	Man-Hrs	Mat'l	Labor	Equip	Total	Price O&P	LV Total	LV Price
Economy, mini perforated	Inst	SF	2C	.026	1.23	.83	---	2.06	2.60	1.71	2.11
Standard, random perforated	Inst	SF	2C	.026	1.73	.83	---	2.56	3.15	2.17	2.62
Designer, swirl perforation	Inst	SF	2C	.026	2.80	.83	---	3.63	4.33	3.16	3.70
Deluxe, sculptured face	Inst	SF	2C	.026	3.20	.83	---	4.03	4.77	3.53	4.11

3/4" thick, 12" x 12" or 12" x 24"

Description	Oper	Unit	Crew Size	Man-Hrs	Mat'l	Labor	Equip	Total	Price O&P	LV Total	LV Price
Economy, mini perforated	Inst	SF	2C	.026	1.37	.83	---	2.20	2.76	1.84	2.25
Standard, random perforated	Inst	SF	2C	.026	1.95	.83	---	2.78	3.40	2.37	2.84
Designer, swirl perforation	Inst	SF	2C	.026	3.17	.83	---	4.00	4.74	3.51	4.09
Deluxe, sculptured face	Inst	SF	2C	.026	3.64	.83	---	4.47	5.26	3.93	4.55

Applied by adhesive to furring strips

Description	Oper	Unit	Crew Size	Man-Hrs	Mat'l	Labor	Equip	Total	Price O&P	LV Total	LV Price
ADD	Inst	SF	2C	.002	---	.06	---	.06	.10	.06	.10

Stapled

Description	Oper	Unit	Crew Size	Man-Hrs	Mat'l	Labor	Equip	Total	Price O&P	LV Total	LV Price
Tile only, no grid system	Demo	SF	LB	.020	---	.53	---	.53	.79	.37	.56
Tile on furring strips	Demo	SF	LB	.015	---	.40	---	.40	.60	.27	.40

Mineral fiber, vinyl coated, tile only

Applied in square pattern by staples, nails, or clips; 5% tile waste

1/2" thick, 12" x 12" or 12" x 24"

Description	Oper	Unit	Crew Size	Man-Hrs	Mat'l	Labor	Equip	Total	Price O&P	LV Total	LV Price
Economy, mini perforated	Inst	SF	2C	.024	.95	.77	---	1.72	2.20	1.42	1.78
Standard, random perforated	Inst	SF	2C	.024	1.41	.77	---	2.18	2.71	1.85	2.25
Designer, swirl perforation	Inst	SF	2C	.024	2.25	.77	---	3.02	3.63	2.63	3.11
Deluxe, sculptured face	Inst	SF	2C	.024	2.56	.77	---	3.33	3.97	2.91	3.41

5/8" thick, 12" x 12" or 12" x 24"

Description	Oper	Unit	Crew Size	Man-Hrs	Mat'l	Labor	Equip	Total	Price O&P	LV Total	LV Price
Economy, mini perforated	Inst	SF	2C	.024	1.13	.77	---	1.90	2.40	1.59	1.96
Standard, random perforated	Inst	SF	2C	.024	1.57	.77	---	2.34	2.88	2.00	2.41
Designer, swirl perforation	Inst	SF	2C	.024	2.56	.77	---	3.33	3.97	2.91	3.41
Deluxe, sculptured face	Inst	SF	2C	.024	2.91	.77	---	3.68	4.36	3.23	3.77

3/4" thick, 12" x 12" or 12" x 24"

Description	Oper	Unit	Crew Size	Man-Hrs	Mat'l	Labor	Equip	Total	Price O&P	LV Total	LV Price
Economy, mini perforated	Inst	SF	2C	.024	1.25	.77	---	2.02	2.53	1.71	2.09
Standard, random perforated	Inst	SF	2C	.024	1.77	.77	---	2.54	3.10	2.18	2.61
Designer, swirl perforation	Inst	SF	2C	.024	2.88	.77	---	3.65	4.32	3.21	3.74
Deluxe, sculptured face	Inst	SF	2C	.024	3.30	.77	---	4.07	4.79	3.59	4.16

Applied by staples, nails or clips to furring strips

Description	Oper	Unit	Crew Size	Man-Hrs	Mat'l	Labor	Equip	Total	Price O&P	LV Total	LV Price
ADD	Inst	SF	2C	.024	---	.77	---	.77	1.16	.55	.82

| Description | Oper | Unit | Crew Size | Man-Hours Per Unit | Costs Based On Small Volume | | | | Large Volume | |
					Avg Mat'l Unit Cost	Avg Labor Unit Cost	Avg Equip Unit Cost	Avg Total Unit Cost	Avg Price Incl O&P	Avg Total Unit Cost	Avg Price Incl O&P
Tile patterns, effect on labor											
Note: The following percentage adjustments for Small Volume also apply to Large											
Herringbone, Increase manhours	Inst	%	2C	25.0	---	---	---	---	---	---	---
Diagonal, Increase manhours	Inst	%	2C	20.0	---	---	---	---	---	---	---
Ashlar, Increase manhours	Inst	%	2C	30.0	---	---	---	---	---	---	---
Furring strips, 8% waste included											
Over wood											
1" x 4", 12" oc	Inst	SF	2C	.014	.29	.45	---	.74	.99	.59	.78
1" x 4", 16" oc	Inst	SF	2C	.012	.26	.39	---	.65	.86	.50	.65
Over plaster											
1" x 4", 12" oc	Inst	SF	2C	.018	.29	.58	---	.87	1.19	.69	.92
1" x 4", 16" oc	Inst	SF	2C	.014	.26	.45	---	.71	.96	.56	.75

Description	Oper	Unit	Crew Size	Man-Hours Per Unit	Costs Based On Small Volume					Large Volume	
					Avg Mat'l Unit Cost	Avg Labor Unit Cost	Avg Equip Unit Cost	Avg Total Unit Cost	Avg Price Incl O&P	Avg Total Unit Cost	Avg Price Incl O&P

Adhesives

Better quality, gun-applied in continuous bead to wood or metal framing or furring members. Per 100 SF of surface area including 6% waste

Panel adhesives

Subfloor adhesive, on floors

12" oc members

Description	Oper	Unit	Crew Size	Man-Hours Per Unit	Avg Mat'l Unit Cost	Avg Labor Unit Cost	Avg Equip Unit Cost	Avg Total Unit Cost	Avg Price Incl O&P	Avg Total Unit Cost	Avg Price Incl O&P
1/8" diameter (11 fl.oz./CSF)	Inst	CSF	CA	.108	2.60	3.47	---	6.07	8.06	4.75	6.18
1/4" diameter (43 fl.oz./CSF)	Inst	CSF	CA	.108	10.40	3.47	---	13.87	16.60	11.73	13.90
3/8" diameter (97 fl.oz./CSF)	Inst	CSF	CA	.108	23.40	3.47	---	26.87	30.90	23.41	26.70
1/2" diameter (172 fl.oz./CSF)	Inst	CSF	CA	.108	41.30	3.47	---	44.77	50.70	39.61	44.50
16" oc members											
1/8" diameter (9 fl.oz./CSF)	Inst	CSF	CA	.080	2.08	2.57	---	4.65	6.14	3.67	4.75
1/4" diameter (34 fl.oz./CSF)	Inst	CSF	CA	.080	8.28	2.57	---	10.85	13.00	9.25	10.90
3/8" diameter (78 fl.oz./CSF)	Inst	CSF	CA	.080	18.70	2.57	---	21.27	24.40	18.60	21.20
1/2" diameter (137 fl.oz./CSF)	Inst	CSF	CA	.080	33.10	2.57	---	35.67	40.20	31.60	35.40
24" oc members											
1/8" diameter (7 fl.oz./CSF)	Inst	CSF	CA	.074	1.56	2.37	---	3.93	5.28	3.07	4.04
1/4" diameter (26 fl.oz./CSF)	Inst	CSF	CA	.074	6.21	2.37	---	8.58	10.40	7.26	8.65
3/8" diameter (58 fl.oz./CSF)	Inst	CSF	CA	.074	14.00	2.37	---	16.37	19.00	14.27	16.40
1/2" diameter (103 fl.oz./CSF)	Inst	CSF	CA	.074	24.80	2.37	---	27.17	30.80	23.97	27.00

Wall sheathing or shear panel wall adhesive on walls, floors or ceilings

Description	Oper	Unit	Crew Size	Man-Hours Per Unit	Avg Mat'l Unit Cost	Avg Labor Unit Cost	Avg Equip Unit Cost	Avg Total Unit Cost	Avg Price Incl O&P	Avg Total Unit Cost	Avg Price Incl O&P
12" oc members											
1/8" diameter (11 fl.oz./CSF)	Inst	CSF	CA	.143	3.34	4.59	---	7.93	10.60	6.22	8.13
1/4" diameter (43 fl.oz./CSF)	Inst	CSF	CA	.143	13.30	4.59	---	17.89	21.50	15.21	18.00
3/8" diameter (97 fl.oz./CSF)	Inst	CSF	CA	.143	30.00	4.59	---	34.59	39.90	30.21	34.60
1/2" diameter (172 fl.oz./CSF)	Inst	CSF	CA	.143	53.20	4.59	---	57.79	65.40	51.01	57.40
16" oc members											
1/8" diameter (9 fl.oz./CSF)	Inst	CSF	CA	.130	2.67	4.17	---	6.84	9.20	5.32	7.02
1/4" diameter (34 fl.oz./CSF)	Inst	CSF	CA	.130	10.70	4.17	---	14.87	18.00	12.51	14.90
3/8" diameter (78 fl.oz./CSF)	Inst	CSF	CA	.130	24.00	4.17	---	28.17	32.70	24.52	28.20
1/2" diameter (137 fl.oz./CSF)	Inst	CSF	CA	.130	42.50	4.17	---	46.67	53.00	41.22	46.50
24" oc members											
1/8" diameter (7 fl.oz./CSF)	Inst	CSF	CA	.120	2.01	3.85	---	5.86	7.99	4.51	6.03
1/4" diameter (26 fl.oz./CSF)	Inst	CSF	CA	.120	7.99	3.85	---	11.84	14.60	9.89	12.00
3/8" diameter (58 fl.oz./CSF)	Inst	CSF	CA	.120	18.00	3.85	---	21.85	25.60	18.90	21.90
1/2" diameter (103 fl.oz./CSF)	Inst	CSF	CA	.120	31.90	3.85	---	35.75	40.90	31.40	35.60

Polystyrene or polyurethane foam panel adhesive, on walls

Description	Oper	Unit	Crew Size	Man-Hours Per Unit	Avg Mat'l Unit Cost	Avg Labor Unit Cost	Avg Equip Unit Cost	Avg Total Unit Cost	Avg Price Incl O&P	Avg Total Unit Cost	Avg Price Incl O&P
12" oc members											
1/8" diameter (11 fl.oz./CSF)	Inst	CSF	CA	.143	4.09	4.59	---	8.68	11.40	6.89	8.86
1/4" diameter (43 fl.oz./CSF)	Inst	CSF	CA	.143	16.30	4.59	---	20.89	24.80	17.91	20.90
3/8" diameter (97 fl.oz./CSF)	Inst	CSF	CA	.143	36.70	4.59	---	41.29	47.30	36.31	41.20
1/2" diameter (172 fl.oz./CSF)	Inst	CSF	CA	.143	65.00	4.59	---	69.59	78.40	61.71	69.20
16" oc members											
1/8" diameter (9 fl.oz./CSF)	Inst	CSF	CA	.130	3.27	4.17	---	7.44	9.86	5.86	7.61
1/4" diameter (34 fl.oz./CSF)	Inst	CSF	CA	.130	13.00	4.17	---	17.17	20.60	14.62	17.30
3/8" diameter (78 fl.oz./CSF)	Inst	CSF	CA	.130	29.40	4.17	---	33.57	38.60	29.32	33.50
1/2" diameter (137 fl.oz./CSF)	Inst	CSF	CA	.130	52.00	4.17	---	56.17	63.50	49.72	55.90
24" oc members											
1/8" diameter (7 fl.oz./CSF)	Inst	CSF	CA	.129	2.45	4.14	---	6.59	8.91	4.91	6.47
1/4" diameter (26 fl.oz./CSF)	Inst	CSF	CA	.129	9.77	4.14	---	13.91	17.00	11.49	13.70
3/8" diameter (58 fl.oz./CSF)	Inst	CSF	CA	.129	22.00	4.14	---	26.14	30.50	22.50	25.90
1/2" diameter (103 fl.oz./CSF)	Inst	CSF	CA	.129	39.00	4.14	---	43.14	49.10	37.80	42.60

Description	Oper	Unit	Crew Size	Man-Hours Per Unit	Costs Based On Small Volume						Large Volume	
					Avg Mat'l Unit Cost	Avg Labor Unit Cost	Avg Equip Unit Cost	Avg Total Unit Cost	Avg Price Incl O&P		Avg Total Unit Cost	Avg Price Incl O&P
Gypsum drywall adhesive, on ceilings												
12" oc members												
1/8" diameter (11 fl.oz./CSF)	Inst	CSF	CA	.143	3.72	4.59	---	8.31	**11.00**		6.56	**8.50**
1/4" diameter (43 fl.oz./CSF)	Inst	CSF	CA	.143	14.80	4.59	---	19.39	**23.20**		16.51	**19.50**
3/8" diameter (97 fl.oz./CSF)	Inst	CSF	CA	.143	33.40	4.59	---	37.99	**43.70**		33.31	**37.90**
1/2" diameter (172 fl.oz./CSF)	Inst	CSF	CA	.143	59.20	4.59	---	63.79	**72.00**		56.41	**63.40**
16" oc members												
1/8" diameter (9 fl.oz./CSF)	Inst	CSF	CA	.130	2.97	4.17	---	7.14	**9.53**		5.59	**7.32**
1/4" diameter (34 fl.oz./CSF)	Inst	CSF	CA	.130	11.90	4.17	---	16.07	**19.30**		13.62	**16.10**
3/8" diameter (78 fl.oz./CSF)	Inst	CSF	CA	.130	26.70	4.17	---	30.87	**35.70**		27.02	**30.90**
1/2" diameter (137 fl.oz./CSF)	Inst	CSF	CA	.130	47.30	4.17	---	51.47	**58.30**		45.52	**51.20**
24" oc members												
1/8" diameter (7 fl.oz./CSF)	Inst	CSF	CA	.120	2.23	3.85	---	6.08	**8.23**		4.71	**6.25**
1/4" diameter (26 fl.oz./CSF)	Inst	CSF	CA	.120	8.89	3.85	---	12.74	**15.60**		10.70	**12.80**
3/8" diameter (58 fl.oz./CSF)	Inst	CSF	CA	.120	20.10	3.85	---	23.95	**27.80**		20.80	**23.90**
1/2" diameter (103 fl.oz./CSF)	Inst	CSF	CA	.120	35.50	3.85	---	39.35	**44.80**		34.60	**39.20**
Gypsum drywall adhesive, on walls												
12" oc members												
1/8" diameter (11 fl.oz./CSF)	Inst	CSF	CA	.143	3.72	4.59	---	8.31	**11.00**		6.56	**8.50**
1/4" diameter (43 fl.oz./CSF)	Inst	CSF	CA	.143	14.80	4.59	---	19.39	**23.20**		16.51	**19.50**
3/8" diameter (97 fl.oz./CSF)	Inst	CSF	CA	.143	33.40	4.59	---	37.99	**43.70**		33.31	**37.90**
1/2" diameter (172 fl.oz./CSF)	Inst	CSF	CA	.143	59.20	4.59	---	63.79	**72.00**		56.41	**63.40**
16" oc members												
1/8" diameter (9 fl.oz./CSF)	Inst	CSF	CA	.130	2.97	4.17	---	7.14	**9.53**		5.59	**7.32**
1/4" diameter (34 fl.oz./CSF)	Inst	CSF	CA	.130	11.90	4.17	---	16.07	**19.30**		13.62	**16.10**
3/8" diameter (78 fl.oz./CSF)	Inst	CSF	CA	.130	26.70	4.17	---	30.87	**35.70**		27.02	**30.90**
1/2" diameter (137 fl.oz./CSF)	Inst	CSF	CA	.130	47.30	4.17	---	51.47	**58.30**		45.52	**51.20**
24" oc members												
1/8" diameter (7 fl.oz./CSF)	Inst	CSF	CA	.120	2.23	3.85	---	6.08	**8.23**		4.71	**6.25**
1/4" diameter (26 fl.oz./CSF)	Inst	CSF	CA	.120	8.89	3.85	---	12.74	**15.60**		10.70	**12.80**
3/8" diameter (58 fl.oz./CSF)	Inst	CSF	CA	.120	20.10	3.85	---	23.95	**27.80**		20.80	**23.90**
1/2" diameter (103 fl.oz./CSF)	Inst	CSF	CA	.120	35.50	3.85	---	39.35	**44.80**		34.60	**39.20**
Hardboard or plastic panel adhesive, on walls												
12" oc members												
1/8" diameter (11 fl.oz./CSF)	Inst	CSF	CA	.143	3.72	4.59	---	8.31	**11.00**		6.56	**8.50**
1/4" diameter (43 fl.oz./CSF)	Inst	CSF	CA	.143	14.80	4.59	---	19.39	**23.20**		16.51	**19.50**
3/8" diameter (97 fl.oz./CSF)	Inst	CSF	CA	.143	33.40	4.59	---	37.99	**43.70**		33.31	**37.90**
1/2" diameter (172 fl.oz./CSF)	Inst	CSF	CA	.143	59.20	4.59	---	63.79	**72.00**		56.41	**63.40**
16" oc members												
1/8" diameter (9 fl.oz./CSF)	Inst	CSF	CA	.130	2.97	4.17	---	7.14	**9.53**		5.59	**7.32**
1/4" diameter (34 fl.oz./CSF)	Inst	CSF	CA	.130	11.90	4.17	---	16.07	**19.30**		13.62	**16.10**
3/8" diameter (78 fl.oz./CSF)	Inst	CSF	CA	.130	26.70	4.17	---	30.87	**35.70**		27.02	**30.90**
1/2" diameter (137 fl.oz./CSF)	Inst	CSF	CA	.130	47.30	4.17	---	51.47	**58.30**		45.52	**51.20**
24" oc members												
1/8" diameter (7 fl.oz./CSF)	Inst	CSF	CA	.120	2.23	3.85	---	6.08	**8.23**		4.71	**6.25**
1/4" diameter (26 fl.oz./CSF)	Inst	CSF	CA	.120	8.89	3.85	---	12.74	**15.60**		10.70	**12.80**
3/8" diameter (58 fl.oz./CSF)	Inst	CSF	CA	.120	20.10	3.85	---	23.95	**27.80**		20.80	**23.90**
1/2" diameter (103 fl.oz./CSF)	Inst	CSF	CA	.120	35.50	3.85	---	39.35	**44.80**		34.60	**39.20**

				Costs Based On Small Volume						Large Volume	
				Man-Hours Per Unit	Avg Mat'l Unit Cost	Avg Labor Unit Cost	Avg Equip Unit Cost	Avg Total Unit Cost	Avg Price Incl O&P	Avg Total Unit Cost	Avg Price Incl O&P
Description	Oper	Unit	Crew Size								

Air conditioning and ventilating systems

System components

Condensing units

Air cooled, compressor, standard controls

1.0 ton	Inst	Ea	SB	10.0	826.00	315.00	---	1141.00	**1290.00**	988.00	**1220.00**
1.5 ton	Inst	Ea	SB	11.4	1010.00	359.00	---	1369.00	**1540.00**	1191.00	**1460.00**
2.0 ton	Inst	Ea	SB	13.3	1160.00	419.00	---	1579.00	**1780.00**	1377.00	**1690.00**
2.5 ton	Inst	Ea	SB	16.0	1390.00	504.00	---	1894.00	**2130.00**	1643.00	**2010.00**
3.0 ton	Inst	Ea	SB	20.0	1640.00	630.00	---	2270.00	**2570.00**	1974.00	**2430.00**
4.0 ton	Inst	Ea	SB	26.7	2160.00	841.00	---	3001.00	**3400.00**	2601.00	**3200.00**
5.0 ton	Inst	Ea	SB	40.0	2640.00	1260.00	---	3900.00	**4500.00**	3370.00	**4190.00**
Minimum Job Charge	Inst	Job	SB	6.67	---	210.00	---	210.00	**309.00**	168.00	**247.00**

Diffusers

Aluminum, opposed blade damper, unless otherwise noted

Ceiling or sidewall, linear

2" wide	Inst	LF	UA	.333	30.30	11.90	---	42.20	**47.90**	36.52	**45.20**
4" wide	Inst	LF	UA	.370	40.60	13.20	---	53.80	**60.20**	46.80	**57.30**
6" wide	Inst	LF	UA	.417	50.30	14.90	---	65.20	**72.30**	56.80	**69.20**
8" wide	Inst	LF	UA	.500	58.70	17.80	---	76.50	**85.10**	66.60	**81.30**

Perforated, 24" x 24" panel size

6" x 6"	Inst	Ea	UA	.625	82.60	22.30	---	104.90	**116.00**	91.40	**111.00**
8" x 8"	Inst	Ea	UA	.714	85.10	25.50	---	110.60	**123.00**	96.30	**117.00**
10" x 10"	Inst	Ea	UA	.833	86.40	29.70	---	116.10	**130.00**	100.90	**124.00**
12" x 12"	Inst	Ea	UA	1.00	89.70	35.70	---	125.40	**142.00**	108.40	**134.00**
18" x 18"	Inst	Ea	UA	1.25	119.00	44.60	---	163.60	**185.00**	141.70	**175.00**

Rectangular, one- to four-way blow

6" x 6"	Inst	Ea	UA	.625	48.40	22.30	---	70.70	**81.40**	60.90	**76.00**
12" x 6"	Inst	Ea	UA	.714	61.90	25.50	---	87.40	**99.60**	75.60	**93.60**
12" x 9"	Inst	Ea	UA	.833	72.90	29.70	---	102.60	**117.00**	88.80	**110.00**
12" x 12"	Inst	Ea	UA	1.00	85.10	35.70	---	120.80	**138.00**	104.40	**130.00**
24" x 12"	Inst	Ea	UA	1.25	144.00	44.60	---	188.60	**210.00**	164.70	**201.00**

T-bar mounting, 24" x 24" lay in frame

6" x 6"	Inst	Ea	UA	.625	78.10	22.30	---	100.40	**111.00**	87.40	**106.00**
9" x 9"	Inst	Ea	UA	.714	87.10	25.50	---	112.60	**125.00**	98.00	**119.00**
12" x 12"	Inst	Ea	UA	.833	114.00	29.70	---	143.70	**158.00**	125.80	**152.00**
15" x 15"	Inst	Ea	UA	1.00	119.00	35.70	---	154.70	**171.00**	134.50	**164.00**
18" x 18"	Inst	Ea	UA	1.25	156.00	44.60	---	200.60	**222.00**	174.70	**213.00**
Minimum Job Charge	Inst	Job	UA	2.86	---	102.00	---	102.00	**151.00**	81.70	**121.00**

Ductwork

Fabricated rectangular, includes fittings, joints, supports

Aluminum alloy

Under 100 lbs	Inst	Lb	UF	.429	3.99	14.00	---	17.99	**24.70**	14.75	**20.70**
100 to 500 lbs	Inst	Lb	UF	.333	2.98	10.90	---	13.88	**19.10**	11.38	**16.00**
500 to 1,000 lbs	Inst	Lb	UF	.273	2.05	8.91	---	10.96	**15.20**	8.95	**12.60**
Over 1,000 lbs	Inst	Lb	UF	.231	1.56	7.54	---	9.10	**12.70**	7.43	**10.50**

Galvanized steel

Under 400 lbs	Inst	Lb	UF	.150	.84	4.90	---	5.74	**8.09**	4.67	**6.66**
400 to 1,000 lbs	Inst	Lb	UF	.140	.71	4.57	---	5.28	**7.48**	4.29	**6.14**
1,000 to 2,000 lbs	Inst	Lb	UF	.130	.61	4.24	---	4.85	**6.89**	3.94	**5.65**
2,000 to 5,000 lbs	Inst	Lb	UF	.125	.52	4.08	---	4.60	**6.56**	3.73	**5.36**
Over 10,000 lbs	Inst	Lb	UF	.120	.45	3.92	---	4.37	**6.25**	3.53	**5.10**

Description	Oper	Unit	Crew Size	Man-Hours Per Unit	Avg Mat'l Unit Cost	Avg Labor Unit Cost	Avg Equip Unit Cost	Avg Total Unit Cost	Avg Price Incl O&P	Avg Total Unit Cost	Avg Price Incl O&P
										Large Volume	

Flexible, coated fabric on spring steel, aluminum, or corrosion-resistant metal

Non-insulated

Description	Oper	Unit	Crew Size	Man-Hours Per Unit	Avg Mat'l Unit Cost	Avg Labor Unit Cost	Avg Equip Unit Cost	Avg Total Unit Cost	Avg Price Incl O&P	Avg Total Unit Cost	Avg Price Incl O&P
3" diameter	Inst	LF	UD	.073	1.01	2.27	---	3.28	**4.38**	2.71	**3.71**
5" diameter	Inst	LF	UD	.089	1.17	2.77	---	3.94	**5.27**	3.26	**4.48**
6" diameter	Inst	LF	UD	.100	1.32	3.12	---	4.44	**5.93**	3.66	**5.03**
7" diameter	Inst	LF	UD	.114	1.73	3.55	---	5.28	**6.99**	4.37	**5.97**
8" diameter	Inst	LF	UD	.133	1.79	4.14	---	5.93	**7.92**	4.93	**6.77**
10" diameter	Inst	LF	UD	.160	2.30	4.98	---	7.28	**9.68**	6.04	**8.26**
12" diameter	Inst	LF	UD	.200	2.72	6.23	---	8.95	**11.90**	7.41	**10.20**

Insulated

Description	Oper	Unit	Crew Size	Man-Hours Per Unit	Avg Mat'l Unit Cost	Avg Labor Unit Cost	Avg Equip Unit Cost	Avg Total Unit Cost	Avg Price Incl O&P	Avg Total Unit Cost	Avg Price Incl O&P
3" diameter	Inst	LF	UD	.080	1.73	2.49	---	4.22	**5.42**	3.53	**4.72**
4" diameter	Inst	LF	UD	.089	1.73	2.77	---	4.50	**5.83**	3.75	**5.04**
5" diameter	Inst	LF	UD	.100	2.03	3.12	---	5.15	**6.64**	4.30	**5.77**
6" diameter	Inst	LF	UD	.114	2.23	3.55	---	5.78	**7.49**	4.82	**6.48**
7" diameter	Inst	LF	UD	.133	2.72	4.14	---	6.86	**8.85**	5.76	**7.73**
8" diameter	Inst	LF	UD	.160	2.79	4.98	---	7.77	**10.20**	6.47	**8.75**
10" diameter	Inst	LF	UD	.200	3.62	6.23	---	9.85	**12.80**	8.21	**11.10**
12" diameter	Inst	LF	UD	.267	4.22	8.32	---	12.54	**16.50**	10.39	**14.10**
14" diameter	Inst	LF	UD	.400	5.11	12.50	---	17.61	**23.60**	14.52	**20.00**

Duct accessories

Air extractors

Description	Oper	Unit	Crew Size	Man-Hours Per Unit	Avg Mat'l Unit Cost	Avg Labor Unit Cost	Avg Equip Unit Cost	Avg Total Unit Cost	Avg Price Incl O&P	Avg Total Unit Cost	Avg Price Incl O&P
12" x 4"	Inst	Ea	UA	.417	15.80	14.90	---	30.70	**37.80**	26.00	**33.80**
8" x 6"	Inst	Ea	UA	.500	15.80	17.80	---	33.60	**42.20**	28.40	**37.30**
20" x 8"	Inst	Ea	UA	.625	35.50	22.30	---	57.80	**68.50**	49.40	**62.80**
18" x 10"	Inst	Ea	UA	.714	34.80	25.50	---	60.30	**72.50**	51.50	**65.80**
24" x 12"	Inst	Ea	UA	1.00	48.40	35.70	---	84.10	**101.00**	71.60	**91.80**

Dampers, fire, curtain-type, vertical

Description	Oper	Unit	Crew Size	Man-Hours Per Unit	Avg Mat'l Unit Cost	Avg Labor Unit Cost	Avg Equip Unit Cost	Avg Total Unit Cost	Avg Price Incl O&P	Avg Total Unit Cost	Avg Price Incl O&P
8" x 4"	Inst	Ea	UA	.417	21.50	14.90	---	36.40	**43.50**	31.10	**39.60**
12" x 4"	Inst	Ea	UA	.455	21.50	16.20	---	37.70	**45.50**	32.20	**41.20**
16" x 14"	Inst	Ea	UA	.833	34.20	29.70	---	63.90	**78.20**	54.30	**70.30**
24" x 20"	Inst	Ea	UA	1.25	43.20	44.60	---	87.80	**109.00**	74.20	**97.10**

Dampers, multi-blade, opposed blade

Description	Oper	Unit	Crew Size	Man-Hours Per Unit	Avg Mat'l Unit Cost	Avg Labor Unit Cost	Avg Equip Unit Cost	Avg Total Unit Cost	Avg Price Incl O&P	Avg Total Unit Cost	Avg Price Incl O&P
12" x 12"	Inst	Ea	UA	.625	31.60	22.30	---	53.90	**64.60**	46.00	**58.80**
12" x 18"	Inst	Ea	UA	.833	41.90	29.70	---	71.60	**85.90**	61.20	**78.20**
24" x 24"	Inst	Ea	UA	1.25	89.00	44.60	---	133.60	**155.00**	115.10	**144.00**
48" x 36"	Inst	Ea	UD	3.33	270.00	104.00	---	374.00	**423.00**	323.20	**399.00**

Dampers, variable volume modulating, motorized

Description	Oper	Unit	Crew Size	Man-Hours Per Unit	Avg Mat'l Unit Cost	Avg Labor Unit Cost	Avg Equip Unit Cost	Avg Total Unit Cost	Avg Price Incl O&P	Avg Total Unit Cost	Avg Price Incl O&P
12" x 12"	Inst	Ea	UA	1.00	163.00	35.70	---	198.70	**215.00**	173.50	**209.00**
24" x 12"	Inst	Ea	UA	1.67	195.00	59.60	---	254.60	**283.00**	221.40	**270.00**
30" x 18"	Inst	Ea	UA	2.50	361.00	89.20	---	450.20	**493.00**	393.30	**476.00**
Thermostat, ADD	Inst	Ea	UA	1.67	38.70	59.60	---	98.30	**127.00**	81.90	**110.00**

Dampers, multi-blade, parallel blade

Description	Oper	Unit	Crew Size	Man-Hours Per Unit	Avg Mat'l Unit Cost	Avg Labor Unit Cost	Avg Equip Unit Cost	Avg Total Unit Cost	Avg Price Incl O&P	Avg Total Unit Cost	Avg Price Incl O&P
8" x 8"	Inst	Ea	UA	.500	65.80	17.80	---	83.60	**92.20**	73.00	**88.60**
16" x 10"	Inst	Ea	UA	.625	69.00	22.30	---	91.30	**102.00**	79.30	**97.20**
18" x 12"	Inst	Ea	UA	.833	71.00	29.70	---	100.70	**115.00**	87.10	**108.00**
28" x 16"	Inst	Ea	UA	1.25	96.10	44.60	---	140.70	**162.00**	121.40	**151.00**

Mixing box, with electric or pneumatic motor, with silencer

Description	Oper	Unit	Crew Size	Man-Hours Per Unit	Avg Mat'l Unit Cost	Avg Labor Unit Cost	Avg Equip Unit Cost	Avg Total Unit Cost	Avg Price Incl O&P	Avg Total Unit Cost	Avg Price Incl O&P
150 to 250 CFM	Inst	Ea	UD	2.00	587.00	62.30	---	649.30	**679.00**	572.80	**676.00**
270 to 600 CFM	Inst	Ea	UD	2.50	606.00	77.90	---	683.90	**722.00**	603.30	**714.00**

				Costs Based On Small Volume					Large Volume	
Description	Oper	Unit	Crew Size	Man-Hours Per Unit	Avg Mat'l Unit Cost	Avg Labor Unit Cost	Avg Equip Unit Cost	Avg Total Unit Cost	Avg Price Incl O&P	Avg Total Unit Cost	Avg Price Incl O&P

Fans

Air conditioning or processed air handling

Axial flow, compact, low sound 2.5" self-propelled

3,800 CFM, 5 HP	Inst	Ea	UE	8.33	4290.00	265.00	---	4555.00	4680.00	4032.00	4710.00
15,600 CFM, 10 HP	Inst	Ea	UE	16.7	7480.00	531.00	---	8011.00	8270.00	7093.00	8300.00

In-line centrifugal, supply/exhaust booster, aluminum wheel/hub, disconnect switch, 1/4" self-propelled

500 CFM, 10" dia. connect	Inst	Ea	UE	8.33	903.00	265.00	---	1168.00	1300.00	1017.00	1240.00
1,520 CFM, 16" dia. connect	Inst	Ea	UE	12.5	1180.00	398.00	---	1578.00	1770.00	1368.00	1680.00
3,480 CFM, 20" dia. connect	Inst	Ea	UE	25.0	1550.00	795.00	---	2345.00	2720.00	2016.00	2530.00

Ceiling fan, right angle, extra quiet 0.10" self-propelled

200 CFM	Inst	Ea	UE	1.56	192.00	49.60	---	241.60	266.00	210.80	256.00
900 CFM	Inst	Ea	UE	2.08	458.00	66.10	---	524.10	556.00	461.10	548.00
3,000 CFM	Inst	Ea	UE	3.13	871.00	99.50	---	970.50	1020.00	855.50	1010.00
Exterior wall or roof cap	Inst	Ea	UA	.833	127.00	29.70	---	156.70	171.00	136.80	165.00

Roof ventilators, corrosive fume resistant, plastic blades

Roof ventilator, centrifugal V belt drive motor, 1/4" self-propelled

250 CFM, 1/4 HP	Inst	Ea	UE	6.25	2680.00	199.00	---	2879.00	2970.00	2549.00	2980.00
900 CFM, 1/3 HP	Inst	Ea	UE	8.33	2900.00	265.00	---	3165.00	3290.00	2802.00	3290.00
1,650 CFM, 1/2 HP	Inst	Ea	UE	10.0	3450.00	318.00	---	3768.00	3920.00	3334.00	3910.00
2,250 CFM, 1 HP	Inst	Ea	UE	12.5	3610.00	398.00	---	4008.00	4200.00	3538.00	4170.00

Utility set centrifugal V belt drive motor, 1/4" self-propelled

1,200 CFM, 1/4 HP	Inst	Ea	UE	6.25	2680.00	199.00	---	2879.00	2970.00	2549.00	2980.00
1,500 CFM, 1/3 HP	Inst	Ea	UE	8.33	2710.00	265.00	---	2975.00	3100.00	2632.00	3090.00
1,850 CFM, 1/2 HP	Inst	Ea	UE	10.0	2710.00	318.00	---	3028.00	3180.00	2674.00	3150.00
2,200 CFM, 3/4 HP	Inst	Ea	UE	12.5	2740.00	398.00	---	3138.00	3330.00	2758.00	3280.00

Direct drive

320 CFM, 11" x 11" damper	Inst	Ea	UE	6.25	324.00	199.00	---	523.00	618.00	448.00	567.00
600 CFM, 11" x 11" damper	Inst	Ea	UE	6.25	356.00	199.00	---	555.00	650.00	476.00	600.00

Ventilation, residential

Attic

Roof-type ventilators

Aluminum dome, damper and curb

6" diameter, 300 CFM	Inst	Ea	EA	.833	271.00	28.70	---	299.70	313.00	264.90	311.00
7" diameter, 450 CFM	Inst	Ea	EA	.909	295.00	31.30	---	326.30	341.00	288.00	339.00
9" diameter, 900 CFM	Inst	Ea	EA	1.00	477.00	34.40	---	511.40	528.00	453.50	529.00
12" diameter, 1,000 CFM	Inst	Ea	EA	1.25	293.00	43.00	---	336.00	356.00	295.40	350.00
16" diameter, 1,500 CFM	Inst	Ea	EA	1.43	353.00	49.20	---	402.20	425.00	354.20	420.00
20" diameter, 2,500 CFM	Inst	Ea	EA	1.67	432.00	57.40	---	489.40	516.00	430.70	510.00
26" diameter, 4,000 CFM	Inst	Ea	EA	2.00	522.00	68.80	---	590.80	623.00	521.00	616.00
32" diameter, 6,500 CFM	Inst	Ea	EA	2.50	722.00	86.00	---	808.00	848.00	712.80	841.00
38" diameter, 8,000 CFM	Inst	Ea	EA	3.33	1070.00	115.00	---	1185.00	1240.00	1046.80	1230.00
50" diameter, 13,000 CFM	Inst	Ea	EA	5.00	1550.00	172.00	---	1722.00	1800.00	1518.00	1790.00

Plastic ABS dome

900 CFM	Inst	Ea	EA	.833	86.40	28.70	---	115.10	128.00	100.00	122.00
1,600 CFM	Inst	Ea	EA	1.00	129.00	34.40	---	163.40	179.00	142.50	172.00

Description	Oper	Unit	Crew Size	Man-Hours Per Unit	Avg Mat'l Unit Cost	Avg Labor Unit Cost	Avg Equip Unit Cost	Avg Total Unit Cost	Avg Price Incl O&P	Avg Total Unit Cost	Avg Price Incl O&P
					Costs Based On Small Volume					**Large Volume**	
Wall-type ventilators, one speed, with shutter											
12" diameter, 1,000 CFM	Inst	Ea	EA	.833	160.00	28.70	---	188.70	**202.00**	165.90	**197.00**
14" diameter, 1,500 CFM	Inst	Ea	EA	1.00	195.00	34.40	---	229.40	**245.00**	201.50	**240.00**
16" diameter, 2,000 CFM	Inst	Ea	EA	1.25	293.00	43.00	---	336.00	**356.00**	295.40	**350.00**
Entire structure, wall-type, one speed, with shutter											
30" diameter, 4,800 CFM	Inst	Ea	EA	2.00	503.00	68.80	---	571.80	**604.00**	504.00	**596.00**
36" diameter, 7,000 CFM	Inst	Ea	EA	2.50	548.00	86.00	---	634.00	**674.00**	557.80	**662.00**
42" diameter, 10,000 CFM	Inst	Ea	EA	3.33	671.00	115.00	---	786.00	**838.00**	689.80	**822.00**
48" diameter, 16,000 CFM	Inst	Ea	EA	5.00	877.00	172.00	---	1049.00	**1130.00**	920.00	**1100.00**
Two speeds, ADD	Inst	Ea	---	---	62.60	---	---	62.60	**72.00**	55.80	**64.20**
Entire structure, lay-down type, one speed, with shutter											
30" diameter, 4,500 CFM	Inst	Ea	EA	1.67	548.00	57.40	---	605.40	**632.00**	534.70	**629.00**
36" diameter, 6,500 CFM	Inst	Ea	EA	2.00	600.00	68.80	---	668.80	**700.00**	590.00	**695.00**
42" diameter, 9,000 CFM	Inst	Ea	EA	2.50	710.00	86.00	---	796.00	**835.00**	701.80	**828.00**
48" diameter, 12,000 CFM	Inst	Ea	EA	3.33	935.00	115.00	---	1050.00	**1100.00**	925.80	**1090.00**
Two speeds, ADD	Inst	Ea	---	---	15.40	---	---	15.40	**17.70**	13.70	**15.70**
12 hour timer, ADD	Inst	Ea	EA	.625	27.70	21.50	---	49.20	**59.10**	41.90	**53.50**
Minimum Job Charge	Inst	Job	EA	2.86	---	98.40	---	98.40	**144.00**	78.80	**115.00**
Grilles											
Aluminum											
Air return											
6" x 6"	Inst	Ea	UA	.417	12.80	14.90	---	27.70	**34.80**	23.30	**30.70**
10" x 6"	Inst	Ea	UA	.455	15.60	16.20	---	31.80	**39.60**	26.90	**35.20**
16" x 8"	Inst	Ea	UA	.500	22.70	17.80	---	40.50	**49.10**	34.50	**44.40**
12" x 12"	Inst	Ea	UA	.625	22.70	22.30	---	45.00	**55.70**	38.00	**49.70**
24" x 12"	Inst	Ea	UA	.714	40.60	25.50	---	66.10	**78.30**	56.60	**71.80**
48" x 24"	Inst	Ea	UA	1.00	156.00	35.70	---	191.70	**209.00**	167.50	**202.00**
Minimum Job Charge	Inst	Job	UA	2.86	---	102.00	---	102.00	**151.00**	81.70	**121.00**
Registers, air supply											
Ceiling/wall, O.B. damper, anodized aluminum											
One- or two-way deflection, adjustable curved face bars											
14" x 8"	Inst	Ea	UA	.625	27.10	22.30	---	49.40	**60.10**	42.00	**54.20**
Baseboard, adjustable damper, enameled steel											
10" x 6"	Inst	Ea	UA	.417	11.70	14.90	---	26.60	**33.80**	22.40	**29.60**
12" x 5"	Inst	Ea	UA	.455	12.80	16.20	---	29.00	**36.80**	24.40	**32.30**
12" x 6"	Inst	Ea	UA	.455	11.70	16.20	---	27.90	**35.80**	23.50	**31.30**
12" x 8"	Inst	Ea	UA	.500	17.10	17.80	---	34.90	**43.50**	29.50	**38.60**
14" x 6"	Inst	Ea	UA	.556	12.80	19.80	---	32.60	**42.10**	27.20	**36.50**
Minimum Job Charge	Inst	Job	UA	2.86	---	102.00	---	102.00	**151.00**	81.70	**121.00**
Ventilators											
Base, damper, screen; 8" neck diameter											
215 CFM @ 5 MPH wind	Inst	Ea	UD	3.33	71.60	104.00	---	175.60	**225.00**	147.00	**196.00**

					Costs Based On Small Volume					Large Volume	
Description	Oper	Unit	Crew Size	Man-Hours Per Unit	Avg Mat'l Unit Cost	Avg Labor Unit Cost	Avg Equip Unit Cost	Avg Total Unit Cost	Avg Price Incl O&P	Avg Total Unit Cost	Avg Price Incl O&P
System units complete											
Fan coil air conditioning											
Cabinet mounted, with filters											
Chilled water											
0.5 ton cooling	Inst	Ea	SB	5.00	929.00	158.00	---	1087.00	1160.00	954.00	1140.00
1 ton cooling	Inst	Ea	SB	6.67	1050.00	210.00	---	1260.00	1360.00	1105.00	1320.00
2.5 ton cooling	Inst	Ea	SB	10.0	1900.00	315.00	---	2215.00	2370.00	1952.00	2320.00
3 ton cooling	Inst	Ea	SB	20.0	2100.00	630.00	---	2730.00	3020.00	2374.00	2890.00
10 ton cooling	Inst	Ea	SF	15.0	3030.00	497.00	---	3527.00	3760.00	3097.00	3690.00
15 ton cooling	Inst	Ea	SF	20.0	4220.00	662.00	---	4882.00	5200.00	4300.00	5110.00
20 ton cooling	Inst	Ea	SF	40.0	5420.00	1320.00	---	6740.00	7370.00	5890.00	7110.00
30 ton cooling	Inst	Ea	SF	60.0	8030.00	1990.00	---	10020.00	11000.00	8750.00	10600.00
Minimum Job Charge	Inst	Job	SB	6.67	---	210.00	---	210.00	309.00	168.00	247.00
Heat pumps											
Air to air, split system, not including curbs or pads											
2 ton cooling, 8.5 MBHP heat	Inst	Ea	SB	20.0	2640.00	630.00	---	3270.00	3570.00	2864.00	3450.00
3 ton cooling, 13 MBHP heat	Inst	Ea	SB	40.0	3420.00	1260.00	---	4680.00	5270.00	4060.00	4990.00
7.5 ton cooling, 33 MBHP heat	Inst	Ea	SB	61.5	8290.00	1940.00	---	10230.00	11100.00	8920.00	10700.00
15 ton cooling, 64 MBHP heat	Inst	Ea	SB	80.0	14800.00	2520.00	---	17320.00	18500.00	15220.00	18200.00
Air to air, single package, not including curbs, pads, or plenums											
2 ton cooling, 6.5 MBHP heat	Inst	Ea	SB	16.0	3190.00	504.00	---	3694.00	3930.00	3253.00	3870.00
3 ton cooling, 10 MBHP heat	Inst	Ea	SB	26.7	3840.00	841.00	---	4681.00	5070.00	4091.00	4920.00
Water source to air, single package											
1 ton cooling, 13 MBHP heat	Inst	Ea	SB	11.4	1230.00	359.00	---	1589.00	1750.00	1378.00	1680.00
2 ton cooling, 19 MBHP heat	Inst	Ea	SB	13.3	1550.00	419.00	---	1969.00	2160.00	1717.00	2080.00
5 ton cooling, 29 MBHP heat	Inst	Ea	SB	26.7	2710.00	841.00	---	3551.00	3950.00	3091.00	3760.00
Minimum Job Charge	Inst	Job	SB	10.0	---	315.00	---	315.00	463.00	252.00	370.00
Roof top air conditioners											
Standard controls, curb, energy economizer											
Single zone, electric-fired cooling, gas-fired heating											
3 ton cooling, 60 MBHP heat	Inst	Ea	SB	16.7	5130.00	526.00	---	5656.00	5900.00	4989.00	5870.00
4 ton cooling, 95 MBHP heat	Inst	Ea	SB	20.0	5320.00	630.00	---	5950.00	6250.00	5244.00	6200.00
10 ton cooling, 200 MBHP heat	Inst	Ea	SF	75.0	12000	2480	---	14480	15600	12690	15200
30 ton cooling, 540 MBHP heat	Inst	Ea	SG	200	36900	6780	---	43680	46900	38330	45800
40 ton cooling, 675 MBHP heat	Inst	Ea	SG	246	48000	8340	---	56340	60300	49580	59200
50 ton cooling, 810 MBHP heat	Inst	Ea	SG	320	58200	10900	---	69100	74100	60960	73000
Single zone, gas-fired cooling and heating											
3 ton cooling, 90 MBHP heat	Inst	Ea	SB	20.0	6390.00	630.00	---	7020.00	7310.00	6194.00	7290.00
Multizone, electric-fired cooling, gas-fired cooling											
20 ton cooling, 360 MBHP heat	Inst	Ea	SG	200	57700	6780	---	64480	67600	56830	67100
30 ton cooling, 540 MBHP heat	Inst	Ea	SG	286	72200	9700	---	81900	86500	72170	85500
70 ton cooling, 1,500 MBHP heat	Inst	Ea	SG	500	135000	17000	---	152000	160000	134600	159000
80 ton cooling, 1,500 MBHP heat	Inst	Ea	SG	571	155000	19400	---	174400	183000	153500	181000
105 ton cooling, 1,500 MBHP heat	Inst	Ea	SG	800	203000	27100	---	230100	243000	202700	240000
Minimum Job Charge	Inst	Job	SB	6.25	---	197.00	---	197.00	289.00	158.00	232.00

Description	Oper	Unit	Crew Size	Man-Hours Per Unit	Avg Mat'l Unit Cost	Avg Labor Unit Cost	Avg Equip Unit Cost	Avg Total Unit Cost	Avg Price Incl O&P	Avg Total Unit Cost	Avg Price Incl O&P
					Costs Based On Small Volume						**Large Volume**	
Window unit air conditioners											
Semi-permanent installation, 3-speed fan, 125-volt GFI receptacle, energy-efficient models											
6,000 BTUH (0.5 ton cooling)	Inst	Ea	EB	5.00	619.00	172.00	---	791.00	**870.00**	690.00	**836.00**
9,000 BTUH (0.75 ton cooling)	Inst	Ea	EB	5.00	684.00	172.00	---	856.00	**935.00**	748.00	**902.00**
12,000 BTUH (1.0 ton cooling)	Inst	Ea	EB	5.00	729.00	172.00	---	901.00	**980.00**	788.00	**948.00**
18,000 BTUH (1.5 ton cooling)	Inst	Ea	EB	6.67	832.00	229.00	---	1061.00	**1170.00**	925.00	**1120.00**
24,000 BTUH (2.0 ton cooling)	Inst	Ea	EB	6.67	909.00	229.00	---	1138.00	**1240.00**	994.00	**1200.00**
36,000 BTUH (3.0 ton cooling)	Inst	Ea	EB	6.67	1170.00	229.00	---	1399.00	**1510.00**	1233.00	**1470.00**
Minimum Job Charge	Inst	Job	EA	3.33	---	115.00	---	115.00	**167.00**	91.80	**134.00**

Awnings. See Canopies, page 55
Backfill. See Excavation, page 109

Bath accessories.
For plumbing fixtures, see individual items.

The material cost of an item includes the fixture, water supply, and trim (includes fittings and faucets). The labor cost of an item includes installation of the fixture and connection of water supply and/or electricity, but no demolition or clean-up is included. Average rough-in of pipe, waste, and vent is an "add" item shown below each major grouping of fixtures, unless noted otherwise.

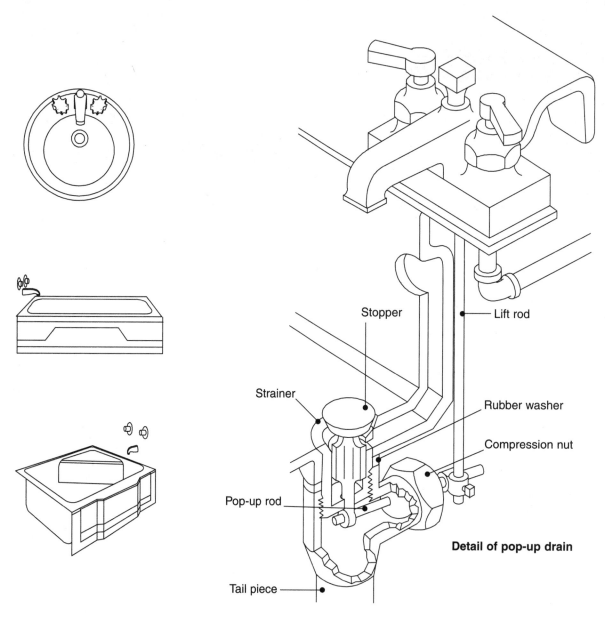

Stopper

Lift rod

Strainer

Rubber washer

Compression nut

Pop-up rod

Detail of pop-up drain

Tail piece

From: *Basic Plumbing with Illustrations* Craftsman Book Company

Bath accessories

Description	Oper	Unit	Crew Size	Man-Hours Per Unit	Avg Mat'l Unit Cost	Avg Labor Unit Cost	Avg Equip Unit Cost	Avg Total Unit Cost	Avg Price Incl O&P	Avg Total Unit Cost	Avg Price Incl O&P
										Large Volume	
Average quality											
Cup holder, surface mounted											
Chrome	Inst	Ea	CA	.444	6.08	14.30	---	20.38	**28.40**	15.71	**21.80**
Cup and toothbrush holder											
Brass	Inst	Ea	CA	.444	8.50	14.30	---	22.80	**31.20**	17.70	**24.10**
Chrome	Inst	Ea	CA	.444	6.80	14.30	---	21.10	**29.20**	16.30	**22.50**
Cup, toothbrush and soapholder, recessed											
Chrome	Inst	Ea	CA	.667	13.70	21.40	---	35.10	**47.90**	27.40	**37.10**
Glass shelf, chrome											
18" long	Inst	Ea	CA	.533	40.80	17.10	---	57.90	**72.50**	46.40	**57.90**
24" long	Inst	Ea	CA	.533	42.50	17.10	---	59.60	**74.50**	47.80	**59.50**
Grab bars, concealed mounting											
Satin stainless steel											
12" long	Inst	Ea	CA	.533	23.20	17.10	---	40.30	**52.40**	31.90	**41.20**
18" long	Inst	Ea	CA	.533	23.20	17.10	---	40.30	**52.40**	31.90	**41.20**
24" long	Inst	Ea	CA	.533	23.20	17.10	---	40.30	**52.40**	31.90	**41.20**
30" long	Inst	Ea	CA	.533	25.00	17.10	---	42.10	**54.40**	33.40	**42.90**
36" long	Inst	Ea	CA	.533	25.30	17.10	---	42.40	**54.70**	33.60	**43.20**
42" long	Inst	Ea	CA	.533	27.30	17.10	---	44.40	**57.00**	35.30	**45.10**
48" long	Inst	Ea	CA	.533	29.80	17.10	---	46.90	**60.00**	37.40	**47.50**
Angle bar, 16" L x 32" H	Inst	Ea	CA	.533	55.30	17.10	---	72.40	**89.20**	58.30	**71.60**
Tub-shower bar, 36" L x 54" H	Inst	Ea	CA	.533	86.70	17.10	---	103.80	**125.00**	84.20	**101.00**
Robe hooks, single or double	Inst	Ea	CA	.333	3.95	10.70	---	14.65	**20.60**	11.28	**15.80**
Shower curtain rod, 1" dia x 5' 5" L											
with adjacent rod holder	Inst	Ea	CA	.333	12.70	10.70	---	23.40	**30.60**	18.42	**24.00**
Soap dish, chrome											
Surface	Inst	Ea	CA	.333	6.08	10.70	---	16.78	**23.00**	13.03	**17.80**
Recessed	Inst	Ea	CA	.333	11.50	10.70	---	22.20	**29.30**	17.51	**23.00**
Soap holder with bar											
Surface	Inst	Ea	CA	.333	6.80	10.70	---	17.50	**23.90**	13.62	**18.50**
Recessed	Inst	Ea	CA	.333	16.00	10.70	---	26.70	**34.40**	21.22	**27.20**
Toilet roll holder											
Surface	Inst	Ea	CA	.333	10.20	10.70	---	20.90	**27.80**	16.42	**21.70**
Recessed	Inst	Ea	CA	.333	20.00	10.70	---	30.70	**39.10**	24.52	**31.00**
Toilet roll holder with hood, recessed											
Surface	Inst	Ea	CA	.444	21.30	14.30	---	35.60	**45.80**	28.20	**36.20**
Recessed	Inst	Ea	CA	.444	34.00	14.30	---	48.30	**60.50**	38.70	**48.20**
Towel bars, square or round											
Stainless steel											
18"	Inst	Ea	CA	.333	10.20	10.70	---	20.90	**27.80**	16.46	**21.70**
24"	Inst	Ea	CA	.333	11.20	10.70	---	21.90	**28.90**	17.26	**22.70**
30"	Inst	Ea	CA	.333	12.20	10.70	---	22.90	**30.00**	18.02	**23.60**
36"	Inst	Ea	CA	.333	13.10	10.70	---	23.80	**31.10**	18.82	**24.50**
Aluminum											
18"	Inst	Ea	CA	.333	9.39	10.70	---	20.09	**26.80**	15.76	**20.90**
24"	Inst	Ea	CA	.333	10.10	10.70	---	20.80	**27.70**	16.35	**21.60**
30"	Inst	Ea	CA	.333	10.80	10.70	---	21.50	**28.50**	16.95	**22.30**
36"	Inst	Ea	CA	.333	11.70	10.70	---	22.40	**29.40**	17.61	**23.10**
Towel pin, chrome	Inst	Ea	CA	.296	6.08	9.50	---	15.58	**21.20**	12.13	**16.50**

Description	Oper	Unit	Crew Size	Man-Hours Per Unit	Avg Mat'l Unit Cost	Avg Labor Unit Cost	Avg Equip Unit Cost	Avg Total Unit Cost	Avg Price Incl O&P	Avg Total Unit Cost	Avg Price Incl O&P
					Costs Based On Small Volume					**Large Volume**	
Towel ring											
Chrome	Inst	Ea	CA	.296	8.33	9.50	---	17.83	**23.80**	13.98	**18.60**
Antique brass	Inst	Ea	CA	.296	8.71	9.50	---	18.21	**24.30**	14.30	**18.90**
Towel ladder, 4 arms											
Antique brass	Inst	Ea	CA	.444	50.20	14.30	---	64.50	**79.10**	52.00	**63.50**
Polished chrome	Inst	Ea	CA	.444	40.30	14.30	---	54.60	**67.70**	43.90	**54.20**
Towel supply shelf, with towel bar											
18" long	Inst	Ea	CA	.333	68.40	10.70	---	79.10	**94.70**	64.32	**76.80**
24" long	Inst	Ea	CA	.333	122.00	10.70	---	132.70	**156.00**	108.02	**127.00**
Wall-to-floor angle bar with flange,											
bolt, washer and screws	Inst	Ea	CA	.593	97.80	19.00	---	116.80	**141.00**	94.80	**114.00**

Medicine cabinets

No electrical work included; for wall outlet cost, see Electrical, page 105

Surface mounting, no wall opening

Description	Oper	Unit	Crew Size	Man-Hours Per Unit	Avg Mat'l Unit Cost	Avg Labor Unit Cost	Avg Equip Unit Cost	Avg Total Unit Cost	Avg Price Incl O&P	Avg Total Unit Cost	Avg Price Incl O&P
Swing door cabinets with reversible mirror door											
16" x 22"	Inst	Ea	CA	1.07	50.60	34.30	---	84.90	**110.00**	67.30	**86.40**
16" x 26"	Inst	Ea	CA	1.07	67.60	34.30	---	101.90	**129.00**	81.40	**103.00**
Swing door, corner cabinets with reversible mirror door											
16" x 36"	Inst	Ea	CA	1.07	88.80	34.30	---	123.10	**154.00**	98.80	**123.00**
Sliding door cabinets, bypassing mirror doors											
Toplighted, stainless steel											
20" x 20"	Inst	Ea	CA	1.07	106.00	34.30	---	140.30	**174.00**	113.20	**139.00**
24" x 20"	Inst	Ea	CA	1.07	128.00	34.30	---	162.30	**198.00**	130.70	**159.00**
28" x 20"	Inst	Ea	CA	1.07	140.00	34.30	---	174.30	**213.00**	141.70	**171.00**
Cosmetic box with framed mirror stainless steel, Builders series											
18" x 26"	Inst	Ea	CA	1.07	70.40	34.30	---	104.70	**132.00**	83.60	**105.00**
24" x 32"	Inst	Ea	CA	1.07	74.30	34.30	---	108.60	**137.00**	86.90	**109.00**
30" x 32"	Inst	Ea	CA	1.07	81.20	34.30	---	115.50	**145.00**	92.60	**115.00**
36" x 32"	Inst	Ea	CA	1.07	85.90	34.30	---	120.20	**150.00**	96.50	**120.00**
48" x 32"	Inst	Ea	CA	1.07	97.80	34.30	---	132.10	**164.00**	106.20	**131.00**

3-way mirror, tri-view

Description	Oper	Unit	Crew Size	Man-Hours Per Unit	Avg Mat'l Unit Cost	Avg Labor Unit Cost	Avg Equip Unit Cost	Avg Total Unit Cost	Avg Price Incl O&P	Avg Total Unit Cost	Avg Price Incl O&P
30" x 30"											
Frameless, beveled mirror	Inst	Ea	CA	1.18	225.00	37.90	---	262.90	**315.00**	213.50	**255.00**
Natural oak	Inst	Ea	CA	1.18	258.00	37.90	---	295.90	**354.00**	241.50	**288.00**
White finish	Inst	Ea	CA	1.18	241.00	37.90	---	278.90	**334.00**	227.50	**271.00**
36" x 30"											
Frameless, beveled mirror	Inst	Ea	CA	1.18	255.00	37.90	---	292.90	**351.00**	238.50	**285.00**
Natural oak	Inst	Ea	CA	1.18	289.00	37.90	---	326.90	**389.00**	266.50	**317.00**
White finish	Inst	Ea	CA	1.18	272.00	37.90	---	309.90	**370.00**	252.50	**300.00**
48" x 30"											
Frameless, beveled mirror	Inst	Ea	CA	1.33	331.00	42.70	---	373.70	**444.00**	304.10	**361.00**
Natural oak	Inst	Ea	CA	1.33	355.00	42.70	---	397.70	**473.00**	325.10	**385.00**
White finish	Inst	Ea	CA	1.33	347.00	42.70	---	389.70	**463.00**	318.10	**377.00**
With matching light fixture											
with 2 lights	Inst	Ea	EA	2.67	59.50	91.80	---	151.30	**202.00**	117.80	**157.00**
with 3 lights	Inst	Ea	EA	2.67	89.30	91.80	---	181.10	**237.00**	142.30	**185.00**
with 4 lights	Inst	Ea	EA	2.67	97.80	91.80	---	189.60	**246.00**	149.30	**193.00**
with 6 lights	Inst	Ea	EA	2.67	119.00	91.80	---	210.80	**271.00**	166.80	**213.00**
With matching light fixture for beveled mirror cabinets only											
with 2 lights	Inst	Ea	EA	2.67	80.80	91.80	---	172.60	**227.00**	135.30	**177.00**
with 3 lights	Inst	Ea	EA	2.67	97.80	91.80	---	189.60	**246.00**	149.30	**193.00**
with 4 lights	Inst	Ea	EA	2.67	128.00	91.80	---	219.80	**281.00**	173.80	**221.00**
with 6 lights	Inst	Ea	EA	2.67	149.00	91.80	---	240.80	**305.00**	191.80	**241.00**

Description	Oper	Unit	Crew Size	Man-Hours Per Unit	Avg Mat'l Unit Cost	Avg Labor Unit Cost	Avg Equip Unit Cost	Avg Total Unit Cost	Avg Price Incl O&P	Avg Total Unit Cost	Avg Price Incl O&P
					Costs Based On Small Volume					Large Volume	
Recessed mounting, overall sizes											
Swing door with mirror											
Builders series											
14" x 18"											
Polished brass strip	Inst	Ea	CA	1.33	51.90	42.70	---	94.60	**124.00**	74.80	**97.30**
Polished edge strip	Inst	Ea	CA	1.33	74.00	42.70	---	116.70	**149.00**	93.00	**118.00**
Stainless steel strip	Inst	Ea	CA	1.33	62.10	42.70	---	104.80	**135.00**	83.20	**107.00**
14" x 24"											
Polished brass strip	Inst	Ea	CA	1.33	57.00	42.70	---	99.70	**130.00**	79.00	**102.00**
Polished edge strip	Inst	Ea	CA	1.33	79.10	42.70	---	121.80	**155.00**	97.20	**123.00**
Stainless steel strip	Inst	Ea	CA	1.33	68.00	42.70	---	110.70	**142.00**	88.10	**113.00**
Decorator series											
14" x 18"											
Frameless bevel mirror	Inst	Ea	CA	1.33	84.20	42.70	---	126.90	**161.00**	101.40	**128.00**
Natural oak	Inst	Ea	CA	1.33	114.00	42.70	---	156.70	**195.00**	125.90	**156.00**
White birch	Inst	Ea	CA	1.33	111.00	42.70	---	153.70	**191.00**	123.10	**153.00**
White finish	Inst	Ea	CA	1.33	136.00	42.70	---	178.70	**220.00**	144.10	**177.00**
14" x 24"											
Frameless bevel mirror	Inst	Ea	CA	1.33	85.00	42.70	---	127.70	**162.00**	102.10	**129.00**
Natural oak	Inst	Ea	CA	1.33	115.00	42.70	---	157.70	**196.00**	126.60	**157.00**
White birch	Inst	Ea	CA	1.33	111.00	42.70	---	153.70	**192.00**	123.80	**154.00**
White finish	Inst	Ea	CA	1.33	140.00	42.70	---	182.70	**225.00**	148.10	**181.00**
Oak framed cabinet with oval mirror											
20" x 36"	Inst	Ea	CA	1.33	170.00	42.70	---	212.70	**260.00**	172.10	**209.00**
Mirrors											
Decorator oval mirrors, antique gold											
16" x 24"	Inst	Ea	CA	.213	55.30	6.84	---	62.14	**73.80**	50.63	**60.00**
16" x 32"	Inst	Ea	CA	.213	68.00	6.84	---	74.84	**88.50**	61.13	**72.10**

Shower equipment. See Shower stalls, page 240
Vanity cabinets. See Cabinets, page 53

			Costs Based On Small Volume						Large Volume	
Description	Oper	Unit	Crew Size	Man-Hours Per Unit	Avg Mat'l Unit Cost	Avg Labor Unit Cost	Avg Equip Unit Cost	Avg Total Unit Cost	Avg Price Incl O&P	Avg Total Unit Cost	Avg Price Incl O&P

Bathtubs (includes whirlpools)

Plumbing fixtures with good quality supply fittings and faucets included in material cost. Labor cost for installation only of fixture, supply fittings, and faucets. For rough-in, see Adjustments in this bathtub section

Frequently encountered applications

Detach & reset operations

Description	Oper	Unit	Crew Size	Man-Hours Per Unit	Avg Mat'l Unit Cost	Avg Labor Unit Cost	Avg Equip Unit Cost	Avg Total Unit Cost	Avg Price Incl O&P	Avg Total Unit Cost	Avg Price Incl O&P
Free standing tub	Reset	Ea	SB	4.57	55.00	144.00	---	199.00	275.00	139.50	192.00
Recessed tub	Reset	Ea	SB	5.71	55.00	180.00	---	235.00	328.00	164.50	229.00
Sunken tub	Reset	Ea	SB	7.62	55.00	240.00	---	295.00	416.00	206.50	291.00

Remove operations

Description	Oper	Unit	Crew Size	Man-Hours Per Unit	Avg Mat'l Unit Cost	Avg Labor Unit Cost	Avg Equip Unit Cost	Avg Total Unit Cost	Avg Price Incl O&P	Avg Total Unit Cost	Avg Price Incl O&P
Free standing tub	Demo	Ea	SB	1.90	15.00	59.90	---	74.90	105.00	52.40	73.70
Recessed tub	Demo	Ea	SB	2.86	30.00	90.10	---	120.10	167.00	84.00	117.00
Sunken tub	Demo	Ea	SB	3.81	45.00	120.00	---	165.00	228.00	115.60	160.00

Replace operations

Recessed into wall (above floor installation)

Description	Oper	Unit	Crew Size	Man-Hours Per Unit	Avg Mat'l Unit Cost	Avg Labor Unit Cost	Avg Equip Unit Cost	Avg Total Unit Cost	Avg Price Incl O&P	Avg Total Unit Cost	Avg Price Incl O&P
Enameled cast iron, colors											
60" L x 30" W x 14" H	Inst	Ea	SB	9.5	900.00	299.00	---	1199.00	1470.00	840.00	1030.00
66" L x 32" W x 16-1/4" H	Inst	Ea	SB	9.5	1910.00	299.00	---	2209.00	2630.00	1540.00	1840.00
Enameled steel, colors											
60" L x 30" W x 17-1/2" H	Inst	Ea	SB	9.5	455.00	299.00	---	754.00	964.00	529.00	675.00

Sunken (below floor installation) or full frame support with full sidewall finish

Description	Oper	Unit	Crew Size	Man-Hours Per Unit	Avg Mat'l Unit Cost	Avg Labor Unit Cost	Avg Equip Unit Cost	Avg Total Unit Cost	Avg Price Incl O&P	Avg Total Unit Cost	Avg Price Incl O&P
Enameled cast iron, colors											
60" L x 32" W x 18-1/4" H	Inst	Ea	SB	9.5	2160.00	299.00	---	2459.00	2920.00	1720.00	2040.00
Bathtub with whirlpool											
Plumbing installation	Inst	Ea	SB	19.0	3760.00	599.00	---	4359.00	5200.00	3049.00	3640.00
Electrical installation	Inst	Ea	EA	8.79	128.00	302.00	---	430.00	588.00	301.30	411.00
71-3/4" L x 36" W x 20-7/8" H	Inst	Ea	SB	9.5	3670.00	299.00	---	3969.00	4660.00	2780.00	3260.00
Bathtub with whirlpool											
Plumbing installation	Inst	Ea	SB	19.0	5760.00	599.00	---	6359.00	7500.00	4449.00	5250.00
Electrical installation	Inst	Ea	EA	8.79	128.00	302.00	---	430.00	588.00	301.30	411.00
Fiberglass, colors											
60" L x 42" W x 20" H	Inst	Ea	SB	9.5	1280.00	299.00	---	1579.00	1920.00	1108.00	1340.00
Bathtub with whirlpool											
Plumbing installation	Inst	Ea	SB	19.0	2410.00	599.00	---	3009.00	3650.00	2099.00	2550.00
Electrical installation	Inst	Ea	EA	8.79	128.00	302.00	---	430.00	588.00	301.30	411.00
72" L x 42" W x 20" H	Inst	Ea	SB	9.5	1340.00	299.00	---	1639.00	1980.00	1146.00	1380.00
Bathtub with whirlpool											
Plumbing installation	Inst	Ea	SB	19.0	2590.00	599.00	---	3189.00	3860.00	2229.00	2700.00
Electrical installation	Inst	Ea	EA	8.79	128.00	302.00	---	430.00	588.00	301.30	411.00
60" L x 60" W x 21" H, corner	Inst	Ea	SB	9.5	1500.00	299.00	---	1799.00	2170.00	1260.00	1520.00
Bathtub with whirlpool											
Plumbing installation	Inst	Ea	SB	19.0	2870.00	599.00	---	3469.00	4180.00	2429.00	2930.00
Electrical installation	Inst	Ea	EA	8.79	128.00	302.00	---	430.00	588.00	301.30	411.00

Adjustments

Description	Oper	Unit	Crew Size	Man-Hours Per Unit	Avg Mat'l Unit Cost	Avg Labor Unit Cost	Avg Equip Unit Cost	Avg Total Unit Cost	Avg Price Incl O&P	Avg Total Unit Cost	Avg Price Incl O&P
Install rough-In	Inst	Ea	SB	14.3	85.00	450.00	---	535.00	760.00	374.50	531.00
Install shower head with mixer valve over tub											
Open wall, ADD	Inst	Ea	SA	2.86	115.00	104.00	---	219.00	285.00	153.20	199.00
Closed wall, ADD	Inst	Ea	SA	5.71	180.00	208.00	---	388.00	512.00	271.00	359.00

Description	Oper	Unit	Crew Size	Man-Hours Per Unit	Avg Mat'l Unit Cost	Avg Labor Unit Cost	Avg Equip Unit Cost	Avg Total Unit Cost	Avg Price Incl O&P	Avg Total Unit Cost	Avg Price Incl O&P
										Large Volume	

Free-standing
American Standard Products, enameled steel
Reminiscence Slipper Soaking Bathtub,
white or black textured exterior
72-1/8" L x 37-1/2" W x 28-3/4" H

Description	Oper	Unit	Crew	MH	Mat'l	Labor	Equip	Total	Price	L.Total	L.Price
White	Inst	Ea	SB	11.4	3210.00	359.00	---	3569.00	4220.00	2492.00	2950.00
Colors	Inst	Ea	SB	11.4	3270.00	359.00	---	3629.00	4280.00	2542.00	3000.00
Premium Colors	Inst	Ea	SB	11.4	3320.00	359.00	---	3679.00	4350.00	2582.00	3040.00
Required accessories											
Antique ball-and-claw legs (four)	Inst	Set	SB	---	479.00	---	---	479.00	551.00	336.00	386.00
Antique bath faucet without riser	Inst	Ea	SB	---	553.00	---	---	553.00	635.00	387.00	445.00
Antique drain, chain & stopper	Inst	Set	SB	---	195.00	---	---	195.00	224.00	136.00	157.00
Antique riser tubes	Inst	Set	SB	---	106.00	---	---	106.00	122.00	74.40	85.50

Kohler Products, enameled cast iron
Birthday Bath
72" L x 37-1/2" W x 21-1/2" H

Description	Oper	Unit	Crew	MH	Mat'l	Labor	Equip	Total	Price	L.Total	L.Price
White	Inst	Ea	SB	11.4	6050.00	359.00	---	6409.00	7490.00	4482.00	5240.00
Colors	Inst	Ea	SB	11.4	7080.00	359.00	---	7439.00	8670.00	5212.00	6070.00
Premium Colors	Inst	Ea	SB	11.4	7860.00	359.00	---	8219.00	9560.00	5752.00	6700.00
Required accessories											
Antique ball-and-claw legs (four)	Inst	Set	SB	---	893.00	---	---	893.00	1030.00	625.00	718.00
Antique bath faucet without riser	Inst	Ea	SB	---	638.00	---	---	638.00	733.00	446.00	513.00
Antique drain, chain & stopper	Inst	Set	SB	---	319.00	---	---	319.00	367.00	223.00	257.00
Antique riser tubes	Inst	Set	SB	---	106.00	---	---	106.00	122.00	74.40	85.50

Vintage Bath
72" L x 42" W x 22" H

Description	Oper	Unit	Crew	MH	Mat'l	Labor	Equip	Total	Price	L.Total	L.Price
White	Inst	Ea	SB	11.4	5770.00	359.00	---	6129.00	7170.00	4292.00	5020.00
Colors	Inst	Ea	SB	11.4	6740.00	359.00	---	7099.00	8280.00	4972.00	5790.00
Required accessories											
Wood base	Inst	Ea	SB	---	2330.00	---	---	2330.00	2690.00	1630.00	1880.00
Ceramic base	Inst	Ea	SB	---	977.00	---	---	977.00	1120.00	684.00	787.00
Adjustable feet	Inst	Set	SB	---	42.50	---	---	42.50	48.90	29.80	34.20

Recessed
American Standard Products, Americast
60" L x 32" W x 17-3/4" H, w/ grab bar (Cambridge)

Description	Oper	Unit	Crew	MH	Mat'l	Labor	Equip	Total	Price	L.Total	L.Price
White	Inst	Ea	SB	9.5	935.00	299.00	---	1234.00	1510.00	865.00	1060.00
Colors	Inst	Ea	SB	9.5	1060.00	299.00	---	1359.00	1660.00	953.00	1160.00
Premium Colors	Inst	Ea	SB	9.5	1150.00	299.00	---	1449.00	1760.00	1015.00	1240.00
Bathtub with whirlpool (System-I)											
Plumbing installation											
White	Inst	Ea	SB	19.0	2900.00	599.00	---	3499.00	4220.00	2449.00	2950.00
Colors	Inst	Ea	SB	19.0	3010.00	599.00	---	3609.00	4340.00	2519.00	3040.00
Premium Colors	Inst	Ea	SB	19.0	3160.00	599.00	---	3759.00	4510.00	2629.00	3160.00
Electrical installation	Inst	Ea	EA	8.79	128.00	302.00	---	430.00	588.00	301.30	411.00

Description	Oper	Unit	Crew Size	Man-Hours Per Unit	Costs Based On Small Volume						Large Volume	
					Avg Mat'l Unit Cost	Avg Labor Unit Cost	Avg Equip Unit Cost	Avg Total Unit Cost	Avg Price Incl O&P		Avg Total Unit Cost	Avg Price Incl O&P
60" L x 34" W x 17-1/2" H, w/ luxury ledge (Princeton)												
White	Inst	Ea	SB	9.5	997.00	299.00	---	1296.00	1590.00		909.00	1110.00
Colors	Inst	Ea	SB	9.5	1080.00	299.00	---	1379.00	1680.00		966.00	1180.00
Premium Colors	Inst	Ea	SB	9.5	1150.00	299.00	---	1449.00	1770.00		1019.00	1240.00
66" L x 32" W x 20" H (Stratford)												
White	Inst	Ea	SB	9.5	1410.00	299.00	---	1709.00	2060.00		1195.00	1440.00
Colors	Inst	Ea	SB	9.5	1630.00	299.00	---	1929.00	2310.00		1351.00	1620.00
Bathtub with whirlpool												
Plumbing installation												
White	Inst	Ea	SB	19.0	3690.00	599.00	---	4289.00	5120.00		2999.00	3580.00
Colors	Inst	Ea	SB	19.0	3910.00	599.00	---	4509.00	5370.00		3159.00	3760.00
Electrical installation	Inst	Ea	EA	8.79	128.00	302.00	---	430.00	588.00		301.30	411.00
Options												
Grab bar kit												
Chrome	Inst	Ea	SB	---	155.00	---	---	155.00	178.00		108.00	125.00
Brass	Inst	Ea	SB	---	263.00	---	---	263.00	302.00		184.00	211.00
White	Inst	Ea	SB	---	263.00	---	---	263.00	302.00		184.00	211.00
Satin	Inst	Ea	SB	---	263.00	---	---	263.00	302.00		184.00	211.00
Drain kit												
Chrome	Inst	Ea	SB	---	106.00	---	---	106.00	122.00		74.40	85.50
Brass	Inst	Ea	SB	---	185.00	---	---	185.00	213.00		130.00	149.00
White	Inst	Ea	SB	---	211.00	---	---	211.00	242.00		148.00	170.00
Satin	Inst	Ea	SB	---	185.00	---	---	185.00	213.00		130.00	149.00

American Standard Products, enameled steel

Description	Oper	Unit	Crew Size	Man-Hours Per Unit	Avg Mat'l Unit Cost	Avg Labor Unit Cost	Avg Equip Unit Cost	Avg Total Unit Cost	Avg Price Incl O&P		Avg Total Unit Cost	Avg Price Incl O&P
60" L x 30" W x 15" H (Salem)												
White, slip resistant	Inst	Ea	SB	9.5	423.00	299.00	---	722.00	926.00		506.00	649.00
Colors, slip resistant	Inst	Ea	SB	9.5	447.00	299.00	---	746.00	954.00		523.00	669.00
60" L x 30" W x 17-1/2" H (Solar)												
White, slip resistant	Inst	Ea	SB	9.5	428.00	299.00	---	727.00	932.00		510.00	653.00
Colors, slip resistant	Inst	Ea	SB	9.5	455.00	299.00	---	754.00	964.00		529.00	675.00

Kohler Products, enameled cast iron

Description	Oper	Unit	Crew Size	Man-Hours Per Unit	Avg Mat'l Unit Cost	Avg Labor Unit Cost	Avg Equip Unit Cost	Avg Total Unit Cost	Avg Price Incl O&P		Avg Total Unit Cost	Avg Price Incl O&P
48" L x 44" W x 14" H, corner (Mayflower)												
White	Inst	Ea	SB	9.5	2290.00	299.00	---	2589.00	3080.00		1810.00	2150.00
Colors	Inst	Ea	SB	9.5	2770.00	299.00	---	3069.00	3620.00		2150.00	2540.00
54" L x 30-1/4" W x 14" H (Seaforth)												
White	Inst	Ea	SB	9.5	1180.00	299.00	---	1479.00	1800.00		1036.00	1260.00
Colors	Inst	Ea	SB	9.5	1400.00	299.00	---	1699.00	2050.00		1188.00	1430.00
60" L x 32" W x 16-1/4" H (Mendota)												
White	Inst	Ea	SB	9.5	1060.00	299.00	---	1359.00	1660.00		955.00	1170.00
Colors	Inst	Ea	SB	9.5	1290.00	299.00	---	1589.00	1920.00		1112.00	1350.00
Premium Colors	Inst	Ea	SB	9.5	1430.00	299.00	---	1729.00	2090.00		1210.00	1460.00

Description	Oper	Unit	Crew Size	Man-Hours Per Unit	Costs Based On Small Volume						Large Volume	
					Avg Mat'l Unit Cost	Avg Labor Unit Cost	Avg Equip Unit Cost	Avg Total Unit Cost	Avg Price Incl O&P		Avg Total Unit Cost	Avg Price Incl O&P
60" L x 34-1/4" W x 14" H (Villager)												
White	Inst	Ea	SB	9.5	765.00	299.00	---	1064.00	1320.00		745.00	925.00
Colors	Inst	Ea	SB	9.5	900.00	299.00	---	1199.00	1470.00		840.00	1030.00
66" L x 32" W x 16-1/4" H (Dynametric)												
White	Inst	Ea	SB	9.5	1590.00	299.00	---	1889.00	2270.00		1320.00	1590.00
Colors	Inst	Ea	SB	9.5	1910.00	299.00	---	2209.00	2630.00		1540.00	1840.00

Sunken

American Standard Products, acrylic

58-1/4" L x 58-1/4" W x 22-1/2" H, corner, with grab bar drilling (Savona)

Bathtub with whirlpool (System-II)

Description	Oper	Unit	Crew Size	Man-Hours Per Unit	Avg Mat'l Unit Cost	Avg Labor Unit Cost	Avg Equip Unit Cost	Avg Total Unit Cost	Avg Price Incl O&P		Avg Total Unit Cost	Avg Price Incl O&P
Plumbing installation												
White	Inst	Ea	SB	19.0	3410.00	599.00	---	4009.00	4800.00		2809.00	3360.00
Colors	Inst	Ea	SB	19.0	3460.00	599.00	---	4059.00	4860.00		2839.00	3400.00
Premium Colors	Inst	Ea	SB	19.0	3520.00	599.00	---	4119.00	4930.00		2879.00	3450.00
Electrical installation	Inst	Ea	EA	8.79	128.00	302.00	---	430.00	588.00		301.30	411.00
Options												
Grab bar kit												
Chrome	Inst	Ea	SB	---	94.40	---	---	94.40	109.00		66.10	76.00
Brass	Inst	Ea	SB	---	132.00	---	---	132.00	152.00		92.20	106.00
White	Inst	Ea	SB	---	118.00	---	---	118.00	136.00		82.70	95.10
Satin	Inst	Ea	SB	---	132.00	---	---	132.00	152.00		92.20	106.00
Drain kit												
Chrome	Inst	Ea	SB	---	106.00	---	---	106.00	122.00		74.40	85.50
Brass	Inst	Ea	SB	---	185.00	---	---	185.00	213.00		130.00	149.00
White	Inst	Ea	SB	---	211.00	---	---	211.00	242.00		148.00	170.00
Satin	Inst	Ea	SB	---	185.00	---	---	185.00	213.00		130.00	149.00
59-1/2" L x 42-1/4" W x 21-1/4" H, oval (Savona)												
White	Inst	Ea	SB	9.5	1090.00	299.00	---	1389.00	1690.00		971.00	1180.00
Colors	Inst	Ea	SB	9.5	1140.00	299.00	---	1439.00	1750.00		1005.00	1220.00
Premium Colors	Inst	Ea	SB	9.5	1180.00	299.00	---	1479.00	1800.00		1038.00	1260.00
Bathtub with whirlpool (System-II)												
Plumbing installation												
White	Inst	Ea	SB	19.0	2060.00	599.00	---	2659.00	3250.00		1859.00	2280.00
Colors	Inst	Ea	SB	19.0	2120.00	599.00	---	2719.00	3320.00		1899.00	2320.00
Premium Colors	Inst	Ea	SB	19.0	2170.00	599.00	---	2769.00	3380.00		1939.00	2360.00
Electrical installation	Inst	Ea	EA	8.79	128.00	302.00	---	430.00	588.00		301.30	411.00
Drain kit												
Chrome	Inst	Ea	SB	---	106.00	---	---	106.00	122.00		74.40	85.50
Brass	Inst	Ea	SB	---	185.00	---	---	185.00	213.00		130.00	149.00
White	Inst	Ea	SB	---	211.00	---	---	211.00	242.00		148.00	170.00
Satin	Inst	Ea	SB	---	185.00	---	---	185.00	213.00		130.00	149.00

Description	Oper	Unit	Crew Size	Man-Hours Per Unit	Costs Based On Small Volume						Large Volume	
					Avg Mat'l Unit Cost	Avg Labor Unit Cost	Avg Equip Unit Cost	Avg Total Unit Cost	Avg Price Incl O&P		Avg Total Unit Cost	Avg Price Incl O&P
60" L x 32" W x 21-1/2" H (Ellisse Built-In)												
Bathtub with whirlpool (System-I)												
Plumbing installation												
White	Inst	Ea	SB	19.0	2760.00	599.00	---	3359.00	4060.00		2349.00	2840.00
Colors	Inst	Ea	SB	19.0	2820.00	599.00	---	3419.00	4120.00		2389.00	2880.00
Electrical installation	Inst	Ea	EA	8.79	128.00	302.00	---	430.00	588.00		301.30	411.00
Bathtub with whirlpool (System-II)												
Plumbing installation												
White	Inst	Ea	SB	19.0	2970.00	599.00	---	3569.00	4290.00		2499.00	3010.00
Colors	Inst	Ea	SB	19.0	3020.00	599.00	---	3619.00	4350.00		2529.00	3050.00
Electrical installation	Inst	Ea	EA	8.79	128.00	302.00	---	430.00	588.00		301.30	411.00
Options												
Grab bar kit												
Chrome	Inst	Ea	SB	---	150.00	---	---	150.00	172.00		105.00	120.00
Brass	Inst	Ea	SB	---	250.00	---	---	250.00	287.00		175.00	201.00
White	Inst	Ea	SB	---	250.00	---	---	250.00	287.00		175.00	201.00
Drain kit												
Chrome	Inst	Ea	SB	---	106.00	---	---	106.00	122.00		74.40	85.50
Brass	Inst	Ea	SB	---	185.00	---	---	185.00	213.00		130.00	149.00
White	Inst	Ea	SB	---	211.00	---	---	211.00	242.00		148.00	170.00
69-1/4" L x 38-1/2" W x 19-3/4" H (Ellisse Oval)												
Bathtub with whirlpool (System-I)												
Plumbing installation												
White	Inst	Ea	SB	19.0	2410.00	599.00	---	3009.00	3650.00		2109.00	2550.00
Colors	Inst	Ea	SB	19.0	2450.00	599.00	---	3049.00	3700.00		2129.00	2590.00
Premium Colors	Inst	Ea	SB	19.0	2490.00	599.00	---	3089.00	3740.00		2159.00	2620.00
Electrical installation	Inst	Ea	EA	8.79	128.00	302.00	---	430.00	588.00		301.30	411.00
Options												
Grab bar kit												
Chrome	Inst	Ea	SB	---	112.00	---	---	112.00	129.00		78.50	90.30
Brass	Inst	Ea	SB	---	187.00	---	---	187.00	216.00		131.00	151.00
White	Inst	Ea	SB	---	187.00	---	---	187.00	216.00		131.00	151.00
Satin	Inst	Ea	SB	---	187.00	---	---	187.00	216.00		131.00	151.00
Drain kit												
Chrome	Inst	Ea	SB	---	79.70	---	---	79.70	91.60		55.80	64.20
Brass	Inst	Ea	SB	---	139.00	---	---	139.00	160.00		97.30	112.00
White	Inst	Ea	SB	---	158.00	---	---	158.00	182.00		111.00	127.00
Satin	Inst	Ea	SB	---	139.00	---	---	139.00	160.00		97.30	112.00

Description	Oper	Unit	Crew Size	Man-Hours Per Unit	Avg Mat'l Unit Cost	Avg Labor Unit Cost	Avg Equip Unit Cost	Avg Total Unit Cost	Avg Price Incl O&P	Avg Total Unit Cost	Avg Price Incl O&P
								Costs Based On Small Volume		Large Volume	
72" L x 48" W x 21-1/2" H (Ellisse)											
Bathtub with whirlpool											
Plumbing installation											
White	Inst	Ea	SB	19.0	3570.00	599.00	---	4169.00	4990.00	2919.00	3490.00
Colors	Inst	Ea	SB	19.0	3620.00	599.00	---	4219.00	5050.00	2959.00	3530.00
Premium Colors	Inst	Ea	SB	19.0	3680.00	599.00	---	4279.00	5110.00	2999.00	3580.00
Electrical installation	Inst	Ea	EA	8.79	128.00	302.00	---	430.00	588.00	301.30	411.00
Drain kit											
Chrome	Inst	Ea	SB	---	106.00	---	---	106.00	122.00	74.40	85.50
Brass	Inst	Ea	SB	---	185.00	---	---	185.00	213.00	130.00	149.00
White	Inst	Ea	SB	---	211.00	---	---	211.00	242.00	148.00	170.00
Satin	Inst	Ea	SB	---	185.00	---	---	185.00	213.00	130.00	149.00
Kohler Products, enameled cast iron											
60" L x 36" W x 20-3/8" H (Steeping Bath)											
White	Inst	Ea	SB	9.5	2370.00	299.00	---	2669.00	3160.00	1870.00	2210.00
Colors	Inst	Ea	SB	9.5	2860.00	299.00	---	3159.00	3730.00	2210.00	2610.00
Premium Colors	Inst	Ea	SB	9.5	3230.00	299.00	---	3529.00	4150.00	2470.00	2910.00
Bathtub with whirlpool (System-III)											
Plumbing installation											
White	Inst	Ea	SB	19.0	5770.00	599.00	---	6369.00	7520.00	4459.00	5260.00
Colors	Inst	Ea	SB	19.0	6260.00	599.00	---	6859.00	8080.00	4799.00	5660.00
Premium Colors	Inst	Ea	SB	19.0	6530.00	599.00	---	7129.00	8390.00	4989.00	5870.00
Electrical installation	Inst	Ea	EA	8.79	128.00	302.00	---	430.00	588.00	301.30	411.00
60" L x 32" W x 18-1/4" H (Tea-for-Two)											
White	Inst	Ea	SB	9.5	1820.00	299.00	---	2119.00	2540.00	1490.00	1780.00
Colors	Inst	Ea	SB	9.5	2160.00	299.00	---	2459.00	2920.00	1720.00	2040.00
Premium Colors	Inst	Ea	SB	9.5	2400.00	299.00	---	2699.00	3200.00	1890.00	2240.00
Bathtub with whirlpool (System-II)											
Plumbing installation											
White	Inst	Ea	SB	19.0	3490.00	599.00	---	4089.00	4900.00	2859.00	3430.00
Colors	Inst	Ea	SB	19.0	3760.00	599.00	---	4359.00	5200.00	3049.00	3640.00
Premium Colors	Inst	Ea	SB	19.0	3900.00	599.00	---	4499.00	5360.00	3149.00	3750.00
Electrical installation	Inst	Ea	EA	8.79	128.00	302.00	---	430.00	588.00	301.30	411.00
66" L x 32" W x 18-1/4" H (Maestro)											
White	Inst	Ea	SB	9.5	1450.00	299.00	---	1749.00	2110.00	1230.00	1480.00
Colors	Inst	Ea	SB	9.5	1700.00	299.00	---	1999.00	2390.00	1400.00	1670.00
Premium Colors	Inst	Ea	SB	9.5	1880.00	299.00	---	2179.00	2600.00	1530.00	1820.00
Bathtub with whirlpool (System-II)											
Plumbing installation											
White	Inst	Ea	SB	19.0	4320.00	599.00	---	4919.00	5850.00	3449.00	4090.00
Colors	Inst	Ea	SB	19.0	4680.00	599.00	---	5279.00	6260.00	3699.00	4390.00
Premium Colors	Inst	Ea	SB	19.0	4880.00	599.00	---	5479.00	6490.00	3839.00	4540.00
Electrical installation	Inst	Ea	EA	8.79	128.00	302.00	---	430.00	588.00	301.30	411.00

				Costs Based On Small Volume					Large Volume	
Description	Oper	Unit	Crew Size	Man-Hours Per Unit	Avg Mat'l Unit Cost	Avg Labor Unit Cost	Avg Equip Unit Cost	Avg Total Unit Cost	Avg Price Incl O&P	Avg Total Unit Cost	Avg Price Incl O&P
66" L x 32" W x 18" H (Tea-for-Two)											
White	Inst	Ea	SB	9.5	1550.00	299.00	---	1849.00	2220.00	1300.00	1560.00
Colors	Inst	Ea	SB	9.5	1820.00	299.00	---	2119.00	2540.00	1490.00	1780.00
Premium Colors	Inst	Ea	SB	9.5	2030.00	299.00	---	2329.00	2770.00	1630.00	1940.00
Bathtub with whirlpool (System-II)											
Plumbing installation											
White	Inst	Ea	SB	19.0	4430.00	599.00	---	5029.00	5980.00	3519.00	4190.00
Colors	Inst	Ea	SB	19.0	4710.00	599.00	---	5309.00	6290.00	3709.00	4400.00
Premium Colors	Inst	Ea	SB	19.0	4910.00	599.00	---	5509.00	6530.00	3859.00	4570.00
Electrical installation	Inst	Ea	EA	8.79	128.00	302.00	---	430.00	588.00	301.30	411.00
71-3/4" L x 36" W x 20-7/8" H (Tea-for-Two)											
White	Inst	Ea	SB	9.5	3030.00	299.00	---	3329.00	3930.00	2330.00	2750.00
Colors	Inst	Ea	SB	9.5	3670.00	299.00	---	3969.00	4660.00	2780.00	3260.00
Premium Colors	Inst	Ea	SB	9.5	4140.00	299.00	---	4439.00	5200.00	3110.00	3640.00
Bathtub with whirlpool (System-III)											
Plumbing installation											
White	Inst	Ea	SB	19.0	5320.00	599.00	---	5919.00	6990.00	4139.00	4900.00
Colors	Inst	Ea	SB	19.0	5760.00	599.00	---	6359.00	7500.00	4449.00	5250.00
Premium Colors	Inst	Ea	SB	19.0	6000.00	599.00	---	6599.00	7780.00	4619.00	5450.00
Electrical installation	Inst	Ea	EA	8.79	128.00	302.00	---	430.00	588.00	301.30	411.00
72" L x 36" W x 18-1/4" H (Caribbean)											
White	Inst	Ea	SB	9.5	2630.00	299.00	---	2929.00	3460.00	2050.00	2430.00
Colors	Inst	Ea	SB	9.5	3120.00	299.00	---	3419.00	4030.00	2390.00	2820.00
Premium Colors	Inst	Ea	SB	9.5	3490.00	299.00	---	3789.00	4450.00	2650.00	3120.00
Bathtub with whirlpool (System-III)											
Plumbing installation											
White	Inst	Ea	SB	19.0	5760.00	599.00	---	6359.00	7510.00	4449.00	5250.00
Colors	Inst	Ea	SB	19.0	6250.00	599.00	---	6849.00	8070.00	4799.00	5650.00
Premium Colors	Inst	Ea	SB	19.0	6520.00	599.00	---	7119.00	8380.00	4979.00	5860.00
Electrical installation	Inst	Ea	EA	8.79	128.00	302.00	---	430.00	588.00	301.30	411.00
72" L x 42" W x 22" H (Seawall Bath) with personal shower system											
White	Inst	Ea	SB	9.5	6540.00	299.00	---	6839.00	7960.00	4790.00	5580.00
Colors	Inst	Ea	SB	9.5	7470.00	299.00	---	7769.00	9030.00	5440.00	6320.00
Premium Colors	Inst	Ea	SB	9.5	8170.00	299.00	---	8469.00	9840.00	5930.00	6890.00
Bathtub with whirlpool (System-III)											
Plumbing installation											
White	Inst	Ea	SB	19.0	10500.00	599.00	---	11099.00	13000.00	7779.00	9080.00
Colors	Inst	Ea	SB	19.0	11600.00	599.00	---	12199.00	14200.00	8549.00	9970.00
Premium Colors	Inst	Ea	SB	19.0	11900.00	599.00	---	12499.00	14600.00	8779.00	10200.00
Electrical installation	Inst	Ea	EA	8.79	128.00	302.00	---	430.00	588.00	301.30	411.00
Options											
Grab bar kit											
Chrome	Inst	Ea	SB	---	184.00	---	---	184.00	212.00	129.00	148.00
Brass	Inst	Ea	SB	---	324.00	---	---	324.00	372.00	227.00	261.00
Nickel	Inst	Ea	SB	---	424.00	---	---	424.00	488.00	297.00	342.00
Drain kit											
Chrome	Inst	Ea	SB	---	264.00	---	---	264.00	304.00	185.00	213.00
Brass	Inst	Ea	SB	---	277.00	---	---	277.00	318.00	194.00	223.00
Nickel	Inst	Ea	SB	---	349.00	---	---	349.00	401.00	244.00	281.00

Description	Oper	Unit	Crew Size	Man-Hours Per Unit	Avg Mat'l Unit Cost	Avg Labor Unit Cost	Avg Equip Unit Cost	Avg Total Unit Cost	Avg Price Incl O&P	Avg Total Unit Cost	Avg Price Incl O&P
										Large Volume	
Kohler Products, fiberglass											
60" L x 36" W x 20" H (Mariposa)											
White	Inst	Ea	SB	9.5	1100.00	299.00	---	1399.00	1700.00	978.00	1190.00
Colors	Inst	Ea	SB	9.5	1130.00	299.00	---	1429.00	1740.00	999.00	1220.00
Bathtub with whirlpool (System-I)											
Plumbing installation											
White	Inst	Ea	SB	19.0	2290.00	599.00	---	2889.00	3510.00	2019.00	2460.00
Colors	Inst	Ea	SB	19.0	2360.00	599.00	---	2959.00	3600.00	2069.00	2520.00
Electrical installation	Inst	Ea	EA	8.79	128.00	302.00	---	430.00	588.00	301.30	411.00
60" L x 42" W x 20" H (Hourglass)											
White	Inst	Ea	SB	9.5	1050.00	299.00	---	1349.00	1650.00	944.00	1150.00
Colors	Inst	Ea	SB	9.5	1080.00	299.00	---	1379.00	1680.00	964.00	1180.00
Bathtub with whirlpool (System-I)											
Plumbing installation											
White	Inst	Ea	SB	19.0	2150.00	599.00	---	2749.00	3350.00	1919.00	2340.00
Colors	Inst	Ea	SB	19.0	2210.00	599.00	---	2809.00	3420.00	1969.00	2390.00
Electrical installation	Inst	Ea	EA	8.79	128.00	302.00	---	430.00	588.00	301.30	411.00
60" L x 42" W x 20" H (Symbio)											
White	Inst	Ea	SB	9.5	1550.00	299.00	---	1849.00	2230.00	1300.00	1560.00
Colors	Inst	Ea	SB	9.5	1610.00	299.00	---	1909.00	2290.00	1340.00	1600.00
Bathtub with whirlpool (System-I)											
Plumbing installation											
White	Inst	Ea	SB	19.0	3700.00	599.00	---	4299.00	5140.00	3009.00	3600.00
Colors	Inst	Ea	SB	19.0	3840.00	599.00	---	4439.00	5300.00	3109.00	3710.00
Electrical installation	Inst	Ea	EA	8.79	128.00	302.00	---	430.00	588.00	301.30	411.00
60" L x 42" W x 20" H (Synchrony)											
White	Inst	Ea	SB	9.5	1250.00	299.00	---	1549.00	1870.00	1082.00	1310.00
Colors	Inst	Ea	SB	9.5	1280.00	299.00	---	1579.00	1920.00	1108.00	1340.00
Bathtub with whirlpool (System-I)											
Plumbing installation											
White	Inst	Ea	SB	19.0	2330.00	599.00	---	2929.00	3560.00	2049.00	2490.00
Colors	Inst	Ea	SB	19.0	2410.00	599.00	---	3009.00	3650.00	2099.00	2550.00
Electrical installation	Inst	Ea	EA	8.79	128.00	302.00	---	430.00	588.00	301.30	411.00
60" L x 42-1/4" W x 18" H (Dockside)											
White	Inst	Ea	SB	9.5	1240.00	299.00	---	1539.00	1860.00	1076.00	1300.00
Colors	Inst	Ea	SB	9.5	1270.00	299.00	---	1569.00	1900.00	1101.00	1330.00
Bathtub with whirlpool (System-I)											
Plumbing installation											
White	Inst	Ea	SB	19.0	2300.00	599.00	---	2899.00	3530.00	2029.00	2470.00
Colors	Inst	Ea	SB	19.0	2380.00	599.00	---	2979.00	3610.00	2079.00	2530.00
Electrical installation	Inst	Ea	EA	8.79	128.00	302.00	---	430.00	588.00	301.30	411.00
60" L x 60" W x 21" H, corner (Sojourn)											
White	Inst	Ea	SB	9.5	1460.00	299.00	---	1759.00	2120.00	1230.00	1480.00
Colors	Inst	Ea	SB	9.5	1500.00	299.00	---	1799.00	2170.00	1260.00	1520.00
Bathtub with whirlpool (System-I)											
Plumbing installation											
White	Inst	Ea	SB	19.0	2780.00	599.00	---	3379.00	4070.00	2359.00	2850.00
Colors	Inst	Ea	SB	19.0	2870.00	599.00	---	3469.00	4180.00	2429.00	2930.00
Electrical installation	Inst	Ea	EA	8.79	128.00	302.00	---	430.00	588.00	301.30	411.00

Description	Oper	Unit	Crew Size	Man-Hours Per Unit	Costs Based On Small Volume					Large Volume	
					Avg Mat'l Unit Cost	Avg Labor Unit Cost	Avg Equip Unit Cost	Avg Total Unit Cost	Avg Price Incl O&P	Avg Total Unit Cost	Avg Price Incl O&P
66" L x 35-7/8" W x 20" H (Mariposa)											
White	Inst	Ea	SB	9.5	1100.00	299.00	---	1399.00	**1710.00**	982.00	**1200.00**
Colors	Inst	Ea	SB	9.5	1130.00	299.00	---	1429.00	**1740.00**	1003.00	**1220.00**
Bathtub with whirlpool (System-I)											
Plumbing installation											
White	Inst	Ea	SB	19.0	2390.00	599.00	---	2989.00	**3630.00**	2089.00	**2540.00**
Colors	Inst	Ea	SB	19.0	2470.00	599.00	---	3069.00	**3720.00**	2149.00	**2600.00**
Electrical installation	Inst	Ea	EA	8.79	128.00	302.00	---	430.00	**588.00**	301.30	**411.00**
66" L x 42" W x 22-5/8" H (Folio)											
White	Inst	Ea	SB	9.5	1250.00	299.00	---	1549.00	**1880.00**	1086.00	**1320.00**
Colors	Inst	Ea	SB	9.5	1290.00	299.00	---	1589.00	**1920.00**	1112.00	**1350.00**
Bathtub with whirlpool (System-I)											
Plumbing installation											
White	Inst	Ea	SB	19.0	2370.00	599.00	---	2969.00	**3610.00**	2079.00	**2530.00**
Colors	Inst	Ea	SB	19.0	2450.00	599.00	---	3049.00	**3700.00**	2129.00	**2590.00**
Electrical installation	Inst	Ea	EA	8.79	128.00	302.00	---	430.00	**588.00**	301.30	**411.00**
72" L x 36" W x 20" H (Mariposa)											
White	Inst	Ea	SB	9.5	1110.00	299.00	---	1409.00	**1710.00**	985.00	**1200.00**
Colors	Inst	Ea	SB	9.5	1140.00	299.00	---	1439.00	**1750.00**	1006.00	**1220.00**
Bathtub with whirlpool (System-I)											
Plumbing installation											
White	Inst	Ea	SB	19.0	2400.00	599.00	---	2999.00	**3640.00**	2099.00	**2550.00**
Colors	Inst	Ea	SB	19.0	2480.00	599.00	---	3079.00	**3730.00**	2149.00	**2610.00**
Electrical installation	Inst	Ea	EA	8.79	128.00	302.00	---	430.00	**588.00**	301.30	**411.00**
72" L x 42" W x 20" H (Synchrony)											
White	Inst	Ea	SB	9.5	1300.00	299.00	---	1599.00	**1930.00**	1118.00	**1350.00**
Colors	Inst	Ea	SB	9.5	1340.00	299.00	---	1639.00	**1980.00**	1146.00	**1380.00**
Bathtub with whirlpool (System-I)											
Plumbing installation											
White	Inst	Ea	SB	19.0	2510.00	599.00	---	3109.00	**3760.00**	2179.00	**2640.00**
Colors	Inst	Ea	SB	19.0	2590.00	599.00	---	3189.00	**3860.00**	2229.00	**2700.00**
Electrical installation	Inst	Ea	EA	8.79	128.00	302.00	---	430.00	**588.00**	301.30	**411.00**

Adjustments

Detach & Reset Tub

Description	Oper	Unit	Crew Size	Man-Hours Per Unit	Avg Mat'l Unit Cost	Avg Labor Unit Cost	Avg Equip Unit Cost	Avg Total Unit Cost	Avg Price Incl O&P	Avg Total Unit Cost	Avg Price Incl O&P
Free standing tub	Reset	Ea	SB	4.57	55.00	144.00	---	199.00	**275.00**	139.50	**192.00**
Recessed tub	Reset	Ea	SB	5.71	55.00	180.00	---	235.00	**328.00**	164.50	**229.00**
Sunken tub	Reset	Ea	SB	7.62	55.00	240.00	---	295.00	**416.00**	206.50	**291.00**
Remove Tub											
Free standing tub	Demo	Ea	SB	1.90	15.00	59.90	---	74.90	**105.00**	52.40	**73.70**
Recessed tub	Demo	Ea	SB	2.86	30.00	90.10	---	120.10	**167.00**	84.00	**117.00**
Sunken tub	Demo	Ea	SB	3.81	45.00	120.00	---	165.00	**228.00**	115.60	**160.00**
Install Rough-In	Inst	Ea	SB	14.3	85.00	450.00	---	535.00	**760.00**	374.50	**531.00**
Install shower head w/ mixer valve over tub											
Open wall, ADD	Inst	Ea	SA	2.86	115.00	104.00	---	219.00	**285.00**	153.20	**199.00**
Closed wall, ADD	Inst	Ea	SA	5.71	180.00	208.00	---	388.00	**512.00**	271.00	**359.00**

Block, concrete. See Masonry, page 181
Brick. See Masonry, page 178

Cabinets

Top quality cabinets are built with the structural stability of fine furniture. Framing stock is kiln dried and a full 1" thick. Cabinets have backs, usually 5-ply $3/16$"-thick plywood, with all backs and interiors finished. Frames should be constructed of hardwood with mortise and tenon joints; corner blocks should be used on all four corners of all base cabinets. Doors are usually of select $7/16$" thick solid core construction using semi-concealed hinges. End panels are $1/2$" thick and attached to frames with mortise and tenon joints, glued and pinned under pressure. Panels should also be dadoed to receive the tops and bottoms of wall cabinets. Shelves are adjustable with veneer faces and front edges. The hardware includes magnetic catches, heavy duty die cast pulls and hinges, and ball-bearing suspension system. The finish is scratch and stain resistant, including a first coat of hand-wiped stain, a sealer coat, and a synthetic varnish with plastic laminate characteristics.

Average quality cabinets feature hardwood frame construction with plywood backs and veneered plywood end panels. Joints are glued mortise and tenon. Doors are solid core attached with exposed self-closing hinges. Shelves are adjustable, and drawers ride on a ball-bearing side suspension glide system. The finish is usually three coats including stain, sealer, and a mar-resistant top coat for easy cleaning.

Economy quality cabinets feature pine construction with joints glued under pressure. Doors, drawer fronts, and side or end panels are constructed of either $1/2$"-thick wood composition board or $1/2$"-thick veneered pine. Face frames are $3/4$"-thick wood composition board or $3/4$"-thick pine. Features include adjustable shelves, hinge straps, and a three-point suspension system on drawers (using nylon rollers). The finish consists of a filler coat, base coat, and final polyester top coat.

Kitchen Cabinet Installation Procedure

To develop a layout plan, measure and write down the following:

1. Floor space.

2. Height and width of all walls.

3. Location of electrical outlets.

4. Size and position of doors, windows, and vents.

5. Location of any posts or pillars. Walls must be prepared if chair rails or baseboards are located where cabinets will be installed.

6. Common height and depth of base cabinets (including 1" for countertops) and wall cabinets.

When you plan the cabinet placement, consider this Rule of Thumb:

Allow between $4^1/2'$ and $5^1/2'$ of counter surface between the refrigerator and sink. Allow between 3' and 4' between the sink and range.

1. What do you have to fit into the available space?

2. Is there enough counter space on both sides of all appliances and sinks? The kitchen has three work centers, each with a major appliance as its hub, and each needing adequate counter space. They are:

   a. Fresh and frozen food center — Refrigerator-freezer

   b. Clean-up center — Sink with disposal-dishwasher

   c. Cooking center — Range-oven

3. Will the sink workspace fit neatly in front of a window?

4. The kitchen triangle is the most efficient kitchen design; it means placing each major center at approximately equidistant triangle points. The ideal triangle is 22 feet total. It should never be less than 13 feet or more than 25 feet.

5. Where are the centers located? A logical working and walking pattern is from refrigerator to sink to range. The refrigerator should be at a triangle point near a door, to minimize the distance to bring in groceries and reduce traffic that could interfere with food preparation. The range should be at a triangle point near the serving and dining area. The sink is located between the two. The refrigerator should be located far enough from the range so that the heat will not affect the refrigerator's cooling efficiency.

6. Does the plan allow for lighting the range and sink work centers and for ventilating the range center?

7. Make sure that open doors (such as cabinet doors or entrance/exit doors) won't interfere with access to an appliance. To clear appliances and cabinets, a door opening should not be less than 30" from the corner, since such equipment is 24" to 28" in depth. A clearance of 48" is necessary when a range is next to a door.

8. Next locate the wall studs with a stud finder or hammer, since all cabinets attach to walls with screws, never nails. Also remove chair rails or baseboards where they conflict with cabinets.

Wall Cabinets: From the highest point on the floor, measure up approximately 84" to determine the top of wall cabinets. Using two #10 x $2^1/2$" wood screws, drill through hanging strips built into the cabinet backs at top and bottom. Use a level to assure that cabinets and doors are aligned, then tighten the screws.

Base Cabinets: Start with a corner unit and a base unit on each side of the corner unit. Place this combination in the corner and work out from both sides. Use "C" clamps when connecting cabinets to draw adjoining units into alignment. With the front face plumb and the unit level from front to back and across the front edge, attach the unit to wall studs by screwing through the hanging strips. To attach adjoining cabinets, drill two holes in the vertical side of one cabinet, inside the door (near top and bottom), and just into the stile of the adjoining cabinet.

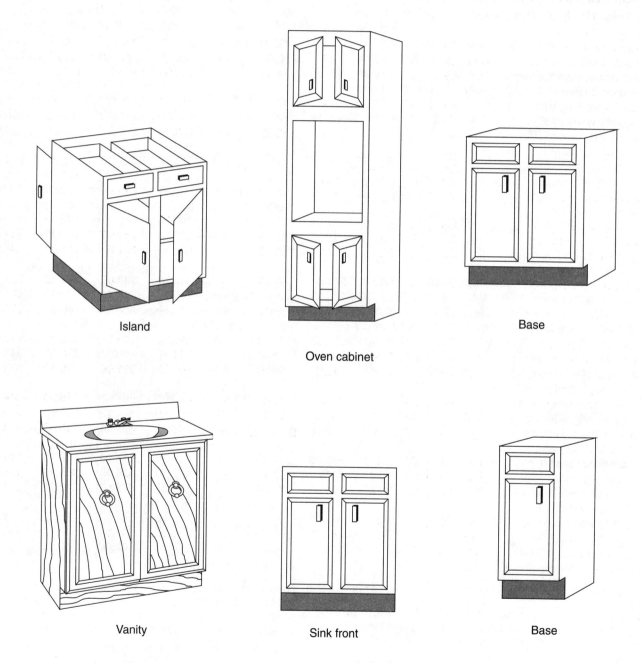

Island

Oven cabinet

Base

Vanity

Sink front

Base

Description	Oper	Unit	Crew Size	Man-Hours Per Unit	Avg Mat'l Unit Cost	Avg Labor Unit Cost	Avg Equip Unit Cost	Avg Total Unit Cost	Avg Price Incl O&P	Avg Total Unit Cost	Avg Price Incl O&P
										Large Volume	
Cabinets											
Labor costs include hanging and fitting of cabinets											
Kitchen											
3' x 4', wood; base, wall, or peninsula	Demo	Ea	LB	1.067	---	28.40	---	28.40	**42.40**	17.10	**25.40**
Kitchen; all hardware included											
Base cabinets, 35" H, 24" D; no tops											
12" W, 1 door, 1 drawer											
High quality workmanship	Inst	Ea	CJ	1.67	191.00	49.10	---	240.10	**302.00**	198.40	**247.00**
Good quality workmanship	Inst	Ea	CJ	1.11	156.00	32.60	---	188.60	**236.00**	157.60	**195.00**
Average quality workmanship	Inst	Ea	CJ	1.11	119.00	32.60	---	151.60	**192.00**	124.60	**156.00**
15" W, 1 door, 1 drawer											
High quality workmanship	Inst	Ea	CJ	1.67	201.00	49.10	---	250.10	**315.00**	207.40	**258.00**
Good quality workmanship	Inst	Ea	CJ	1.11	166.00	32.60	---	198.60	**248.00**	166.60	**206.00**
Average quality workmanship	Inst	Ea	CJ	1.11	126.00	32.60	---	158.60	**200.00**	131.60	**163.00**
18" W, 1 door, 1 drawer											
High quality workmanship	Inst	Ea	CJ	1.67	208.00	49.10	---	257.10	**323.00**	213.40	**265.00**
Good quality workmanship	Inst	Ea	CJ	1.11	172.00	32.60	---	204.60	**255.00**	171.60	**212.00**
Average quality workmanship	Inst	Ea	CJ	1.11	131.00	32.60	---	163.60	**206.00**	135.60	**169.00**
21" W, 1 door, 1 drawer											
High quality workmanship	Inst	Ea	CJ	1.93	231.00	56.70	---	287.70	**362.00**	239.10	**297.00**
Good quality workmanship	Inst	Ea	CJ	1.23	191.00	36.10	---	227.10	**283.00**	190.80	**235.00**
Average quality workmanship	Inst	Ea	CJ	1.23	147.00	36.10	---	183.10	**231.00**	151.80	**189.00**
24" W, 1 door, 1 drawer											
High quality workmanship	Inst	Ea	CJ	1.93	254.00	56.70	---	310.70	**390.00**	259.10	**321.00**
Good quality workmanship	Inst	Ea	CJ	1.23	208.00	36.10	---	244.10	**304.00**	205.80	**254.00**
Average quality workmanship	Inst	Ea	CJ	1.23	184.00	36.10	---	220.10	**275.00**	184.80	**228.00**
30" W, 2 doors, 2 drawers											
High quality workmanship	Inst	Ea	CJ	2.58	324.00	75.80	---	399.80	**502.00**	332.20	**412.00**
Good quality workmanship	Inst	Ea	CJ	1.74	280.00	51.10	---	331.10	**413.00**	278.50	**343.00**
Average quality workmanship	Inst	Ea	CJ	1.74	212.00	51.10	---	263.10	**331.00**	218.50	**271.00**
36" W, 2 doors, 2 drawers											
High quality workmanship	Inst	Ea	CJ	2.58	373.00	75.80	---	448.80	**561.00**	375.20	**464.00**
Good quality workmanship	Inst	Ea	CJ	1.74	308.00	51.10	---	359.10	**446.00**	303.50	**373.00**
Average quality workmanship	Inst	Ea	CJ	1.74	235.00	51.10	---	286.10	**358.00**	238.50	**295.00**
42" W, 2 doors, 2 drawers											
High quality workmanship	Inst	Ea	CJ	3.33	431.00	97.80	---	528.80	**663.00**	439.70	**546.00**
Good quality workmanship	Inst	Ea	CJ	2.22	357.00	65.20	---	422.20	**526.00**	355.10	**438.00**
Average quality workmanship	Inst	Ea	CJ	2.22	271.00	65.20	---	336.20	**423.00**	279.10	**347.00**
48" W, 4 doors, 2 drawers											
High quality workmanship	Inst	Ea	CJ	3.33	415.00	97.80	---	512.80	**644.00**	425.70	**529.00**
Good quality workmanship	Inst	Ea	CJ	2.22	343.00	65.20	---	408.20	**509.00**	343.10	**423.00**
Average quality workmanship	Inst	Ea	CJ	2.22	259.00	65.20	---	324.20	**409.00**	268.10	**334.00**

				Costs Based On Small Volume						Large Volume	
Description	Oper	Unit	Crew Size	Man-Hours Per Unit	Avg Mat'l Unit Cost	Avg Labor Unit Cost	Avg Equip Unit Cost	Avg Total Unit Cost	Avg Price Incl O&P	Avg Total Unit Cost	Avg Price Incl O&P
Base corner cabinet, blind, 35" H, 24" D, no tops											
36" W, 1 door, 1 drawer											
High quality workmanship	Inst	Ea	CJ	2.39	271.00	70.20	---	341.20	**431.00**	282.00	**351.00**
Good quality workmanship	Inst	Ea	CJ	1.65	217.00	48.50	---	265.50	**333.00**	221.00	**274.00**
Average quality workmanship	Inst	Ea	CJ	1.65	173.00	48.50	---	221.50	**281.00**	182.00	**228.00**
42" W, 1 door, 1 drawer											
High quality workmanship	Inst	Ea	CJ	3.02	299.00	88.70	---	387.70	**492.00**	318.50	**398.00**
Good quality workmanship	Inst	Ea	CJ	1.93	247.00	56.70	---	303.70	**381.00**	253.10	**313.00**
Average quality workmanship	Inst	Ea	CJ	1.93	187.00	56.70	---	243.70	**310.00**	200.10	**250.00**
36" x 36" (lazy Susan)											
High quality workmanship	Inst	Ea	CJ	2.39	404.00	70.20	---	474.20	**590.00**	400.00	**493.00**
Good quality workmanship	Inst	Ea	CJ	1.65	334.00	48.50	---	382.50	**474.00**	325.00	**399.00**
Average quality workmanship	Inst	Ea	CJ	1.65	256.00	48.50	---	304.50	**379.00**	255.00	**315.00**
Utility closets, 81" to 85" H, 24" W, 24" D											
High quality workmanship	Inst	Ea	CJ	3.33	620.00	97.80	---	717.80	**890.00**	607.70	**747.00**
Good quality workmanship	Inst	Ea	CJ	2.22	483.00	65.20	---	548.20	**677.00**	467.10	**572.00**
Average quality workmanship	Inst	Ea	CJ	2.22	368.00	65.20	---	433.20	**539.00**	365.10	**449.00**
Drawer base cabinets, 35" H, 24" D											
4 drawers, no tops											
18" W											
High quality workmanship	Inst	Ea	CJ	1.67	311.00	49.10	---	360.10	**447.00**	304.40	**374.00**
Good quality workmanship	Inst	Ea	CJ	1.11	267.00	32.60	---	299.60	**369.00**	255.60	**313.00**
Average quality workmanship	Inst	Ea	CJ	1.11	204.00	32.60	---	236.60	**293.00**	199.60	**246.00**
24" W											
High quality workmanship	Inst	Ea	CJ	1.93	349.00	56.70	---	405.70	**503.00**	343.10	**422.00**
Good quality workmanship	Inst	Ea	CJ	1.23	288.00	36.10	---	324.10	**399.00**	276.80	**338.00**
Average quality workmanship	Inst	Ea	CJ	1.23	221.00	36.10	---	257.10	**319.00**	216.80	**267.00**
Island cabinets											
Island base cabinets, 35" H, 24" D											
24" W, 1 door both sides											
High quality workmanship	Inst	Ea	CJ	2.05	452.00	60.20	---	512.20	**632.00**	436.10	**534.00**
Good quality workmanship	Inst	Ea	CJ	1.37	371.00	40.20	---	411.20	**506.00**	353.10	**430.00**
Average quality workmanship	Inst	Ea	CJ	1.37	285.00	40.20	---	325.20	**403.00**	277.10	**339.00**
30" W, 2 doors both sides											
High quality workmanship	Inst	Ea	CJ	2.67	478.00	78.40	---	556.40	**691.00**	470.00	**578.00**
Good quality workmanship	Inst	Ea	CJ	1.78	394.00	52.30	---	446.30	**551.00**	380.40	**466.00**
Average quality workmanship	Inst	Ea	CJ	1.78	299.00	52.30	---	351.30	**438.00**	296.40	**365.00**
36" W, 2 doors both sides											
High quality workmanship	Inst	Ea	CJ	2.67	502.00	78.40	---	580.40	**720.00**	492.00	**604.00**
Good quality workmanship	Inst	Ea	CJ	1.78	415.00	52.30	---	467.30	**576.00**	398.40	**488.00**
Average quality workmanship	Inst	Ea	CJ	1.78	315.00	52.30	---	367.30	**456.00**	310.40	**382.00**
48" W, 4 doors both sides											
High quality workmanship	Inst	Ea	CJ	3.56	539.00	105.00	---	644.00	**804.00**	539.60	**667.00**
Good quality workmanship	Inst	Ea	CJ	2.32	445.00	68.10	---	513.10	**636.00**	434.80	**534.00**
Average quality workmanship	Inst	Ea	CJ	2.32	338.00	68.10	---	406.10	**508.00**	339.80	**420.00**

Description	Oper	Unit	Crew Size	Man-Hours Per Unit	Avg Mat'l Unit Cost	Avg Labor Unit Cost	Avg Equip Unit Cost	Avg Total Unit Cost	Avg Price Incl O&P	Avg Total Unit Cost	Avg Price Incl O&P
								Costs Based On Small Volume		**Large Volume**	
Corner island base cabinets, 35" H, 24" D											
42" W, 4 doors, 2 drawers											
High quality workmanship	Inst	Ea	CJ	3.56	478.00	105.00	---	583.00	**730.00**	485.60	**602.00**
Good quality workmanship	Inst	Ea	CJ	2.32	390.00	68.10	---	458.10	**571.00**	386.80	**476.00**
Average quality workmanship	Inst	Ea	CJ	2.32	294.00	68.10	---	362.10	**455.00**	300.80	**374.00**
48" W, 6 doors, 2 drawers											
High quality workmanship	Inst	Ea	CJ	4.44	501.00	130.00	---	631.00	**796.00**	521.40	**650.00**
Good quality workmanship	Inst	Ea	CJ	2.96	413.00	86.90	---	499.90	**626.00**	418.30	**517.00**
Average quality workmanship	Inst	Ea	CJ	2.96	313.00	86.90	---	399.90	**506.00**	329.30	**411.00**
Hanging corner island wall cabinets, 18" H, 24" D											
18" W, 3 doors											
High quality workmanship	Inst	Ea	CJ	2.42	347.00	71.10	---	418.10	**522.00**	349.60	**432.00**
Good quality workmanship	Inst	Ea	CJ	1.62	254.00	47.60	---	301.60	**376.00**	253.50	**312.00**
Average quality workmanship	Inst	Ea	CJ	1.62	194.00	47.60	---	241.60	**304.00**	200.50	**249.00**
24" W, 3 doors											
High quality workmanship	Inst	Ea	CJ	2.42	357.00	71.10	---	428.10	**535.00**	358.60	**443.00**
Good quality workmanship	Inst	Ea	CJ	1.62	264.00	47.60	---	311.60	**388.00**	262.50	**324.00**
Average quality workmanship	Inst	Ea	CJ	1.62	200.00	47.60	---	247.60	**311.00**	205.50	**255.00**
30" W, 3 doors											
High quality workmanship	Inst	Ea	CJ	3.14	368.00	92.20	---	460.20	**579.00**	381.20	**473.00**
Good quality workmanship	Inst	Ea	CJ	2.11	275.00	62.00	---	337.00	**423.00**	280.00	**348.00**
Average quality workmanship	Inst	Ea	CJ	2.11	210.00	62.00	---	272.00	**345.00**	223.00	**279.00**
Hanging island cabinets, 18" H, 12" D											
30" W, 2 doors both sides											
High quality workmanship	Inst	Ea	CJ	2.86	368.00	84.00	---	452.00	**567.00**	375.90	**465.00**
Good quality workmanship	Inst	Ea	CJ	1.90	270.00	55.80	---	325.80	**407.00**	272.50	**337.00**
Average quality workmanship	Inst	Ea	CJ	1.90	205.00	55.80	---	260.80	**329.00**	214.50	**268.00**
36" W, 2 doors both sides											
High quality workmanship	Inst	Ea	CJ	2.86	378.00	84.00	---	462.00	**580.00**	384.90	**477.00**
Good quality workmanship	Inst	Ea	CJ	1.90	278.00	55.80	---	333.80	**418.00**	279.50	**346.00**
Average quality workmanship	Inst	Ea	CJ	1.90	212.00	55.80	---	267.80	**338.00**	221.50	**275.00**
Hanging island cabinets, 24" H, 12" D											
24" W, 2 doors both sides											
High quality workmanship	Inst	Ea	CJ	2.42	347.00	71.10	---	418.10	**522.00**	349.60	**432.00**
Good quality workmanship	Inst	Ea	CJ	1.62	242.00	47.60	---	289.60	**361.00**	242.50	**299.00**
Average quality workmanship	Inst	Ea	CJ	1.62	179.00	47.60	---	226.60	**286.00**	186.50	**232.00**
30" W, 2 doors both sides											
High quality workmanship	Inst	Ea	CJ	3.14	361.00	92.20	---	453.20	**571.00**	374.20	**466.00**
Good quality workmanship	Inst	Ea	CJ	2.11	261.00	62.00	---	323.00	**406.00**	268.00	**333.00**
Average quality workmanship	Inst	Ea	CJ	2.11	193.00	62.00	---	255.00	**324.00**	208.00	**260.00**
36" W, 2 doors both sides											
High quality workmanship	Inst	Ea	CJ	3.14	378.00	92.20	---	470.20	**592.00**	390.20	**485.00**
Good quality workmanship	Inst	Ea	CJ	2.11	278.00	62.00	---	340.00	**427.00**	283.00	**351.00**
Average quality workmanship	Inst	Ea	CJ	2.11	212.00	62.00	---	274.00	**347.00**	225.00	**281.00**
42" W, 2 doors both sides											
High quality workmanship	Inst	Ea	CJ	4.10	457.00	120.00	---	577.00	**729.00**	477.30	**594.00**
Good quality workmanship	Inst	Ea	CJ	2.71	336.00	79.60	---	415.60	**523.00**	345.90	**429.00**
Average quality workmanship	Inst	Ea	CJ	2.71	257.00	79.60	---	336.60	**428.00**	275.90	**345.00**
48" W, 4 doors both sides											
High quality workmanship	Inst	Ea	CJ	4.10	548.00	120.00	---	668.00	**838.00**	557.30	**691.00**
Good quality workmanship	Inst	Ea	CJ	2.71	404.00	79.60	---	483.60	**604.00**	405.90	**501.00**
Average quality workmanship	Inst	Ea	CJ	2.71	308.00	79.60	---	387.60	**489.00**	320.90	**399.00**

Description	Oper	Unit	Crew Size	Man-Hours Per Unit	Costs Based On Small Volume						Large Volume	
					Avg Mat'l Unit Cost	Avg Labor Unit Cost	Avg Equip Unit Cost	Avg Total Unit Cost	Avg Price Incl O&P		Avg Total Unit Cost	Avg Price Incl O&P

Oven cabinets

81" to 85" H, 24" D, 27" W

Description	Oper	Unit	Crew Size	Man-Hours	Mat'l	Labor	Equip	Total	Price O&P	Total LV	Price LV
High quality workmanship	Inst	Ea	CJ	3.33	662.00	97.80	---	759.80	**940.00**	644.70	**791.00**
Good quality workmanship	Inst	Ea	CJ	2.22	515.00	65.20	---	580.20	**715.00**	495.10	**605.00**
Average quality workmanship	Inst	Ea	CJ	2.22	394.00	65.20	---	459.20	**570.00**	388.10	**477.00**

Sink/range

Base cabinets, 35" H, 24" D, 2 doors, no tops included

30" W

Description	Oper	Unit	Crew Size	Man-Hours	Mat'l	Labor	Equip	Total	Price O&P	Total LV	Price LV
High quality workmanship	Inst	Ea	CJ	2.22	258.00	65.20	---	323.20	**408.00**	268.10	**333.00**
Good quality workmanship	Inst	Ea	CJ	1.48	212.00	43.50	---	255.50	**320.00**	214.10	**265.00**
Average quality workmanship	Inst	Ea	CJ	1.48	162.00	43.50	---	205.50	**259.00**	169.10	**211.00**
36" W											
High quality workmanship	Inst	Ea	CJ	2.22	330.00	65.20	---	395.20	**493.00**	331.10	**409.00**
Good quality workmanship	Inst	Ea	CJ	1.48	273.00	43.50	---	316.50	**393.00**	268.10	**329.00**
Average quality workmanship	Inst	Ea	CJ	1.48	208.00	43.50	---	251.50	**315.00**	210.10	**260.00**
42" W											
High quality workmanship	Inst	Ea	CJ	2.91	365.00	85.50	---	450.50	**567.00**	375.10	**465.00**
Good quality workmanship	Inst	Ea	CJ	1.93	302.00	56.70	---	358.70	**448.00**	302.10	**373.00**
Average quality workmanship	Inst	Ea	CJ	1.93	229.00	56.70	---	285.70	**360.00**	237.10	**294.00**
48" W											
High quality workmanship	Inst	Ea	CJ	2.91	414.00	85.50	---	499.50	**625.00**	417.10	**516.00**
Good quality workmanship	Inst	Ea	CJ	1.93	340.00	56.70	---	396.70	**493.00**	335.10	**413.00**
Average quality workmanship	Inst	Ea	CJ	1.93	260.00	56.70	---	316.70	**398.00**	265.10	**328.00**

Sink/range front cabinets, 35" H, 2 doors

30" W

Description	Oper	Unit	Crew Size	Man-Hours	Mat'l	Labor	Equip	Total	Price O&P	Total LV	Price LV
High quality workmanship	Inst	Ea	CJ	1.93	162.00	56.70	---	218.70	**279.00**	177.10	**223.00**
Good quality workmanship	Inst	Ea	CJ	1.33	134.00	39.10	---	173.10	**220.00**	142.50	**178.00**
Average quality workmanship	Inst	Ea	CJ	1.33	103.00	39.10	---	142.10	**182.00**	114.60	**145.00**
36" W											
High quality workmanship	Inst	Ea	CJ	1.93	218.00	56.70	---	274.70	**347.00**	227.10	**283.00**
Good quality workmanship	Inst	Ea	CJ	1.33	181.00	39.10	---	220.10	**275.00**	183.50	**227.00**
Average quality workmanship	Inst	Ea	CJ	1.33	141.00	39.10	---	180.10	**227.00**	148.50	**185.00**
42" W											
High quality workmanship	Inst	Ea	CJ	2.50	239.00	73.40	---	312.40	**397.00**	256.40	**321.00**
Good quality workmanship	Inst	Ea	CJ	1.67	200.00	49.10	---	249.10	**313.00**	206.40	**256.00**
Average quality workmanship	Inst	Ea	CJ	1.67	151.00	49.10	---	200.10	**255.00**	163.40	**205.00**
48" W											
High quality workmanship	Inst	Ea	CJ	2.50	252.00	73.40	---	325.40	**413.00**	267.40	**334.00**
Good quality workmanship	Inst	Ea	CJ	1.67	216.00	49.10	---	265.10	**333.00**	221.40	**274.00**
Average quality workmanship	Inst	Ea	CJ	1.67	164.00	49.10	---	213.10	**270.00**	174.40	**218.00**

Wall cabinets

Description	Oper	Unit	Crew Size	Man-Hours Per Unit	Costs Based On Small Volume						Large Volume	
					Avg Mat'l Unit Cost	Avg Labor Unit Cost	Avg Equip Unit Cost	Avg Total Unit Cost	Avg Price Incl O&P		Avg Total Unit Cost	Avg Price Incl O&P

Refrigerator cabinets, 15" H, 12" D, 2 doors

Description	Oper	Unit	Crew Size	M-H	Mat'l	Labor	Equip	Total	Price O&P	Total LV	Price LV
30" W											
High quality workmanship	Inst	Ea	CJ	2.58	182.00	75.80	---	257.80	**332.00**	206.20	**261.00**
Good quality workmanship	Inst	Ea	CJ	1.70	135.00	49.90	---	184.90	**237.00**	149.30	**189.00**
Average quality workmanship	Inst	Ea	CJ	1.70	103.00	49.90	---	152.90	**199.00**	121.80	**155.00**
33" W											
High quality workmanship	Inst	Ea	CJ	2.58	189.00	75.80	---	264.80	**340.00**	212.20	**269.00**
Good quality workmanship	Inst	Ea	CJ	1.70	138.00	49.90	---	187.90	**241.00**	152.30	**192.00**
Average quality workmanship	Inst	Ea	CJ	1.70	105.00	49.90	---	154.90	**201.00**	123.30	**157.00**
36" W											
High quality workmanship	Inst	Ea	CJ	2.58	200.00	75.80	---	275.80	**353.00**	222.20	**280.00**
Good quality workmanship	Inst	Ea	CJ	1.70	147.00	49.90	---	196.90	**251.00**	160.30	**202.00**
Average quality workmanship	Inst	Ea	CJ	1.70	112.00	49.90	---	161.90	**209.00**	129.50	**164.00**

Range cabinets, 21" H, 12" D

Description	Oper	Unit	Crew Size	M-H	Mat'l	Labor	Equip	Total	Price O&P	Total LV	Price LV
24" W, 1 door											
High quality workmanship	Inst	Ea	CJ	2.50	156.00	73.40	---	229.40	**297.00**	182.40	**232.00**
Good quality workmanship	Inst	Ea	CJ	1.68	116.00	49.30	---	165.30	**213.00**	131.70	**167.00**
Average quality workmanship	Inst	Ea	CJ	1.68	87.50	49.30	---	136.80	**179.00**	107.20	**137.00**
30" W, 2 doors											
High quality workmanship	Inst	Ea	CJ	2.50	173.00	73.40	---	246.40	**318.00**	197.40	**251.00**
Good quality workmanship	Inst	Ea	CJ	1.68	128.00	49.30	---	177.30	**227.00**	142.70	**180.00**
Average quality workmanship	Inst	Ea	CJ	1.68	98.00	49.30	---	147.30	**192.00**	116.50	**149.00**
33" W, 2 doors											
High quality workmanship	Inst	Ea	CJ	2.67	180.00	78.40	---	258.40	**334.00**	207.00	**262.00**
Good quality workmanship	Inst	Ea	CJ	1.78	133.00	52.30	---	185.30	**238.00**	149.40	**188.00**
Average quality workmanship	Inst	Ea	CJ	1.78	102.00	52.30	---	154.30	**200.00**	121.30	**155.00**
36" W, 2 doors											
High quality workmanship	Inst	Ea	CJ	2.67	193.00	78.40	---	271.40	**349.00**	218.00	**275.00**
Good quality workmanship	Inst	Ea	CJ	1.78	140.00	52.30	---	192.30	**246.00**	155.40	**196.00**
Average quality workmanship	Inst	Ea	CJ	1.78	107.00	52.30	---	159.30	**207.00**	126.00	**161.00**
42" W, 2 doors											
High quality workmanship	Inst	Ea	CJ	2.67	210.00	78.40	---	288.40	**370.00**	233.00	**294.00**
Good quality workmanship	Inst	Ea	CJ	1.78	156.00	52.30	---	208.30	**265.00**	169.40	**213.00**
Average quality workmanship	Inst	Ea	CJ	1.78	119.00	52.30	---	171.30	**221.00**	136.40	**174.00**
48" W, 2 doors											
High quality workmanship	Inst	Ea	CJ	3.33	238.00	97.80	---	335.80	**432.00**	269.70	**341.00**
Good quality workmanship	Inst	Ea	CJ	2.22	177.00	65.20	---	242.20	**310.00**	196.10	**246.00**
Average quality workmanship	Inst	Ea	CJ	2.22	135.00	65.20	---	200.20	**259.00**	158.10	**202.00**

Description	Oper	Unit	Crew Size	Man-Hours Per Unit	Costs Based On Small Volume						Large Volume	
					Avg Mat'l Unit Cost	Avg Labor Unit Cost	Avg Equip Unit Cost	Avg Total Unit Cost	Avg Price Incl O&P		Avg Total Unit Cost	Avg Price Incl O&P
Range cabinets, 30" H, 12" D												
12" W, 1 door												
High quality workmanship	Inst	Ea	CJ	1.95	137.00	57.30	---	194.30	**250.00**		155.70	**197.00**
Good quality workmanship	Inst	Ea	CJ	1.30	99.80	38.20	---	138.00	**177.00**		111.30	**140.00**
Average quality workmanship	Inst	Ea	CJ	1.30	75.30	38.20	---	113.50	**148.00**		89.60	**114.00**
15" W, 1 door												
High quality workmanship	Inst	Ea	CJ	1.95	151.00	57.30	---	208.30	**267.00**		167.70	**212.00**
Good quality workmanship	Inst	Ea	CJ	1.30	110.00	38.20	---	148.20	**190.00**		120.60	**152.00**
Average quality workmanship	Inst	Ea	CJ	1.30	84.00	38.20	---	122.20	**158.00**		97.30	**124.00**
18" W, 1 door												
High quality workmanship	Inst	Ea	CJ	1.95	158.00	57.30	---	215.30	**275.00**		174.70	**219.00**
Good quality workmanship	Inst	Ea	CJ	1.30	116.00	38.20	---	154.20	**196.00**		124.90	**157.00**
Average quality workmanship	Inst	Ea	CJ	1.30	87.50	38.20	---	125.70	**162.00**		100.40	**127.00**
21" W, 1 door												
High quality workmanship	Inst	Ea	CJ	2.22	166.00	65.20	---	231.20	**297.00**		186.10	**235.00**
Good quality workmanship	Inst	Ea	CJ	1.48	123.00	43.50	---	166.50	**212.00**		135.10	**169.00**
Average quality workmanship	Inst	Ea	CJ	1.48	92.80	43.50	---	136.30	**176.00**		108.30	**138.00**
24" W, 1 door												
High quality workmanship	Inst	Ea	CJ	2.22	186.00	65.20	---	251.20	**320.00**		203.10	**256.00**
Good quality workmanship	Inst	Ea	CJ	1.48	137.00	43.50	---	180.50	**229.00**		147.10	**184.00**
Average quality workmanship	Inst	Ea	CJ	1.48	103.00	43.50	---	146.50	**189.00**		117.60	**149.00**
27" W, 2 doors												
High quality workmanship	Inst	Ea	CJ	2.22	205.00	65.20	---	270.20	**343.00**		220.10	**276.00**
Good quality workmanship	Inst	Ea	CJ	1.48	151.00	43.50	---	194.50	**246.00**		159.10	**199.00**
Average quality workmanship	Inst	Ea	CJ	1.48	116.00	43.50	---	159.50	**204.00**		128.10	**162.00**
30" W, 2 doors												
High quality workmanship	Inst	Ea	CJ	2.96	217.00	86.90	---	303.90	**391.00**		244.30	**309.00**
Good quality workmanship	Inst	Ea	CJ	1.98	161.00	58.20	---	219.20	**280.00**		178.00	**224.00**
Average quality workmanship	Inst	Ea	CJ	1.98	121.00	58.20	---	179.20	**232.00**		142.00	**181.00**
33" W, 2 doors												
High quality workmanship	Inst	Ea	CJ	2.96	229.00	86.90	---	315.90	**405.00**		255.30	**322.00**
Good quality workmanship	Inst	Ea	CJ	1.98	170.00	58.20	---	228.20	**291.00**		185.00	**233.00**
Average quality workmanship	Inst	Ea	CJ	1.98	130.00	58.20	---	188.20	**243.00**		150.00	**190.00**
36" W, 2 doors												
High quality workmanship	Inst	Ea	CJ	2.96	238.00	86.90	---	324.90	**416.00**		263.30	**331.00**
Good quality workmanship	Inst	Ea	CJ	1.98	177.00	58.20	---	235.20	**299.00**		192.00	**240.00**
Average quality workmanship	Inst	Ea	CJ	1.98	135.00	58.20	---	193.20	**249.00**		154.00	**196.00**
42" W, 2 doors												
High quality workmanship	Inst	Ea	CJ	3.81	259.00	112.00	---	371.00	**479.00**		296.30	**376.00**
Good quality workmanship	Inst	Ea	CJ	2.54	191.00	74.60	---	265.60	**341.00**		213.60	**270.00**
Average quality workmanship	Inst	Ea	CJ	2.54	145.00	74.60	---	219.60	**286.00**		173.60	**221.00**
48" W, 2 doors												
High quality workmanship	Inst	Ea	CJ	3.81	306.00	112.00	---	418.00	**535.00**		338.30	**426.00**
Good quality workmanship	Inst	Ea	CJ	2.54	226.00	74.60	---	300.60	**383.00**		244.60	**307.00**
Average quality workmanship	Inst	Ea	CJ	2.54	172.00	74.60	---	246.60	**318.00**		196.60	**249.00**
24" W, blind corner unit, 1 door												
High quality workmanship	Inst	Ea	CJ	2.81	166.00	82.50	---	248.50	**323.00**		196.30	**251.00**
Good quality workmanship	Inst	Ea	CJ	1.90	131.00	55.80	---	186.80	**241.00**		149.50	**190.00**
Average quality workmanship	Inst	Ea	CJ	1.90	128.00	55.80	---	183.80	**237.00**		146.50	**186.00**

Description	Oper	Unit	Crew Size	Man-Hours Per Unit	Avg Mat'l Unit Cost	Avg Labor Unit Cost	Avg Equip Unit Cost	Avg Total Unit Cost	Avg Price Incl O&P	Avg Total Unit Cost	Avg Price Incl O&P
								Costs Based On Small Volume		**Large Volume**	
24" x 24" angle corner unit, stationary											
High quality workmanship	Inst	Ea	CJ	2.81	238.00	82.50	---	320.50	**409.00**	260.30	**327.00**
Good quality workmanship	Inst	Ea	CJ	1.90	175.00	55.80	---	230.80	**294.00**	188.50	**236.00**
Average quality workmanship	Inst	Ea	CJ	1.90	137.00	55.80	---	192.80	**248.00**	154.50	**195.00**
24" x 24" angle corner unit, lazy Susan											
High quality workmanship	Inst	Ea	CJ	2.81	354.00	82.50	---	436.50	**548.00**	362.30	**450.00**
Good quality workmanship	Inst	Ea	CJ	1.90	287.00	55.80	---	342.80	**428.00**	287.50	**355.00**
Average quality workmanship	Inst	Ea	CJ	1.90	196.00	55.80	---	251.80	**319.00**	207.50	**259.00**

Vanity cabinets and sink top

Description	Oper	Unit	Crew Size	Man-Hours Per Unit	Avg Mat'l Unit Cost	Avg Labor Unit Cost	Avg Equip Unit Cost	Avg Total Unit Cost	Avg Price Incl O&P	Avg Total Unit Cost	Avg Price Incl O&P
Disconnect plumbing and remove to dumpster	Demo	Ea	LB	1.67	---	44.50	---	44.50	**66.30**	26.60	**39.70**
Remove old unit, replace with new unit, reconnect plumbing	Reset	Ea	SB	3.81	---	120.00	---	120.00	**176.00**	72.10	**106.00**

Vanity units, with marble tops, good quality fittings and faucets; hardware, deluxe, finished models

Stained ash and birch primed composition construction

Labor costs include fitting and hanging only of vanity units

For rough-in costs, see Adjustments in this cabinets section, page 54

2-door units

Description	Oper	Unit	Crew Size	Man-Hours Per Unit	Avg Mat'l Unit Cost	Avg Labor Unit Cost	Avg Equip Unit Cost	Avg Total Unit Cost	Avg Price Incl O&P	Avg Total Unit Cost	Avg Price Incl O&P
20" x 16"											
Ash	Inst	Ea	SB	8.42	396.00	265.00	---	661.00	**864.00**	508.00	**652.00**
Birch	Inst	Ea	SB	8.42	361.00	265.00	---	626.00	**822.00**	477.00	**615.00**
Composition construction	Inst	Ea	SB	8.42	273.00	265.00	---	538.00	**717.00**	400.00	**522.00**
25" x 19"											
Ash	Inst	Ea	SB	8.42	460.00	265.00	---	725.00	**942.00**	566.00	**721.00**
Birch	Inst	Ea	SB	8.42	420.00	265.00	---	685.00	**894.00**	530.00	**678.00**
Composition construction	Inst	Ea	SB	8.42	299.00	265.00	---	564.00	**749.00**	423.00	**550.00**
31" x 19"											
Ash	Inst	Ea	SB	8.42	548.00	265.00	---	813.00	**1050.00**	643.00	**814.00**
Birch	Inst	Ea	SB	8.42	497.00	265.00	---	762.00	**986.00**	598.00	**760.00**
Composition construction	Inst	Ea	SB	8.42	355.00	265.00	---	620.00	**816.00**	473.00	**609.00**
35" x 19"											
Ash	Inst	Ea	SB	8.89	592.00	280.00	---	872.00	**1120.00**	692.00	**875.00**
Birch	Inst	Ea	SB	8.89	536.00	280.00	---	816.00	**1050.00**	642.00	**816.00**
Composition construction	Inst	Ea	SB	8.89	387.00	280.00	---	667.00	**876.00**	511.00	**658.00**
For drawers in any above unit, ADD per drawer	Inst	Ea	SB	---	52.50	---	---	52.50	**63.00**	46.50	**55.80**

2-door cutback units with 3 drawers

Description	Oper	Unit	Crew Size	Man-Hours Per Unit	Avg Mat'l Unit Cost	Avg Labor Unit Cost	Avg Equip Unit Cost	Avg Total Unit Cost	Avg Price Incl O&P	Avg Total Unit Cost	Avg Price Incl O&P
36" x 19"											
Ash	Inst	Ea	SB	8.89	686.00	280.00	---	966.00	**1230.00**	776.00	**976.00**
Birch	Inst	Ea	SB	8.89	616.00	280.00	---	896.00	**1150.00**	714.00	**902.00**
Composition construction	Inst	Ea	SB	8.89	460.00	280.00	---	740.00	**964.00**	576.00	**736.00**
49" x 19											
Ash	Inst	Ea	SB	8.89	858.00	280.00	---	1138.00	**1440.00**	928.00	**1160.00**
Birch	Inst	Ea	SB	8.89	742.00	280.00	---	1022.00	**1300.00**	825.00	**1040.00**
Composition construction	Inst	Ea	SB	8.89	555.00	280.00	---	835.00	**1080.00**	659.00	**836.00**
60" x 19"											
Ash	Inst	Ea	SB	11.4	1080.00	359.00	---	1439.00	**1830.00**	1168.00	**1460.00**
Birch	Inst	Ea	SB	11.4	898.00	359.00	---	1257.00	**1610.00**	1005.00	**1260.00**
Composition construction	Inst	Ea	SB	11.4	698.00	359.00	---	1057.00	**1370.00**	828.00	**1050.00**

Description	Oper	Unit	Crew Size	Man-Hours Per Unit	Avg Mat'l Unit Cost	Avg Labor Unit Cost	Avg Equip Unit Cost	Avg Total Unit Cost	Avg Price Incl O&P	Avg Total Unit Cost	Avg Price Incl O&P
					Costs Based On Small Volume					**Large Volume**	
Corner unit, 1 door											
22" x 22"											
Ash	Inst	Ea	SB	8.42	509.00	265.00	---	774.00	**1000.00**	609.00	**773.00**
Birch	Inst	Ea	SB	8.42	481.00	265.00	---	746.00	**967.00**	584.00	**743.00**
Composition construction	Inst	Ea	SB	8.42	392.00	265.00	---	657.00	**860.00**	505.00	**648.00**
Adjustments											
To remove and reset											
Vanity units with tops	Reset	Ea	SA	5.71	---	208.00	---	208.00	**305.00**	121.00	**178.00**
Top only	Reset	Ea	SA	3.33	---	121.00	---	121.00	**178.00**	72.70	**107.00**
To install rough-in	Inst	Ea	SB	6.67	---	210.00	---	210.00	**309.00**	126.00	**185.00**

Description	Oper	Unit	Crew Size	Man-Hours Per Unit	Avg Mat'l Unit Cost	Avg Labor Unit Cost	Avg Equip Unit Cost	Avg Total Unit Cost	Avg Price Incl O&P	Avg Total Unit Cost	Avg Price Incl O&P
						Costs Based On Small Volume					**Large Volume**	

Canopies
Costs for awnings include all hardware

Residential prefabricated

Aluminum

Description	Oper	Unit	Crew Size	Man-Hours Per Unit	Avg Mat'l Unit Cost	Avg Labor Unit Cost	Avg Equip Unit Cost	Avg Total Unit Cost	Avg Price Incl O&P	Avg Total Unit Cost	Avg Price Incl O&P
Carport, freestanding											
16' x 8'	Inst	Ea	2C	8.21	1290.00	263.00	---	1553.00	1880.00	1271.00	1520.00
20' x 10'	Inst	Ea	2C	12.3	1370.00	395.00	---	1765.00	2160.00	1417.00	1720.00
Door canopies, 36" projection											
4' wide	Inst	Ea	CA	1.54	216.00	49.40	---	265.40	322.00	216.10	259.00
5' wide	Inst	Ea	CA	2.05	242.00	65.80	---	307.80	376.00	248.70	301.00
6' wide	Inst	Ea	CA	3.08	261.00	98.80	---	359.80	448.00	286.20	352.00
8' wide	Inst	Ea	2C	4.92	357.00	158.00	---	515.00	647.00	407.00	503.00
10' wide	Inst	Ea	2C	8.21	408.00	263.00	---	671.00	865.00	519.00	657.00
12' wide	Inst	Ea	2C	12.3	485.00	395.00	---	880.00	1150.00	671.00	861.00
Door canopies, 42" projection											
4' wide	Inst	Ea	CA	1.54	247.00	49.40	---	296.40	358.00	243.10	290.00
5' wide	Inst	Ea	CA	2.05	270.00	65.80	---	335.80	409.00	272.70	329.00
6' wide	Inst	Ea	CA	3.08	342.00	98.80	---	440.80	541.00	355.20	431.00
8' wide	Inst	Ea	2C	4.92	399.00	158.00	---	557.00	696.00	443.00	545.00
10' wide	Inst	Ea	2C	8.21	462.00	263.00	---	725.00	927.00	565.00	710.00
12' wide	Inst	Ea	2C	12.3	543.00	395.00	---	938.00	1220.00	720.00	917.00
Door canopies, 48" projection											
4' wide	Inst	Ea	CA	1.54	270.00	49.40	---	319.40	385.00	262.10	313.00
5' wide	Inst	Ea	CA	2.05	316.00	65.80	---	381.80	462.00	312.70	374.00
6' wide	Inst	Ea	CA	3.08	359.00	98.80	---	457.80	561.00	370.20	448.00
8' wide	Inst	Ea	2C	4.92	489.00	158.00	---	647.00	799.00	520.00	633.00
10' wide	Inst	Ea	2C	8.21	546.00	263.00	---	809.00	1020.00	637.00	792.00
12' wide	Inst	Ea	2C	12.3	615.00	395.00	---	1010.00	1300.00	781.00	988.00
Door canopies, 54" projection											
4' wide	Inst	Ea	CA	1.54	311.00	49.40	---	360.40	431.00	297.10	352.00
5' wide	Inst	Ea	CA	2.05	330.00	65.80	---	395.80	478.00	323.70	387.00
6' wide	Inst	Ea	CA	3.08	411.00	98.80	---	509.80	620.00	414.20	499.00
8' wide	Inst	Ea	2C	4.92	560.00	158.00	---	718.00	881.00	580.00	703.00
10' wide	Inst	Ea	2C	8.21	633.00	263.00	---	896.00	1120.00	710.00	876.00
12' wide	Inst	Ea	2C	12.3	719.00	395.00	---	1114.00	1420.00	870.00	1090.00
Patio cover											
16' x 8'	Inst	Ea	2C	8.21	690.00	263.00	---	953.00	1190.00	759.00	933.00
20' x 10'	Inst	Ea	2C	12.3	934.00	395.00	---	1329.00	1670.00	1053.00	1300.00
Window awnings, 3' high											
4' wide	Inst	Ea	CA	1.76	71.90	56.50	---	128.40	167.00	97.90	125.00
6' wide	Inst	Ea	CA	2.46	86.80	78.90	---	165.70	218.00	125.30	162.00
9' wide	Inst	Ea	2C	4.10	119.00	132.00	---	251.00	334.00	186.70	245.00
12' wide	Inst	Ea	2C	6.15	150.00	197.00	---	347.00	468.00	255.00	339.00
Window awnings, 4' high											
4' wide	Inst	Ea	CA	1.76	82.80	56.50	---	139.30	180.00	107.20	136.00
6' wide	Inst	Ea	CA	2.46	109.00	78.90	---	187.90	244.00	144.40	184.00
9' wide	Inst	Ea	2C	4.10	150.00	132.00	---	282.00	369.00	212.70	275.00
12' wide	Inst	Ea	2C	6.15	196.00	197.00	---	393.00	521.00	295.00	384.00

| | | | Costs Based On Small Volume | | | | | | Large Volume | |
| | | | | | | | | | | |
Description	Oper	Unit	Crew Size	Man-Hours Per Unit	Avg Mat'l Unit Cost	Avg Labor Unit Cost	Avg Equip Unit Cost	Avg Total Unit Cost	Avg Price Incl O&P	Avg Total Unit Cost	Avg Price Incl O&P
Window awnings, 6' high											
4' wide	Inst	Ea	CA	1.76	136.00	56.50	---	192.50	**241.00**	152.60	**188.00**
6' wide	Inst	Ea	CA	2.46	184.00	78.90	---	262.90	**330.00**	208.30	**257.00**
9' wide	Inst	Ea	2C	4.10	230.00	132.00	---	362.00	**462.00**	281.70	**354.00**
12' wide	Inst	Ea	2C	6.15	259.00	197.00	---	456.00	**594.00**	349.00	**446.00**
Roll-up											
3' wide	Inst	Ea	CA	1.54	56.40	49.40	---	105.80	**139.00**	80.10	**103.00**
4' wide	Inst	Ea	CA	1.76	72.50	56.50	---	129.00	**168.00**	98.30	**126.00**
6' wide	Inst	Ea	CA	2.46	89.70	78.90	---	168.60	**222.00**	127.70	**165.00**
9' wide	Inst	Ea	2C	4.10	130.00	132.00	---	262.00	**347.00**	196.70	**256.00**
12' wide	Inst	Ea	2C	6.15	159.00	197.00	---	356.00	**479.00**	263.00	**348.00**

Canvas

Traditional fabric awning, with waterproof, colorfast acrylic duck, double-stitched seams, tubular metal framing and all hardware included

Description	Oper	Unit	Crew Size	Man-Hours Per Unit	Avg Mat'l Unit Cost	Avg Labor Unit Cost	Avg Equip Unit Cost	Avg Total Unit Cost	Avg Price Incl O&P	Avg Total Unit Cost	Avg Price Incl O&P
Window awning, 24" drop											
3' wide	Inst	Ea	CA	1.54	184.00	49.40	---	233.40	**286.00**	189.10	**228.00**
4' wide	Inst	Ea	CA	1.76	225.00	56.50	---	281.50	**344.00**	228.60	**276.00**
Window awning, 30" drop											
3' wide	Inst	Ea	CA	1.54	210.00	49.40	---	259.40	**316.00**	211.10	**254.00**
4' wide	Inst	Ea	CA	1.76	253.00	56.50	---	309.50	**376.00**	252.60	**303.00**
5' wide	Inst	Ea	CA	2.05	288.00	65.80	---	353.80	**429.00**	287.70	**346.00**
6' wide	Inst	Ea	CA	2.46	322.00	78.90	---	400.90	**489.00**	325.30	**393.00**
8' wide	Inst	Ea	2C	4.92	359.00	158.00	---	517.00	**649.00**	409.00	**506.00**
10' wide	Inst	Ea	2C	6.15	408.00	197.00	---	605.00	**766.00**	476.00	**593.00**
Minimum Job Charge	Inst	Ea	CA	3.08	---	98.80	---	98.80	**148.00**	64.20	**96.30**

Carpet

Description	Oper	Unit	Crew Size	Man-Hours Per Unit	Avg Mat'l Unit Cost	Avg Labor Unit Cost	Avg Equip Unit Cost	Avg Total Unit Cost	Avg Price Incl O&P	Avg Total Unit Cost	Avg Price Incl O&P
						Costs Based On Small Volume					Large Volume	

Detach & reset operations

Description	Oper	Unit	Crew Size	Man-Hrs	Mat'l	Labor	Equip	Total	Price O&P	LV Total	LV Price O&P
Detach and relay											
Existing carpet only	Reset	SY	FA	.159	.20	4.70	.42	5.32	**7.55**	3.06	**4.37**
Existing carpet w/ new pad	Reset	SY	FA	.185	3.04	5.46	---	8.50	**11.50**	5.71	**7.61**

Remove operations

Description	Oper	Unit	Crew Size	Man-Hrs	Mat'l	Labor	Equip	Total	Price O&P	LV Total	LV Price O&P
Carpet only, tacked	Demo	SY	FA	.058	---	1.71	---	1.71	**2.52**	1.03	**1.52**
Pad, minimal glue	Demo	SY	FA	.029	---	.86	---	.86	**1.26**	.53	**.78**
Scrape up backing residue	Demo	SY	FA	.139	---	4.11	---	4.11	**6.04**	2.45	**3.60**

Replace operations

Price includes consultation, measurement, pad separate, carpet separate, installation using tack strips and hot melt tape on seams

Description	Oper	Unit	Crew Size	Man-Hrs	Mat'l	Labor	Equip	Total	Price O&P	LV Total	LV Price O&P
Standard quality											
Poly, thin pile density	Inst	SY	FA	.139	13.80	4.11	.42	18.33	**22.30**	13.70	**16.60**
Loop pile, 20 oz.	Inst	SY	FA	.139	9.20	4.11	.42	13.73	**17.00**	10.06	**12.30**
Pad	Inst	SY	FA	.029	2.47	.86	---	3.33	**4.10**	2.51	**3.06**
Average quality											
Poly, medium pile density	Inst	SY	FA	.139	18.40	4.11	.42	22.93	**27.60**	17.40	**20.80**
Loop pile, 26 oz.	Inst	SY	FA	.139	12.70	4.11	.42	17.23	**21.00**	12.80	**15.50**
Cut pile, 36 oz.	Inst	SY	FA	.139	15.00	4.11	.42	19.53	**23.70**	14.70	**17.60**
Wool	Inst	SY	FA	.139	23.00	4.11	.42	27.53	**32.90**	21.10	**25.00**
Pad	Inst	SY	FA	.029	3.04	.86	---	3.90	**4.76**	2.96	**3.58**
High quality											
Nylon, medium pile density	Inst	SY	FA	.139	27.60	4.11	.42	32.13	**38.20**	24.80	**29.30**
Loop pile, 36 oz.	Inst	SY	FA	.139	15.00	4.11	.42	19.53	**23.70**	14.70	**17.60**
Cut pile, 46 oz.	Inst	SY	FA	.139	20.70	4.11	.42	25.23	**30.30**	19.30	**22.90**
Wool	Inst	SY	FA	.139	46.00	4.11	.42	50.53	**59.40**	39.50	**46.20**
Pad	Inst	SY	FA	.029	6.39	.86	---	7.25	**8.61**	5.64	**6.66**
Premium quality											
Nylon, thick pile density	Inst	SY	FA	.139	31.10	4.11	.42	35.63	**42.20**	27.50	**32.40**
Cut pile, 54 oz.	Inst	SY	FA	.139	28.80	4.11	.42	33.33	**39.50**	25.70	**30.30**
Wool	Inst	SY	FA	.139	92.00	4.11	.42	96.53	**112.00**	76.30	**88.50**
Decorator, floral or design	Inst	SY	FA	.267	109.00	7.89	.42	117.31	**138.00**	92.38	**108.00**
Pad	Inst	SY	FA	.029	8.14	.86	---	9.00	**10.60**	7.04	**8.27**
Steps											
Waterfall (wrap over nose)	Inst	Ea	FA	.167	9.20	4.93	---	14.13	**17.80**	10.31	**12.80**
Tucked (under tread nose)	Inst	Ea	FA	.267	9.20	7.89	---	17.09	**22.20**	12.09	**15.40**
Cove or wall wrap											
4" high	Inst	LF	FA	.103	2.05	3.04	---	5.09	**6.83**	3.47	**4.58**
6" high	Inst	LF	FA	.111	3.07	3.28	---	6.35	**8.35**	4.44	**5.74**
8" high	Inst	LF	FA	.121	4.60	3.57	---	8.17	**10.50**	5.84	**7.40**

Description	Oper	Unit	Crew Size	Man-Hours Per Unit	Avg Mat'l Unit Cost	Avg Labor Unit Cost	Avg Equip Unit Cost	Avg Total Unit Cost	Avg Price Incl O&P	Avg Total Unit Cost	Avg Price Incl O&P
								Costs Based On Small Volume		**Large Volume**	

Caulking

Material costs are typical costs for the listed bead diameters. Figures in parentheses, following bead diameter, indicate approximate coverage including 5% waste. Labor costs are per LF of bead length and assume good quality application on smooth to slightly irregular surfaces

Description	Oper	Unit	Crew Size	Man-Hours Per Unit	Avg Mat'l Unit Cost	Avg Labor Unit Cost	Avg Equip Unit Cost	Avg Total Unit Cost	Avg Price Incl O&P	Avg Total Unit Cost	Avg Price Incl O&P
Multi-purpose caulk, good quality											
1/8" (11.6 LF/fluid oz.)	Inst	LF	CA	.026	.01	.83	---	.84	**1.26**	.59	**.88**
1/4" (2.91 LF/fluid oz.)	Inst	LF	CA	.036	.06	1.16	---	1.22	**1.81**	.85	**1.26**
3/8" (1.29 LF/fluid oz.)	Inst	LF	CA	.043	.13	1.38	---	1.51	**2.23**	1.06	**1.56**
1/2" (.729 LF/fluid oz.)	Inst	LF	CA	.048	.24	1.54	---	1.78	**2.60**	1.25	**1.82**
Butyl flex caulk, premium quality											
1/8" (11.6 LF/fluid oz.)	Inst	LF	CA	.026	.02	.83	---	.85	**1.28**	.60	**.89**
1/4" (2.91 LF/fluid oz.)	Inst	LF	CA	.036	.09	1.16	---	1.25	**1.84**	.87	**1.29**
3/8" (1.29 LF/fluid oz.)	Inst	LF	CA	.043	.21	1.38	---	1.59	**2.32**	1.13	**1.65**
1/2" (.729 LF/fluid oz.)	Inst	LF	CA	.048	.38	1.54	---	1.92	**2.77**	1.36	**1.95**
Latex, premium quality											
1/8" (11.6 LF/fluid oz.)	Inst	LF	CA	.026	.03	.83	---	.86	**1.29**	.60	**.89**
1/4" (2.91 LF/fluid oz.)	Inst	LF	CA	.036	.13	1.16	---	1.29	**1.89**	.90	**1.32**
3/8" (1.29 LF/fluid oz.)	Inst	LF	CA	.043	.29	1.38	---	1.67	**2.42**	1.19	**1.72**
1/2" (.729 LF/fluid oz.)	Inst	LF	CA	.048	.52	1.54	---	2.06	**2.93**	1.48	**2.09**
Latex caulk, good quality											
1/8" (11.6 LF/fluid oz.)	Inst	LF	CA	.026	.02	.83	---	.85	**1.28**	.60	**.89**
1/4" (2.91 LF/fluid oz.)	Inst	LF	CA	.036	.09	1.16	---	1.25	**1.84**	.87	**1.29**
3/8" (1.29 LF/fluid oz.)	Inst	LF	CA	.043	.21	1.38	---	1.59	**2.32**	1.13	**1.65**
1/2" (.729 LF/fluid oz.)	Inst	LF	CA	.048	.38	1.54	---	1.92	**2.77**	1.36	**1.95**
Oil base caulk, good quality											
1/8" (11.6 LF/fluid oz.)	Inst	LF	CA	.026	.03	.83	---	.86	**1.29**	.60	**.89**
1/4" (2.91 LF/fluid oz.)	Inst	LF	CA	.036	.12	1.16	---	1.28	**1.88**	.90	**1.32**
3/8" (1.29 LF/fluid oz.)	Inst	LF	CA	.043	.27	1.38	---	1.65	**2.39**	1.18	**1.71**
1/2" (.729 LF/fluid oz.)	Inst	LF	CA	.048	.47	1.54	---	2.01	**2.87**	1.44	**2.04**
Oil base caulk, economy quality											
1/8" (11.6 LF/fluid oz.)	Inst	LF	CA	.026	.01	.83	---	.84	**1.26**	.59	**.88**
1/4" (2.91 LF/fluid oz.)	Inst	LF	CA	.036	.06	1.16	---	1.22	**1.81**	.85	**1.26**
3/8" (1.29 LF/fluid oz.)	Inst	LF	CA	.043	.13	1.38	---	1.51	**2.23**	1.06	**1.56**
1/2" (.729 LF/fluid oz.)	Inst	LF	CA	.048	.24	1.54	---	1.78	**2.60**	1.25	**1.82**
Silicone caulk, good quality											
1/8" (11.6 LF/fluid oz.)	Inst	LF	CA	.026	.04	.83	---	.87	**1.30**	.61	**.90**
1/4" (2.91 LF/fluid oz.)	Inst	LF	CA	.036	.14	1.16	---	1.30	**1.90**	.91	**1.34**
3/8" (1.29 LF/fluid oz.)	Inst	LF	CA	.043	.32	1.38	---	1.70	**2.45**	1.22	**1.76**
1/2" (.729 LF/fluid oz.)	Inst	LF	CA	.048	.57	1.54	---	2.11	**2.99**	1.52	**2.14**
Silicone caulk, premium quality											
1/8" (11.6 LF/fluid oz.)	Inst	LF	CA	.026	.04	.83	---	.87	**1.30**	.61	**.90**
1/4" (2.91 LF/fluid oz.)	Inst	LF	CA	.036	.17	1.16	---	1.33	**1.94**	.94	**1.37**
3/8" (1.29 LF/fluid oz.)	Inst	LF	CA	.043	.39	1.38	---	1.77	**2.54**	1.27	**1.82**
1/2" (.729 LF/fluid oz.)	Inst	LF	CA	.048	.69	1.54	---	2.23	**3.14**	1.61	**2.25**
Tub caulk, white siliconized											
1/8" (11.6 LF/fluid oz.)	Inst	LF	CA	.026	.01	.83	---	.84	**1.26**	.59	**.88**
1/4" (2.91 LF/fluid oz.)	Inst	LF	CA	.036	.06	1.16	---	1.22	**1.81**	.85	**1.26**
3/8" (1.29 LF/fluid oz.)	Inst	LF	CA	.043	.13	1.38	---	1.51	**2.23**	1.06	**1.56**
1/2" (.729 LF/fluid oz.)	Inst	LF	CA	.048	.24	1.54	---	1.78	**2.60**	1.25	**1.82**

Description	Oper	Unit	Crew Size	Man-Hours Per Unit	Avg Mat'l Unit Cost	Avg Labor Unit Cost	Avg Equip Unit Cost	Avg Total Unit Cost	Avg Price Incl O&P	Avg Total Unit Cost	Avg Price Incl O&P
								Costs Based On Small Volume		Large Volume	
Anti-algae and mildew-resistant tub caulk, premium quality white or clear silicone											
1/8" (11.6 LF/fluid oz.)	Inst	LF	CA	.026	.04	.83	---	.87	**1.30**	.61	**.90**
1/4" (2.91 LF/fluid oz.)	Inst	LF	CA	.036	.18	1.16	---	1.34	**1.95**	.94	**1.37**
3/8" (1.29 LF/fluid oz.)	Inst	LF	CA	.043	.40	1.38	---	1.78	**2.55**	1.28	**1.83**
1/2" (.729 LF/fluid oz.)	Inst	LF	CA	.048	.71	1.54	---	2.25	**3.16**	1.63	**2.27**
Elastomeric caulk, premium quality											
1/8" (11.6 LF/fluid oz.)	Inst	LF	CA	.026	.04	.83	---	.87	**1.30**	.61	**.90**
1/4" (2.91 LF/fluid oz.)	Inst	LF	CA	.036	.17	1.16	---	1.33	**1.94**	.94	**1.37**
3/8" (1.29 LF/fluid oz.)	Inst	LF	CA	.043	.37	1.38	---	1.75	**2.51**	1.26	**1.80**
1/2" (.729 LF/fluid oz.)	Inst	LF	CA	.048	.66	1.54	---	2.20	**3.10**	1.59	**2.22**

Note: The following percentage adjustment for Small Volume also applies to Large

Description	Oper	Unit	Crew Size	Man-Hours Per Unit	Avg Mat'l Unit Cost	Avg Labor Unit Cost	Avg Equip Unit Cost	Avg Total Unit Cost	Avg Price Incl O&P	Avg Total Unit Cost	Avg Price Incl O&P
Add for irregular surfaces such as vertical masonry or lap siding	Inst	%	CA	---	5.0	---	---	---	---	---	---
Caulking gun, heavy duty, professional type	Inst	Ea	CA	---	30.00	---	---	30.00	**36.00**	24.00	**28.80**
Caulking gun, economy grade											
11-oz. cartridge	Inst	Ea	CA	---	5.68	---	---	5.68	**6.82**	4.54	**5.45**
29-oz. cartridge	Inst	Ea	CA	---	19.90	---	---	19.90	**23.80**	15.90	**19.10**

Ceramic tile

1. **Dimensions**. There are many sizes of ceramic tile. Only $4\frac{1}{4}$" x $4\frac{1}{4}$" and 1" x 1" will be discussed here.

   a. $4\frac{1}{4}$" x $4\frac{1}{4}$" tile is furnished both unmounted and back-mounted. Back-mounted tile are usually furnished in sheets of 12 tile.

   b. 1" x 1" mosaic tile is furnished face-mounted and back-mounted in sheets; normally, 2'-0" x 1'-0".

2. **Installation**. There are three methods:

   a. Conventional, which uses portland cement, sand and wet tile grout.

   b. Dry-set, which uses dry-set mix and dry tile grout mix.

   c. Organic adhesive, which uses adhesive and dry tile grout mix.

   The conventional method is the most expensive and is used less frequently than the other methods.

3. **Estimating Technique**. For tile, determine the area and add 5% to 10% for waste. For cove, base or trim, determine the length in linear feet and add 5% to 10% for waste.

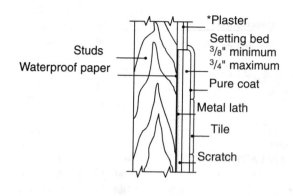

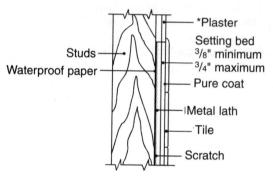

*This installation would be similar if wall finish above tile wainscot were of other material such as wallboard, plywood, etc.

Wood or steel construction with plaster above tile wainscot

Courtesy: *Ceramic Tile Institute of America*
700 N. Virgil Ave., Ste 300, Los Angeles, CA 90029

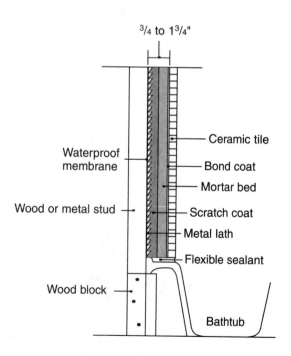

Cross section of bathtub wall using cement mortar

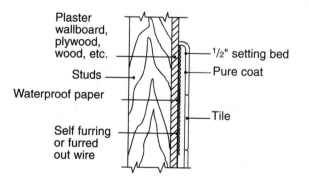

Wood or steel construction with solid covered backing

Ceramic tile

Description	Oper	Unit	Crew Size	Man-Hours Per Unit	Avg Mat'l Unit Cost	Avg Labor Unit Cost	Avg Equip Unit Cost	Avg Total Unit Cost	Avg Price Incl O&P	Avg Total Unit Cost	Avg Price Incl O&P
Countertop/backsplash											
Adhesive set with backmounted tile											
1" x 1"	Inst	SF	TB	.381	7.41	10.80	---	18.21	**24.40**	13.00	**17.00**
4-1/4" x 4-1/4"	Inst	SF	TB	.333	5.92	9.43	---	15.35	**20.70**	10.87	**14.30**
Cove/base											
Adhesive set with unmounted tile											
4-1/4" x 4-1/4"	Inst	LF	TB	.333	5.69	9.43	---	15.12	**20.40**	10.67	**14.10**
6" x 4-1/4"	Inst	LF	TB	.281	5.56	7.96	---	13.52	**18.10**	9.65	**12.60**
Conventional mortar set with unmounted tile											
4-1/4" x 4-1/4"	Inst	LF	TB	.593	5.57	16.80	---	22.37	**31.10**	15.00	**20.50**
6" x 4-1/4"	Inst	LF	TB	.533	5.39	15.10	---	20.49	**28.40**	13.80	**18.80**
Dry-set mortar with unmounted tile											
4-1/4" x 4-1/4"	Inst	LF	TB	.410	5.60	11.60	---	17.20	**23.50**	11.89	**15.90**
6" x 4-1/4"	Inst	LF	TB	.356	5.43	10.10	---	15.53	**21.10**	10.80	**14.40**
Floors											
1-1/2" x 1-1/2"											
Adhesive or dry-set base	Demo	SF	LB	.048	---	1.28	---	1.28	**1.91**	.77	**1.15**
Conventional mortar base	Demo	SF	LB	.056	---	1.49	---	1.49	**2.22**	.91	**1.35**
Adhesive set with backmounted tile											
1-1/2" x 1-1/2"	Inst	SF	TB	.222	7.41	6.28	---	13.69	**17.80**	10.29	**13.00**
4-1/4" x 4-1/4"	Inst	SF	TB	.205	5.92	5.80	---	11.72	**15.30**	8.69	**11.10**
Conventional mortar set with backmounted tile											
1-1/2" x 1-1/2"	Inst	SF	TB	.444	7.06	12.60	---	19.66	**26.60**	13.77	**18.30**
4-1/4" x 4-1/4"	Inst	SF	TB	.381	5.59	10.80	---	16.39	**22.30**	11.39	**15.20**
Dry-set mortar with backmounted tile											
1-1/2" x 1-1/2"	Inst	SF	TB	.281	7.14	7.96	---	15.10	**19.90**	11.04	**14.20**
4-1/4" x 4-1/4"	Inst	SF	TB	.254	5.65	7.19	---	12.84	**17.10**	9.27	**12.00**
Wainscot cap											
Adhesive set with unmounted tile											
2" x 6"	Inst	LF	TB	.222	2.90	6.28	---	9.18	**12.60**	6.32	**8.47**
Conventional mortar set with unmounted tile											
2" x 6"	Inst	LF	TB	.356	2.85	10.10	---	12.95	**18.10**	8.53	**11.70**
Dry-set mortar with unmounted tile											
2" x 6"	Inst	LF	TB	.267	2.85	7.56	---	10.41	**14.40**	7.03	**9.53**
Walls											
1-1/2" x 1-1/2" or 4-1/4" x 4-1/4"											
Adhesive or dry-set base	Demo	SF	LB	.056	---	1.49	---	1.49	**2.22**	.88	**1.31**
Conventional mortar base	Demo	SF	LB	.067	---	1.78	---	1.78	**2.66**	1.07	**1.59**
Adhesive set with backmounted tile											
1-1/2" x 1-1/2"	Inst	SF	TB	.267	7.41	7.56	---	14.97	**19.60**	11.05	**14.20**
4-1/4" x 4-1/4"	Inst	SF	TB	.242	5.92	6.85	---	12.77	**16.90**	9.31	**12.00**
Conventional mortar set with backmounted tile											
1-1/2" x 1-1/2"	Inst	SF	TB	.533	7.06	15.10	---	22.16	**30.30**	15.27	**20.50**
4-1/4" x 4-1/4"	Inst	SF	TB	.444	5.59	12.60	---	18.19	**24.90**	12.47	**16.80**
Dry-set mortar with backmounted tile											
1-1/2" x 1-1/2"	Inst	SF	TB	.333	7.14	9.43	---	16.57	**22.10**	11.94	**15.60**
4-1/4" x 4-1/4"	Inst	SF	TB	.296	5.65	8.38	---	14.03	**18.80**	10.01	**13.10**

Description	Oper	Unit	Crew Size	Man-Hours Per Unit	Costs Based On Small Volume						Large Volume	
					Avg Mat'l Unit Cost	Avg Labor Unit Cost	Avg Equip Unit Cost	Avg Total Unit Cost	Avg Price Incl O&P		Avg Total Unit Cost	Avg Price Incl O&P

Closet door systems

Labor costs include hanging and fitting of doors and hardware

Bi-folding units

Includes hardware and pine fascia trim

Unfinished

Description	Oper	Unit	Crew Size	Man-Hours Per Unit	Avg Mat'l Unit Cost	Avg Labor Unit Cost	Avg Equip Unit Cost	Avg Total Unit Cost	Avg Price Incl O&P	Avg Total Unit Cost	Avg Price Incl O&P
Birch, flush face, 1-3/8" T, hollow core											
2'-0" x 6'-8", 2 doors	Inst	Set	2C	2.67	96.40	85.70	---	182.10	239.00	133.40	171.00
2'-6" x 6'-8", 2 doors	Inst	Set	2C	2.67	116.00	85.70	---	201.70	262.00	150.30	191.00
3'-0" x 6'-8", 2 doors	Inst	Set	2C	2.67	116.00	85.70	---	201.70	262.00	150.30	191.00
4'-0" x 6'-8", 4 doors	Inst	Set	2C	3.33	166.00	107.00	---	273.00	351.00	205.20	259.00
6'-0" x 6'-8", 4 doors	Inst	Set	2C	3.33	216.00	107.00	---	323.00	409.00	248.20	308.00
8'-0" x 6'-8", 4 doors	Inst	Set	2C	3.33	297.00	107.00	---	404.00	502.00	317.20	387.00
4'-0" x 8'-0", 4 doors	Inst	Set	2C	3.33	278.00	107.00	---	385.00	480.00	301.20	369.00
6'-0" x 8'-0", 4 doors	Inst	Set	2C	3.33	352.00	107.00	---	459.00	565.00	364.20	441.00
8'-0" x 8'-0", 4 doors	Inst	Set	2C	3.33	443.00	107.00	---	550.00	669.00	441.20	530.00
Lauan, flush face, 1-3/8" T, hollow core											
2'-0" x 6'-8", 2 doors	Inst	Set	2C	2.67	74.10	85.70	---	159.80	214.00	114.40	150.00
2'-6" x 6'-8", 2 doors	Inst	Set	2C	2.67	79.10	85.70	---	164.80	220.00	118.70	155.00
3'-0" x 6'-8", 2 doors	Inst	Set	2C	2.67	87.10	85.70	---	172.80	229.00	125.50	162.00
4'-0" x 6'-8", 4 doors	Inst	Set	2C	3.33	131.00	107.00	---	238.00	311.00	176.20	225.00
6'-0" x 6'-8", 4 doors	Inst	Set	2C	3.33	159.00	107.00	---	266.00	343.00	199.20	252.00
8'-0" x 6'-8", 4 doors	Inst	Set	2C	3.33	214.00	107.00	---	321.00	406.00	246.20	306.00
4'-0" x 8'-0", 4 doors	Inst	Set	2C	3.33	207.00	107.00	---	314.00	398.00	240.20	299.00
6'-0" x 8'-0", 4 doors	Inst	Set	2C	3.33	260.00	107.00	---	367.00	459.00	285.20	351.00
8'-0" x 8'-0", 4 doors	Inst	Set	2C	3.33	325.00	107.00	---	432.00	535.00	341.20	415.00
Ponderosa pine, colonial raised panel (2/door), 1-3/8" T, solid core											
2'-0" x 6'-8", 2 doors	Inst	Set	2C	2.67	101.00	85.70	---	186.70	245.00	137.50	176.00
4'-0" x 6'-8", 4 doors	Inst	Set	2C	3.33	204.00	107.00	---	311.00	394.00	237.20	296.00
6'-0" x 6'-8", 4 doors	Inst	Set	2C	3.33	263.00	107.00	---	370.00	463.00	288.20	354.00
8'-0" x 6'-8", 4 doors	Inst	Set	2C	3.33	328.00	107.00	---	435.00	537.00	343.20	417.00
Ponderosa pine, raised panel louver, 1-3/8" T, solid core											
2'-0" x 6'-8", 2 doors	Inst	Set	2C	2.67	128.00	85.70	---	213.70	275.00	160.30	202.00
4'-0" x 6'-8", 4 doors	Inst	Set	2C	3.33	250.00	107.00	---	357.00	447.00	277.20	341.00
6'-0" x 6'-8", 4 doors	Inst	Set	2C	3.33	329.00	107.00	---	436.00	539.00	344.20	419.00
Hardboard, flush face, 1-3/8" T, hollow core											
2'-0" x 6'-8", 2 doors	Inst	Set	2C	2.67	64.20	85.70	---	149.90	202.00	106.00	140.00
2'-6" x 6'-8", 2 doors	Inst	Set	2C	2.67	68.40	85.70	---	154.10	207.00	109.60	144.00
3'-0" x 6'-8", 2 doors	Inst	Set	2C	2.67	74.00	85.70	---	159.70	214.00	114.30	149.00
4'-0" x 6'-8", 4 doors	Inst	Set	2C	3.33	122.00	107.00	---	229.00	300.00	168.20	216.00
6'-0" x 6'-8", 4 doors	Inst	Set	2C	3.33	144.00	107.00	---	251.00	326.00	187.20	237.00
8'-0" x 6'-8", 4 doors	Inst	Set	2C	3.33	216.00	107.00	---	323.00	409.00	248.20	308.00
4'-0" x 8'-0", 4 doors	Inst	Set	2C	3.33	221.00	107.00	---	328.00	414.00	252.20	313.00
6'-0" x 8'-0", 4 doors	Inst	Set	2C	3.33	266.00	107.00	---	373.00	466.00	290.20	357.00
8'-0" x 8'-0", 4 doors	Inst	Set	2C	3.33	370.00	107.00	---	477.00	586.00	380.20	459.00
Sen (ash), flush face, 1-3/8" T, hollow core											
2'-0" x 6'-8", 2 doors	Inst	Set	2C	2.67	116.00	85.70	---	201.70	262.00	150.30	191.00
2'-6" x 6'-8", 2 doors	Inst	Set	2C	2.67	129.00	85.70	---	214.70	277.00	161.30	203.00
3'-0" x 6'-8", 2 doors	Inst	Set	2C	2.67	140.00	85.70	---	225.70	290.00	171.30	215.00
4'-0" x 6'-8", 4 doors	Inst	Set	2C	3.33	201.00	107.00	---	308.00	392.00	236.20	294.00
6'-0" x 6'-8", 4 doors	Inst	Set	2C	3.33	247.00	107.00	---	354.00	445.00	275.20	339.00
8'-0" x 6'-8", 4 doors	Inst	Set	2C	3.33	339.00	107.00	---	446.00	550.00	353.20	429.00
4'-0" x 8'-0", 4 doors	Inst	Set	2C	3.33	159.00	107.00	---	266.00	343.00	199.20	252.00
6'-0" x 8'-0", 4 doors	Inst	Set	2C	3.33	460.00	107.00	---	567.00	689.00	456.20	547.00
8'-0" x 8'-0", 4 doors	Inst	Set	2C	3.33	656.00	107.00	---	763.00	914.00	623.20	739.00

Description	Oper	Unit	Crew Size	Man-Hours Per Unit	Costs Based On Small Volume						Large Volume	
					Avg Mat'l Unit Cost	Avg Labor Unit Cost	Avg Equip Unit Cost	Avg Total Unit Cost	Avg Price Incl O&P		Avg Total Unit Cost	Avg Price Incl O&P

Prefinished

Walnut tone, mar-resistant finish

Embossed (distressed wood appearance) lauan, flush face, 1-3/8" T, hollow core

Description	Oper	Unit	Crew Size	MH	Mat'l	Labor	Equip	Total	Price O&P	LV Total	LV Price
2'-0" x 6'-8", 2 doors	Inst	Set	2C	2.67	67.90	85.70	---	153.60	**207.00**	109.10	**144.00**
4'-0" x 6'-8", 4 doors	Inst	Set	2C	3.33	136.00	107.00	---	243.00	**316.00**	180.20	**229.00**
6'-0" x 6'-8", 4 doors	Inst	Set	2C	3.33	178.00	107.00	---	285.00	**365.00**	216.20	**271.00**
Lauan, flush face, 1-3/8" T, hollow core											
2'-0" x 6'-8", 2 doors	Inst	Set	2C	2.67	75.90	85.70	---	161.60	**216.00**	116.00	**151.00**
4'-0" x 6'-8", 4 doors	Inst	Set	2C	3.33	151.00	107.00	---	258.00	**334.00**	192.20	**244.00**
6'-0" x 6'-8", 4 doors	Inst	Set	2C	3.33	202.00	107.00	---	309.00	**393.00**	236.20	**295.00**
Ponderosa pine, full louver, 1-3/8" T, hollow core				---							
2'-0" x 6'-8", 2 doors	Inst	Set	2C	2.67	158.00	85.70	---	243.70	**310.00**	185.30	**231.00**
4'-0" x 6'-8", 4 doors	Inst	Set	2C	3.33	308.00	107.00	---	415.00	**515.00**	327.20	**398.00**
6'-0" x 6'-8", 4 doors	Inst	Set	2C	3.33	392.00	107.00	---	499.00	**611.00**	398.20	**481.00**
Ponderosa pine, raised louver, 1-3/8" T, hollow core				---							
2'-0" x 6'-8", 2 doors	Inst	Set	2C	2.67	167.00	85.70	---	252.70	**320.00**	193.30	**240.00**
4'-0" x 6'-8", 4 doors	Inst	Set	2C	3.33	325.00	107.00	---	432.00	**535.00**	341.20	**415.00**
6'-0" x 6'-8", 4 doors	Inst	Set	2C	3.33	392.00	107.00	---	499.00	**611.00**	398.20	**481.00**

Sliding or bypassing units

Includes hardware, 4-5/8 " jambs, header, and fascia

Wood inserts, 1-3/8" T, hollow core

Unfinished birch

Description	Oper	Unit	Crew Size	MH	Mat'l	Labor	Equip	Total	Price O&P	LV Total	LV Price
4'-0" x 6'-8", 2 doors	Inst	Set	2C	3.81	191.00	122.00	---	313.00	**403.00**	236.50	**297.00**
6'-0" x 6'-8", 2 doors	Inst	Set	2C	3.81	235.00	122.00	---	357.00	**453.00**	273.50	**340.00**
8'-0" x 6'-8", 2 doors	Inst	Set	2C	3.81	396.00	122.00	---	518.00	**638.00**	410.50	**498.00**
4'-0" x 8'-0", 2 doors	Inst	Set	2C	3.81	331.00	122.00	---	453.00	**564.00**	355.50	**435.00**
6'-0" x 8'-0", 2 doors	Inst	Set	2C	3.81	384.00	122.00	---	506.00	**625.00**	400.50	**487.00**
8'-0" x 8'-0", 2 doors	Inst	Set	2C	3.81	531.00	122.00	---	653.00	**794.00**	526.50	**631.00**
10'-0" x 6'-8", 3 doors	Inst	Set	2C	5.33	592.00	171.00	---	763.00	**938.00**	608.00	**734.00**
12'-0" x 6'-8", 3 doors	Inst	Set	2C	5.33	638.00	171.00	---	809.00	**991.00**	647.00	**780.00**
10'-0" x 8'-0", 3 doors	Inst	Set	2C	5.33	737.00	171.00	---	908.00	**1100.00**	731.00	**876.00**
12'-0" x 8'-0", 3 doors	Inst	Set	2C	5.33	802.00	171.00	---	973.00	**1180.00**	786.00	**940.00**
Unfinished hardboard											
4'-0" x 6'-8", 2 doors	Inst	Set	2C	3.81	100.00	122.00	---	222.00	**298.00**	158.80	**208.00**
6'-0" x 6'-8", 2 doors	Inst	Set	2C	3.81	123.00	122.00	---	245.00	**325.00**	178.50	**231.00**
8'-0" x 6'-8", 2 doors	Inst	Set	2C	3.81	225.00	122.00	---	347.00	**443.00**	265.50	**331.00**
4'-0" x 8'-0", 2 doors	Inst	Set	2C	3.81	155.00	122.00	---	277.00	**362.00**	205.50	**262.00**
6'-0" x 8'-0", 2 doors	Inst	Set	2C	3.81	178.00	122.00	---	300.00	**388.00**	225.50	**285.00**
8'-0" x 8'-0", 2 doors	Inst	Set	2C	3.81	230.00	122.00	---	352.00	**448.00**	269.50	**336.00**
10'-0" x 6'-8", 3 doors	Inst	Set	2C	5.33	278.00	171.00	---	449.00	**577.00**	340.00	**427.00**
12'-0" x 6'-8", 3 doors	Inst	Set	2C	5.33	340.00	171.00	---	511.00	**648.00**	393.00	**488.00**
10'-0" x 8'-0", 3 doors	Inst	Set	2C	5.33	380.00	171.00	---	551.00	**693.00**	426.00	**526.00**
12'-0" x 8'-0", 3 doors	Inst	Set	2C	5.33	345.00	171.00	---	516.00	**653.00**	397.00	**492.00**
Unfinished lauan											
4'-0" x 6'-8", 2 doors	Inst	Set	2C	3.81	138.00	122.00	---	260.00	**342.00**	191.50	**245.00**
6'-0" x 6'-8", 2 doors	Inst	Set	2C	3.81	162.00	122.00	---	284.00	**370.00**	211.50	**269.00**
8'-0" x 6'-8", 2 doors	Inst	Set	2C	3.81	235.00	122.00	---	357.00	**453.00**	273.50	**340.00**
4'-0" x 8'-0", 2 doors	Inst	Set	2C	3.81	224.00	122.00	---	346.00	**441.00**	264.50	**330.00**
6'-0" x 8'-0", 2 doors	Inst	Set	2C	3.81	265.00	122.00	---	387.00	**488.00**	298.50	**369.00**
8'-0" x 8'-0", 2 doors	Inst	Set	2C	3.81	357.00	122.00	---	479.00	**593.00**	377.50	**460.00**
10'-0" x 6'-8", 3 doors	Inst	Set	2C	5.33	322.00	171.00	---	493.00	**627.00**	377.00	**470.00**
12'-0" x 6'-8", 3 doors	Inst	Set	2C	5.33	346.00	171.00	---	517.00	**655.00**	398.00	**493.00**
10'-0" x 8'-0", 3 doors	Inst	Set	2C	5.33	460.00	171.00	---	631.00	**786.00**	495.00	**605.00**
12'-0" x 8'-0", 3 doors	Inst	Set	2C	5.33	504.00	171.00	---	675.00	**836.00**	532.00	**648.00**

					Costs Based On Small Volume						Large Volume	
Description	Oper	Unit	Crew Size	Man-Hours Per Unit	Avg Mat'l Unit Cost	Avg Labor Unit Cost	Avg Equip Unit Cost	Avg Total Unit Cost	Avg Price Incl O&P		Avg Total Unit Cost	Avg Price Incl O&P
Unfinished red oak												
4'-0" x 6'-8", 2 doors	Inst	Set	2C	3.81	207.00	122.00	---	329.00	421.00		249.50	313.00
6'-0" x 6'-8", 2 doors	Inst	Set	2C	3.81	254.00	122.00	---	376.00	476.00		290.50	359.00
8'-0" x 6'-8", 2 doors	Inst	Set	2C	3.81	419.00	122.00	---	541.00	665.00		430.50	520.00
4'-0" x 8'-0", 2 doors	Inst	Set	2C	3.81	357.00	122.00	---	479.00	593.00		377.50	460.00
6'-0" x 8'-0", 2 doors	Inst	Set	2C	3.81	391.00	122.00	---	513.00	633.00		406.50	493.00
8'-0" x 8'-0", 2 doors	Inst	Set	2C	3.81	537.00	122.00	---	659.00	801.00		531.50	637.00
10'-0" x 6'-8", 3 doors	Inst	Set	2C	5.33	534.00	171.00	---	705.00	870.00		558.00	677.00
12'-0" x 6'-8", 3 doors	Inst	Set	2C	5.33	538.00	171.00	---	709.00	876.00		562.00	681.00
10'-0" x 8'-0", 3 doors	Inst	Set	2C	5.33	736.00	171.00	---	907.00	1100.00		730.00	875.00
12'-0" x 8'-0", 3 doors	Inst	Set	2C	5.33	753.00	171.00	---	924.00	1120.00		745.00	892.00

Mirror bypass units

Frameless unit with 1/2" beveled mirror edges												
4'-0" x 6'-8", 2 doors	Inst	Set	2C	4.44	244.00	142.00	---	386.00	494.00		293.70	367.00
6'-0" x 6'-8", 2 doors	Inst	Set	2C	4.44	314.00	142.00	---	456.00	575.00		353.70	436.00
8'-0" x 6'-8", 2 doors	Inst	Set	2C	4.44	436.00	142.00	---	578.00	715.00		456.70	556.00
4'-0" x 8'-0", 2 doors	Inst	Set	2C	4.44	298.00	142.00	---	440.00	556.00		339.70	420.00
6'-0" x 8'-0", 2 doors	Inst	Set	2C	4.44	353.00	142.00	---	495.00	620.00		386.70	475.00
8'-0" x 8'-0", 2 doors	Inst	Set	2C	4.44	462.00	142.00	---	604.00	745.00		479.70	582.00
10'-0" x 6'-8", 3 doors	Inst	Set	2C	5.33	587.00	171.00	---	758.00	931.00		603.00	729.00
12'-0" x 6'-8", 3 doors	Inst	Set	2C	5.33	635.00	171.00	---	806.00	987.00		644.00	776.00
10'-0" x 8'-0", 3 doors	Inst	Set	2C	5.33	669.00	171.00	---	840.00	1030.00		673.00	810.00
12'-0" x 8'-0", 3 doors	Inst	Set	2C	5.33	733.00	171.00	---	904.00	1100.00		727.00	872.00
Aluminum frame unit												
4'-0" x 6'-8", 2 doors	Inst	Set	2C	4.44	250.00	142.00	---	392.00	501.00		298.70	373.00
6'-0" x 6'-8", 2 doors	Inst	Set	2C	4.44	323.00	142.00	---	465.00	585.00		360.70	445.00
8'-0" x 6'-8", 2 doors	Inst	Set	2C	4.44	391.00	142.00	---	533.00	663.00		418.70	512.00
4'-0" x 8'-0", 2 doors	Inst	Set	2C	4.44	283.00	142.00	---	425.00	539.00		326.70	406.00
6'-0" x 8'-0", 2 doors	Inst	Set	2C	4.44	366.00	142.00	---	508.00	634.00		397.70	487.00
8'-0" x 8'-0", 2 doors	Inst	Set	2C	4.44	454.00	142.00	---	596.00	736.00		472.70	574.00
10'-0" x 6'-8", 3 doors	Inst	Set	2C	5.33	552.00	171.00	---	723.00	891.00		573.00	695.00
12'-0" x 6'-8", 3 doors	Inst	Set	2C	5.33	601.00	171.00	---	772.00	948.00		616.00	743.00
10'-0" x 8'-0", 3 doors	Inst	Set	2C	5.33	634.00	171.00	---	805.00	985.00		643.00	775.00
12'-0" x 8'-0", 3 doors	Inst	Set	2C	5.33	693.00	171.00	---	864.00	1050.00		694.00	834.00
Golden oak frame unit												
4'-0" x 6'-8", 2 doors	Inst	Set	2C	4.44	296.00	142.00	---	438.00	554.00		337.70	418.00
6'-0" x 6'-8", 2 doors	Inst	Set	2C	4.44	384.00	142.00	---	526.00	655.00		412.70	505.00
8'-0" x 6'-8", 2 doors	Inst	Set	2C	4.44	463.00	142.00	---	605.00	747.00		480.70	583.00
4'-0" x 8'-0", 2 doors	Inst	Set	2C	4.44	335.00	142.00	---	477.00	599.00		370.70	456.00
6'-0" x 8'-0", 2 doors	Inst	Set	2C	4.44	434.00	142.00	---	576.00	712.00		454.70	553.00
8'-0" x 8'-0", 2 doors	Inst	Set	2C	4.44	538.00	142.00	---	680.00	833.00		544.70	656.00
10'-0" x 6'-8", 3 doors	Inst	Set	2C	5.33	654.00	171.00	---	825.00	1010.00		661.00	795.00
12'-0" x 6'-8", 3 doors	Inst	Set	2C	5.33	713.00	171.00	---	884.00	1080.00		711.00	853.00
10'-0" x 8'-0", 3 doors	Inst	Set	2C	5.33	752.00	171.00	---	923.00	1120.00		744.00	891.00
12'-0" x 8'-0", 3 doors	Inst	Set	2C	5.33	822.00	171.00	---	993.00	1200.00		804.00	960.00

Description	Oper	Unit	Crew Size	Man-Hours Per Unit	Costs Based On Small Volume						Large Volume	
					Avg Mat'l Unit Cost	Avg Labor Unit Cost	Avg Equip Unit Cost	Avg Total Unit Cost	Avg Price Incl O&P		Avg Total Unit Cost	Avg Price Incl O&P
Steel frame unit												
4'-0" x 6'-8", 2 doors	Inst	Set	2C	4.44	175.00	142.00	---	317.00	415.00		234.70	300.00
6'-0" x 6'-8", 2 doors	Inst	Set	2C	4.44	229.00	142.00	---	371.00	477.00		280.70	353.00
8'-0" x 6'-8", 2 doors	Inst	Set	2C	4.44	281.00	142.00	---	423.00	536.00		324.70	404.00
4'-0" x 8'-0", 2 doors	Inst	Set	2C	4.44	192.00	142.00	---	334.00	435.00		249.70	317.00
6'-0" x 8'-0", 2 doors	Inst	Set	2C	4.44	260.00	142.00	---	402.00	513.00		306.70	383.00
8'-0" x 8'-0", 2 doors	Inst	Set	2C	4.44	327.00	142.00	---	469.00	589.00		363.70	449.00
10'-0" x 6'-8", 3 doors	Inst	Set	2C	5.33	412.00	171.00	---	583.00	730.00		454.00	558.00
12'-0" x 6'-8", 3 doors	Inst	Set	2C	5.33	460.00	171.00	---	631.00	786.00		495.00	605.00
10'-0" x 8'-0", 3 doors	Inst	Set	2C	5.33	468.00	171.00	---	639.00	795.00		502.00	613.00
12'-0" x 8'-0", 3 doors	Inst	Set	2C	5.33	526.00	171.00	---	697.00	861.00		551.00	669.00

Accordion doors

Description	Oper	Unit	Crew Size	Man-Hours Per Unit	Avg Mat'l Unit Cost	Avg Labor Unit Cost	Avg Equip Unit Cost	Avg Total Unit Cost	Avg Price Incl O&P	Avg Total Unit Cost	Avg Price Incl O&P
Custom prefinished woodgrain print											
2'-0" x 6'-8"	Inst	Set	2C	2.67	282.00	85.70	---	367.70	453.00	291.30	353.00
3'-0" x 6'-8"	Inst	Set	2C	2.67	382.00	85.70	---	467.70	568.00	376.30	451.00
4'-0" x 6'-8"	Inst	Set	2C	2.67	495.00	85.70	---	580.70	697.00	472.30	562.00
5'-0" x 6'-8"	Inst	Set	2C	2.67	572.00	85.70	---	657.70	786.00	538.30	637.00
6'-0" x 6'-8"	Inst	Set	2C	2.67	698.00	85.70	---	783.70	931.00	646.30	761.00
7'-0" x 6'-8"	Inst	Set	2C	3.33	830.00	107.00	---	937.00	1120.00	772.20	910.00
8'-0" x 6'-8"	Inst	Set	2C	3.33	905.00	107.00	---	1012.00	1200.00	835.20	983.00
9'-0" x 6'-8"	Inst	Set	2C	3.33	1010.00	107.00	---	1117.00	1330.00	927.20	1090.00
10'-0" x 6'-8"	Inst	Set	2C	3.33	1170.00	107.00	---	1277.00	1500.00	1058.20	1240.00
Note: The following percentage adjustment for Small Volume also applies to Large											
For 8'-0"H, ADD	Inst	%	2C	---	12.0	---	---	---	---	---	---
Heritage prefinished real wood veneer											
2'-0" x 6'-8"	Inst	Set	2C	2.67	550.00	85.70	---	635.70	761.00	519.30	616.00
3'-0" x 6'-8"	Inst	Set	2C	2.67	773.00	85.70	---	858.70	1020.00	710.30	834.00
4'-0" x 6'-8"	Inst	Set	2C	2.67	1000.00	85.70	---	1085.70	1280.00	905.30	1060.00
5'-0" x 6'-8"	Inst	Set	2C	2.67	1150.00	85.70	---	1235.70	1460.00	1034.30	1210.00
6'-0" x 6'-8"	Inst	Set	2C	2.67	1390.00	85.70	---	1475.70	1720.00	1231.30	1440.00
7'-0" x 6'-8"	Inst	Set	2C	3.33	1610.00	107.00	---	1717.00	2010.00	1434.20	1680.00
8'-0" x 6'-8"	Inst	Set	2C	3.33	1770.00	107.00	---	1877.00	2190.00	1564.20	1830.00
9'-0" x 6'-8"	Inst	Set	2C	3.33	2000.00	107.00	---	2107.00	2460.00	1764.20	2050.00
10'-0" x 6'-8"	Inst	Set	2C	3.33	2220.00	107.00	---	2327.00	2720.00	1954.20	2270.00
Note: The following percentage adjustment for Small Volume also applies to Large											
For 8'-0"H, ADD	Inst	%	2C	---	20.0	---	---	---	---	---	---

Track and hardware only

Description	Oper	Unit	Crew Size	Man-Hours Per Unit	Avg Mat'l Unit Cost	Avg Labor Unit Cost	Avg Equip Unit Cost	Avg Total Unit Cost	Avg Price Incl O&P	Avg Total Unit Cost	Avg Price Incl O&P
4'-0" x 6'-8", 2 doors	Inst	Set	---	---	20.70	---	---	20.70	23.80	17.60	20.30
6'-0" x 6'-8", 2 doors	Inst	Set	---	---	27.60	---	---	27.60	31.70	23.50	27.10
8'-0" x 6'-8", 2 doors	Inst	Set	---	---	34.50	---	---	34.50	39.70	29.40	33.80
4'-0" x 8'-0", 2 doors	Inst	Set	---	---	20.70	---	---	20.70	23.80	17.60	20.30
6'-0" x 8'-0", 2 doors	Inst	Set	---	---	26.50	---	---	26.50	30.40	22.50	25.90
8'-0" x 8'-0", 2 doors	Inst	Set	---	---	34.50	---	---	34.50	39.70	29.40	33.80
10'-0" x 6'-8", 3 doors	Inst	Set	---	---	47.20	---	---	47.20	54.20	40.20	46.20
12'-0" x 6'-8", 3 doors	Inst	Set	---	---	58.70	---	---	58.70	67.50	50.00	57.50
10'-0" x 8'-0", 3 doors	Inst	Set	---	---	47.20	---	---	47.20	54.20	40.20	46.20
12'-0" x 8'-0", 3 doors	Inst	Set	---	---	58.70	---	---	58.70	67.50	50.00	57.50

Description	Oper	Unit	Crew Size	Man-Hours Per Unit	Avg Mat'l Unit Cost	Avg Labor Unit Cost	Avg Equip Unit Cost	Avg Total Unit Cost	Avg Price Incl O&P	Avg Total Unit Cost	Avg Price Incl O&P
					Costs Based On Small Volume					Large Volume	

Columns

See also Framing, page 123

Aluminum, extruded; self supporting

Designed as decorative, loadbearing elements for porches, entrances, colonnades, etc.; primed, knocked-down, and carton packed complete with cap and base

Description	Oper	Unit	Crew Size	Man-Hours Per Unit	Avg Mat'l Unit Cost	Avg Labor Unit Cost	Avg Equip Unit Cost	Avg Total Unit Cost	Avg Price Incl O&P	Avg Total Unit Cost	Avg Price Incl O&P
Column with standard cap and base											
8" dia. x 8' to 12' H	Inst	Ea	CS	6.67	173.00	202.00	89.40	464.40	600.00	322.60	413.00
10" dia. x 8' to 12' H	Inst	Ea	CS	6.67	231.00	202.00	89.40	522.40	670.00	372.60	473.00
10" dia. x 16' to 20' H	Inst	Ea	CS	8.89	261.00	269.00	119.00	649.00	836.00	456.50	582.00
12" dia. x 9' to 12' H	Inst	Ea	CS	6.67	419.00	202.00	89.40	710.40	895.00	533.60	666.00
12" dia. x 16' to 24' H	Inst	Ea	CS	8.89	483.00	269.00	119.00	871.00	1100.00	645.50	809.00
Column with Corinthian cap and decorative base											
8" dia. x 8' to 12' H	Inst	Ea	CS	6.67	301.00	202.00	89.40	592.40	753.00	432.60	544.00
10" dia. x 8' to 12' H	Inst	Ea	CS	6.67	378.00	202.00	89.40	669.40	845.00	497.60	623.00
10" dia. x 16' to 20' H	Inst	Ea	CS	8.89	413.00	269.00	119.00	801.00	1020.00	586.50	738.00
12" dia. x 9' to 12' H	Inst	Ea	CS	6.67	618.00	202.00	89.40	909.40	1130.00	703.60	870.00
12" dia. x 16' to 24' H	Inst	Ea	CS	8.89	694.00	269.00	119.00	1082.00	1360.00	826.50	1030.00

Brick. See Masonry, page 178

Wood, treated, No. 1 common and better white pine, T&G construction

Designed as decorative, loadbearing elements for porches, entrances, colonnades, etc.; primed, knocked-down, and carton packed complete with cap and base

Description	Oper	Unit	Crew Size	Man-Hours Per Unit	Avg Mat'l Unit Cost	Avg Labor Unit Cost	Avg Equip Unit Cost	Avg Total Unit Cost	Avg Price Incl O&P	Avg Total Unit Cost	Avg Price Incl O&P
Plain column with standard cap and base											
8" dia. x 8' to 12' H	Inst	Ea	CS	8.00	268.00	242.00	107.00	617.00	792.00	438.30	557.00
10" dia. x 8' to 12' H	Inst	Ea	CS	8.00	321.00	242.00	107.00	670.00	856.00	484.30	612.00
12" dia. x 8' to 12' H	Inst	Ea	CS	8.89	372.00	269.00	119.00	760.00	969.00	550.50	695.00
14" dia. x 12' to 16' H	Inst	Ea	CS	11.4	607.00	345.00	153.00	1105.00	1400.00	818.90	1030.00
16" dia. x 18' to 20' H	Inst	Ea	CS	12.5	891.00	378.00	168.00	1437.00	1800.00	1091.00	1360.00
Plain column with Corinthian cap and decorative base											
8" dia. x 8' to 12' H	Inst	Ea	CS	8.00	307.00	242.00	107.00	656.00	839.00	472.30	597.00
10" dia. x 8' to 12' H	Inst	Ea	CS	8.00	398.00	242.00	107.00	747.00	948.00	549.30	691.00
12" dia. x 8' to 12' H	Inst	Ea	CS	8.89	448.00	269.00	119.00	836.00	1060.00	616.50	774.00
14" dia. x 12' to 16' H	Inst	Ea	CS	11.4	675.00	345.00	153.00	1173.00	1480.00	877.90	1100.00
16" dia. x 18' to 20' H	Inst	Ea	CS	12.5	950.00	378.00	168.00	1496.00	1880.00	1141.00	1420.00

Concrete

Concrete Footings

1. **Dimensions.** 6" T, 8" T, 12" T x 12" W, 16" W, 20" W; 12" T x 24" W.

2. **Installation**

   a. Forms. 2" side forms equal in height to the thickness of the footing. 2" x 4" stakes 4'-0" oc, no less than the thickness of the footing. 2" x 4" bracing for stakes 8'-0" oc for 6" and 8" thick footings and 4'-0" oc for 12" thick footings. 1" x 2" or 1" x 3" spreaders 4'-0" oc

   b. Concrete. 1-2-4 mix is used in this section.

   c. Reinforcing steel. Various sizes, but usually only #3, #4, or #5 straight rods with end ties are used.

3. **Notes on Labor**

   a. Forming. Output based on a crew of two carpenters and one laborer.

   b. Grading, finish. Output based on what one laborer can do in one day.

   c. Reinforcing steel. Output based on what one laborer or one ironworker can do in one day.

   d. Concrete. Output based on a crew of two laborers and one carpenter.

   e. Forms, wrecking and cleaning. Output based on what one laborer can do in one day.

4. **Estimating Technique.** Determine the linear feet of footing.

Concrete Foundations

1. **Dimensions.** 8" T, 12" T x 4' H, 8' H, or 12' H.

2. **Installation**

   a. Forms. 4' x 8' panels made of 3/4" form grade plywood backed with 2" x 4" studs and sills (studs approximately 16" oc), three sets and six sets of 2" x 4" wales for 4', 8' and 12' high walls. 2" x 4" wales for 4', 8' and 12' high walls. 2" x 4" diagonal braces (with stakes) 12'-0" oc one side. Snap ties spaced 22" oc, 20" oc and 17" oc along each wale for 4', 8' and 12' high walls. Paraffin oil coating for forms. Twelve uses are estimated for panels; twenty uses for wales and braces; snap ties are used only once.

   b. Concrete. 1-2-4 mix.

   c. Reinforcing steel. Sizes #3 to #7. Bars are straight except dowels which may on occasion be bent rods.

3. **Notes on Labor**

   a. Concrete, placing. Output based on a crew of one carpenter and five laborers.

   b. Forming. Output based on a crew of four carpenters and one laborer.

   c. Reinforcing steel rods. Output based on what two laborers or two ironworkers can do in one day.

   d. Wrecking and cleaning forms. Output based on what two laborers can do in one day.

4. **Estimating Technique**

   a. Determine linear feet of wall if wall is 8" or 12" x 4', 8' or 12', or determine square feet of wall. Then calculate and add the linear feet of rods.

Concrete Interior Floor Finishes

1. **Dimensions.** $3^{1}/_{2}$", 4", 5", 6" thick x various areas.

2. **Installation**

   a. Forms. A wood form may or may not be required. A foundation wall may serve as a form for both basement and first floor slabs. In this section, only 2" x 4" and 2" x 6" side forms with stakes 4'-0" oc are considered.

   b. Finish grading. Dirt or gravel.

   c. Screeds (wood strips placed in area where concrete is to be placed). The concrete when placed will be finished even with top of the screeds. Screeds must be pulled before concrete sets up and the voids filled with concrete. 2" x 2" and 2" x 4" screeds with 2" x 2" stakes 6'-0" oc will be covered in this section.

   d. Steel reinforcing. Items to be covered are: #3, #4, #5 rods; 6 x 6/10-10 and 6 x 6/6-6 welded wire mesh.

   e. Concrete. 1-2-4 mix.

3. **Notes on Labor**

   a. Forms and screeds. Output based on a crew of two masons and one laborer.

   b. Finish grading. Output based on what one laborer can do in one day.

   c. Reinforcing. Output based on what two laborers or two ironworkers can do in one day.

   d. Concrete, place and finish. Output based on three cement masons and five laborers as a crew.

   e. Wrecking and cleaning forms. Output based on what one laborer can do in one day.

4. **Estimating Technique**

   a. Finish grading, mesh, and concrete. Determine the area and add waste.

   b. Forms, screeds and rods. Determine the linear feet.

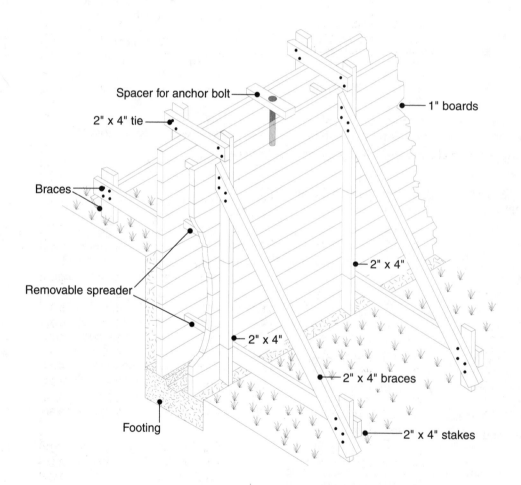

Spacer for anchor bolt

2" x 4" tie

1" boards

Braces

Removable spreader

2" x 4"

2" x 4"

2" x 4" braces

Footing

2" x 4" stakes

Concrete, cast in place

Description	Oper	Unit	Crew Size	Man-Hours Per Unit	Avg Mat'l Unit Cost	Avg Labor Unit Cost	Avg Equip Unit Cost	Avg Total Unit Cost	Avg Price Incl O&P	Avg Total Unit Cost	Avg Price Incl O&P
										Large Volume	

Footings

Demo with air tools, reinforced

Description	Oper	Unit	Crew Size	Man-Hours Per Unit	Avg Mat'l Unit Cost	Avg Labor Unit Cost	Avg Equip Unit Cost	Avg Total Unit Cost	Avg Price Incl O&P	Avg Total Unit Cost	Avg Price Incl O&P
8" T x 12" W	Demo	LF	AD	.202	---	5.92	.78	6.70	**9.54**	5.67	**8.08**
8" T x 16" W	Demo	LF	AD	.214	---	6.27	.83	7.10	**10.10**	6.03	**8.59**
8" T x 20" W	Demo	LF	AD	.229	---	6.71	.88	7.59	**10.80**	6.44	**9.16**
12" T x 12" W	Demo	LF	AD	.224	---	6.57	.87	7.44	**10.60**	6.31	**8.98**
12" T x 16" W	Demo	LF	AD	.235	---	6.89	.91	7.80	**11.10**	6.63	**9.44**
12" T x 20" W	Demo	LF	AD	.247	---	7.24	.96	8.20	**11.70**	6.99	**9.96**
12" T x 24" W	Demo	LF	AD	.261	---	7.65	1.01	8.66	**12.30**	7.37	**10.50**

Place per LF poured footing

Forming, 4 uses

Description	Oper	Unit	Crew Size	Man-Hours Per Unit	Avg Mat'l Unit Cost	Avg Labor Unit Cost	Avg Equip Unit Cost	Avg Total Unit Cost	Avg Price Incl O&P	Avg Total Unit Cost	Avg Price Incl O&P
2" x 6"	Inst	LF	CS	.116	.91	3.51	---	4.42	**6.18**	3.93	**5.55**
2" x 8"	Inst	LF	CS	.130	1.06	3.94	---	5.00	**6.96**	4.43	**6.25**
2" x 12"	Inst	LF	CS	.145	1.63	4.39	---	6.02	**8.21**	5.26	**7.27**

Grading, finish by hand

Description	Oper	Unit	Crew Size	Man-Hours Per Unit	Avg Mat'l Unit Cost	Avg Labor Unit Cost	Avg Equip Unit Cost	Avg Total Unit Cost	Avg Price Incl O&P	Avg Total Unit Cost	Avg Price Incl O&P
6", 8", 12" T x 12" W	Inst	LF	1L	.026	---	.69	---	.69	**1.03**	.61	**.91**
6", 8", 12" T x 16" W	Inst	LF	1L	.030	---	.80	---	.80	**1.19**	.72	**1.07**
6", 8", 12" T x 20" W	Inst	LF	1L	.034	---	.91	---	.91	**1.35**	.83	**1.23**
12" T x 24" W	Inst	LF	1L	.039	---	1.04	---	1.04	**1.55**	.93	**1.39**

Reinforcing steel in place, material costs include lap, waste, and tie wire

Description	Oper	Unit	Crew Size	Man-Hours Per Unit	Avg Mat'l Unit Cost	Avg Labor Unit Cost	Avg Equip Unit Cost	Avg Total Unit Cost	Avg Price Incl O&P	Avg Total Unit Cost	Avg Price Incl O&P
Two (No. 3) 3/8" rods	Inst	LF	1L	.015	.58	.40	---	.98	**1.18**	.81	**1.00**
Two (No. 4) 1/2" rods	Inst	LF	1L	.016	.66	.43	---	1.09	**1.30**	.87	**1.06**
Two (No. 5) 5/8" rods	Inst	LF	1L	.016	1.03	.43	---	1.46	**1.67**	1.18	**1.38**
Three (No. 3) 3/8" rods	Inst	LF	1L	.019	.87	.51	---	1.38	**1.62**	1.11	**1.33**
Three (No. 4) 1/2" rods	Inst	LF	1L	.020	.99	.53	---	1.52	**1.78**	1.23	**1.46**
Three (No. 5) 5/8" rods	Inst	LF	1L	.021	1.54	.56	---	2.10	**2.37**	1.68	**1.92**

Concrete, pour from truck into forms, using 3,000 PSI, 1-1/2" aggregate, 5.7 sack mix

Description	Oper	Unit	Crew Size	Man-Hours Per Unit	Avg Mat'l Unit Cost	Avg Labor Unit Cost	Avg Equip Unit Cost	Avg Total Unit Cost	Avg Price Incl O&P	Avg Total Unit Cost	Avg Price Incl O&P
6" T x 12" W (54.0 LF/CY)	Inst	LF	LC	.043	2.34	1.22	---	3.56	**4.16**	2.89	**3.43**
6" T x 16" W (40.5 LF/CY)	Inst	LF	LC	.048	2.34	1.37	---	3.71	**4.38**	3.00	**3.60**
6" T x 20" W (32.3 LF/CY)	Inst	LF	LC	.049	3.52	1.39	---	4.91	**5.60**	3.91	**4.53**
8" T x 12" W (40.5 LF/CY)	Inst	LF	LC	.048	2.34	1.37	---	3.71	**4.38**	3.00	**3.60**
8" T x 16" W (30.4 LF/CY)	Inst	LF	LC	.052	3.52	1.48	---	5.00	**5.72**	4.00	**4.65**
8" T x 20" W (24.3 LF/CY)	Inst	LF	LC	.055	4.69	1.57	---	6.26	**7.02**	4.94	**5.63**
12" T x 12" W (27.0 LF/CY)	Inst	LF	LC	.050	4.69	1.42	---	6.11	**6.81**	4.83	**5.46**
12" T x 16" W (20.3 LF/CY)	Inst	LF	LC	.058	5.86	1.65	---	7.51	**8.32**	5.92	**6.64**
12" T x 20" W (16.2 LF/CY)	Inst	LF	LC	.063	7.03	1.79	---	8.82	**9.70**	6.95	**7.75**
12" T x 24" W (13.5 LF/CY)	Inst	LF	LC	.055	8.20	1.57	---	9.77	**10.50**	7.60	**8.29**

Forms, wreck, remove and clean

Description	Oper	Unit	Crew Size	Man-Hours Per Unit	Avg Mat'l Unit Cost	Avg Labor Unit Cost	Avg Equip Unit Cost	Avg Total Unit Cost	Avg Price Incl O&P	Avg Total Unit Cost	Avg Price Incl O&P
2" x 6"	Inst	LF	1L	.051	---	1.36	---	1.36	**2.02**	1.23	**1.83**
2" x 8"	Inst	LF	1L	.056	---	1.49	---	1.49	**2.22**	1.33	**1.98**
2" x 12"	Inst	LF	1L	.061	---	1.63	---	1.63	**2.42**	1.47	**2.18**

Description	Oper	Unit	Crew Size	Man-Hours Per Unit	Avg Mat'l Unit Cost	Avg Labor Unit Cost	Avg Equip Unit Cost	Avg Total Unit Cost	Avg Price Incl O&P	Avg Total Unit Cost	Avg Price Incl O&P
						Costs Based On Small Volume				Large Volume	

Foundations and retaining walls, demo with air tools, per LF wall

With reinforcing

4'-0" H

| 8" T | Demo | LF | AD | .353 | --- | 10.40 | 1.36 | 11.76 | **16.70** | 9.95 | **14.20** |
| 12" T | Demo | LF | AD | .490 | --- | 14.40 | 1.89 | 16.29 | **23.10** | 13.70 | **19.60** |

8'-0" H

| 8" T | Demo | LF | AD | .381 | --- | 11.20 | 1.47 | 12.67 | **18.00** | 10.75 | **15.30** |
| 12" T | Demo | LF | AD | .545 | --- | 16.00 | 2.11 | 18.11 | **25.80** | 15.28 | **21.80** |

12"-0" H

| 8" T | Demo | LF | AD | .414 | --- | 12.10 | 1.60 | 13.70 | **19.60** | 11.76 | **16.70** |
| 12" T | Demo | LF | AD | .649 | --- | 19.00 | 2.50 | 21.50 | **30.70** | 18.11 | **25.80** |

Without reinforcing

4'-0" H

| 8" T | Demo | LF | AD | .308 | --- | 9.03 | 1.19 | 10.22 | **14.60** | 8.66 | **12.30** |
| 12" T | Demo | LF | AD | .414 | --- | 12.10 | 1.60 | 13.70 | **19.60** | 11.76 | **16.70** |

8'-0" H

| 8" T | Demo | LF | AD | .338 | --- | 9.91 | 1.31 | 11.22 | **16.00** | 9.48 | **13.50** |
| 12" T | Demo | LF | AD | .471 | --- | 13.80 | 1.82 | 15.62 | **22.30** | 13.24 | **18.90** |

12"-0" H

| 8" T | Demo | LF | AD | .369 | --- | 10.80 | 1.43 | 12.23 | **17.40** | 10.48 | **14.90** |
| 12" T | Demo | LF | AD | .545 | --- | 16.00 | 2.11 | 18.11 | **25.80** | 15.28 | **21.80** |

Place foundation or wall

Forming only. Material price includes panel forms, wales, braces, snap ties, paraffin oil and nails

8" or 12" T x 4'-0" H

Make (@12 uses)	Inst	SF	CU	.063	1.44	1.95	---	3.39	**4.37**	2.89	**3.79**
Erect and coat	Inst	SF	CU	.106	.03	3.29	---	3.32	**4.96**	3.06	**4.58**
Wreck and clean	Inst	SF	LB	.048	---	1.28	---	1.28	**1.91**	1.15	**1.71**

8" or 12" T x 8'-0" H

Make (@12 uses)	Inst	SF	CU	.066	1.32	2.05	---	3.37	**4.39**	2.89	**3.84**
Erect and coat	Inst	SF	CU	.128	.03	3.97	---	4.00	**5.98**	3.68	**5.51**
Wreck and clean	Inst	SF	LB	.058	---	1.55	---	1.55	**2.30**	1.39	**2.06**

8" or 12" T x 12'-0" H

Make (@12 uses)	Inst	SF	CU	.065	1.38	2.02	---	3.40	**4.40**	2.91	**3.84**
Erect and coat	Inst	SF	CU	.150	.03	4.65	---	4.68	**7.01**	4.30	**6.44**
Wreck and clean	Inst	SF	LB	.068	---	1.81	---	1.81	**2.70**	1.65	**2.46**

Reinforcing steel rods. 5% waste included, pricing based on LF of rod

No. 3 (3/8" rod)	Inst	LF	LB	.013	.28	.35	---	.63	**.80**	.53	**.69**
No. 4 (1/2" rod)	Inst	LF	LB	.013	.32	.35	---	.67	**.84**	.56	**.72**
No. 5 (5/8" rod)	Inst	LF	LB	.014	.49	.37	---	.86	**1.05**	.72	**.89**
No. 6 (3/4" rod)	Inst	LF	LB	.016	.71	.43	---	1.14	**1.35**	.91	**1.10**
No. 7 (7/8" rod)	Inst	LF	LB	.018	.96	.48	---	1.44	**1.67**	1.16	**1.37**

Description	Oper	Unit	Crew Size	Man-Hours Per Unit	Costs Based On Small Volume					Large Volume	
					Avg Mat'l Unit Cost	Avg Labor Unit Cost	Avg Equip Unit Cost	Avg Total Unit Cost	Avg Price Incl O&P	Avg Total Unit Cost	Avg Price Incl O&P
Concrete, placed from trucks into forms, material cost includes 5% waste											
and assumes use of 3,000 PSI, 1-1/2" aggregate, 5.7 sack mix											
8" T x 4'-0" H (10.12 LF/CY)	Inst	LF	LG	.077	11.70	2.12	---	13.82	**14.90**	10.81	**11.80**
8" T x 4'-0" H (40.5 SF/CY)	Inst	SF	LG	.019	2.34	.52	---	2.86	**3.12**	2.25	**2.48**
8" T x 8'-0" H (5.06 LF/CY)	Inst	LF	LG	.154	23.40	4.24	---	27.64	**29.80**	21.63	**23.50**
8" T x 8'-0" H (40.5 SF/CY)	Inst	SF	LG	.019	2.34	.52	---	2.86	**3.12**	2.25	**2.48**
8" T x 12'-0" H (3.37 LF/CY)	Inst	LF	LG	.232	35.20	6.39	---	41.59	**44.70**	32.36	**35.20**
8" T x 12'-0" H (40.5 SF/CY)	Inst	SF	LG	.019	2.34	.52	---	2.86	**3.12**	2.25	**2.48**
12' T x 4'-0" H (6.75 LF/CY)	Inst	LF	LG	.115	17.60	3.17	---	20.77	**22.30**	16.17	**17.60**
12" T x 4'-0" H (27.0 LF/CY)	Inst	SF	LG	.029	4.69	.80	---	5.49	**5.88**	4.27	**4.62**
12" T x 8'-0" H (3.38 LF/CY)	Inst	LF	LG	.231	35.20	6.36	---	41.56	**44.60**	32.33	**35.20**
12" T x 8'-0" H (27.0 SF/CY)	Inst	SF	LG	.026	4.69	.72	---	5.41	**5.76**	4.27	**4.62**
12" T x 12'-0" H (2.25 LF/CY)	Inst	LF	LG	.312	51.60	8.60	---	60.20	**64.40**	47.70	**51.90**
12" T x 12'-0" H (27.0 SF/CY)	Inst	SF	LG	.026	4.69	.72	---	5.41	**5.76**	4.27	**4.62**

Slabs, sidewalks and driveways, demo with air tools

Description	Oper	Unit	Crew Size	Man-Hours Per Unit	Avg Mat'l Unit Cost	Avg Labor Unit Cost	Avg Equip Unit Cost	Avg Total Unit Cost	Avg Price Incl O&P	Avg Total Unit Cost	Avg Price Incl O&P
With reinforcing (6 x 6 / 10 x 10)											
4" T	Demo	SF	AD	.074	---	2.17	.20	2.37	**3.41**	2.02	**2.90**
5" T	Demo	SF	AD	.085	---	2.49	.23	2.72	**3.92**	2.34	**3.37**
6" T	Demo	SF	AD	.097	---	2.84	.26	3.10	**4.47**	2.65	**3.82**
Without reinforcing											
4" T	Demo	SF	AD	.051	---	1.49	.14	1.63	**2.35**	1.41	**2.03**
5" T	Demo	SF	AD	.059	---	1.73	.16	1.89	**2.72**	1.63	**2.35**
6" T	Demo	SF	AD	.067	---	1.96	.18	2.14	**3.09**	1.82	**2.62**

Place concrete

Description	Oper	Unit	Crew Size	Man-Hours Per Unit	Avg Mat'l Unit Cost	Avg Labor Unit Cost	Avg Equip Unit Cost	Avg Total Unit Cost	Avg Price Incl O&P	Avg Total Unit Cost	Avg Price Incl O&P
Forming, 4 uses											
2" x 4" (4" T slab)	Inst	LF	DD	.045	.31	1.31	---	1.62	**2.25**	1.44	**2.01**
2" x 6" (5" T and 6" T slab)	Inst	LF	DD	.045	.39	1.31	---	1.70	**2.33**	1.50	**2.07**
Grading											
Dirt, cut and fill, +/- 1/10 ft	Inst	SF	1L	.011	---	.29	---	.29	**.44**	.27	**.40**
Gravel, 3/4" to 1-1/2" stone	Inst	SF	1L	.012	.46	.32	---	.78	**.94**	.64	**.79**
Screeds, 3 uses											
2" x 2"	Inst	LF	DD	.031	.13	.90	---	1.03	**1.47**	.95	**1.35**
2" x 4"	Inst	LF	DD	.034	.24	.99	---	1.23	**1.71**	1.11	**1.56**
Reinforcing steel rods, includes 5% waste, costs are per LF of rod or SF of mesh											
No. 3 (3/8" rod)	Inst	LF	LB	.012	.28	.32	---	.60	**.76**	.50	**.65**
No. 4 (1/2" rod)	Inst	LF	LB	.013	.32	.35	---	.67	**.84**	.53	**.68**
No. 5 (5/8" rod)	Inst	LF	LB	.013	.49	.35	---	.84	**1.01**	.69	**.85**
6 x 6 / 10-10 @ 21 lbs/CSF	Inst	SF	LB	.003	.32	.08	---	.40	**.44**	.32	**.36**
6 x 6 / 6-6 @ 42 lbs/CSF	Inst	SF	LB	.004	.52	.11	---	.63	**.68**	.50	**.55**
Concrete, pour and finish (steel trowel), material cost includes 5% waste											
and assumes use of 2500 PSI, 1" aggregate, 5.5 sack mix											
3-1/2" T (92.57 SF/CY)	Inst	SF	DF	.025	1.17	.70	---	1.87	**2.21**	1.54	**1.85**
4" T (81.00 SF/CY)	Inst	SF	DF	.026	1.17	.73	---	1.90	**2.25**	1.56	**1.89**
5" T (64.80 SF/CY)	Inst	SF	DF	.027	2.34	.76	---	3.10	**3.46**	2.48	**2.82**
6" T (54.00 SF/CY)	Inst	SF	DF	.028	2.34	.79	---	3.13	**3.50**	2.51	**2.86**
Forms, wreck and clean											
2" x 4"	Inst	LF	1L	.025	---	.67	---	.67	**.99**	.59	**.87**
2" x 6"	Inst	LF	1L	.026	---	.69	---	.69	**1.03**	.64	**.95**

Description	Oper	Unit	Crew Size	Man-Hours Per Unit	Avg Mat'l Unit Cost	Avg Labor Unit Cost	Avg Equip Unit Cost	Avg Total Unit Cost	Avg Price Incl O&P	Avg Total Unit Cost	Avg Price Incl O&P
					Costs Based On Small Volume					Large Volume	

Material information
Ready mix delivered by truck

Material costs only, prices are typical for most cities and assumes delivery up to 20 miles for 10 CY or more, 3 " to 4 " slump

Footing and foundation, using 1-1/2" aggregate

Description	Oper	Unit	Crew Size	Man-Hours	Mat'l	Labor	Equip	Total	Price O&P	Total	Price O&P
2000 PSI, 4.8 sack mix	Inst	CY	---	---	104.00	---	---	104.00	**104.00**	78.60	78.60
2500 PSI, 5.2 sack mix	Inst	CY	---	---	107.00	---	---	107.00	**107.00**	81.20	81.20
3000 PSI, 5.7 sack mix	Inst	CY	---	---	112.00	---	---	112.00	**112.00**	84.60	84.60
3500 PSI, 6.3 sack mix	Inst	CY	---	---	114.00	---	---	114.00	**114.00**	86.20	86.20
4000 PSI, 6.9 sack mix	Inst	CY	---	---	122.00	---	---	122.00	**122.00**	92.20	92.20

Slab, sidewalk and driveway, using 1" aggregate

2000 PSI, 5.0 sack mix	Inst	CY	---	---	106.00	---	---	106.00	**106.00**	80.30	80.30
2500 PSI, 5.5 sack mix	Inst	CY	---	---	110.00	---	---	110.00	**110.00**	83.70	83.70
3000 PSI, 6.0 sack mix	Inst	CY	---	---	114.00	---	---	114.00	**114.00**	86.20	86.20
3500 PSI, 6.6 sack mix	Inst	CY	---	---	119.00	---	---	119.00	**119.00**	90.50	90.50
4000 PSI, 7.1 sack mix	Inst	CY	---	---	123.00	---	---	123.00	**123.00**	93.00	93.00

Pump & gout mix, using pea gravel, 3/8" aggregate

2000 PSI, 6.0 sack mix	Inst	CY	---	---	114.00	---	---	114.00	**114.00**	86.20	86.20
2500 PSI, 6.5 sack mix	Inst	CY	---	---	118.00	---	---	118.00	**118.00**	89.60	89.60
3000 PSI, 7.2 sack mix	Inst	CY	---	---	124.00	---	---	124.00	**124.00**	93.90	93.90
3500 PSI, 7.9 sack mix	Inst	CY	---	---	129.00	---	---	129.00	**129.00**	98.10	98.10
4000 PSI, 8.5 sack mix	Inst	CY	---	---	137.00	---	---	137.00	**137.00**	104.00	104.00

Adjustments
Extra delivery costs for ready-mix concrete

Delivery over 20 miles, ADD	Inst	Mi	---	---	2.14	---	---	2.14	**2.14**	2.14	2.14
Standby charge in excess of 5 minutes per CY delivered											
per minute extra time, ADD	Inst	Min	---	---	1.07	---	---	1.07	**1.07**	1.07	1.07
Loads 7.25 CY or less, ADD											
7.25 CY	Inst	LS	---	---	6.33	---	---	6.33	**6.33**	6.33	6.33
6.0 CY	Inst	LS	---	---	22.60	---	---	22.60	**22.60**	22.60	22.60
5.0 CY	Inst	LS	---	---	36.20	---	---	36.20	**36.20**	36.20	36.20
4.0 CY	Inst	LS	---	---	49.80	---	---	49.80	**49.80**	49.80	49.80
3.0 CY	Inst	LS	---	---	63.30	---	---	63.30	**63.30**	63.30	63.30
2.0 CY	Inst	LS	---	---	76.90	---	---	76.90	**76.90**	76.90	76.90
1.0 CY	Inst	LS	---	---	90.50	---	---	90.50	**90.50**	90.50	90.50

Extra costs for non-standard mix additives

High early strength											
5.0 sack mix	Inst	CY	---	---	10.20	---	---	10.20	**10.20**	10.20	10.20
6.0 sack mix	Inst	CY	---	---	13.20	---	---	13.20	**13.20**	13.20	13.20
White cement (architectural)	Inst	CY	---	---	45.80	---	---	45.80	**45.80**	45.80	45.80
For 1% calcium chloride	Inst	CY	---	---	1.53	---	---	1.53	**1.53**	1.53	1.53
For 2% calcium chloride	Inst	CY	---	---	2.14	---	---	2.14	**2.14**	2.14	2.14
Chemical compensated shrinkage (WRDA Admix)	Inst	CY	---	---	13.70	---	---	13.70	**13.70**	13.70	13.70

Note: The following percentage adjustment for Small Volume also applies to Large

Super plasticized mix, with 7"-8" slump	Inst	%	---	---	8.0	---	---	8.0	**8.0**	8.0	8.0
Coloring of concrete, ADD											
Light or sand colors	Inst	CY	---	---	16.90	---	---	16.90	**16.90**	16.90	16.90
Medium or buff colors	Inst	CY	---	---	24.90	---	---	24.90	**24.90**	24.90	24.90
Dark colors	Inst	CY	---	---	37.50	---	---	37.50	**37.50**	37.50	37.50
Green	Inst	CY	---	---	41.70	---	---	41.70	**41.70**	41.70	41.70

Description	Oper	Unit	Costs Based On Small Volume							Large Volume	
			Crew Size	Man-Hours Per Unit	Avg Mat'l Unit Cost	Avg Labor Unit Cost	Avg Equip Unit Cost	Avg Total Unit Cost	Avg Price Incl O&P	Avg Total Unit Cost	Avg Price Incl O&P
Extra costs for non-standard aggregates											
Lightweight aggregate, ADD											
Mix from truck to forms	Inst	CY	---	---	36.40	---	---	36.40	**36.40**	36.40	**36.40**
Pump mix	Inst	CY	---	---	37.50	---	---	37.50	**37.50**	37.50	**37.50**
Granite aggregate, ADD	Inst	CY	---	---	5.35	---	---	5.35	**5.35**	5.35	**5.35**

Concrete block. See Masonry, page 181

				Costs Based On Small Volume						Large Volume	
Description	Oper	Unit	Crew Size	Man-Hours Per Unit	Avg Mat'l Unit Cost	Avg Labor Unit Cost	Avg Equip Unit Cost	Avg Total Unit Cost	Avg Price Incl O&P	Avg Total Unit Cost	Avg Price Incl O&P

Countertops

Formica

One-piece tops; straight, "L", or "U" shapes; surfaced with laminated plastic cemented to particleboard base

Post formed countertop with raised front drip edge

25" W, 1-1/2" H front edge, 4" H coved backsplash

Description	Oper	Unit	Crew Size	Man-Hours Per Unit	Avg Mat'l Unit Cost	Avg Labor Unit Cost	Avg Equip Unit Cost	Avg Total Unit Cost	Avg Price Incl O&P	Avg Total Unit Cost	Avg Price Incl O&P
Satin/suede patterns	Inst	LF	2C	.808	25.90	25.90	---	51.80	68.70	39.20	50.50
Solid patterns	Inst	LF	2C	.808	28.80	25.90	---	54.70	72.00	41.80	53.50
Specialty finish patterns	Inst	LF	2C	.808	34.60	25.90	---	60.50	78.70	47.10	59.60
Wood tone patterns	Inst	LF	2C	.808	36.00	25.90	---	61.90	80.30	48.40	61.10
Double roll top, 1-1/2" H front and back edges, no backsplash											
25" W											
Satin/suede patterns	Inst	LF	2C	.762	37.60	24.50	---	62.10	80.00	49.00	61.40
Solid patterns	Inst	LF	2C	.762	41.80	24.50	---	66.30	84.80	52.80	65.80
Specialty finish patterns	Inst	LF	2C	.762	50.20	24.50	---	74.70	94.40	60.40	74.50
Wood tone patterns	Inst	LF	2C	.762	52.30	24.50	---	76.80	96.80	62.30	76.70
36" W											
Satin/suede patterns	Inst	LF	2C	.833	46.30	26.70	---	73.00	93.40	58.30	72.60
Solid patterns	Inst	LF	2C	.833	51.50	26.70	---	78.20	99.30	63.00	78.00
Specialty finish patterns	Inst	LF	2C	.833	61.80	26.70	---	88.50	111.00	72.30	88.70
Wood tone patterns	Inst	LF	2C	.833	64.30	26.70	---	91.00	114.00	74.70	91.40
Post formed countertop with square edge veneer front											
25" W, 1-1/2" H front edge, 4" H coved backsplash											
Satin/suede patterns	Inst	LF	2C	.808	32.40	25.90	---	58.30	76.20	45.10	57.30
Solid patterns	Inst	LF	2C	.808	36.10	25.90	---	62.00	80.40	48.40	61.10
Specialty finish patterns	Inst	LF	2C	.808	43.30	25.90	---	69.20	88.70	55.00	68.60
Wood tone patterns	Inst	LF	2C	.808	45.10	25.90	---	71.00	90.70	56.60	70.50
Self-edge countertop with square edge veneer front											
25" W, 1-1/2" H front edge, 4" H coved backsplash (@ 90-degree angle to deck)											
Satin/suede patterns	Inst	LF	2C	.808	77.90	25.90	---	103.80	128.00	86.50	105.00
Solid patterns	Inst	LF	2C	.808	86.60	25.90	---	112.50	138.00	94.40	114.00
Specialty finish patterns	Inst	LF	2C	.808	104.00	25.90	---	129.90	158.00	110.20	132.00
Wood tone patterns	Inst	LF	2C	.808	108.00	25.90	---	133.90	163.00	114.10	137.00
Material cost adjustments, ADD											
Diagonal corner cut											
Standard	Inst	Ea	---	---	24.10	---	---	24.10	24.10	22.00	22.00
With plateau shelf	Inst	Ea	---	---	114.00	---	---	114.00	114.00	104.00	104.00
Radius corner cut											
3" or 6"	Inst	Ea	---	---	16.10	---	---	16.10	16.10	14.60	14.60
12"	Inst	Ea	---	---	16.10	---	---	16.10	16.10	14.60	14.60
Quarter radius end	Inst	Ea	---	---	16.10	---	---	16.10	16.10	14.60	14.60
Half radius end	Inst	Ea	---	---	29.50	---	---	29.50	29.50	26.80	26.80
Miter corner, shop assembled	Inst	Ea	---	---	24.30	---	---	24.30	24.30	22.10	22.10
Splicing any top or leg 12' or longer, shop assembled	Inst	Ea	---	---	28.10	---	---	28.10	28.10	25.60	25.60
Endsplash with finished sides and edges	Inst	Ea	---	---	21.40	---	---	21.40	21.40	19.50	19.50
Sink or range cutout	Inst	Ea	---	---	9.38	---	---	9.38	9.38	8.54	8.54

Description	Oper	Unit	Crew Size	Man-Hours Per Unit	Costs Based On Small Volume						Large Volume	
					Avg Mat'l Unit Cost	Avg Labor Unit Cost	Avg Equip Unit Cost	Avg Total Unit Cost	Avg Price Incl O&P		Avg Total Unit Cost	Avg Price Incl O&P

Ceramic tile

Countertop/backsplash

Adhesive set with backmounted tile

Description	Oper	Unit	Crew Size	Man-Hours Per Unit	Avg Mat'l Unit Cost	Avg Labor Unit Cost	Avg Equip Unit Cost	Avg Total Unit Cost	Avg Price Incl O&P	Avg Total Unit Cost	Avg Price Incl O&P
1" x 1"	Inst	SF	TB	.381	7.41	10.80	---	18.21	**24.40**	13.00	**18.50**
4-1/4" x 4-1/4"	Inst	SF	TB	.333	5.92	9.43	---	15.35	**20.70**	10.87	**15.60**

Cove/base

Adhesive set with unmounted tile

Description	Oper	Unit	Crew Size	Man-Hours Per Unit	Avg Mat'l Unit Cost	Avg Labor Unit Cost	Avg Equip Unit Cost	Avg Total Unit Cost	Avg Price Incl O&P	Avg Total Unit Cost	Avg Price Incl O&P
4-1/4" x 4-1/4"	Inst	LF	TB	.333	5.69	9.43	---	15.12	**20.40**	10.67	**14.10**
6" x 4-1/4"	Inst	LF	TB	.281	5.56	7.96	---	13.52	**18.10**	9.65	**12.60**
Conventional/mortar set with unmounted tile											
4-1/4" x 4-1/4"	Inst	LF	TB	.593	5.57	16.80	---	22.37	**31.10**	15.00	**20.50**
6" x 4-1/4"	Inst	LF	TB	.533	5.39	15.10	---	20.49	**28.40**	13.80	**18.80**
Dry-set mortar with unmounted tile											
4-1/4" x 4-1/4"	Inst	LF	TB	.410	5.60	11.60	---	17.20	**23.50**	11.89	**15.90**
6" x 4-1/4"	Inst	LF	TB	.356	5.43	10.10	---	15.53	**21.10**	10.80	**14.40**

Wood

Butcher block construction throughout top; custom, straight, "L", or "U" shapes

Self-edge top; 26" W, with 4" H backsplash, 1-1/2" H front and back edges

Description	Oper	Unit	Crew Size	Man-Hours Per Unit	Avg Mat'l Unit Cost	Avg Labor Unit Cost	Avg Equip Unit Cost	Avg Total Unit Cost	Avg Price Incl O&P	Avg Total Unit Cost	Avg Price Incl O&P
	Inst	LF	2C	.746	74.00	23.90	---	97.90	**121.00**	81.60	**99.30**
Material cost adjustments, ADD											
Miter corner	Inst	Ea	---	---	18.50	---	---	18.50	**18.50**	16.50	**16.50**
Sink or surface saver cutout	Inst	Ea	---	---	55.50	---	---	55.50	**55.50**	49.50	**49.50**
45-degree plateau corner	Inst	Ea	---	---	278.00	---	---	278.00	**278.00**	248.00	**248.00**
End splash	Inst	Ea	---	---	64.80	---	---	64.80	**64.80**	57.80	**57.80**

					Costs Based On Small Volume					Large Volume	
Description	Oper	Unit	Crew Size	Man-Hours Per Unit	Avg Mat'l Unit Cost	Avg Labor Unit Cost	Avg Equip Unit Cost	Avg Total Unit Cost	Avg Price Incl O&P	Avg Total Unit Cost	Avg Price Incl O&P

Cupolas

Natural finish redwood, aluminum roof

Description	Oper	Unit	Crew Size	Man-Hours Per Unit	Avg Mat'l Unit Cost	Avg Labor Unit Cost	Avg Equip Unit Cost	Avg Total Unit Cost	Avg Price Incl O&P	Avg Total Unit Cost	Avg Price Incl O&P
24" x 24" x 25" H	Inst	Ea	CA	1.90	297.00	61.00	---	358.00	463.00	291.60	373.00
30" x 30" x 30" H	Inst	Ea	CA	2.22	372.00	71.20	---	443.20	572.00	360.70	462.00
35" x 35" x 33" H	Inst	Ea	CA	2.67	509.00	85.70	---	594.70	764.00	486.30	621.00
Natural finish redwood, copper roof											
22" x 22" x 33" H	Inst	Ea	CA	1.90	348.00	61.00	---	409.00	527.00	334.60	427.00
25" x 25" x 39" H	Inst	Ea	CA	2.22	421.00	71.20	---	492.20	633.00	403.70	515.00
35" x 35" x 33" H	Inst	Ea	CA	2.67	596.00	85.70	---	681.70	873.00	561.30	715.00
Natural finish redwood, octagonal shape copper roof											
31" W, 37" H	Inst	Ea	CA	2.67	608.00	85.70	---	693.70	888.00	571.30	727.00
35" W, 43" H	Inst	Ea	CA	3.33	773.00	107.00	---	880.00	1130.00	726.20	923.00
Weathervanes for above, aluminum											
18" H, black finish	Inst	Ea	---	---	67.30	---	---	67.30	67.30	57.60	57.60
24" H, black finish	Inst	Ea	---	---	88.50	---	---	88.50	88.50	75.80	75.80
36" H, black and gold finish	Inst	Ea	---	---	198.00	---	---	198.00	198.00	170.00	170.00

Demolition

Average Manhours per Unit is based on what the designated crew can do in one day. In this section, the crew might be a laborer, a carpenter, a floor layer, etc. Who does the wrecking depends on the quantity of wrecking to be done.

A contractor might use laborers exclusively on a large volume job, but use a carpenter or a carpenter and a laborer on a small volume job.

The choice of tools and equipment greatly affects the quantity of work that a worker can accomplish in a day. A person can remove more brick with a compressor and pneumatic tool than with a sledgehammer or pry bar. In this section, the phrase "by hand"

includes the use of hand tools, i.e., sledgehammers, wrecking bars, claw hammers, etc. When the Average Manhours per Unit is based on the use of equipment (not hand tools), a description of the equipment is provided.

The Average Manhours per Unit is not based on the use of "heavy" equipment such as bulldozers, cranes with wrecking balls, etc. The Average Manhours per Unit does include the labor involved in hauling wrecked material or debris to a dumpster located at the site. Average rental costs for dumpsters are: $450.00 for 40 CY (20' x 8' x 8' H); $365.00 for 30 CY (20' x 8' x 6' H). Rental period includes delivery and pickup when full.

				Costs Based On Small Volume						Large Volume	
Description	Oper	Unit	Crew Size	Man-Hours Per Unit	Avg Mat'l Unit Cost	Avg Labor Unit Cost	Avg Equip Unit Cost	Avg Total Unit Cost	Avg Price Incl O&P	Avg Total Unit Cost	Avg Price Incl O&P

Demolition

Wreck and remove to dumpster

Concrete

Footings, with air tools; reinforced

Description	Oper	Unit	Crew Size	Man-Hours Per Unit	Avg Mat'l Unit Cost	Avg Labor Unit Cost	Avg Equip Unit Cost	Avg Total Unit Cost	Avg Price Incl O&P	Avg Total Unit Cost	Avg Price Incl O&P
8" T x 12" W (.67 CF/LF)	Demo	LF	AB	.205	---	5.87	1.19	7.06	9.88	4.58	6.41
8" T x 16" W (.89 CF/LF)	Demo	LF	AB	.246	---	7.05	1.43	8.48	11.90	5.51	7.71
8" T x 20" W (1.11 CF/LF)	Demo	LF	AB	.274	---	7.85	1.58	9.43	13.20	6.13	8.58
12" T x 12" W (1.00 CF/LF)	Demo	LF	AB	.352	---	10.10	2.04	12.14	17.00	7.88	11.06
12" T x 16" W (1.33 CF/LF)	Demo	LF	AB	.410	---	11.70	2.38	14.08	19.80	9.19	12.96
12" T x 20" W (1.67 CF/LF)	Demo	LF	AB	.448	---	12.80	2.59	15.39	21.60	10.01	14.00
12" T x 24" W (2.00 CF/LF)	Demo	LF	AB	.492	---	14.10	2.85	16.95	23.70	11.01	15.40

Foundations and retaining walls, with air tools, per LF wall

With reinforcing
4'-0" H

Description	Oper	Unit	Crew Size	Man-Hours Per Unit	Avg Mat'l Unit Cost	Avg Labor Unit Cost	Avg Equip Unit Cost	Avg Total Unit Cost	Avg Price Incl O&P	Avg Total Unit Cost	Avg Price Incl O&P
8" T (2.67 CF/LF)	Demo	SF	AB	.410	---	11.70	2.38	14.08	19.80	9.19	12.90
12" T (4.00 CF/LF)	Demo	SF	AB	.547	---	15.70	3.17	18.87	26.40	12.26	17.20

8'-0" H

| 8" T (5.33 CF/LF) | Demo | SF | AB | .448 | --- | 12.80 | 2.59 | 15.39 | 21.60 | 10.01 | 14.00 |
| 12" T (8.00 CF/LF) | Demo | SF | AB | .615 | --- | 17.60 | 3.56 | 21.16 | 29.60 | 13.82 | 19.30 |

12'-0" H

| 8" T (8.00 CF/LF) | Demo | SF | AB | .492 | --- | 14.10 | 2.85 | 16.95 | 23.70 | 11.01 | 15.40 |
| 12" T (12.00 CF/LF) | Demo | SF | AB | .703 | --- | 20.10 | 4.07 | 24.17 | 33.90 | 15.75 | 22.00 |

Without reinforcing
4'-0" H

| 8" T (2.67 CF/LF) | Demo | SF | AB | .308 | --- | 8.82 | 1.78 | 10.60 | 14.80 | 6.89 | 9.64 |
| 12" T (4.00 CF/LF) | Demo | SF | AB | .410 | --- | 11.70 | 2.38 | 14.08 | 19.80 | 9.19 | 12.90 |

8'-0" H

| 8" T (5.33 CF/LF) | Demo | SF | AB | .352 | --- | 10.10 | 2.04 | 12.14 | 17.00 | 7.88 | 11.00 |
| 12" T (8.00 CF/LF) | Demo | SF | AB | .492 | --- | 14.10 | 2.85 | 16.95 | 23.70 | 11.01 | 15.40 |

12'-0" H

| 8" T (8.00 CF/LF) | Demo | SF | AB | .410 | --- | 11.70 | 2.38 | 14.08 | 19.80 | 9.19 | 12.90 |
| 12" T (12.00 CF/LF) | Demo | SF | AB | .615 | --- | 17.60 | 3.56 | 21.16 | 29.60 | 13.82 | 19.30 |

Slabs with air tools

With reinforcing

Description	Oper	Unit	Crew Size	Man-Hours Per Unit	Avg Mat'l Unit Cost	Avg Labor Unit Cost	Avg Equip Unit Cost	Avg Total Unit Cost	Avg Price Incl O&P	Avg Total Unit Cost	Avg Price Incl O&P
4" T	Demo	SF	AB	.058	---	1.66	.34	2.00	2.80	1.31	1.83
5" T	Demo	SF	AB	.067	---	1.92	.39	2.31	3.23	1.48	2.07
6" T	Demo	SF	AB	.070	---	2.00	.41	2.41	3.38	1.58	2.21

Without reinforcing

4" T	Demo	SF	AB	.049	---	1.40	.29	1.69	2.37	1.11	1.55
5" T	Demo	SF	AB	.058	---	1.66	.34	2.00	2.80	1.31	1.83
6" T	Demo	SF	AB	.067	---	1.92	.39	2.31	3.23	1.48	2.07

Masonry

Description	Oper	Unit	Crew Size	Man-Hours Per Unit	Costs Based On Small Volume					Large Volume	
					Avg Mat'l Unit Cost	Avg Labor Unit Cost	Avg Equip Unit Cost	Avg Total Unit Cost	Avg Price Incl O&P	Avg Total Unit Cost	Avg Price Incl O&P

Brick

Description	Oper	Unit	Crew Size	Man-Hours Per Unit	Avg Mat'l Unit Cost	Avg Labor Unit Cost	Avg Equip Unit Cost	Avg Total Unit Cost	Avg Price Incl O&P	Avg Total Unit Cost	Avg Price Incl O&P
Chimneys											
4" T wall	Demo	VLF	LB	.985	---	26.20	---	26.20	**39.10**	17.10	**25.40**
8" T wall	Demo	VLF	LB	2.46	---	65.50	---	65.50	**97.60**	42.60	**63.50**
Columns, 12" x 12" o.d.	Demo	VLF	LB	.492	---	13.10	---	13.10	**19.50**	8.52	**12.70**
Veneer, 4" T, with air tools	Demo	SF	AB	.091	---	2.61	.53	3.14	**4.39**	2.03	**2.84**
Walls, with air tools											
8" T wall	Demo	SF	AB	.182	---	5.21	1.06	6.27	**8.77**	4.10	**5.73**
12" T wall	Demo	SF	AB	.259	---	7.42	1.50	8.92	**12.50**	5.79	**8.10**

Concrete block; lightweight (haydite), standard or heavyweight

Foundations and retaining walls; no excavation included

Without reinforcing or with only lateral reinforcing

Description	Oper	Unit	Crew Size	Man-Hours Per Unit	Avg Mat'l Unit Cost	Avg Labor Unit Cost	Avg Equip Unit Cost	Avg Total Unit Cost	Avg Price Incl O&P	Avg Total Unit Cost	Avg Price Incl O&P
With air tools											
8" W x 8" H x 16" L	Demo	SF	AB	.088	---	2.52	.51	3.03	**4.24**	1.96	**2.75**
12" W x 8" H x 16" L	Demo	SF	AB	.103	---	2.95	.59	3.54	**4.96**	2.31	**3.23**
Without air tools											
8" W x 8" H x 16" L	Demo	SF	LB	.109	---	2.90	---	2.90	**4.33**	1.89	**2.82**
12" W x 8" H x 16" L	Demo	SF	LB	.130	---	3.46	---	3.46	**5.16**	2.24	**3.33**

With vertical reinforcing in every other core (2 cores per block) with cores filled

Description	Oper	Unit	Crew Size	Man-Hours Per Unit	Avg Mat'l Unit Cost	Avg Labor Unit Cost	Avg Equip Unit Cost	Avg Total Unit Cost	Avg Price Incl O&P	Avg Total Unit Cost	Avg Price Incl O&P
With air tools											
8" W x 8" H x 16" L	Demo	SF	AB	.145	---	4.15	.84	4.99	**6.99**	3.24	**4.53**
12" W x 8" H x 16" L	Demo	SF	AB	.170	---	4.87	.98	5.85	**8.19**	3.79	**5.30**

Exterior walls (above grade) and partitions, no shoring included

Without reinforcing or with only lateral reinforcing

Description	Oper	Unit	Crew Size	Man-Hours Per Unit	Avg Mat'l Unit Cost	Avg Labor Unit Cost	Avg Equip Unit Cost	Avg Total Unit Cost	Avg Price Incl O&P	Avg Total Unit Cost	Avg Price Incl O&P
With air tools											
8" W x 8" H x 16" L	Demo	SF	AB	.070	---	2.00	.41	2.41	**3.38**	1.58	**2.21**
12" W x 8" H x 16" L	Demo	SF	AB	.082	---	2.35	.48	2.83	**3.96**	1.83	**2.56**
Without air tools											
8" W x 8" H x 16" L	Demo	SF	LB	.088	---	2.34	---	2.34	**3.49**	1.52	**2.26**
12" W x 8" H x 16" L	Demo	SF	LB	.103	---	2.74	---	2.74	**4.09**	1.78	**2.66**

Fences

Without reinforcing or with only lateral reinforcing

Description	Oper	Unit	Crew Size	Man-Hours Per Unit	Avg Mat'l Unit Cost	Avg Labor Unit Cost	Avg Equip Unit Cost	Avg Total Unit Cost	Avg Price Incl O&P	Avg Total Unit Cost	Avg Price Incl O&P
With air tools											
6" W x 4" H x 16" L	Demo	SF	AB	.063	---	1.80	.37	2.17	**3.04**	1.41	**1.98**
6" W x 6" H x 16" L	Demo	SF	AB	.067	---	1.92	.39	2.31	**3.23**	1.48	**2.07**
8" W x 8" H x 16" L	Demo	SF	AB	.070	---	2.00	.41	2.41	**3.38**	1.58	**2.21**
12" W x 8" H x 16" L	Demo	SF	AB	.082	---	2.35	.48	2.83	**3.96**	1.83	**2.56**
Without air tools											
6" W x 4" H x 16" L	Demo	SF	LB	.079	---	2.10	---	2.10	**3.14**	1.39	**2.06**
6" W x 6" H x 16" L	Demo	SF	LB	.083	---	2.21	---	2.21	**3.29**	1.44	**2.14**
8" W x 8" H x 16" L	Demo	SF	LB	.088	---	2.34	---	2.34	**3.49**	1.52	**2.26**
12" W x 8" H x 16" L	Demo	SF	LB	.103	---	2.74	---	2.74	**4.09**	1.78	**2.66**

Quarry tile, 6" or 9" squares

Floors

Description	Oper	Unit	Crew Size	Man-Hours Per Unit	Avg Mat'l Unit Cost	Avg Labor Unit Cost	Avg Equip Unit Cost	Avg Total Unit Cost	Avg Price Incl O&P	Avg Total Unit Cost	Avg Price Incl O&P
Conventional mortar set	Demo	SF	LB	.055	---	1.47	---	1.47	**2.18**	.96	**1.43**
Dry-set mortar	Demo	SF	LB	.048	---	1.28	---	1.28	**1.91**	.83	**1.23**

Description	Oper	Unit	Crew Size	Man-Hours Per Unit	Avg Mat'l Unit Cost	Avg Labor Unit Cost	Avg Equip Unit Cost	Avg Total Unit Cost	Avg Price Incl O&P	Avg Total Unit Cost	Avg Price Incl O&P
								Costs Based On Small Volume		**Large Volume**	

Rough carpentry (framing)

Dimension lumber

Beams, set on steel columns

Built-up from 2" lumber

Description	Oper	Unit	Crew Size	Man-Hours Per Unit	Avg Mat'l Unit Cost	Avg Labor Unit Cost	Avg Equip Unit Cost	Avg Total Unit Cost	Avg Price Incl O&P	Avg Total Unit Cost	Avg Price Incl O&P
4" T x 10" W - 10' L (2 pcs)	Demo	LF	LB	.029	---	.77	---	.77	**1.15**	.51	**.75**
4" T x 12" W - 12' L (2 pcs)	Demo	LF	LB	.024	---	.64	---	.64	**.95**	.43	**.64**
6" T x 10" W - 10' L (3 pcs)	Demo	LF	LB	.029	---	.77	---	.77	**1.15**	.51	**.75**
6" T x 12" W - 12' L (3 pcs)	Demo	LF	LB	.024	---	.64	---	.64	**.95**	.43	**.64**
Single member (solid lumber)											
3" T x 12" W - 12' L	Demo	LF	LB	.024	---	.64	---	.64	**.95**	.43	**.64**
4" T x 12" W - 12' L	Demo	LF	LB	.024	---	.64	---	.64	**.95**	.43	**.64**
Bracing, diagonal, notched-in, studs oc											
1" x 6" - 10'	Demo	LF	LB	.041	---	1.09	---	1.09	**1.63**	.69	**1.03**
Bridging, "X" type, 1" x 3", (8", 10", 12" T)											
16" oc	Demo	LF	LB	.070	---	1.86	---	1.86	**2.78**	1.23	**1.83**
Columns or posts											
4" x 4" -8' L	Demo	LF	LB	.032	---	.85	---	.85	**1.27**	.56	**.83**
6" x 6" -8' L	Demo	LF	LB	.032	---	.85	---	.85	**1.27**	.56	**.83**
6" x 8" -8' L	Demo	LF	LB	.038	---	1.01	---	1.01	**1.51**	.67	**.99**
8" x 8" -8' L	Demo	LF	LB	.040	---	1.07	---	1.07	**1.59**	.69	**1.03**
Fascia											
1" x 4" - 12" L	Demo	LF	LB	.019	---	.51	---	.51	**.75**	.32	**.48**
Firestops or stiffeners											
2" x 4" - 16"	Demo	LF	LB	.052	---	1.39	---	1.39	**2.06**	.91	**1.35**
2" x 6" - 16"	Demo	LF	LB	.052	---	1.39	---	1.39	**2.06**	.91	**1.35**
Furring strips, 1" x 4" - 8' L											
Walls; strips 12" oc											
Studs 16" oc	Demo	SF	LB	.020	---	.53	---	.53	**.79**	.35	**.52**
Studs 24" oc	Demo	SF	LB	.018	---	.48	---	.48	**.71**	.32	**.48**
Masonry (concrete blocks)	Demo	SF	LB	.022	---	.59	---	.59	**.87**	.40	**.60**
Concrete	Demo	SF	LB	.038	---	1.01	---	1.01	**1.51**	.67	**.99**
Ceiling; joists 16" oc											
Strips 12" oc	Demo	SF	LB	.029	---	.77	---	.77	**1.15**	.51	**.75**
Strips 16" oc	Demo	SF	LB	.023	---	.61	---	.61	**.91**	.40	**.60**
Headers or lintels, over openings											
Built-up or single member											
4" T x 6" W - 4' L	Demo	LF	LB	.044	---	1.17	---	1.17	**1.75**	.77	**1.15**
4" T x 8" W - 8' L	Demo	LF	LB	.034	---	.91	---	.91	**1.35**	.59	**.87**
4" T x 10" W - 10' L	Demo	LF	LB	.029	---	.77	---	.77	**1.15**	.51	**.75**
4" T x 12" W - 12' L	Demo	LF	LB	.024	---	.64	---	.64	**.95**	.43	**.64**
4" T x 14" W - 14' L	Demo	LF	LB	.024	---	.64	---	.64	**.95**	.43	**.64**

Description	Oper	Unit	Crew Size	Man-Hours Per Unit	Avg Mat'l Unit Cost	Avg Labor Unit Cost	Avg Equip Unit Cost	Avg Total Unit Cost	Avg Price Incl O&P	Avg Total Unit Cost	Avg Price Incl O&P
Joists											
Ceiling											
2" x 4" - 8' L	Demo	LF	LB	.020	---	.53	---	.53	.79	.35	.52
2" x 4" - 10' L	Demo	LF	LB	.018	---	.48	---	.48	.71	.29	.44
2" x 8" - 12' L	Demo	LF	LB	.017	---	.45	---	.45	.67	.29	.44
2" x 10" - 14' L	Demo	LF	LB	.016	---	.43	---	.43	.64	.27	.40
2" x 12" - 16' L	Demo	LF	LB	.016	---	.43	---	.43	.64	.27	.40
Floor; seated on sill plate											
2" x 8" - 12' L	Demo	LF	LB	.015	---	.40	---	.40	.60	.27	.40
2" x 10" - 14' L	Demo	LF	LB	.014	---	.37	---	.37	.56	.24	.36
2" x 12" - 16' L	Demo	LF	LB	.014	---	.37	---	.37	.56	.24	.36
Ledgers											
Nailed, 2" x 6" - 12' L	Demo	LF	LB	.035	---	.93	---	.93	1.39	.61	.91
Bolted, 3" x 8" - 12' L	Demo	LF	LB	.025	---	.67	---	.67	.99	.43	.64
Plates; 2" x 4" or 2" x 6"											
Double top nailed	Demo	LF	LB	.030	---	.80	---	.80	1.19	.53	.79
Sill, nailed	Demo	LF	LB	.018	---	.48	---	.48	.71	.32	.48
Sill or bottom, bolted	Demo	LF	LB	.036	---	.96	---	.96	1.43	.61	.91
Rafters											
Common											
2" x 4" - 14' L	Demo	LF	LB	.014	---	.37	---	.37	.56	.24	.36
2" x 6" - 14' L	Demo	LF	LB	.015	---	.40	---	.40	.60	.27	.40
2" x 8" - 14' L	Demo	LF	LB	.017	---	.45	---	.45	.67	.29	.44
2" x 10" - 14' L	Demo	LF	LB	.021	---	.56	---	.56	.83	.35	.52
Hip and/or valley											
2" x 4" - 16' L	Demo	LF	LB	.012	---	.32	---	.32	.48	.21	.32
2" x 6" - 16' L	Demo	LF	LB	.013	---	.35	---	.35	.52	.21	.32
2" x 8" - 16' L	Demo	LF	LB	.015	---	.40	---	.40	.60	.27	.40
2" x 10" - 16' L	Demo	LF	LB	.018	---	.48	---	.48	.71	.32	.48
Jack											
2" x 4" - 6' L	Demo	LF	LB	.031	---	.83	---	.83	1.23	.53	.79
2" x 6" - 6' L	Demo	LF	LB	.033	---	.88	---	.88	1.31	.59	.87
2" x 8" - 6' L	Demo	LF	LB	.038	---	1.01	---	1.01	1.51	.64	.95
2" x 10" - 6' L	Demo	LF	LB	.046	---	1.23	---	1.23	1.83	.80	1.19
Roof decking, solid T&G											
2" x 6" - 12' L	Demo	LF	LB	.038	---	1.01	---	1.01	1.51	.67	.99
2" x 8" - 12' L	Demo	LF	LB	.027	---	.72	---	.72	1.07	.48	.71
Studs											
2" x 4" - 8' L	Demo	LF	LB	.018	---	.48	---	.48	.71	.32	.48
2" x 6" - 8' L	Demo	LF	LB	.018	---	.48	---	.48	.71	.32	.48
Stud partitions, studs 16" oc with bottom plate and double top plates and firestops; per LF partition											
2" x 4" - 8' L	Demo	LF	LB	.246	---	6.55	---	6.55	9.76	4.26	6.35
2" x 6" - 8' L	Demo	LF	LB	.246	---	6.55	---	6.55	9.76	4.26	6.35

Description	Oper	Unit	Crew Size	Man-Hours Per Unit	Avg Mat'l Unit Cost	Avg Labor Unit Cost	Avg Equip Unit Cost	Avg Total Unit Cost	Avg Price Incl O&P	Avg Total Unit Cost	Avg Price Incl O&P
Boards											
Sheathing, regular or diagonal											
1" x 8"											
Roof	Demo	SF	LB	.018	---	.48	---	.48	**.71**	.32	**.48**
Sidewall	Demo	SF	LB	.015	---	.40	---	.40	**.60**	.27	**.40**
Subflooring, regular or diagonal											
1" x 8" - 16' L	Demo	SF	LB	.016	---	.43	---	.43	**.64**	.27	**.40**
1" x 10" - 16' L	Demo	SF	LB	.013	---	.35	---	.35	**.52**	.21	**.32**
Plywood											
Sheathing											
Roof											
1/2" T, CDX	Demo	SF	LB	.012	---	.32	---	.32	**.48**	.21	**.32**
5/8" T, CDX	Demo	SF	LB	.013	---	.35	---	.35	**.52**	.21	**.32**
Wall											
3/8" or 1/2" T, CDX	Demo	SF	LB	.010	---	.27	---	.27	**.40**	.16	**.24**
5/8" T, CDX	Demo	SF	LB	.010	---	.27	---	.27	**.40**	.19	**.28**
Subflooring											
5/8", 3/4" T, CDX	Demo	SF	LB	.011	---	.29	---	.29	**.44**	.19	**.28**
1-1/8" T, 2-4-1, T&G long edges	Demo	SF	LB	.015	---	.40	---	.40	**.60**	.27	**.40**
Trusses, "W" pattern with gin pole, 24' to 30' spans											
3 - in - 12 slope	Demo	Ea	LB	.947	---	25.20	---	25.20	**37.60**	16.40	**24.40**
5 - in - 12 slope	Demo	Ea	LB	.947	---	25.20	---	25.20	**37.60**	16.40	**24.40**
Finish carpentry											
Bath accessories, screwed	Demo	Ea	LB	.197	---	5.25	---	5.25	**7.82**	3.41	**5.08**
Cabinets											
Kitchen, to 3' x 4', wood; base, wall, or peninsula	Demo	Ea	LB	.985	---	26.20	---	26.20	**39.10**	17.10	**25.40**
Medicine, metal	Demo	Ea	LB	.821	---	21.90	---	21.90	**32.60**	14.20	**21.20**
Vanity, cabinet and sink top											
Disconnect plumbing and remove to dumpster	Demo	Ea	LB	1.54	---	41.00	---	41.00	**61.10**	26.60	**39.70**
Remove old unit, replace with new unit, reconnect plumbing	Demo	Ea	SB	3.52	---	111.00	---	111	**163**	72.10	**106.00**
Hardwood flooring (over wood subfloors)											
Block, set in mastic	Demo	SF	LB	.021	---	.56	---	.56	**.83**	.35	**.52**
Strip, nailed	Demo	SF	LB	.027	---	.72	---	.72	**1.07**	.48	**.71**
Marlite panels, 4' x 8',											
adhesive set	Demo	SF	LB	.029	---	.77	---	.77	**1.15**	.51	**.75**
Molding and trim											
At base (floor)	Demo	LF	LB	.023	---	.61	---	.61	**.91**	.40	**.60**
At ceiling	Demo	LF	LB	.021	---	.56	---	.56	**.83**	.35	**.52**
On walls or cabinets	Demo	LF	LB	.015	---	.40	---	.40	**.60**	.27	**.40**
Paneling											
Plywood, prefinished	Demo	SF	LB	.013	---	.35	---	.35	**.52**	.24	**.36**
Wood	Demo	SF	LB	.015	---	.40	---	.40	**.60**	.27	**.40**

Description	Oper	Unit	Crew Size	Man-Hours Per Unit	Avg Mat'l Unit Cost	Avg Labor Unit Cost	Avg Equip Unit Cost	Avg Total Unit Cost	Avg Price Incl O&P	Avg Total Unit Cost	Avg Price Incl O&P
								Costs Based On Small Volume		Large Volume	

Weather protection
Insulation
Batt/roll, with wall or ceiling finish already removed
Description	Oper	Unit	Crew Size	M-H/Unit	Mat'l	Labor	Equip	Total	Price O&P	Total (LV)	Price O&P (LV)
Joists, 16" or 24" oc	Demo SF	LB		.008	---	.21	---	.21	.32	.13	.20
Rafters, 16" or 24" oc	Demo SF	LB		.010	---	.27	---	.27	.40	.16	.24
Studs, 16" or 24" oc	Demo SF	LB		.007	---	.19	---	.19	.28	.13	.20

Loose, with ceiling finish already removed
Joists, 16" or 24" oc											
4" T	Demo SF	LB		.006	---	.16	---	.16	.24	.11	.16
6" T	Demo SF	LB		.011	---	.29	---	.29	.44	.19	.28

Rigid
Roofs											
1/2" T	Demo Sq	LB		1.45	---	38.60	---	38.60	57.60	25.10	37.40
1" T	Demo Sq	LB		1.64	---	43.70	---	43.70	65.10	28.50	42.50
Walls, 1/2" T	Demo SF	LB		.012	---	.32	---	.32	.48	.19	.28

Sheet metal
Gutter and downspouts
Aluminum	Demo LF	LB		.029	---	.77	---	.77	1.15	.51	.75
Galvanized	Demo LF	LB		.038	---	1.01	---	1.01	1.51	.67	.99

Roofing and siding
Aluminum
Roofing, nailed to wood
Corrugated (2-1/2"), 26" W with 3-3/4" side lap and 6" end lap	Demo Sq	LB		2.46	---	65.50	---	65.50	97.60	42.60	63.50

Siding, nailed to wood
Clapboard (i.e., lap drop)											
8" exposure	Demo SF	LB		.024	---	.64	---	.64	.95	.43	.64
10" exposure	Demo SF	LB		.019	---	.51	---	.51	.75	.35	.52
Corrugated (2-1/2"), 26" W with 2-1/2" side lap and 4" end lap	Demo SF	LB		.021	---	.56	---	.56	.83	.35	.52
Panels, 4' x 8'	Demo SF	LB		.010	---	.27	---	.27	.40	.19	.28
Shingle, 24" L with 12" exposure	Demo SF	LB		.017	---	.45	---	.45	.67	.29	.44

Asphalt shingle roofing
240 lb/Sq, strip, 3 tab 5" exposure	Demo Sq	LB		1.54	---	41.00	---	41.00	61.10	26.60	39.70

Built-up/hot roofing (to wood deck)
3 ply											
With gravel	Demo Sq	LB		2.24	---	59.70	---	59.70	88.90	38.60	57.60
Without gravel	Demo Sq	LB		1.76	---	46.90	---	46.90	69.90	30.40	45.30
5 ply											
With gravel	Demo Sq	LB		2.46	---	65.50	---	65.50	97.60	42.60	63.50
Without gravel	Demo Sq	LB		1.89	---	50.40	---	50.40	75.00	32.80	48.80

Clay tile roofing
2 piece interlocking	Demo Sq	LB		2.46	---	65.50	---	65.50	97.60	42.60	63.50
1 piece	Demo Sq	LB		2.24	---	59.70	---	59.70	88.90	38.60	57.60

Description	Oper	Unit	Crew Size	Man-Hours Per Unit	Avg Mat'l Unit Cost	Avg Labor Unit Cost	Avg Equip Unit Cost	Avg Total Unit Cost	Avg Price Incl O&P	Large Volume Avg Total Unit Cost	Large Volume Avg Price Incl O&P
Hardboard siding											
Lap, 1/2" T x 12" W x 16' L, with 11" exposure	Demo	SF	LB	.018	---	.48	---	.48	.71	.32	.48
Panels, 7/16" T x 4' W x 8' H	Demo	SF	LB	.010	---	.27	---	.27	.40	.16	.24
Mineral surfaced roll roofing											
Single coverage 90 lb/Sq roll with 6" end lap and 2" headlap	Demo	Sq	LB	.769	---	20.50	---	20.50	30.50	13.30	19.90
Double coverage selvage roll, with 6" end lap and 17" exposure	Demo	Sq	LB	1.12	---	29.80	---	29.80	44.50	19.40	28.90
Wood											
Roofing											
Shakes											
24" L with 10" exposure											
1/2" to 3/4" T	Demo	Sq	LB	.985	---	26.20	---	26.20	39.10	17.10	25.40
3/4" to 5/4" T	Demo	Sq	LB	1.07	---	28.50	---	28.50	42.50	18.50	27.60
Shingles											
16" L with 5" exposure	Demo	Sq	LB	2.05	---	54.60	---	54.60	81.40	35.40	52.80
18" L with 5-1/2" exposure	Demo	Sq	LB	1.89	---	50.40	---	50.40	75.00	32.80	48.80
24" L with 7-1/2" exposure	Demo	Sq	LB	1.37	---	36.50	---	36.50	54.40	23.70	35.30
Siding											
Bevel											
1/2" x 8" with 6-3/4" exposure	Demo	SF	LB	.027	---	.72	---	.72	1.07	.48	.71
5/8" x 10" with 8-3/4" exposure	Demo	SF	LB	.021	---	.56	---	.56	.83	.37	.56
3/4" x 12" with 10-3/4" exposure	Demo	SF	LB	.017	---	.45	---	.45	.67	.29	.44
Drop (horizontal), 1/4" T&G											
1" x 8" with 7" exposure	Demo	SF	LB	.026	---	.69	---	.69	1.03	.45	.67
1" x 10" with 9" exposure	Demo	SF	LB	.020	---	.53	---	.53	.79	.35	.52
Board (1" x 12") and batten (1" x 2") @ 12" oc											
Horizontal	Demo	SF	LB	.019	---	.51	---	.51	.75	.35	.52
Vertical											
Standard	Demo	SF	LB	.024	---	.64	---	.64	.95	.43	.64
Reverse	Demo	SF	LB	.023	---	.61	---	.61	.91	.40	.60
Board on board (1" x 12" with 1-1/2" overlap), vertical	Demo	SF	LB	.026	---	.69	---	.69	1.03	.45	.67
Plywood (1/2" T) with battens (1" x 2")											
16" oc battens	Demo	SF	LB	.011	---	.29	---	.29	.44	.19	.28
24" oc battens	Demo	SF	LB	.010	---	.27	---	.27	.40	.19	.28
Shakes											
24" L with 11-1/2" exposure											
1/2" to 3/4" T	Demo	SF	LB	.015	---	.40	---	.40	.60	.27	.40
3/4" to 5/4" T	Demo	SF	LB	.017	---	.45	---	.45	.67	.29	.44
Shingles											
16" L with 7-1/2" exposure	Demo	SF	LB	.025	---	.67	---	.67	.99	.43	.64
18" L with 8-1/2" exposure	Demo	SF	LB	.022	---	.59	---	.59	.87	.37	.56
24" L with 11-1/2" exposure	Demo	SF	LB	.016	---	.43	---	.43	.64	.27	.40

Doors, windows and glazing
Doors with related trim and frame

Description	Oper	Unit	Crew Size	Man-Hours Per Unit	Avg Mat'l Unit Cost	Avg Labor Unit Cost	Avg Equip Unit Cost	Avg Total Unit Cost	Avg Price Incl O&P	Avg Total Unit Cost	Avg Price Incl O&P
Closet, with track											
Folding, 4 doors	Demo Set	LB	2.05	---	54.60	---	54.60	**81.40**	35.40	**52.80**	
Sliding, 2 or 3 doors	Demo Set	LB	2.05	---	54.60	---	54.60	**81.40**	35.40	**52.80**	
Entry, 3' x 7'	Demo Ea	LB	1.76	---	46.90	---	46.90	**69.90**	30.40	**45.30**	
Fire, 3' x 7'	Demo Ea	LB	1.76	---	46.90	---	46.90	**69.90**	30.40	**45.30**	
Garage											
Wood, aluminum, or hardboard											
Single	Demo Ea	LB	3.08	---	82.10	---	82.10	**122**	53.30	**79.40**	
Double	Demo Ea	LB	4.10	---	109.00	---	109	**163**	71.10	**106.00**	
Steel											
Single	Demo Ea	LB	3.52	---	93.80	---	93.80	**140**	61.00	**90.90**	
Double	Demo Ea	LB	4.92	---	131.00	---	131	**195**	85.30	**127.00**	
Glass sliding, with track											
2 lites wide	Demo Set	LB	3.08	---	82.10	---	82.10	**122**	53.30	**79.40**	
3 lites wide	Demo Set	LB	4.10	---	109.00	---	109	**163**	71.10	**106.00**	
4 lites wide	Demo Set	LB	6.15	---	164.00	---	164	**244**	107.00	**159.00**	
Interior, 3' x 7'	Demo Ea	LB	1.54	---	41.00	---	41.00	**61.10**	26.60	**39.70**	
Screen, 3' x 7'	Demo Ea	LB	1.23	---	32.80	---	32.80	**48.80**	21.30	**31.80**	
Storm combination, 3' x 7'	Demo Ea	LB	1.23	---	32.80	---	32.80	**48.80**	21.30	**31.80**	

Windows, with related trim and frame

Description	Oper	Unit	Crew Size	Man-Hours Per Unit	Avg Mat'l Unit Cost	Avg Labor Unit Cost	Avg Equip Unit Cost	Avg Total Unit Cost	Avg Price Incl O&P	Avg Total Unit Cost	Avg Price Incl O&P
To 12 SF											
Aluminum	Demo Ea	LB	1.17	---	31.20	---	31.20	**46.40**	20.30	**30.20**	
Wood	Demo Ea	LB	1.54	---	41.00	---	41.00	**61.10**	26.60	**39.70**	
13 SF to 50 SF											
Aluminum	Demo Ea	LB	1.89	---	50.40	---	50.40	**75.00**	32.80	**48.80**	
Wood	Demo Ea	LB	2.46	---	65.50	---	65.50	**97.60**	42.60	**63.50**	

Glazing, clean sash and remove old putty or rubber

Description	Oper	Unit	Crew Size	Man-Hours Per Unit	Avg Mat'l Unit Cost	Avg Labor Unit Cost	Avg Equip Unit Cost	Avg Total Unit Cost	Avg Price Incl O&P	Avg Total Unit Cost	Avg Price Incl O&P
3/32" T float, putty or rubber											
8" x 12" (0.667 SF)	Demo SF	GA	.492	---	14.70	---	14.70	**21.70**	9.53	**14.10**	
12" x 16" (1.333 SF)	Demo SF	GA	.274	---	8.16	---	8.16	**12.10**	5.30	**7.84**	
14" x 20" (1.944 SF)	Demo SF	GA	.224	---	6.67	---	6.67	**9.87**	4.32	**6.39**	
16" x 24" (2.667 SF)	Demo SF	GA	.176	---	5.24	---	5.24	**7.76**	3.39	**5.02**	
24" x 26" (4.333 SF)	Demo SF	GA	.130	---	3.87	---	3.87	**5.73**	2.50	**3.70**	
36" x 24" (6.000 SF)	Demo SF	GA	.098	---	2.92	---	2.92	**4.32**	1.91	**2.82**	
1/8" T float, putty, steel sash											
12" x 16" (1.333 SF)	Demo SF	GA	.274	---	8.16	---	8.16	**12.10**	5.30	**7.84**	
16" x 20" (2.222 SF)	Demo SF	GA	.189	---	5.63	---	5.63	**8.33**	3.66	**5.42**	
16" x 24" (2.667 SF)	Demo SF	GA	.176	---	5.24	---	5.24	**7.76**	3.39	**5.02**	
24" x 26" (4.333 SF)	Demo SF	GA	.130	---	3.87	---	3.87	**5.73**	2.50	**3.70**	
28" x 32" (6.222 SF)	Demo SF	GA	.095	---	2.83	---	2.83	**4.19**	1.85	**2.73**	
36" x 36" (9.000 SF)	Demo SF	GA	.077	---	2.29	---	2.29	**3.39**	1.49	**2.20**	
36" x 48" (12.000 SF)	Demo SF	GA	.065	---	1.94	---	1.94	**2.86**	1.25	**1.85**	
1/4" T float											
Wood sash with putty											
72" x 48" (24.0 SF)	Demo SF	GA	.067	---	2.00	---	2.00	**2.95**	1.28	**1.90**	
Aluminum sash with aluminum channel and rigid neoprene rubber											
48" x 96" (32.0 SF)	Demo SF	GA	.070	---	2.08	---	2.08	**3.08**	1.37	**2.03**	
96" x 96" (64.0 SF)	Demo SF	GA	.068	---	2.03	---	2.03	**3.00**	1.31	**1.94**	

Description	Oper	Unit	Crew Size	Man-Hours Per Unit	Avg Mat'l Unit Cost	Avg Labor Unit Cost	Avg Equip Unit Cost	Avg Total Unit Cost	Avg Price Incl O&P	Avg Total Unit Cost	Avg Price Incl O&P
						Costs Based On Small Volume				Large Volume	
1" T insulating glass; with 2 pieces 1/4" float and 1/2" air space											
To 6.0 SF	Demo	SF	GA	.246	---	7.33	---	7.33	**10.80**	4.76	**7.05**
6.1 to 12.0 SF	Demo	SF	GA	.112	---	3.34	---	3.34	**4.94**	2.17	**3.22**
12.1 to 18.0 SF	Demo	SF	GA	.085	---	2.53	---	2.53	**3.75**	1.64	**2.42**
18.1 to 24.0 SF	Demo	SF	GA	.082	---	2.44	---	2.44	**3.61**	1.58	**2.34**

Aluminum sliding door glass with aluminum channel and rigid neoprene rubber

Description	Oper	Unit	Crew Size	Man-Hours Per Unit	Avg Mat'l Unit Cost	Avg Labor Unit Cost	Avg Equip Unit Cost	Avg Total Unit Cost	Avg Price Incl O&P	Avg Total Unit Cost	Avg Price Incl O&P
34" x 76" (17.944 SF)											
5/8" T insulating glass with 2 pieces 5/32" T (tempered with 1-1/4" air space)	Demo	SF	GA	.072	---	2.14	---	2.14	**3.17**	1.40	**2.07**
5/32" T tempered	Demo	SF	GA	.063	---	1.88	---	1.88	**2.78**	1.22	**1.81**
46" x 76" (24.278 SF)											
5/8" T insulating glass with 2 pieces 5/32" T (tempered with 1-1/4" air space)	Demo	SF	GA	.057	---	1.70	---	1.70	**2.51**	1.10	**1.63**
5/32" T tempered	Demo	SF	GA	.050	---	1.49	---	1.49	**2.20**	.98	**1.45**

Finishes

Exterior and interior, with hand tools; no insulation removal included

Plaster and stucco; remove to studs or sheathing

Description	Oper	Unit	Crew Size	Man-Hours Per Unit	Avg Mat'l Unit Cost	Avg Labor Unit Cost	Avg Equip Unit Cost	Avg Total Unit Cost	Avg Price Incl O&P	Avg Total Unit Cost	Avg Price Incl O&P
Lath (wood or metal) and plaster, walls and ceiling											
2 coats	Demo	SY	LB	.189	---	5.03	---	5.03	**7.50**	3.28	**4.88**
3 coats	Demo	SY	LB	.205	---	5.46	---	5.46	**8.14**	3.54	**5.28**
Stucco and metal netting											
2 coats	Demo	SY	LB	.259	---	6.90	---	6.90	**10.30**	4.48	**6.67**
3 coats	Demo	SY	LB	.308	---	8.21	---	8.21	**12.20**	5.33	**7.94**

Wallboard, gypsum (drywall),

Description	Oper	Unit	Crew Size	Man-Hours Per Unit	Avg Mat'l Unit Cost	Avg Labor Unit Cost	Avg Equip Unit Cost	Avg Total Unit Cost	Avg Price Incl O&P	Avg Total Unit Cost	Avg Price Incl O&P
walls and ceilings	Demo	SF	LB	.016	---	.43	---	.43	**.64**	.27	**.40**

Ceramic, metal, plastic tile

Description	Oper	Unit	Crew Size	Man-Hours Per Unit	Avg Mat'l Unit Cost	Avg Labor Unit Cost	Avg Equip Unit Cost	Avg Total Unit Cost	Avg Price Incl O&P	Avg Total Unit Cost	Avg Price Incl O&P
Floors, 1" x 1"											
Adhesive or dry-set base	Demo	SF	LB	.045	---	1.20	---	1.20	**1.79**	.77	**1.15**
Conventional mortar base	Demo	SF	LB	.052	---	1.39	---	1.39	**2.06**	.91	**1.35**
Walls, 1" x 1" or 4-1/4" x 4-1/4"											
Adhesive or dry-set base	Demo	SF	LB	.051	---	1.36	---	1.36	**2.02**	.88	**1.31**
Conventional mortar base	Demo	SF	LB	.062	---	1.65	---	1.65	**2.46**	1.07	**1.59**

Acoustical or insulating ceiling tile

Description	Oper	Unit	Crew Size	Man-Hours Per Unit	Avg Mat'l Unit Cost	Avg Labor Unit Cost	Avg Equip Unit Cost	Avg Total Unit Cost	Avg Price Incl O&P	Avg Total Unit Cost	Avg Price Incl O&P
Adhesive set, tile only	Demo	SF	LB	.019	---	.51	---	.51	**.75**	.32	**.48**
Stapled, tile only	Demo	SF	LB	.021	---	.56	---	.56	**.83**	.37	**.56**
Stapled, tile and furring strips	Demo	SF	LB	.016	---	.43	---	.43	**.64**	.27	**.40**

Suspended ceiling system;

Description	Oper	Unit	Crew Size	Man-Hours Per Unit	Avg Mat'l Unit Cost	Avg Labor Unit Cost	Avg Equip Unit Cost	Avg Total Unit Cost	Avg Price Incl O&P	Avg Total Unit Cost	Avg Price Incl O&P
panels and grid system	Demo	SF	LB	.014	---	.37	---	.37	**.56**	.24	**.36**

Resilient flooring, adhesive set

Description	Oper	Unit	Crew Size	Man-Hours Per Unit	Avg Mat'l Unit Cost	Avg Labor Unit Cost	Avg Equip Unit Cost	Avg Total Unit Cost	Avg Price Incl O&P	Avg Total Unit Cost	Avg Price Incl O&P
Sheet products	Demo	SY	LB	.154	---	4.10	---	4.10	**6.11**	2.66	**3.97**
Tile products	Demo	SF	LB	.016	---	.43	---	.43	**.64**	.29	**.44**

Wallpaper

Average output is expressed in rolls (36.0 SF/single roll)

Description	Oper	Unit	Crew Size	Man-Hours Per Unit	Avg Mat'l Unit Cost	Avg Labor Unit Cost	Avg Equip Unit Cost	Avg Total Unit Cost	Avg Price Incl O&P	Avg Total Unit Cost	Avg Price Incl O&P
Single layer of paper from plaster with steaming equipment	Demo	Roll	1L	.615	---	16.40	---	16.40	24.40	10.70	15.90
Several layers of paper from plaster with steaming equipment	Demo	Roll	1L	1.03	---	27.40	---	27.40	40.90	17.80	26.50
Vinyls (with non-woven, woven, or synthetic fiber backings) from plaster with steaming equipment	Demo	Roll	1L	.410	---	10.90	---	10.90	16.30	7.11	10.60
Single layer of paper from drywall with steaming equipment	Demo	Roll	1L	.615	---	16.40	---	16.40	24.40	10.70	15.90
Several layers of paper from drywall with steaming equipment	Demo	Roll	1L	1.03	---	27.40	---	27.40	40.90	17.80	26.50
Vinyls (with synthetic fiber backing) from drywall with steaming equipment	Demo	Roll	1L	.410	---	10.90	---	10.90	16.30	7.11	10.60
Vinyls (with other backings), from drywall with steaming equipment	Demo	Roll	1L	.615	---	16.40	---	16.40	24.40	10.70	15.90

				Costs Based On Small Volume					Large Volume	
Description	Oper	Unit	Crew Size	Man-Hours Per Unit	Avg Mat'l Unit Cost	Avg Labor Unit Cost	Avg Equip Unit Cost	Avg Total Unit Cost	Avg Price Incl O&P	Avg Total Unit Cost	Avg Price Incl O&P

Dishwashers

High quality units. Labor cost includes rough-in

Frequently encountered applications

Description	Oper	Unit	Crew Size	Man-Hours Per Unit	Avg Mat'l	Avg Labor	Avg Equip	Avg Total	Avg Price O&P	Lg Total	Lg Price O&P
Detach & reset unit	Reset	Ea	SA	2.05	---	74.50	---	74.50	**110.00**	48.40	**71.10**
Remove unit	Demo	Ea	SA	1.23	---	44.70	---	44.70	**65.70**	29.10	**42.70**
Install unit											
Standard	Inst	Ea	SC	10.3	553.00	364.00	---	917.00	**1170.00**	691.00	**870.00**
Average	Inst	Ea	SC	10.3	680.00	364.00	---	1044.00	**1320.00**	796.00	**991.00**
High	Inst	Ea	SC	10.3	850.00	364.00	---	1214.00	**1510.00**	936.00	**1150.00**
Premium	Inst	Ea	SC	10.3	978.00	364.00	---	1342.00	**1660.00**	1041.00	**1270.00**

Built-in front loading

Description	Oper	Unit	Crew Size	Man-Hours Per Unit	Avg Mat'l	Avg Labor	Avg Equip	Avg Total	Avg Price O&P	Lg Total	Lg Price O&P
Six cycles	Inst	Ea	SC	10.3	978.00	364.00	---	1342.00	**1660.00**	1041.00	**1270.00**
Five cycles	Inst	Ea	SC	10.3	850.00	364.00	---	1214.00	**1510.00**	936.00	**1150.00**
Four cycles	Inst	Ea	SC	10.3	680.00	364.00	---	1044.00	**1320.00**	796.00	**991.00**
Three cycles	Inst	Ea	SC	10.3	553.00	364.00	---	917.00	**1170.00**	691.00	**870.00**
Two cycles	Inst	Ea	SC	10.3	327.00	364.00	---	691.00	**912.00**	506.00	**657.00**

Adjustments

Description	Oper	Unit	Crew Size	Man-Hours Per Unit	Avg Mat'l	Avg Labor	Avg Equip	Avg Total	Avg Price O&P	Lg Total	Lg Price O&P
Remove and reset dishwasher only	Reset	Ea	SA	2.05	---	74.50	---	74.50	**110.00**	48.40	**71.10**
Remove dishwasher	Demo	Ea	SA	1.23	---	44.70	---	44.70	**65.70**	29.10	**42.70**
Front and side panel kits											
White or prime finish	Inst	Ea	---	---	42.50	---	---	42.50	**42.50**	35.00	**35.00**
Regular color finish	Inst	Ea	---	---	42.50	---	---	42.50	**42.50**	35.00	**35.00**
Brushed chrome finish	Inst	Ea	---	---	51.00	---	---	51.00	**51.00**	42.00	**42.00**
Stainless steel finish	Inst	Ea	---	---	63.80	---	---	63.80	**63.80**	52.50	**52.50**
Black-glass acrylic finish with trim kit	Inst	Ea	---	---	85.00	---	---	85.00	**85.00**	70.00	**70.00**
Stainless steel trim kit	Inst	Ea	---	---	38.30	---	---	38.30	**38.30**	31.50	**31.50**

Portable front loading

Convertible, front loading, portable dishwashers; hardwood top, front & side panels

Description	Oper	Unit	Crew Size	Man-Hours Per Unit	Avg Mat'l	Avg Labor	Avg Equip	Avg Total	Avg Price O&P	Lg Total	Lg Price O&P
Six cycles											
White	Inst	Ea	SC	6.15	680.00	218.00	---	898.00	**1100.00**	701.00	**852.00**
Colors	Inst	Ea	SC	6.15	723.00	218.00	---	941.00	**1150.00**	736.00	**892.00**
Four cycles											
White	Inst	Ea	SC	6.15	595.00	218.00	---	813.00	**1000.00**	631.00	**771.00**
Colors	Inst	Ea	SC	6.15	638.00	218.00	---	856.00	**1050.00**	666.00	**812.00**
Three cycles											
White	Inst	Ea	SC	6.15	510.00	218.00	---	728.00	**906.00**	561.00	**691.00**
Colors	Inst	Ea	SC	6.15	553.00	218.00	---	771.00	**955.00**	596.00	**731.00**

Front loading portables

Description	Oper	Unit	Crew Size	Man-Hours Per Unit	Avg Mat'l	Avg Labor	Avg Equip	Avg Total	Avg Price O&P	Lg Total	Lg Price O&P
Three cycles with hardwood top											
White	Inst	Ea	SC	6.15	510.00	218.00	---	728.00	**906.00**	561.00	**691.00**
Colors	Inst	Ea	SC	6.15	553.00	218.00	---	771.00	**955.00**	596.00	**731.00**
Two cycles with porcelain top											
White	Inst	Ea	SC	6.15	468.00	218.00	---	686.00	**857.00**	526.00	**651.00**
Colors	Inst	Ea	SC	6.15	510.00	218.00	---	728.00	**906.00**	561.00	**691.00**

Description	Oper	Unit	Crew Size	Man-Hours Per Unit	Costs Based On Small Volume						Large Volume	
					Avg Mat'l Unit Cost	Avg Labor Unit Cost	Avg Equip Unit Cost	Avg Total Unit Cost	Avg Price Incl O&P		Avg Total Unit Cost	Avg Price Incl O&P
Dishwasher - sink combination, includes good quality fittings, 48" cabinet, faucets and water supply kit												
Six cycles												
White	Inst	Ea	SC	16.3	1530.00	577.00	---	2107.00	**2610.00**		1638.00	**2010.00**
Colors	Inst	Ea	SC	16.3	1590.00	577.00	---	2167.00	**2680.00**		1688.00	**2070.00**
Three cycles												
White	Inst	Ea	SC	16.3	1280.00	577.00	---	1857.00	**2310.00**		1428.00	**1760.00**
Remove and reset combination dishwasher - sink only	Reset	Ea	SA	3.42	---	124.00	---	124.00	**183.00**		80.70	**119.00**
Remove only, d/w - sink combo	Demo	Ea	SA	2.05	---	74.50	---	74.50	**110.00**		48.40	**71.10**
Material adjustments												
Note: The following percentage adjustments for Small Volume also apply to Large												
For standard quality, DEDUCT	Inst	%	---	---	-20.0	---	---	---	---		---	---
For economy quality, DEDUCT	Inst	%	---	---	-30.0	---	---	---	---		---	---

Description	Oper	Unit	Crew Size	Man-Hours Per Unit	Costs Based On Small Volume					Large Volume	
					Avg Mat'l Unit Cost	Avg Labor Unit Cost	Avg Equip Unit Cost	Avg Total Unit Cost	Avg Price Incl O&P	Avg Total Unit Cost	Avg Price Incl O&P

Door frames

Exterior wood door frames
Exterior frame with exterior trim

Description	Oper	Unit	Crew Size	Man-Hours Per Unit	Avg Mat'l Unit Cost	Avg Labor Unit Cost	Avg Equip Unit Cost	Avg Total Unit Cost	Avg Price Incl O&P	Avg Total Unit Cost	Avg Price Incl O&P
5/4 x 4-9/16" deep											
Pine	Inst	LF	2C	.076	3.83	2.44	---	6.27	**8.06**	5.02	**6.29**
Oak	Inst	LF	2C	.082	6.55	2.63	---	9.18	**11.50**	7.62	**9.32**
Walnut	Inst	LF	2C	.089	8.18	2.86	---	11.04	**13.70**	9.26	**11.30**
5/4 x 5-3/16" deep											
Pine	Inst	LF	2C	.076	4.16	2.44	---	6.60	**8.44**	5.32	**6.63**
Oak	Inst	LF	2C	.082	7.06	2.63	---	9.69	**12.10**	8.10	**9.87**
Walnut	Inst	LF	2C	.089	10.60	2.86	---	13.46	**16.50**	11.52	**13.80**
5/4 x 6-9/16" deep											
Pine	Inst	LF	2C	.076	4.86	2.44	---	7.30	**9.25**	5.97	**7.38**
Oak	Inst	LF	2C	.082	8.32	2.63	---	10.95	**13.50**	9.26	**11.20**
Walnut	Inst	LF	2C	.089	12.50	2.86	---	15.36	**18.60**	13.20	**15.80**

Exterior sills

Description	Oper	Unit	Crew Size	Man-Hours Per Unit	Avg Mat'l Unit Cost	Avg Labor Unit Cost	Avg Equip Unit Cost	Avg Total Unit Cost	Avg Price Incl O&P	Avg Total Unit Cost	Avg Price Incl O&P
8/4 x 8" deep											
No horns	Inst	LF	2C	.333	10.90	10.70	---	21.60	**28.60**	16.52	**21.20**
2" horns	Inst	LF	2C	.333	11.70	10.70	---	22.40	**29.50**	17.22	**22.10**
3" horns	Inst	LF	2C	.333	12.70	10.70	---	23.40	**30.60**	18.12	**23.10**
8/4 x 10" deep											
No horns	Inst	LF	2C	.444	14.10	14.30	---	28.40	**37.60**	21.67	**27.90**
2" horns	Inst	LF	2C	.444	15.50	14.30	---	29.80	**39.20**	22.87	**29.30**
3" horns	Inst	LF	2C	.444	17.00	14.30	---	31.30	**40.90**	24.27	**30.90**

Exterior, colonial frame and trim

Description	Oper	Unit	Crew Size	Man-Hours Per Unit	Avg Mat'l Unit Cost	Avg Labor Unit Cost	Avg Equip Unit Cost	Avg Total Unit Cost	Avg Price Incl O&P	Avg Total Unit Cost	Avg Price Incl O&P
3' opening, in swing	Inst	Ea	2C	1.33	548.00	42.70	---	590.70	**694.00**	531.70	**621.00**
5' 4" opening, in/out swing	Inst	Ea	2C	1.78	884.00	57.10	---	941.10	**1100.00**	851.30	**992.00**
6' opening, in/out swing	Inst	Ea	2C	2.67	1100.00	85.70	---	1185.70	**1390.00**	1061.30	**1240.00**

Interior wood door frames
Interior frame

Description	Oper	Unit	Crew Size	Man-Hours Per Unit	Avg Mat'l Unit Cost	Avg Labor Unit Cost	Avg Equip Unit Cost	Avg Total Unit Cost	Avg Price Incl O&P	Avg Total Unit Cost	Avg Price Incl O&P
11/16" x 3-5/8" deep											
Pine	Inst	LF	2C	.076	3.87	2.44	---	6.31	**8.11**	5.05	**6.32**
Oak	Inst	LF	2C	.082	4.75	2.63	---	7.38	**9.41**	5.96	**7.41**
Walnut	Inst	LF	2C	.089	7.13	2.86	---	9.99	**12.50**	8.29	**10.10**
11/16" x 4-9/16" deep											
Pine	Inst	LF	2C	.076	3.99	2.44	---	6.43	**8.25**	5.16	**6.45**
Oak	Inst	LF	2C	.082	4.75	2.63	---	7.38	**9.41**	5.96	**7.41**
Walnut	Inst	LF	2C	.089	7.13	2.86	---	9.99	**12.50**	8.29	**10.10**
11/16" x 5-3/16" deep											
Pine	Inst	LF	2C	.076	4.51	2.44	---	6.95	**8.85**	5.65	**7.01**
Oak	Inst	LF	2C	.082	4.88	2.63	---	7.51	**9.56**	6.08	**7.55**
Walnut	Inst	LF	2C	.089	7.39	2.86	---	10.25	**12.80**	8.53	**10.40**

Pocket door frame

Description	Oper	Unit	Crew Size	Man-Hours Per Unit	Avg Mat'l Unit Cost	Avg Labor Unit Cost	Avg Equip Unit Cost	Avg Total Unit Cost	Avg Price Incl O&P	Avg Total Unit Cost	Avg Price Incl O&P
2'-0", to 3'-0" x 6'-8"	Inst	Ea	2C	2.29	99.00	73.50	---	172.50	**224.00**	134.20	**169.00**
3'-6" x 6'-8"	Inst	Ea	2C	2.29	182.00	73.50	---	255.50	**320.00**	210.70	**258.00**
4'-0" x 6'-8"	Inst	Ea	2C	2.29	185.00	73.50	---	258.50	**323.00**	213.70	**260.00**

Threshold, oak

Description	Oper	Unit	Crew Size	Man-Hours Per Unit	Avg Mat'l Unit Cost	Avg Labor Unit Cost	Avg Equip Unit Cost	Avg Total Unit Cost	Avg Price Incl O&P	Avg Total Unit Cost	Avg Price Incl O&P
5/8" x 3-5/8" deep	Inst	LF	2C	.167	2.59	5.36	---	7.95	**11.00**	5.60	**7.56**
5/8" x 4-5/8" deep	Inst	LF	2C	.178	3.37	5.71	---	9.08	**12.40**	6.54	**8.73**
5/8" x 5-5/8" deep	Inst	LF	2C	.190	4.01	6.10	---	10.11	**13.80**	7.37	**9.75**

				Costs Based On Small Volume						Large Volume	
Description	**Oper**	**Unit**	**Crew Size**	**Man-Hours Per Unit**	**Avg Mat'l Unit Cost**	**Avg Labor Unit Cost**	**Avg Equip Unit Cost**	**Avg Total Unit Cost**	**Avg Price Incl O&P**	**Avg Total Unit Cost**	**Avg Price Incl O&P**

Door hardware

Locksets

Outside locks

Knobs (pin tumbler)

Description	Oper	Unit	Crew	MH	Mat'l	Labor	Equip	Total	Price O&P	LV Total	LV Price
Excellent quality	Inst	Ea	CA	1.00	159.00	30.00	---	189.00	**228.00**	156.00	**187.00**
Good quality	Inst	Ea	CA	.800	118.00	24.00	---	142.00	**172.00**	116.00	**139.00**
Average quality	Inst	Ea	CA	.800	82.60	24.00	---	106.60	**131.00**	85.70	**104.00**
Handlesets											
Excellent quality	Inst	Ea	CA	1.00	207.00	30.00	---	237.00	**282.00**	197.00	**233.00**
Good quality	Inst	Ea	CA	.800	130.00	24.00	---	154.00	**185.00**	126.00	**150.00**
Average quality	Inst	Ea	CA	.800	94.40	24.00	---	118.40	**145.00**	95.80	**115.00**
Bath or bedroom locks											
Excellent quality	Inst	Ea	CA	.800	118.00	24.00	---	142.00	**172.00**	117.00	**140.00**
Good quality	Inst	Ea	CA	.615	100.00	18.40	---	118.40	**143.00**	97.90	**117.00**
Average quality	Inst	Ea	CA	.615	76.70	18.40	---	95.10	**116.00**	77.70	**93.50**
Passage latches											
Excellent quality	Inst	Ea	CA	.615	136.00	18.40	---	154.40	**184.00**	128.00	**152.00**
Good quality	Inst	Ea	CA	.471	101.00	14.10	---	115.10	**138.00**	96.14	**114.00**
Average quality	Inst	Ea	CA	.471	70.80	14.10	---	84.90	**103.00**	69.84	**83.60**
Dead locks (double key)											
Excellent quality	Inst	Ea	CA	.800	136.00	24.00	---	160.00	**192.00**	132.00	**158.00**
Good quality	Inst	Ea	CA	.615	106.00	18.40	---	124.40	**150.00**	102.90	**123.00**
Average quality	Inst	Ea	CA	.615	72.00	18.40	---	90.40	**110.00**	73.60	**88.90**

Kickplates, 8" x 30"

Description	Oper	Unit	Crew	MH	Mat'l	Labor	Equip	Total	Price O&P	LV Total	LV Price
Aluminum	Inst	Ea	CA	.471	18.50	14.10	---	32.60	**42.50**	25.14	**32.10**
Brass or bronze	Inst	Ea	CA	.471	26.40	14.10	---	40.50	**51.60**	31.84	**39.90**

Thresholds

Aluminum, pre-notched, draft-proof, standard

3/4" high

Description	Oper	Unit	Crew	MH	Mat'l	Labor	Equip	Total	Price O&P	LV Total	LV Price
32" long	Inst	Ea	CA	.381	12.20	11.40	---	23.60	**31.10**	17.90	**23.20**
36" long	Inst	Ea	CA	.381	13.90	11.40	---	25.30	**33.20**	19.40	**25.00**
42" long	Inst	Ea	CA	.381	16.60	11.40	---	28.00	**36.20**	21.70	**27.60**
48" long	Inst	Ea	CA	.381	19.00	11.40	---	30.40	**39.00**	23.80	**30.00**
60" long	Inst	Ea	CA	.381	23.60	11.40	---	35.00	**44.30**	27.70	**34.50**
72" long	Inst	Ea	CA	.381	28.00	11.40	---	39.40	**49.30**	31.40	**38.80**

Note: The following percentage adjustment for Small Volume also applies to Large

For high rug-type, 1-1/8"

Description	Oper	Unit	Crew	MH	Mat'l	Labor	Equip	Total	Price O&P	LV Total	LV Price
ADD	Inst	%	CA	---	25.0	---	---	---	**---**	---	**---**

Wood, oak threshold with vinyl weather seal

5/8" x 3-1/2"

Description	Oper	Unit	Crew	MH	Mat'l	Labor	Equip	Total	Price O&P	LV Total	LV Price
33" long	Inst	Ea	CA	.381	14.20	11.40	---	25.60	**33.40**	19.60	**25.20**
37" long	Inst	Ea	CA	.381	16.00	11.40	---	27.40	**35.50**	21.20	**27.00**
43" long	Inst	Ea	CA	.381	17.20	11.40	---	28.60	**36.90**	22.20	**28.20**
49" long	Inst	Ea	CA	.381	19.60	11.40	---	31.00	**39.70**	24.30	**30.60**
61" long	Inst	Ea	CA	.381	25.00	11.40	---	36.40	**45.80**	28.90	**35.80**
73" long	Inst	Ea	CA	.381	29.90	11.40	---	41.30	**51.50**	33.10	**40.60**

3/4" x 3-1/2"

Description	Oper	Unit	Crew	MH	Mat'l	Labor	Equip	Total	Price O&P	LV Total	LV Price
33" long	Inst	Ea	CA	.381	15.00	11.40	---	26.40	**34.40**	20.40	**26.00**
37" long	Inst	Ea	CA	.381	16.80	11.40	---	28.20	**36.50**	21.90	**27.80**

Doors

Interior Door Systems

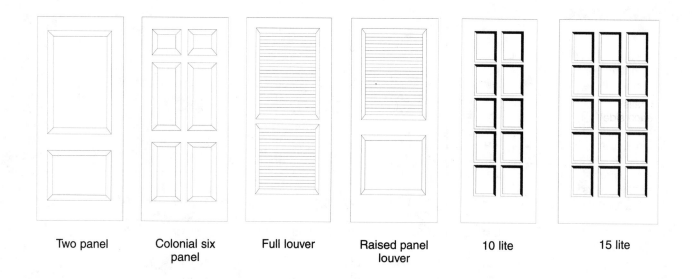

| Two panel | Colonial six panel | Full louver | Raised panel louver | 10 lite | 15 lite |

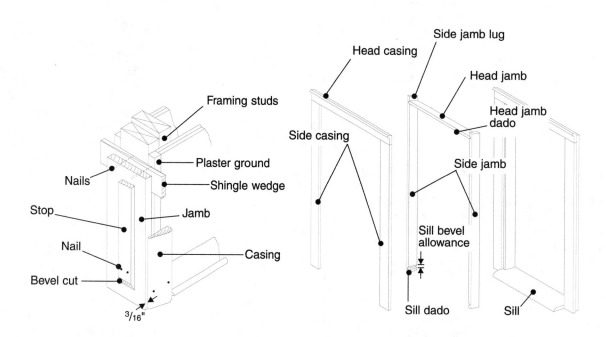

Parts of Exterior Door Frame

Assembled Package Door Units (Interior)

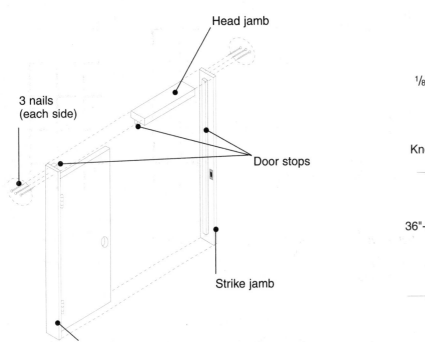

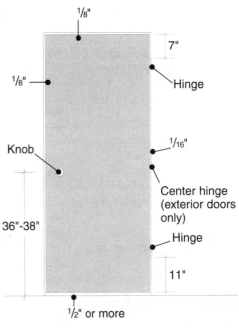

Unit Includes:

Jamb: $4^9/_{16}"$ finger joint pine w/ stops applied.

Butts: 1 pair $3^1/_2" \times 3^1/_2"$ applied to door and jamb.

Door: 3 degree bevel one side $^3/_{16}"$ under std. width - net 80"
H center bored - $2^1/_8"$ bore
$2^3/_8"$ backset - 1" edge bore

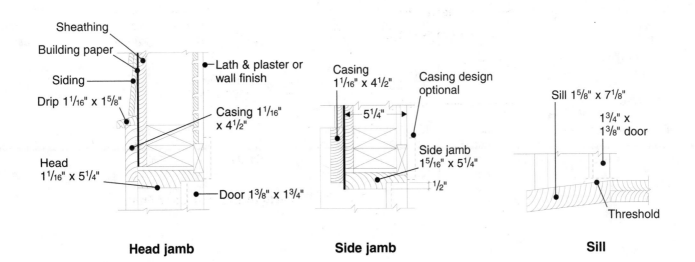

Head jamb   **Side jamb**   **Sill**

				Costs Based On Small Volume					Large Volume	
Description	**Oper**	**Unit**	**Crew Size**	**Man-Hours Per Unit**	**Avg Mat'l Unit Cost**	**Avg Labor Unit Cost**	**Avg Equip Unit Cost**	**Avg Total Unit Cost**	**Avg Price Incl O&P**	**Avg Total Unit Cost**	**Avg Price Incl O&P**

Doors

Entrance doors

Decorative glass doors and matching sidelights

Lafayette collection, 1-3/8" panels

Astoria style

Description	Oper	Unit	Crew Size	Man-Hrs	Mat'l	Labor	Equip	Total	Price O&P	LV Total	LV Price
2'-6", 2'-8", 3'-0" W x 6'-8" H doors											
Fir & pine	Inst	Ea	2C	2.05	1370.00	65.80	---	1435.80	**1670.00**	1212.70	**1410.00**
Oak	Inst	Ea	2C	2.05	1790.00	65.80	---	1855.80	**2160.00**	1572.70	**1830.00**
1'-0", 1'-2" x 6'-8-1/2" sidelights											
Fir & pine	Inst	Ea	2C	1.54	604.00	49.40	---	653.40	**769.00**	549.10	**643.00**
Oak	Inst	Ea	2C	1.54	768.00	49.40	---	817.40	**958.00**	690.10	**804.00**
Bourbon Royale style											
2'-6", 2'-8", 3'-0" W x 6'-8" H doors											
Fir & pine	Inst	Ea	2C	2.05	1340.00	65.80	---	1405.80	**1640.00**	1192.70	**1380.00**
Oak	Inst	Ea	2C	2.05	1770.00	65.80	---	1835.80	**2130.00**	1562.70	**1810.00**
1'-0", 1'-2" x 6'-8-1/2" sidelights											
Fir & pine	Inst	Ea	2C	1.54	636.00	49.40	---	685.40	**806.00**	576.10	**674.00**
Oak	Inst	Ea	2C	1.54	788.00	49.40	---	837.40	**981.00**	707.10	**824.00**
Jubilee style											
2'-6", 2'-8", 3'-0" W x 6'-8" H doors											
Fir & pine	Inst	Ea	2C	2.05	1410.00	65.80	---	1475.80	**1720.00**	1252.70	**1450.00**
Oak	Inst	Ea	2C	2.05	1840.00	65.80	---	1905.80	**2220.00**	1622.70	**1880.00**
1'-0", 1'-2" x 6'-8-1/2" sidelights											
Fir & pine	Inst	Ea	2C	1.54	592.00	49.40	---	641.40	**755.00**	539.10	**631.00**
Oak	Inst	Ea	2C	1.54	900.00	49.40	---	949.40	**1110.00**	803.10	**934.00**
Lexington style											
2'-6", 2'-8", 3'-0" W x 6'-8" H doors											
Fir & pine	Inst	Ea	2C	2.05	1410.00	65.80	---	1475.80	**1720.00**	1242.70	**1450.00**
Oak	Inst	Ea	2C	2.05	1840.00	65.80	---	1905.80	**2210.00**	1612.70	**1870.00**
1'-0", 1'-2" x 6'-8-1/2" sidelights											
Fir & pine	Inst	Ea	2C	1.54	631.00	49.40	---	680.40	**800.00**	572.10	**670.00**
Oak	Inst	Ea	2C	1.54	784.00	49.40	---	833.40	**975.00**	703.10	**819.00**
Marquis style											
2'-6", 2'-8", 3'-0" W x 6'-8" H doors											
Fir & pine	Inst	Ea	2C	2.05	1470.00	65.80	---	1535.80	**1790.00**	1302.70	**1510.00**
Oak	Inst	Ea	2C	2.05	1900.00	65.80	---	1965.80	**2290.00**	1672.70	**1940.00**
1'-0", 1'-2" x 6'-8-1/2" sidelights											
Fir & pine	Inst	Ea	2C	1.54	664.00	49.40	---	713.40	**838.00**	601.10	**702.00**
Oak	Inst	Ea	2C	1.54	740.00	49.40	---	789.40	**925.00**	665.10	**776.00**
Monaco style											
2'-6", 2'-8", 3'-0" W x 6'-8" H doors											
Fir & pine	Inst	Ea	2C	2.05	1730.00	65.80	---	1795.80	**2090.00**	1522.70	**1770.00**
Oak	Inst	Ea	2C	2.05	2150.00	65.80	---	2215.80	**2570.00**	1882.70	**2180.00**
1'-0", 1'-2" x 6'-8-1/2" sidelights											
Fir & pine	Inst	Ea	2C	1.54	792.00	49.40	---	841.40	**985.00**	710.10	**828.00**
Oak	Inst	Ea	2C	1.54	944.00	49.40	---	993.40	**1160.00**	840.10	**977.00**

Description	Oper	Unit	Crew Size	Man-Hours Per Unit	Avg Mat'l Unit Cost	Avg Labor Unit Cost	Avg Equip Unit Cost	Avg Total Unit Cost	Avg Price Incl O&P	Avg Total Unit Cost	Avg Price Incl O&P
								Costs Based On Small Volume		**Large Volume**	
Windsor style											
2'-6", 2'-8", 3'-0" W x 6'-8" H doors											
Fir & pine	Inst	Ea	2C	2.05	1450.00	65.80	---	1515.80	**1770.00**	1292.70	**1500.00**
Oak	Inst	Ea	2C	2.05	1880.00	65.80	---	1945.80	**2260.00**	1652.70	**1910.00**
1'-0", 1'-2" x 6'-8-1/2" sidelights											
Fir & pine	Inst	Ea	2C	1.54	624.00	49.40	---	673.40	**792.00**	566.10	**663.00**
Oak	Inst	Ea	2C	1.54	780.00	49.40	---	829.40	**971.00**	700.10	**816.00**

Stile and rail raised panels

No lites

Description	Oper	Unit	Crew Size	Man-Hours Per Unit	Avg Mat'l Unit Cost	Avg Labor Unit Cost	Avg Equip Unit Cost	Avg Total Unit Cost	Avg Price Incl O&P	Avg Total Unit Cost	Avg Price Incl O&P
2 raised panels											
2'-6" to 3'-0" x 6'-8" x 1-3/4" T	Inst	Ea	2C	2.05	382.00	65.80	---	447.80	**538.00**	369.70	**440.00**
3'-6" x 6'-8" x 1-3/4" T	Inst	Ea	2C	2.22	382.00	71.20	---	453.20	**547.00**	373.50	**446.00**
4 raised panels											
2'-6" to 3'-0" x 6'-8" x 1-3/4" T	Inst	Ea	2C	2.05	348.00	65.80	---	413.80	**499.00**	340.70	**407.00**
3'-6" x 6'-8" x 1-3/4" T	Inst	Ea	2C	2.22	511.00	71.20	---	582.20	**694.00**	483.50	**573.00**
6 raised panels											
2'-0" to 2'-8" x 6'-8" x 1-3/4" T	Inst	Ea	2C	2.05	336.00	65.80	---	401.80	**485.00**	330.70	**395.00**
3'-0" x 6'-8" x 1-3/4" T	Inst	Ea	2C	2.22	346.00	71.20	---	417.20	**504.00**	342.50	**410.00**
3'-6" x 6'-8" x 1-3/4" T	Inst	Ea	2C	2.46	506.00	78.90	---	584.90	**701.00**	484.30	**575.00**
8 raised panels											
2'-6" to 2'-8" x 6'-8" x 1-3/4" T	Inst	Ea	2C	2.05	337.00	65.80	---	402.80	**487.00**	331.70	**396.00**
3'-0" x 6'-8" x 1-3/4" T	Inst	Ea	2C	2.22	346.00	71.20	---	417.20	**504.00**	342.50	**410.00**
3'-6" x 6'-8" x 1-3/4" T	Inst	Ea	2C	2.46	506.00	78.90	---	584.90	**701.00**	484.30	**575.00**
2'-6" to 2'-8" x 7'-0" x 1-3/4" T	Inst	Ea	2C	2.05	365.00	65.80	---	430.80	**518.00**	354.70	**423.00**
3'-0" x 7'-0" x 1-3/4" T	Inst	Ea	2C	2.22	393.00	71.20	---	464.20	**559.00**	382.50	**457.00**
3'-6" x 7'-0" x 1-3/4" T	Inst	Ea	2C	2.46	532.00	78.90	---	610.90	**730.00**	507.30	**601.00**
10 raised panels											
2'-6" to 2'-8" x 8'-0" x 1-3/4" T	Inst	Ea	2C	2.22	443.00	71.20	---	514.20	**616.00**	425.50	**505.00**
3'-0" x 8'-0" x 1-3/4" T	Inst	Ea	2C	2.46	464.00	78.90	---	542.90	**652.00**	448.30	**533.00**
3'-6" x 8'-0" x 1-3/4" T	Inst	Ea	2C	2.71	588.00	87.00	---	675.00	**806.00**	560.10	**664.00**

Lites

Description	Oper	Unit	Crew Size	Man-Hours Per Unit	Avg Mat'l Unit Cost	Avg Labor Unit Cost	Avg Equip Unit Cost	Avg Total Unit Cost	Avg Price Incl O&P	Avg Total Unit Cost	Avg Price Incl O&P
1 raised panel, 1 lite											
2'-6" to 3'-0" x 6'-8" x 1-3/4" T											
Open w/stops	Inst	Ea	2C	2.05	334.00	65.80	---	399.80	**483.00**	328.70	**393.00**
Tempered clear	Inst	Ea	2C	2.05	348.00	65.80	---	413.80	**499.00**	340.70	**407.00**
Acrylic or tempered amber	Inst	Ea	2C	2.05	418.00	65.80	---	483.80	**579.00**	400.70	**475.00**
1 raised panel, 6 lites											
2'-6" to 3'-0" x 6'-8" x 1-3/4" T											
Open w/stops	Inst	Ea	2C	2.05	360.00	65.80	---	425.80	**513.00**	350.70	**418.00**
Tempered clear	Inst	Ea	2C	2.05	385.00	65.80	---	450.80	**541.00**	371.70	**443.00**
Acrylic or tempered amber	Inst	Ea	2C	2.05	437.00	65.80	---	502.80	**601.00**	416.70	**494.00**
1 raised panel, 9 lites											
2'-6" to 3'-0" x 6'-8" x 1-3/4" T											
Open w/stops	Inst	Ea	2C	2.05	366.00	65.80	---	431.80	**519.00**	355.70	**424.00**
Tempered clear	Inst	Ea	2C	2.05	394.00	65.80	---	459.80	**552.00**	379.70	**452.00**
Acrylic or tempered amber	Inst	Ea	2C	2.05	478.00	65.80	---	543.80	**648.00**	451.70	**534.00**
1 raised panel, 12 diamond lites											
2'-6" to 3'-0" x 6'-8" x 1-3/4" T											
Open w/stops	Inst	Ea	2C	2.05	375.00	65.80	---	440.80	**530.00**	363.70	**433.00**
Tempered clear	Inst	Ea	2C	2.05	478.00	65.80	---	543.80	**648.00**	451.70	**534.00**
Acrylic or tempered amber	Inst	Ea	2C	2.05	596.00	65.80	---	661.80	**784.00**	552.70	**651.00**

Description	Oper	Unit	Costs Based On Small Volume							Large Volume	
			Crew Size	Man-Hours Per Unit	Avg Mat'l Unit Cost	Avg Labor Unit Cost	Avg Equip Unit Cost	Avg Total Unit Cost	Avg Price Incl O&P	Avg Total Unit Cost	Avg Price Incl O&P
Dutch doors, country style, fir, stile and rail											
Two raised panels in lower door											
1 lite in upper door											
2'-6" to 3'-0" x 6'-8" x 1-3/4" T											
Empty	Inst	Ea	2C	2.05	460.00	65.80	---	525.80	**628.00**	436.70	**517.00**
Tempered clear	Inst	Ea	2C	2.05	478.00	65.80	---	543.80	**648.00**	451.70	**534.00**
Tempered amber	Inst	Ea	2C	2.05	596.00	65.80	---	661.80	**784.00**	552.70	**651.00**
4 lites in upper door											
2'-6" to 3'-0" x 6'-8" x 1-3/4" T											
Empty	Inst	Ea	2C	2.05	471.00	65.80	---	536.80	**640.00**	445.70	**527.00**
Tempered clear	Inst	Ea	2C	2.05	514.00	65.80	---	579.80	**690.00**	482.70	**570.00**
Tempered amber	Inst	Ea	2C	2.05	543.00	65.80	---	608.80	**723.00**	507.70	**598.00**
9 lites in upper door											
2'-6" to 3'-0" x 6'-8" x 1-3/4" T											
Empty	Inst	Ea	2C	2.05	466.00	65.80	---	531.80	**635.00**	441.70	**523.00**
Tempered clear	Inst	Ea	2C	2.05	484.00	65.80	---	549.80	**655.00**	456.70	**540.00**
Tempered amber	Inst	Ea	2C	2.05	601.00	65.80	---	666.80	**789.00**	556.70	**655.00**
12 diamond lites in upper door											
2'-6" to 3'-0" x 6'-8" x 1-3/4" T											
Empty	Inst	Ea	2C	2.05	512.00	65.80	---	577.80	**688.00**	480.70	**568.00**
Tempered clear	Inst	Ea	2C	2.05	550.00	65.80	---	615.80	**731.00**	513.70	**605.00**
Tempered amber	Inst	Ea	2C	2.05	680.00	65.80	---	745.80	**880.00**	624.70	**733.00**
Four diamond-shaped raised panels in lower door											
1 lite in upper door											
2'-6" to 3'-0" x 6'-8" x 1-3/4" T											
Empty	Inst	Ea	2C	2.05	464.00	65.80	---	529.80	**632.00**	439.70	**521.00**
Tempered clear	Inst	Ea	2C	2.05	481.00	65.80	---	546.80	**652.00**	454.70	**538.00**
Tempered amber	Inst	Ea	2C	2.05	599.00	65.80	---	664.80	**788.00**	555.70	**654.00**
6 lites in upper door											
2'-6" to 3'-0" x 6'-8" x 1-3/4" T											
Empty	Inst	Ea	2C	2.05	502.00	65.80	---	567.80	**675.00**	471.70	**558.00**
Tempered clear	Inst	Ea	2C	2.05	520.00	65.80	---	585.80	**697.00**	487.70	**576.00**
Tempered amber	Inst	Ea	2C	2.05	644.00	65.80	---	709.80	**840.00**	593.70	**698.00**
9 lites in upper door											
2'-6" to 3'-0" x 6'-8" x 1-3/4" T											
Empty	Inst	Ea	2C	2.05	467.00	65.80	---	532.80	**636.00**	442.70	**524.00**
Tempered clear	Inst	Ea	2C	2.05	487.00	65.80	---	552.80	**659.00**	459.70	**544.00**
Tempered amber	Inst	Ea	2C	2.05	604.00	65.80	---	669.80	**793.00**	559.70	**659.00**
12 diamond lites in upper door											
2'-6" to 3'-0" x 6'-8" x 1-3/4" T											
Empty	Inst	Ea	2C	2.05	516.00	65.80	---	581.80	**692.00**	483.70	**572.00**
Tempered clear	Inst	Ea	2C	2.05	555.00	65.80	---	620.80	**736.00**	517.70	**610.00**
Tempered amber	Inst	Ea	2C	2.05	684.00	65.80	---	749.80	**886.00**	628.70	**738.00**

Description	Oper	Unit	Crew Size	Man-Hours Per Unit	Avg Mat'l Unit Cost	Avg Labor Unit Cost	Avg Equip Unit Cost	Avg Total Unit Cost	Avg Price Incl O&P	Avg Total Unit Cost	Avg Price Incl O&P
						Costs Based On Small Volume					Large Volume	

French doors

Douglas fir or hemlock 4-1/2 " stiles and top rail, 9-1/2 " bottom rail, tempered glass with stops

Description	Oper	Unit	Crew Size	Man-Hours Per Unit	Avg Mat'l Unit Cost	Avg Labor Unit Cost	Avg Equip Unit Cost	Avg Total Unit Cost	Avg Price Incl O&P	Avg Total Unit Cost	Avg Price Incl O&P
1 lite											
1-3/8" T and 1-3/4" T											
2'-0" x 6'-8"	Inst	Ea	2C	1.76	251.00	56.50	---	307.50	374.00	251.60	302.00
2'-4", 2'-6", 2'-8" x 6'-8"	Inst	Ea	2C	2.05	257.00	65.80	---	322.80	395.00	262.70	317.00
3'-0" x 6'-8"	Inst	Ea	2C	2.46	261.00	78.90	---	339.90	418.00	274.30	334.00
5 lites, 5 high											
1-3/8" T and 1-3/4" T											
2'-0" x 6'-8"	Inst	Ea	2C	1.76	222.00	56.50	---	278.50	340.00	226.60	273.00
2'-4", 2'-6", 2'-8" x 6'-8"	Inst	Ea	2C	2.05	224.00	65.80	---	289.80	357.00	234.70	285.00
3'-0" x 6'-8"	Inst	Ea	2C	2.46	228.00	78.90	---	306.90	380.00	246.30	301.00
10 lites, 5 high											
1-3/8" T and 1-3/4" T											
2'-0" x 6'-8"	Inst	Ea	2C	1.76	247.00	56.50	---	303.50	368.00	247.60	298.00
2'-4", 2'-6", 2'-8" x 6'-8"	Inst	Ea	2C	2.05	254.00	65.80	---	319.80	390.00	259.70	314.00
3'-0" x 6'-8"	Inst	Ea	2C	2.46	260.00	78.90	---	338.90	417.00	273.30	333.00
15 lites, 5 high											
1-3/8" T and 1-3/4" T											
2'-6", 2'-8" x 6'-8"	Inst	Ea	2C	2.05	283.00	65.80	---	348.80	424.00	284.70	343.00
3'-0" x 6'-8"	Inst	Ea	2C	2.46	295.00	78.90	---	373.90	458.00	304.30	367.00

French sidelites

Douglas fir or hemlock 4-1/2 " stiles and top rail, 9-1/2 " bottom rail, tempered glass with stops

Description	Oper	Unit	Crew Size	Man-Hours Per Unit	Avg Mat'l Unit Cost	Avg Labor Unit Cost	Avg Equip Unit Cost	Avg Total Unit Cost	Avg Price Incl O&P	Avg Total Unit Cost	Avg Price Incl O&P
1 lite											
1-3/4" T											
1'-0", 1'-2" x 6'-8"	Inst	Ea	2C	1.54	171.00	49.40	---	220.40	271.00	178.10	217.00
1'-4", 1'-6" x 6'-8"	Inst	Ea	2C	1.54	183.00	49.40	---	232.40	284.00	189.10	228.00
5 lites, 5 high											
1-3/4" T											
1'-0", 1'-2" x 6'-8"	Inst	Ea	2C	1.54	165.00	49.40	---	214.40	264.00	173.10	211.00
1'-4", 1'-6" x 6'-8"	Inst	Ea	2C	1.54	179.00	49.40	---	228.40	280.00	186.10	225.00

Garden doors

Description	Oper	Unit	Crew Size	Man-Hours Per Unit	Avg Mat'l Unit Cost	Avg Labor Unit Cost	Avg Equip Unit Cost	Avg Total Unit Cost	Avg Price Incl O&P	Avg Total Unit Cost	Avg Price Incl O&P
1 door, "X" unit											
2'-6" x 6'-8"	Inst	LS	2C	2.05	330.00	65.80	---	395.80	479.00	325.70	389.00
3'-0" x 6'-8"	Inst	LS	2C	2.46	354.00	78.90	---	432.90	526.00	354.30	425.00
2 doors, "XO/OX" unit											
5'-0" x 6'-8"	Inst	LS	2C	3.72	620.00	119.00	---	739.00	892.00	607.70	726.00
6'-0" x 6'-8"	Inst	LS	2C	4.44	661.00	142.00	---	803.00	974.00	659.40	791.00
3 doors, "XOX" unit											
7'-6" x 6'-8"	Inst	LS	2C	5.52	879.00	177.00	---	1056.00	1280.00	869.00	1040.00
9'-0" x 6'-8"	Inst	LS	2C	6.67	944.00	214.00	---	1158.00	1410.00	947.00	1140.00

Description	Oper	Unit	Crew Size	Man-Hours Per Unit	Avg Mat'l Unit Cost	Avg Labor Unit Cost	Avg Equip Unit Cost	Avg Total Unit Cost	Avg Price Incl O&P	Avg Total Unit Cost	Avg Price Incl O&P
										Large Volume	
Sliding doors											
Glass sliding doors, with 5-1/2" anodized aluminum frame, trim, weatherstripping, tempered 3/16" T clear single-glazed glass, with screen											
6'-8" H											
5' W, 2 lites, 1 sliding	Inst	Set	2C	6.15	484.00	197.00	---	681.00	852.00	542.00	669.00
6' W, 2 lites, 1 sliding	Inst	Set	2C	6.15	539.00	197.00	---	736.00	916.00	590.00	723.00
7' W, 2 lites, 1 sliding	Inst	Set	2C	6.15	579.00	197.00	---	776.00	962.00	624.00	763.00
8' W, 2 lites, 1 sliding	Inst	Set	2C	6.15	622.00	197.00	---	819.00	1010.00	660.00	805.00
10' W, 2 lites, 1 sliding	Inst	Set	2C	6.15	808.00	197.00	---	1005.00	1230.00	820.00	988.00
9' W, 3 lites, 1 sliding	Inst	Set	2C	8.00	716.00	257.00	---	973.00	1210.00	784.00	962.00
12' W, 3 lites, 1 sliding	Inst	Set	2C	8.00	840.00	257.00	---	1097.00	1350.00	890.00	1080.00
15' W, 3 lites, 1 sliding	Inst	Set	2C	8.00	1120.00	257.00	---	1377.00	1680.00	1132.00	1360.00
8'-0" H											
5' W, 2 lites, 1 sliding	Inst	Set	2C	6.15	570.00	197.00	---	767.00	951.00	616.00	754.00
6' W, 2 lites, 1 sliding	Inst	Set	2C	6.15	612.00	197.00	---	809.00	1000.00	652.00	795.00
7' W, 2 lites, 1 sliding	Inst	Set	2C	6.15	725.00	197.00	---	922.00	1130.00	748.00	906.00
8' W, 2 lites, 1 sliding	Inst	Set	2C	6.15	838.00	197.00	---	1035.00	1260.00	845.00	1020.00
10' W, 2 lites, 1 sliding	Inst	Set	2C	6.15	966.00	197.00	---	1163.00	1410.00	955.00	1140.00
9' W, 3 lites, 1 sliding	Inst	Set	2C	8.00	843.00	257.00	---	1100.00	1350.00	892.00	1090.00
12' W, 3 lites, 1 sliding	Inst	Set	2C	8.00	1130.00	257.00	---	1387.00	1680.00	1136.00	1370.00
15' W, 3 lites, 1 sliding	Inst	Set	2C	8.00	1310.00	257.00	---	1567.00	1890.00	1291.00	1550.00
French sliding door units											
Douglas fir doors, solid brass hardware, screen, frames, and exterior molding included											
1 lite, 2 panels, single glazed											
1-3/4" T											
5'-0" x 6'-8"	Inst	Ea	2C	8.00	2600.00	257.00	---	2857.00	3380.00	2401.00	2820.00
6'-0" x 6'-8"	Inst	Ea	2C	8.00	2770.00	257.00	---	3027.00	3570.00	2541.00	2990.00
7'-0" x 6'-8"	Inst	Ea	2C	8.00	3000.00	257.00	---	3257.00	3830.00	2741.00	3210.00
8'-0" x 6'-8"	Inst	Ea	2C	8.00	3160.00	257.00	---	3417.00	4020.00	2871.00	3360.00
5'-0" x 7'-0"	Inst	Ea	2C	8.00	2660.00	257.00	---	2917.00	3450.00	2451.00	2880.00
6'-0" x 7'-0"	Inst	Ea	2C	8.00	2800.00	257.00	---	3057.00	3610.00	2571.00	3010.00
7'-0" x 7'-0"	Inst	Ea	2C	8.00	3090.00	257.00	---	3347.00	3930.00	2811.00	3290.00
8'-0" x 7'-0"	Inst	Ea	2C	8.00	3280.00	257.00	---	3537.00	4150.00	2971.00	3480.00
5'-0" x 8'-0"	Inst	Ea	2C	8.00	3010.00	257.00	---	3267.00	3850.00	2751.00	3220.00
6'-0" x 8'-0"	Inst	Ea	2C	8.00	3230.00	257.00	---	3487.00	4100.00	2931.00	3430.00
7'-0" x 8'-0"	Inst	Ea	2C	8.00	3470.00	257.00	---	3727.00	4370.00	3141.00	3670.00
8'-0" x 8'-0"	Inst	Ea	2C	8.00	3690.00	257.00	---	3947.00	4630.00	3331.00	3890.00
10 lites, 2 panels, single glazed											
1-3/4" T											
5'-0" x 6'-8"	Inst	Ea	2C	8.00	2810.00	257.00	---	3067.00	3620.00	2581.00	3020.00
6'-0" x 6'-8"	Inst	Ea	2C	8.00	2930.00	257.00	---	3187.00	3760.00	2681.00	3140.00
5'-0" x 7'-0"	Inst	Ea	2C	8.00	2860.00	257.00	---	3117.00	3670.00	2621.00	3070.00
6'-0" x 7'-0"	Inst	Ea	2C	8.00	3010.00	257.00	---	3267.00	3850.00	2751.00	3220.00
5'-0" x 8'-0"	Inst	Ea	2C	8.00	3180.00	257.00	---	3437.00	4040.00	2891.00	3390.00
6'-0" x 8'-0"	Inst	Ea	2C	8.00	3350.00	257.00	---	3607.00	4240.00	3041.00	3550.00

Description	Oper	Unit	Crew Size	Man-Hours Per Unit	Avg Mat'l Unit Cost	Avg Labor Unit Cost	Avg Equip Unit Cost	Avg Total Unit Cost	Avg Price Incl O&P	Avg Total Unit Cost	Avg Price Incl O&P
					Costs Based On Small Volume					**Large Volume**	
15 lites, 2 panels, single glazed											
1-3/4" T											
5'-0" x 6'-8"	Inst	Ea	2C	8.00	2950.00	257.00	---	3207.00	**3780.00**	2701.00	**3160.00**
6'-0" x 6'-8"	Inst	Ea	2C	8.00	3090.00	257.00	---	3347.00	**3940.00**	2811.00	**3300.00**
7'-0" x 6'-8"	Inst	Ea	2C	8.00	3290.00	257.00	---	3547.00	**4170.00**	2981.00	**3490.00**
8'-0" x 6'-8"	Inst	Ea	2C	8.00	3430.00	257.00	---	3687.00	**4330.00**	3111.00	**3640.00**
5'-0" x 7'-0"	Inst	Ea	2C	8.00	3010.00	257.00	---	3267.00	**3850.00**	2751.00	**3220.00**
6'-0" x 7'-0"	Inst	Ea	2C	8.00	3170.00	257.00	---	3427.00	**4030.00**	2881.00	**3370.00**
7'-0" x 7'-0"	Inst	Ea	2C	8.00	3370.00	257.00	---	3627.00	**4270.00**	3061.00	**3580.00**
8'-0" x 7'-0"	Inst	Ea	2C	8.00	3540.00	257.00	---	3797.00	**4460.00**	3201.00	**3740.00**
5'-0" x 8'-0"	Inst	Ea	2C	8.00	3340.00	257.00	---	3597.00	**4220.00**	3031.00	**3540.00**
6'-0" x 8'-0"	Inst	Ea	2C	8.00	3500.00	257.00	---	3757.00	**4420.00**	3171.00	**3710.00**
7'-0" x 8'-0"	Inst	Ea	2C	8.00	3760.00	257.00	---	4017.00	**4710.00**	3391.00	**3950.00**
8'-0" x 8'-0"	Inst	Ea	2C	8.00	3900.00	257.00	---	4157.00	**4870.00**	3511.00	**4100.00**
Fire doors, natural birch											
One hour rating											
2'-6" x 6'-8" x 1-3/4" T	Inst	Ea	2C	1.88	253.00	60.30	---	313.30	**381.00**	255.50	**308.00**
3'-6" x 6'-8" x 1-3/4" T	Inst	Ea	2C	2.22	315.00	71.20	---	386.20	**469.00**	316.50	**380.00**
2'-6" x 7'-0" x 1-3/4" T	Inst	Ea	2C	1.88	264.00	60.30	---	324.30	**394.00**	265.50	**319.00**
3'-6" x 7'-0" x 1-3/4" T	Inst	Ea	2C	2.22	329.00	71.20	---	400.20	**485.00**	328.50	**394.00**
1.5 hour rating											
2'-6" x 6'-8" x 1-3/4" T	Inst	Ea	2C	1.88	359.00	60.30	---	419.30	**503.00**	346.50	**412.00**
3'-6" x 6'-8" x 1-3/4" T	Inst	Ea	2C	2.22	400.00	71.20	---	471.20	**567.00**	388.50	**464.00**
2'-6" x 7'-0" x 1-3/4" T	Inst	Ea	2C	1.88	378.00	60.30	---	438.30	**525.00**	362.50	**431.00**
3'-6" x 7'-0" x 1-3/4" T	Inst	Ea	2C	2.22	472.00	71.20	---	543.20	**650.00**	450.50	**534.00**
Interior doors											
Passage doors, flush face, includes hinge hardware											
Hollow core prehung door units											
Hardboard											
2'-6" x 6'-8" x 1-3/8" T	Inst	Ea	2C	1.63	45.70	52.30	---	98.00	**131.00**	73.40	**96.50**
2'-8" x 6'-8" x 1-3/8" T	Inst	Ea	2C	1.76	48.60	56.50	---	105.10	**141.00**	78.20	**103.00**
3'-0" x 6'-8" x 1-3/8" T	Inst	Ea	2C	1.76	51.30	56.50	---	107.80	**144.00**	80.50	**105.00**
2'-6" x 7'-0" x 1-3/8" T	Inst	Ea	2C	1.63	58.20	52.30	---	110.50	**145.00**	84.10	**109.00**
2'-8" x 7'-0" x 1-3/8" T	Inst	Ea	2C	1.76	61.10	56.50	---	117.60	**155.00**	88.90	**115.00**
3'-0" x 7'-0" x 1-3/8" T	Inst	Ea	2C	1.76	63.80	56.50	---	120.30	**158.00**	91.20	**118.00**
Lauan											
2'-6" x 6'-8" x 1-3/8" T	Inst	Ea	2C	1.63	60.60	52.30	---	112.90	**148.00**	86.20	**111.00**
2'-8" x 6'-8" x 1-3/8" T	Inst	Ea	2C	1.76	62.90	56.50	---	119.40	**157.00**	90.40	**117.00**
3'-0" x 6'-8" x 1-3/8" T	Inst	Ea	2C	1.76	67.40	56.50	---	123.90	**162.00**	94.30	**121.00**
2'-6" x 7'-0" x 1-3/8" T	Inst	Ea	2C	1.63	71.70	52.30	---	124.00	**161.00**	95.70	**122.00**
2'-8" x 7'-0" x 1-3/8" T	Inst	Ea	2C	1.76	74.30	56.50	---	130.80	**170.00**	100.20	**128.00**
3'-0" x 7'-0" x 1-3/8" T	Inst	Ea	2C	1.76	79.00	56.50	---	135.50	**176.00**	104.20	**133.00**
Birch											
2'-6" x 6'-8" x 1-3/8" T	Inst	Ea	2C	1.63	73.20	52.30	---	125.50	**163.00**	97.00	**124.00**
2'-8" x 6'-8" x 1-3/8" T	Inst	Ea	2C	1.76	75.80	56.50	---	132.30	**172.00**	101.40	**129.00**
3'-0" x 6'-8" x 1-3/8" T	Inst	Ea	2C	1.76	80.20	56.50	---	136.70	**177.00**	105.20	**134.00**
2'-6" x 7'-0" x 1-3/8" T	Inst	Ea	2C	1.63	91.10	52.30	---	143.40	**183.00**	112.30	**141.00**
2'-8" x 7'-0" x 1-3/8" T	Inst	Ea	2C	1.76	94.50	56.50	---	151.00	**193.00**	117.50	**148.00**
3'-0" x 7'-0" x 1-3/8" T	Inst	Ea	2C	1.76	99.70	56.50	---	156.20	**199.00**	122.00	**153.00**

Description	Oper	Unit	Crew Size	Man-Hours Per Unit	Costs Based On Small Volume						Large Volume	
					Avg Mat'l Unit Cost	Avg Labor Unit Cost	Avg Equip Unit Cost	Avg Total Unit Cost	Avg Price Incl O&P		Avg Total Unit Cost	Avg Price Incl O&P

Solid core prehung door units
Hardboard
2'-6" x 6'-8" x 1-3/8" T	Inst	Ea	2C	1.76	93.30	56.50	---	149.80	192.00		116.40	147.00
2'-8" x 6'-8" x 1-3/8" T	Inst	Ea	2C	1.88	96.80	60.30	---	157.10	202.00		122.30	154.00
3'-0" x 6'-8" x 1-3/8" T	Inst	Ea	2C	1.88	99.00	60.30	---	159.30	204.00		124.20	157.00
2'-6" x 7'-0" x 1-3/8" T	Inst	Ea	2C	1.76	103.00	56.50	---	159.50	203.00		125.00	157.00
2'-8" x 7'-0" x 1-3/8" T	Inst	Ea	2C	1.88	106.00	60.30	---	166.30	213.00		130.50	164.00
3'-0" x 7'-0" x 1-3/8" T	Inst	Ea	2C	1.88	106.00	60.30	---	166.30	213.00		130.50	164.00

Lauan
2'-6" x 6'-8" x 1-3/8" T	Inst	Ea	2C	1.76	107.00	56.50	---	163.50	207.00		127.80	160.00
2'-8" x 6'-8" x 1-3/8" T	Inst	Ea	2C	1.88	110.00	60.30	---	170.30	217.00		133.70	168.00
3'-0" x 6'-8" x 1-3/8" T	Inst	Ea	2C	1.88	114.00	60.30	---	174.30	222.00		137.40	172.00
2'-6" x 7'-0" x 1-3/8" T	Inst	Ea	2C	1.76	117.00	56.50	---	173.50	219.00		136.60	170.00
2'-8" x 7'-0" x 1-3/8" T	Inst	Ea	2C	1.88	120.00	60.30	---	180.30	228.00		141.50	177.00
3'-0" x 7'-0" x 1-3/8" T	Inst	Ea	2C	1.88	124.00	60.30	---	184.30	233.00		145.50	181.00

Birch
2'-6" x 6'-8" x 1-3/8" T	Inst	Ea	2C	1.76	116.00	56.50	---	172.50	219.00		136.30	170.00
2'-8" x 6'-8" x 1-3/8" T	Inst	Ea	2C	1.88	119.00	60.30	---	179.30	227.00		141.50	176.00
3'-0" x 6'-8" x 1-3/8" T	Inst	Ea	2C	1.88	124.00	60.30	---	184.30	233.00		145.50	181.00
2'-6" x 7'-0" x 1-3/8" T	Inst	Ea	2C	1.76	127.00	56.50	---	183.50	230.00		144.60	179.00
2'-8" x 7'-0" x 1-3/8" T	Inst	Ea	2C	1.88	129.00	60.30	---	189.30	239.00		150.50	186.00
3'-0" x 7'-0" x 1-3/8" T	Inst	Ea	2C	1.88	134.00	60.30	---	194.30	245.00		154.50	191.00

Prehung package door units
Hollow core, flush face door units
Hardboard
2'-6" x 6'-8" x 1-3/8" T	Inst	Ea	2C	2.05	143.00	65.80	---	208.80	263.00		164.70	205.00
2'-8" x 6'-8" x 1-3/8" T	Inst	Ea	2C	2.05	146.00	65.80	---	211.80	266.00		167.70	208.00
3'-0" x 6'-8" x 1-3/8" T	Inst	Ea	2C	2.05	149.00	65.80	---	214.80	270.00		169.70	210.00

Lauan
2'-6" x 6'-8" x 1-3/8" T	Inst	Ea	2C	2.05	158.00	65.80	---	223.80	280.00		177.70	219.00
2'-8" x 6'-8" x 1-3/8" T	Inst	Ea	2C	2.05	160.00	65.80	---	225.80	283.00		179.70	222.00
3'-0" x 6'-8" x 1-3/8" T	Inst	Ea	2C	2.05	165.00	65.80	---	230.80	288.00		183.70	226.00

Birch
2'-6" x 6'-8" x 1-3/8" T	Inst	Ea	2C	2.05	170.00	65.80	---	235.80	295.00		188.70	232.00
2'-8" x 6'-8" x 1-3/8" T	Inst	Ea	2C	2.05	173.00	65.80	---	238.80	298.00		190.70	234.00
3'-0" x 6'-8" x 1-3/8" T	Inst	Ea	2C	2.05	177.00	65.80	---	242.80	303.00		194.70	239.00

Ash
2'-6" x 6'-8" x 1-3/8" T	Inst	Ea	2C	2.05	192.00	65.80	---	257.80	319.00		206.70	253.00
2'-8" x 6'-8" x 1-3/8" T	Inst	Ea	2C	2.05	201.00	65.80	---	266.80	330.00		214.70	262.00
3'-0" x 6'-8" x 1-3/8" T	Inst	Ea	2C	2.05	206.00	65.80	---	271.80	336.00		218.70	267.00

Solid core, flush face door units
Hardboard
2'-6" x 6'-8" x 1-3/8" T	Inst	Ea	2C	2.22	193.00	71.20	---	264.20	328.00		211.50	259.00
2'-8" x 6'-8" x 1-3/8" T	Inst	Ea	2C	2.22	196.00	71.20	---	267.20	332.00		214.50	263.00
3'-0" x 6'-8" x 1-3/8" T	Inst	Ea	2C	2.22	198.00	71.20	---	269.20	335.00		216.50	265.00

Description	Oper	Unit	Crew Size	Man-Hours Per Unit	Costs Based On Small Volume						Large Volume	
					Avg Mat'l Unit Cost	Avg Labor Unit Cost	Avg Equip Unit Cost	Avg Total Unit Cost	Avg Price Incl O&P		Avg Total Unit Cost	Avg Price Incl O&P
Lauan												
2'-6" x 6'-8" x 1-3/8" T	Inst	Ea	2C	2.22	206.00	71.20	---	277.20	344.00		222.50	272.00
2'-8" x 6'-8" x 1-3/8" T	Inst	Ea	2C	2.22	209.00	71.20	---	280.20	348.00		225.50	276.00
3'-0" x 6'-8" x 1-3/8" T	Inst	Ea	2C	2.22	214.00	71.20	---	285.20	353.00		229.50	280.00
Birch												
2'-6" x 6'-8" x 1-3/8" T	Inst	Ea	2C	2.22	216.00	71.20	---	287.20	355.00		231.50	282.00
2'-8" x 6'-8" x 1-3/8" T	Inst	Ea	2C	2.22	218.00	71.20	---	289.20	358.00		233.50	285.00
3'-0" x 6'-8" x 1-3/8" T	Inst	Ea	2C	2.22	223.00	71.20	---	294.20	363.00		237.50	289.00
Ash												
2'-6" x 6'-8" x 1-3/8" T	Inst	Ea	2C	2.22	234.00	71.20	---	305.20	375.00		246.50	300.00
2'-8" x 6'-8" x 1-3/8" T	Inst	Ea	2C	2.22	238.00	71.20	---	309.20	381.00		250.50	304.00
3'-0" x 6'-8" x 1-3/8" T	Inst	Ea	2C	2.22	247.00	71.20	---	318.20	390.00		257.50	312.00

Adjustments

For mitered casing, both sides, supplied but not applied

Description	Oper	Unit	Crew Size	Man-Hours Per Unit	Avg Mat'l Unit Cost	Avg Labor Unit Cost	Avg Equip Unit Cost	Avg Total Unit Cost	Avg Price Incl O&P		Avg Total Unit Cost	Avg Price Incl O&P
9/16" x 1-1/2" finger joint, ADD	Inst	Ea	---	---	29.50	---	---	29.50	33.90		25.00	28.80
9/16" x 2-1/4" solid pine, ADD	Inst	Ea	---	---	41.30	---	---	41.30	47.50		35.40	40.70
9/16" x 2-1/2" solid pine, ADD	Inst	Ea	---	---	38.90	---	---	38.90	44.80		33.30	38.30
For solid fir jamb, 4-9/16", ADD	Inst	Ea	---	---	38.00	---	---	38.00	43.70		32.50	37.40

Screen doors

Aluminum frame with fiberglass wire screen, and plain grille, includes hardware, hinges, closer and latch

Description	Oper	Unit	Crew Size	Man-Hours Per Unit	Avg Mat'l Unit Cost	Avg Labor Unit Cost	Avg Equip Unit Cost	Avg Total Unit Cost	Avg Price Incl O&P		Avg Total Unit Cost	Avg Price Incl O&P
3'-0" x 7'-0"	Inst	Ea	2C	1.54	70.80	49.40	---	120.20	156.00		92.70	118.00

Wood screen doors, pine, aluminum wire screen, 1-1/8" T

Description	Oper	Unit	Crew Size	Man-Hours Per Unit	Avg Mat'l Unit Cost	Avg Labor Unit Cost	Avg Equip Unit Cost	Avg Total Unit Cost	Avg Price Incl O&P		Avg Total Unit Cost	Avg Price Incl O&P
2'-6" x 6'-9"	Inst	Ea	2C	1.54	118.00	49.40	---	167.40	210.00		133.10	164.00
2'-8" x 6'-9"	Inst	Ea	2C	1.54	118.00	49.40	---	167.40	210.00		133.10	164.00
3'-0" x 6'-9"	Inst	Ea	2C	1.54	130.00	49.40	---	179.40	223.00		143.10	176.00
Half screen												
2'-6" x 6'-9"	Inst	Ea	2C	1.54	148.00	49.40	---	197.40	244.00		158.10	193.00
2'-8" x 6'-9"	Inst	Ea	2C	1.54	148.00	49.40	---	197.40	244.00		158.10	193.00
3'-0" x 6'-9"	Inst	Ea	2C	1.54	177.00	49.40	---	226.40	278.00		184.10	222.00

Steel doors

24-gauge steel faces, fully primed, reversible, prices are for a standard 2-3/8" diameter bore

Decorative doors

Description	Oper	Unit	Crew Size	Man-Hours Per Unit	Avg Mat'l Unit Cost	Avg Labor Unit Cost	Avg Equip Unit Cost	Avg Total Unit Cost	Avg Price Incl O&P		Avg Total Unit Cost	Avg Price Incl O&P
Dual glazed, tempered clear												
2'-6", 2'-8", 3'-0" x 6'-8" x 1-3/4" T	Inst	Ea	2C	2.22	280.00	71.20	---	351.20	428.00		285.50	345.00
Dual glazed, tempered clear, two bevel												
2'-6", 2'-8", 3'-0" x 6'-8" x 1-3/4" T	Inst	Ea	2C	2.22	328.00	71.20	---	399.20	484.00		327.50	393.00

Embossed raised panel doors

Description	Oper	Unit	Crew Size	Man-Hours Per Unit	Avg Mat'l Unit Cost	Avg Labor Unit Cost	Avg Equip Unit Cost	Avg Total Unit Cost	Avg Price Incl O&P		Avg Total Unit Cost	Avg Price Incl O&P
6 raised panels												
2'-6", 2'-8", 3'-0" x 6'-8" x 1-3/4" T	Inst	Ea	2C	2.22	158.00	71.20	---	229.20	288.00		181.50	225.00
2'-8" x 8'-0" x 1-3/4" T	Inst	Ea	2C	2.46	266.00	78.90	---	344.90	424.00		278.30	338.00
3'-0" x 8'-0" x 1-3/4" T	Inst	Ea	2C	2.46	290.00	78.90	---	368.90	452.00		299.30	363.00
6 raised panels, dual glazed, tempered clear												
2'-6", 2'-8", 3'-0" x 6'-8" x 1-3/4" T	Inst	Ea	2C	2.22	245.00	71.20	---	316.20	389.00		256.50	311.00
8 raised panels												
2'-8", 3'-0" x 6'-8" x 1-3/4" T	Inst	Ea	2C	2.22	160.00	71.20	---	231.20	291.00		183.50	228.00

					Costs Based On Small Volume					Large Volume	
Description	Oper	Unit	Crew Size	Man-Hours Per Unit	Avg Mat'l Unit Cost	Avg Labor Unit Cost	Avg Equip Unit Cost	Avg Total Unit Cost	Avg Price Incl O&P	Avg Total Unit Cost	Avg Price Incl O&P
Embossed raised panels and dual glazed tempered clear glass											
2 raised panels with 1 lite											
2'-6", 2'-8", 3'-0" x 6'-8" x 1-3/4" T	Inst	Ea	2C	2.22	265.00	71.20	---	336.20	412.00	273.50	331.00
2 raised panels with 9 lites											
2'-6", 2'-8", 3'-0" x 6'-8" x 1-3/4" T	Inst	Ea	2C	2.22	270.00	71.20	---	341.20	417.00	277.50	335.00
2 raised panels with 12 diamond lites											
2'-6", 2'-8", 3'-0" x 6'-8" x 1-3/4" T	Inst	Ea	2C	2.22	280.00	71.20	---	351.20	428.00	285.50	345.00
4 diamond raised panels with 1 lite											
2'-6", 2'-8", 3'-0" x 6'-8" x 1-3/4" T	Inst	Ea	2C	2.22	265.00	71.20	---	336.20	412.00	273.50	331.00
4 diamond raised panels with 9 lites											
2'-6", 2'-8", 3'-0" x 6'-8" x 1-3/4" T	Inst	Ea	2C	2.22	270.00	71.20	---	341.20	417.00	277.50	335.00
4 diamond raised panels with 12 diamond lites											
2'-6", 2'-8", 3'-0" x 6'-8" x 1-3/4" T	Inst	Ea	2C	2.22	280.00	71.20	---	351.20	428.00	285.50	345.00
Flush doors											
2'-0" to 3'-0" x 6'-8" x 1-3/4" T	Inst	Ea	2C	2.22	151.00	71.20	---	222.20	281.00	175.50	218.00
2'-0" to 3'-0" x 8'-0" x 1-3/4" T	Inst	Ea	2C	2.05	274.00	65.80	---	339.80	414.00	276.70	333.00
Dual glazed, tempered clear											
2'-6", 2'-8", 3'-0" x 6'-8" x 1-3/4" T	Inst	Ea	2C	2.22	231.00	71.20	---	302.20	373.00	244.50	297.00
1-1/2 hour fire label											
2'-0" to 3'-0" x 6'-8" x 1-3/4" T	Inst	Ea	2C	2.22	165.00	71.20	---	236.20	297.00	187.50	232.00
2'-0" to 3'-0" x 7'-0" x 1-3/4" T	Inst	Ea	2C	2.05	201.00	65.80	---	266.80	329.00	214.70	261.00
French doors, flush face											
1 lite											
2'-6", 2'-8", 3'-0" x 6'-8" x 1-3/4" T	Inst	Ea	2C	2.22	301.00	71.20	---	372.20	453.00	304.50	366.00
2'-6", 2'-8", 3'-0" x 8,-0" x 1-3/4" T	Inst	Ea	2C	2.46	496.00	78.90	---	574.90	688.00	475.30	565.00
10 lite											
2'-6", 2'-8", 3'-0" x 6'-8" x 1-3/4" T	Inst	Ea	2C	2.22	319.00	71.20	---	390.20	473.00	319.50	383.00
2'-6", 2'-8", 3'-0" x 8,-0" x 1-3/4" T	Inst	Ea	2C	2.46	525.00	78.90	---	603.90	722.00	500.30	594.00
15 lite											
2'-6", 2'-8", 3'-0" x 6'-8" x 1-3/4" T	Inst	Ea	2C	2.22	325.00	71.20	---	396.20	480.00	324.50	389.00
2'-6", 2'-8", 3'-0" x 8,-0" x 1-3/4" T	Inst	Ea	2C	2.46	537.00	78.90	---	615.90	736.00	511.30	606.00
French sidelites											
Embossed raised panel											
3 lite											
1'-0", 1'-2" x 6'-8"	Inst	Ea	2C	1.54	165.00	49.40	---	214.40	264.00	173.10	211.00
Flush face											
1'-0", 1'-2" x 6'-8"	Inst	Ea	2C	1.54	81.40	49.40	---	130.80	168.00	101.80	128.00
Dual glazed, tempered clear											
1'-0", 1'-2" x 6'-8"	Inst	Ea	2C	1.54	171.00	49.40	---	220.40	271.00	178.10	217.00
Prehung package door units											
Flush steel door											
2'-6" x 6'-8" x 1-3/8" T	Inst	Ea	2C	2.05	281.00	65.80	---	346.80	422.00	282.70	340.00

Drywall (Sheetrock or wallboard)

The types of drywall are Standard, Fire Resistant, Water Resistant, and Fire and Water Resistant.

1. **Dimensions.** 1/4", 3/8" x 6' to 12'; 1/2", 5/8" x 6' to 14'.

2. **Installation**

   a. One ply. Sheets are nailed to studs and joists using 1 1/4" or 1 3/8" x .101 type 500 nails. Nails are usually spaced 8" oc on walls and 7" oc on ceilings. The joints between sheets are filled, taped and sanded. Tape and compound are available in kits containing 250 LF of tape and 18 lbs of compound or 75 LF of tape and 5 lbs of compound.

   b. Two plies. Initial ply is nailed to studs or joists. Second ply is laminated to first with taping compounds as the adhesive. Only the joints in the second ply are filled, taped and finished.

3. **Estimating Technique.** Determine area, deduct openings and add approximately 10% for waste.

Ceiling application

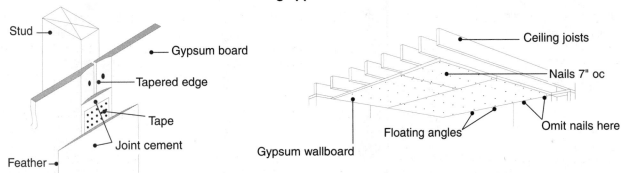

Wall application

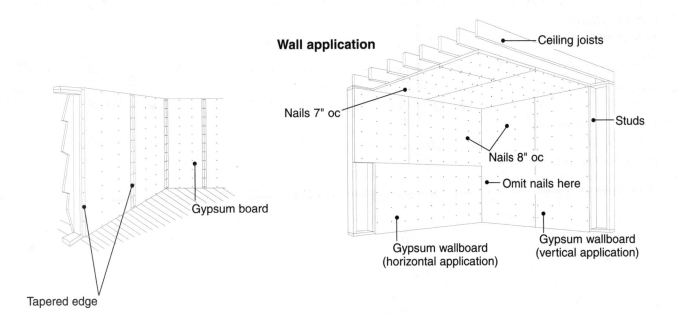

				Costs Based On Small Volume					Large Volume	
Description	Oper	Unit	Crew Size	Man-Hours Per Unit	Avg Mat'l Unit Cost	Avg Labor Unit Cost	Avg Equip Unit Cost	Avg Total Unit Cost	Avg Price Incl O&P	Avg Total Unit Cost	Avg Price Incl O&P

Drywall

Demo Drywall or Sheetrock

Walls and ceilings	Demo	SF	1L	.023	---	.61	---	.61	.91	.27	.4

Install Drywall or Sheetrock (gypsum plasterboard)

Includes tape, finish, nails or screws, and 5% waste

Applied to wood or metal frame

Walls only, 4' x 8', 10', 12'											
3/8" T, standard	Inst	SF	CA	.024	.43	.77	---	1.20	1.65	1.09	1.5
1/2" T, standard	Inst	SF	CA	.024	.43	.77	---	1.20	1.65	1.09	1.5
5/8" T, standard	Inst	SF	CA	.025	.50	.80	---	1.30	1.78	1.19	1.6
Laminated 3/8" T sheets, std.	Inst	SF	CA	.033	.67	1.06	---	1.73	2.36	1.65	2.2
Ceilings only, 4' x 8', 10', 12'											
3/8" T, standard	Inst	SF	CA	.029	.43	.93	---	1.36	1.89	1.21	1.6
1/2" T, standard	Inst	SF	CA	.029	.43	.93	---	1.36	1.89	1.21	1.6
5/8" T, standard	Inst	SF	CA	.031	.50	.99	---	1.49	2.07	1.35	1.8
Laminated 3/8" T sheets, std.	Inst	SF	CA	.040	.67	1.28	---	1.95	2.70	1.88	2.5
Average for ceilings and walls											
3/8" T, standard	Inst	SF	CA	.025	.43	.80	---	1.23	1.70	1.12	1.5
1/2" T, standard	Inst	SF	CA	.025	.43	.80	---	1.23	1.70	1.12	1.5
5/8" T, standard	Inst	SF	CA	.026	.50	.83	---	1.33	1.83	1.22	1.6
Laminated 3/8" T sheets, std.	Inst	SF	CA	.034	.67	1.09	---	1.76	2.41	1.68	2.2
Tape and bed joints on repaired sheetrock											
Walls	Inst	SF	CA	.024	.14	.77	---	.91	1.32	.83	1.2
Ceilings	Inst	SF	CA	.028	.14	.90	---	1.04	1.51	.92	1.3
Thin coat plaster	Inst	SF	CA	.016	.27	.51	---	.78	1.08	.69	.9
Material and labor adjustments											
No tape and finish, DEDUCT	Inst	SF	CA	-.013	-.14	-.42	---	-.56	---	-.47	---
For ceilings 9' H											
Ceiling only, ADD	Inst	SF	CA	.009	---	.29	---	.29	.43	.26	.39
Walls & ceilings (average), ADD	Inst	SF	CA	.004	---	.13	---	.13	.19	.10	.14
Material adjustments											
For fire resistant (1/2" and 5/8" T)											
ADD	Inst	SF	CA	---	.13	---	---	.13	.15	.11	.13
For water resistant (1/2" and 5/8" T)											
ADD	Inst	SF	CA	---	.14	---	---	.14	.16	.12	.14
For fire and water resistant (5/8" T only)											
ADD	Inst	SF	CA	---	.27	---	---	.27	.31	.23	.26

Electrical

General work

Residential service, single phase system. Prices given on a cost per each basis for a unit price system which includes a weathercap, service entrance cable, meter socket, entrance disconnect switch, ground rod with clamp, ground cable, EMT, and panelboard

Description	Oper	Unit	Crew Size	Man-Hours Per Unit	Avg Mat'l Unit Cost	Avg Labor Unit Cost	Avg Equip Unit Cost	Avg Total Unit Cost	Avg Price Incl O&P	Avg Total Unit Cost	Avg Price Incl O&P
								Costs Based On Small Volume		Large Volume	
Weathercap											
100 AMP service	Inst	Ea	EA	1.43	7.03	49.20	---	56.23	79.90	40.86	57.60
200 AMP service	Inst	Ea	EA	2.29	20.70	78.80	---	99.50	139.00	74.00	102.00
Service entrance cable (typical allowance is 20 LF)											
100 AMP service	Inst	LF	EA	.176	.16	6.05	---	6.21	9.02	4.38	6.35
200 AMP service	Inst	LF	EA	.254	.36	8.74	---	9.10	13.20	6.45	9.32
Meter socket											
100 AMP service	Inst	Ea	EA	5.71	38.90	196.00	---	234.90	331.00	173.70	242.00
200 AMP service	Inst	Ea	EA	8.89	60.70	306.00	---	366.70	516.00	275.80	385.00
Entrance disconnect switch											
100 AMP service	Inst	Ea	EA	8.89	192.00	306.00	---	498.00	668.00	397.00	525.00
200 AMP service	Inst	Ea	EA	13.3	385.00	457.00	---	842.00	1110.00	678.00	879.00
Ground rod, with clamp											
100 AMP service	Inst	Ea	EA	3.81	20.70	131.00	---	151.70	215.00	110.80	156.00
200 AMP service	Inst	Ea	EA	3.81	40.00	131.00	---	171.00	237.00	128.50	176.00
Ground cable (typical allowance is 10 LF)											
100 AMP service	Inst	LF	EA	.114	.19	3.92	---	4.11	5.94	2.93	4.22
200 AMP service	Inst	LF	EA	.143	.31	4.92	---	5.23	7.54	3.73	5.35
3/4" EMT (typical allowance is 10 LF)											
200 AMP service	Inst	LF	EA	.152	.77	5.23	---	6.00	8.52	4.39	6.19
Panelboard											
100 AMP service - 12-circuit	Inst	Ea	EA	2.7	170.00	92.90	---	262.90	331.00	211.00	260.00
200 AMP service - 24-circuit	Inst	Ea	EA	4.0	407.00	138.00	---	545.00	669.00	453.10	546.00
Adjustments for other than normal working situations											
Cut and patch, ADD	Inst	%	EA	---	---	---	---	50.0	50.0	20.0	20.0
Dust protection, ADD	Inst	%	EA	---	---	---	---	40.0	40.0	10.0	10.0
Protect existing work, ADD	Inst	%	EA	---	---	---	---	50.0	50.0	20.0	20.0

Wiring per outlet or switch; wall or ceiling

Description	Oper	Unit	Crew Size	Man-Hours Per Unit	Avg Mat'l Unit Cost	Avg Labor Unit Cost	Avg Equip Unit Cost	Avg Total Unit Cost	Avg Price Incl O&P	Avg Total Unit Cost	Avg Price Incl O&P
Romex, non-metallic sheathed cable, 600 volt, copper with ground wire	Inst	Ea	EA	.714	10.40	24.60	---	35.00	47.80	26.72	36.10
BX, flexible armored cable, 600 volt, copper	Inst	Ea	EA	.952	17.80	32.70	---	50.50	68.20	39.20	52.30
EMT with wire, electric metallic thinwall, 1/2"	Inst	Ea	EA	1.90	22.20	65.30	---	87.50	121.00	66.10	90.20
Rigid with wire, 1/2"	Inst	Ea	EA	2.86	29.60	98.40	---	128.00	178.00	96.00	132.00
Wiring, connection, and installation in closed wall structure, ADD	Inst	Ea	EA	1.43	---	49.20	---	49.20	71.80	34.40	50.20

Lighting. See Lighting fixtures, page 172

					Costs Based On Small Volume					Large Volume	
Description	Oper	Unit	Crew Size	Man-Hours Per Unit	Avg Mat'l Unit Cost	Avg Labor Unit Cost	Avg Equip Unit Cost	Avg Total Unit Cost	Avg Price Incl O&P	Avg Total Unit Cost	Avg Price Incl O&P

Special systems
Burglary detection systems

Description	Oper	Unit	Crew Size	Man-Hours Per Unit	Avg Mat'l Unit Cost	Avg Labor Unit Cost	Avg Equip Unit Cost	Avg Total Unit Cost	Avg Price Incl O&P	Avg Total Unit Cost	Avg Price Incl O&P
Alarm bell	Inst	Ea	EA	2.86	88.80	98.40	---	187.20	246.00	150.40	194.0
Burglar alarm, battery operated											
Mechanical trigger	Inst	Ea	EA	2.86	318.00	98.40	---	416.40	510.00	360.80	437.0
Electrical trigger	Inst	Ea	EA	2.86	385.00	98.40	---	483.40	586.00	422.80	507.0
Adjustments, ADD											
Outside key control	Inst	Ea	---	---	88.80	---	---	88.80	102.00	81.60	93.8
Remote signaling circuitry	Inst	Ea	---	---	144.00	---	---	144.00	165.00	132.00	152.0
Card reader											
Standard	Inst	Ea	EA	4.44	1070.00	153.00	---	1223.00	1460.00	1096.00	1290.0
Multi-code	Inst	Ea	EA	4.44	1380.00	153.00	---	1533.00	1810.00	1380.00	1620.0
Detectors											
Motion											
Infrared photoelectric	Inst	Ea	EA	5.71	222.00	196.00	---	418.00	542.00	342.00	435.0
Passive infrared	Inst	Ea	EA	5.71	296.00	196.00	---	492.00	627.00	410.00	514.0
Ultrasonic, 12 volt	Inst	Ea	EA	5.71	266.00	196.00	---	462.00	593.00	383.00	482.0
Microwave											
10' to 200'	Inst	Ea	EA	5.71	770.00	196.00	---	966.00	1170.00	845.00	1010.0
10' to 350'	Inst	Ea	EA	5.71	2220.00	196.00	---	2416.00	2840.00	2178.00	2550.0
Door switches											
Hinge switch	Inst	Ea	EA	2.29	68.10	78.80	---	146.90	193.00	117.60	152.0
Magnetic switch	Inst	Ea	EA	2.29	81.40	78.80	---	160.20	209.00	129.80	166.0
Exit control locks											
Horn alarm	Inst	Ea	EA	2.86	400.00	98.40	---	498.40	603.00	435.80	523.0
Flashing light alarm	Inst	Ea	EA	2.86	451.00	98.40	---	549.40	663.00	483.80	577.0
Glass break alarm switch	Inst	Ea	EA	1.43	54.80	49.20	---	104.00	135.00	84.70	108.0
Indicating panels											
1 channel	Inst	Ea	EA	4.44	422.00	153.00	---	575.00	708.00	498.00	606.0
10 channel	Inst	Ea	EA	7.27	1460.00	250.00	---	1710.00	2040.00	1523.00	1810.0
20 channel	Inst	Ea	EA	11.4	2810.00	392.00	---	3202.00	3810.00	2855.00	3370.0
40 channel	Inst	Ea	EA	20.0	5180.00	688.00	---	5868.00	6960.00	5310.00	6280.0
Police connect panel	Inst	Ea	EA	2.86	281.00	98.40	---	379.40	467.00	326.80	398.0
Siren	Inst	Ea	EA	2.86	169.00	98.40	---	267.40	338.00	223.80	279.0
Switchmats											
30" x 5'	Inst	Ea	EA	2.29	97.70	78.80	---	176.50	227.00	144.80	184.0
30" x 25'	Inst	Ea	EA	2.86	234.00	98.40	---	332.40	413.00	283.80	348.0
Telephone dialer	Inst	Ea	EA	2.29	444.00	78.80	---	522.80	626.00	463.00	550.00

Doorbell systems
Includes transformer, button, bell

Description	Oper	Unit	Crew Size	Man-Hours Per Unit	Avg Mat'l Unit Cost	Avg Labor Unit Cost	Avg Equip Unit Cost	Avg Total Unit Cost	Avg Price Incl O&P	Avg Total Unit Cost	Avg Price Incl O&P
Door chimes, 2 notes	Inst	Ea	EA	1.14	88.80	39.20	---	128.00	159.00	109.10	134.00
Tube-type chimes	Inst	Ea	EA	1.43	222.00	49.20	---	271.20	327.00	238.40	285.00
Transformer and button only	Inst	Ea	EA	.952	51.80	32.70	---	84.50	107.00	70.50	88.20
Push button only	Inst	Ea	EA	.714	22.20	24.60	---	46.80	61.40	37.60	48.60

Description	Oper	Unit	Crew Size	Man-Hours Per Unit	Avg Mat'l Unit Cost	Avg Labor Unit Cost	Avg Equip Unit Cost	Avg Total Unit Cost	Avg Price Incl O&P	Avg Total Unit Cost	Avg Price Incl O&P
										Large Volume	
Fire alarm systems											
Battery and rack	Inst	Ea	EA	3.81	888.00	131.00	---	1019.00	1210.00	907.80	1070.00
Automatic charger	Inst	Ea	EA	1.90	570.00	65.30	---	635.30	751.00	569.70	669.00
Detector											
Fixed temperature	Inst	Ea	EA	1.90	35.50	65.30	---	100.80	136.00	78.30	104.00
Rate of rise	Inst	Ea	EA	1.90	44.40	65.30	---	109.70	146.00	86.50	114.00
Door holder											
Electro-magnetic	Inst	Ea	EA	2.86	99.20	98.40	---	197.60	258.00	159.90	205.00
Combination holder/closer	Inst	Ea	EA	3.81	555.00	131.00	---	686.00	830.00	601.80	721.00
Fire drill switch	Inst	Ea	EA	1.90	111.00	65.30	---	176.30	223.00	147.70	184.00
Glass break alarm switch	Inst	Ea	EA	1.43	63.60	49.20	---	112.80	145.00	92.90	117.00
Signal bell	Inst	Ea	EA	1.90	63.60	65.30	---	128.90	169.00	104.20	134.00
Smoke detector											
Ceiling type	Inst	Ea	EA	1.90	82.90	65.30	---	148.20	191.00	121.90	154.00
Duct type	Inst	Ea	EA	3.81	326.00	131.00	---	457.00	566.00	390.80	478.00
Intercom systems											
Master stations, with digital AM/FM receiver, up to 20 remote stations, & telephone interface											
Solid state with antenna, with 200' wire, 4 speakers	Inst	Set	EA	5.71	1780.00	196.00	---	1976.00	2330.00	1768.00	2080.00
Solid state with antenna, with 1000' wire, 8 speakers	Inst	Set	EA	11.4	2370.00	392.00	---	2762.00	3300.00	2455.00	2900.00
Room to door intercoms											
Master station, door station, with transformer, wire	Inst	Set	EA	2.86	444.00	98.40	---	542.40	654.00	476.80	570.00
Second door station, ADD	Inst	Ea	EA	1.43	59.20	49.20	---	108.40	140.00	88.80	113.00
Installation, wiring, and transformer, wire											
ADD	Inst	LS	EA	2.29	---	78.80	---	78.80	115.00	55.00	80.30
Telephone, phone-jack wiring											
Pre-wiring, per outlet or jack	Inst	Ea	EA	.952	29.60	32.70	---	62.30	81.80	50.10	64.80
Wiring, connection, and installation in closed wall structure, ADD	Inst	Ea	EA	1.43	---	49.20	---	49.20	71.80	34.40	50.20
Television antenna											
Television antenna outlet, with 300 OHM	Inst	Ea	EA	.952	17.80	32.70	---	50.50	68.20	39.20	52.30
Wiring, connection, and installation in closed wall structure, ADD	Inst	Ea	EA	1.43	---	49.20	---	49.20	71.80	34.40	50.20
Thermostat wiring											
For heating units located on the first floor											
Thermostat on first floor	Inst	Ea	EA	1.43	29.60	49.20	---	78.80	106.00	61.60	81.50
Wiring, connection, and installation in closed wall structure, ADD	Inst	Ea	EA	1.43	---	49.20	---	49.20	71.80	34.40	50.20
Thermostat on second floor	Inst	Ea	EA	2.29	37.00	78.80	---	115.80	158.00	89.00	119.00
Wiring, connection, and installation in closed wall structure, ADD	Inst	Ea	EA	1.90	---	65.30	---	65.30	95.40	45.70	66.80

Costs Based On Small Volume

					Costs Based On Small Volume					Large Volume	
Description	Oper	Unit	Crew Size	Man-Hours Per Unit	Avg Mat'l Unit Cost	Avg Labor Unit Cost	Avg Equip Unit Cost	Avg Total Unit Cost	Avg Price Incl O&P	Avg Total Unit Cost	Avg Price Incl O&P

Entrances

Single and double door entrances	Demo	Ea	LB	.889	---	23.70	---	23.70	**35.30**	17.10	**25.40**

Colonial design

White pine includes frames, pediments, and pilasters

Plain carved archway

Single door units

3'-0" W x 6'-8" H	Inst	Ea	2C	3.08	353.00	98.80	---	451.80	**501.00**	366.20	**398.00**

Double door units

Two - 2'-6" W x 6'-8" H	Inst	Ea	2C	4.10	379.00	132.00	---	511.00	**576.00**	409.70	**453.00**
Two - 2'-8" W x 6'-8" H	Inst	Ea	2C	4.10	411.00	132.00	---	543.00	**608.00**	436.70	**480.00**
Two - 3'-0" W x 6'-8" H	Inst	Ea	2C	4.10	537.00	132.00	---	669.00	**734.00**	545.70	**588.00**

Decorative carved archway

Single door units

3'-0" W x 6'-8" H	Inst	Ea	2C	3.08	443.00	98.80	---	541.80	**591.00**	443.20	**475.00**

Double door units

Two - 2'-6" W x 6'-8" H	Inst	Ea	2C	4.10	379.00	132.00	---	511.00	**576.00**	409.70	**453.00**
Two - 2'-8" W x 6'-8" H	Inst	Ea	2C	4.10	402.00	132.00	---	534.00	**600.00**	429.70	**473.00**
Two - 3 -0" W x 6'-8" H	Inst	Ea	2C	4.10	419.00	132.00	---	551.00	**616.00**	444.70	**487.00**

Description	Oper	Unit	Crew Size	Man-Hours Per Unit	Costs Based On Small Volume					Large Volume	
					Avg Mat'l Unit Cost	Avg Labor Unit Cost	Avg Equip Unit Cost	Avg Total Unit Cost	Avg Price Incl O&P	Avg Total Unit Cost	Avg Price Incl O&P

Excavation

Description	Oper	Unit	Crew Size	Man-Hours Per Unit	Avg Mat'l Unit Cost	Avg Labor Unit Cost	Avg Equip Unit Cost	Avg Total Unit Cost	Avg Price Incl O&P	Avg Total Unit Cost	Avg Price Incl O&P
Digging out or trenching	Demo	CY	LB	.056	---	1.49	---	1.49	**2.22**	.91	**1.35**
Pits, medium earth, piled											
With front end loader, track mounted, 1-1/2 CY capacity;											
55 CY per hour	Demo	CY	VB	.061	---	1.91	.88	2.79	**3.71**	1.66	**2.20**
By hand											
To 4'-0" D	Demo	CY	LB	1.78	---	47.40	---	47.40	**70.70**	28.50	**42.50**
4'-0" to 6'-0" D	Demo	CY	LB	2.67	---	71.10	---	71.10	**106.00**	42.60	**63.50**
6'-0" to 8'-0" D	Demo	CY	LB	4.00	---	107.00	---	107.00	**159.00**	71.10	**106.00**
Continuous footing or trench, medium earth, piled											
With tractor backhoe, 3/8 CY capacity (48 HP);											
15 CY per hour	Demo	CY	VB	.222	---	6.95	3.21	10.16	**13.50**	6.09	**8.09**
By hand, to 4'-0" D	Demo	CY	LB	1.78	---	47.40	---	47.40	**70.70**	28.50	**42.50**
Backfilling, by hand, medium soil											
Without compaction	Demo	CY	LB	.941	---	25.10	---	25.10	**37.40**	15.20	**22.70**
With hand compaction											
6" layers	Demo	CY	LB	1.60	---	42.60	---	42.60	**63.50**	25.10	**37.40**
12" layers	Demo	CY	LB	1.23	---	32.80	---	32.80	**48.80**	19.40	**28.90**
With vibrating plate compaction											
6" layers	Demo	CY	AB	1.33	---	38.10	1.86	39.96	**58.20**	24.02	**35.00**
12" layers	Demo	CY	AB	1.14	---	32.70	1.60	34.30	**49.90**	20.03	**29.20**

Facebrick. See Masonry, page 178

Description	Oper	Unit	Crew Size	Man-Hours Per Unit	Costs Based On Small Volume						Large Volume	
					Avg Mat'l Unit Cost	Avg Labor Unit Cost	Avg Equip Unit Cost	Avg Total Unit Cost	Avg Price Incl O&P		Avg Total Unit Cost	Avg Price Incl O&P

Fences

Basketweave

Redwood, preassembled, 8' L panels, includes 4" x 4" line posts, horizontal or vertical weave

Description	Oper	Unit	Crew Size	Man-Hours	Avg Mat'l	Avg Labor	Avg Equip	Avg Total	Avg Price O&P	Avg Total LV	Avg Price LV
5' H	Inst	LF	CS	.133	13.40	4.03	---	17.43	21.50	14.93	18.2(
6' H	Inst	LF	CS	.133	20.10	4.03	---	24.13	29.20	20.93	25.1(

Adjustments

Corner or end posts, 8' H	Inst	Ea	CA	.800	26.80	25.70	---	52.50	69.30	43.50	57.0(
3-1/2' W x 5' H gate, with hardware	Inst	Ea	CA	1.00	114.00	32.10	---	146.10	179.00	126.70	155.0(

Board, per LF complete fence system

6' H boards nailed to wood frame, on 1 side only; 8' L redwood 4" x 4" (milled) posts, set 2' D in concrete filled holes @ 6' oc; frame members 2" x 4" (milled) as 2 rails between posts per 6' L fence section; costs are per LF of fence

Douglas fir frame members with redwood posts

Milled boards, "dog-eared" one end

Cedar

1" x 6" - 6' H	Inst	LF	CS	.533	42.60	16.10	---	58.70	73.20	49.90	61.70
1" x 8" - 6' H	Inst	LF	CS	.453	43.40	13.70	---	57.10	70.50	49.00	59.90
1" x 10" - 6' H	Inst	LF	CS	.400	42.50	12.10	---	54.60	67.10	46.88	57.10

Douglas fir

1" x 6" - 6' H	Inst	LF	CS	.533	41.00	16.10	---	57.10	71.40	48.60	60.10
1" x 8" - 6' H	Inst	LF	CS	.453	41.40	13.70	---	55.10	68.20	47.20	57.90
1" x 10" - 6' H	Inst	LF	CS	.400	41.00	12.10	---	53.10	65.40	45.58	55.50

Redwood

1" x 6" - 6' H	Inst	LF	CS	.533	42.60	16.10	---	58.70	73.20	49.90	61.70
1" x 8" - 6' H	Inst	LF	CS	.453	43.40	13.70	---	57.10	70.50	49.00	59.90
1" x 10" - 6' H	Inst	LF	CS	.400	45.00	12.10	---	57.10	69.90	48.98	59.50

Rough boards, both ends squared

Cedar

1" x 6" - 6' H	Inst	LF	CS	.490	41.70	14.80	---	56.50	70.20	48.30	59.40
1" x 8" - 6' H	Inst	LF	CS	.429	42.50	13.00	---	55.50	68.40	47.49	58.00
1" x 10" - 6' H	Inst	LF	CS	.375	41.60	11.40	---	53.00	64.90	45.54	55.30

Douglas fir

1" x 6" - 6' H	Inst	LF	CS	.490	40.40	14.80	---	55.20	68.70	47.10	58.00
1" x 8" - 6' H	Inst	LF	CS	.429	40.80	13.00	---	53.80	66.40	45.89	56.20
1" x 10" - 6' H	Inst	LF	CS	.375	40.60	11.40	---	52.00	63.70	44.54	54.20

Redwood

1" x 6" - 6' H	Inst	LF	CS	.490	41.70	14.80	---	56.50	70.20	48.30	59.40
1" x 8" - 6' H	Inst	LF	CS	.429	42.50	13.00	---	55.50	68.40	47.49	58.00
1" x 10" - 6' H	Inst	LF	CS	.375	44.10	11.40	---	55.50	67.70	47.74	57.80

				Costs Based On Small Volume						Large Volume	
Description	Oper	Unit	Crew Size	Man-Hours Per Unit	Avg Mat'l Unit Cost	Avg Labor Unit Cost	Avg Equip Unit Cost	Avg Total Unit Cost	Avg Price Incl O&P	Avg Total Unit Cost	Avg Price Incl O&P

Redwood frame members with redwood posts

Milled boards, "dog-eared" one end

Cedar

1" x 6" - 6' H	Inst	LF	CS	.533	42.70	16.10	---	58.80	73.30	50.00	61.70
1" x 8" - 6' H	Inst	LF	CS	.453	43.50	13.70	---	57.20	70.60	49.00	60.00
1" x 10" - 6' H	Inst	LF	CS	.400	42.60	12.10	---	54.70	67.10	46.88	57.10

Douglas fir

1" x 6" - 6' H	Inst	LF	CS	.533	41.10	16.10	---	57.20	71.50	48.60	60.10
1" x 8" - 6' H	Inst	LF	CS	.453	41.50	13.70	---	55.20	68.20	47.20	57.90
1" x 10" - 6' H	Inst	LF	CS	.400	41.10	12.10	---	53.20	65.40	45.58	55.60

Redwood

1" x 6" - 6' H	Inst	LF	CS	.533	42.70	16.10	---	58.80	73.30	50.00	61.70
1" x 8" - 6' H	Inst	LF	CS	.453	43.50	13.70	---	57.20	70.60	49.00	60.00
1" x 10" - 6' H	Inst	LF	CS	.400	45.00	12.10	---	57.10	69.90	49.08	59.60

Rough boards, both ends squared

Cedar

1" x 6" - 6' H	Inst	LF	CS	.490	41.80	14.80	---	56.60	70.30	48.30	59.40
1" x 8" - 6' H	Inst	LF	CS	.429	42.60	13.00	---	55.60	68.50	47.49	58.00
1" x 10" - 6' H	Inst	LF	CS	.375	41.70	11.40	---	53.10	65.00	45.54	55.40

Douglas fir

1" x 6" - 6' H	Inst	LF	CS	.490	40.40	14.80	---	55.20	68.70	47.10	58.00
1" x 8" - 6' H	Inst	LF	CS	.429	40.80	13.00	---	53.80	66.40	45.99	56.20
1" x 10" - 6' H	Inst	LF	CS	.375	40.60	11.40	---	52.00	63.70	44.64	54.30

Redwood

1" x 6" - 6' H	Inst	LF	CS	.490	41.80	14.80	---	56.60	70.30	48.30	59.40
1" x 8" - 6' H	Inst	LF	CS	.429	42.60	13.00	---	55.60	68.50	47.49	58.00
1" x 10" - 6' H	Inst	LF	CS	.375	44.10	11.40	---	55.50	67.80	47.74	57.90

Chain link

9 gauge galvanized steel, includes top rail (1-5/8" o.d.), line posts (2" o.d.) @ 10' oc and sleeves

36" H	Inst	LF	HB	.145	4.93	4.46	---	9.39	12.30	7.73	10.00
42" H	Inst	LF	HB	.152	5.28	4.67	---	9.95	13.00	8.20	10.60
48" H	Inst	LF	HB	.160	5.53	4.92	---	10.45	13.70	8.60	11.10
60" H	Inst	LF	HB	.178	6.38	5.47	---	11.85	15.50	9.75	12.60
72" H	Inst	LF	HB	.200	7.24	6.15	---	13.39	17.50	11.04	14.30

Adjustments

Note: The following percentage adjustments for Small Volume also apply to Large

11-1/2"-gauge galvanized steel fabric DEDUCT	Inst	%	---	---	-27.0	---	---	---	---	---	---
12-gauge galvanized steel fabric DEDUCT	Inst	%	---	---	-29.0	---	---	---	---	---	---
9-gauge green vinyl-coated fabric DEDUCT	Inst	%	---	---	-4.0	---	---	---	---	---	---
11-gauge green vinyl-coated fabric DEDUCT	Inst	%	---	---	-3.8	---	---	---	---	---	---

					Costs Based On Small Volume					Large Volume	
Description	Oper	Unit	Crew Size	Man-Hours Per Unit	Avg Mat'l Unit Cost	Avg Labor Unit Cost	Avg Equip Unit Cost	Avg Total Unit Cost	Avg Price Incl O&P	Avg Total Unit Cost	Avg Price Incl O&P
Filler strips, ADD											
Aluminum, baked-on enamel finish											
Diagonal, 1-7/8" W											
48" H	Inst	LF	LB	.213	4.53	5.67	---	10.20	13.70	8.28	11.00
60" H	Inst	LF	LB	.213	5.49	5.67	---	11.16	14.80	9.14	12.00
72" H	Inst	LF	LB	.213	6.43	5.67	---	12.10	15.90	9.97	12.90
Vertical, 1-1/4" W											
48" H	Inst	LF	LB	.213	4.53	5.67	---	10.20	13.70	8.28	11.00
60" H	Inst	LF	LB	.213	5.33	5.67	---	11.00	14.60	9.00	11.80
72" H	Inst	LF	LB	.213	6.27	5.67	---	11.94	15.70	9.83	12.80
Wood, redwood stain											
Vertical, 1-1/4" W											
48" H	Inst	LF	LB	.213	4.53	5.67	---	10.20	13.70	8.28	11.00
60" H	Inst	LF	LB	.213	5.49	5.67	---	11.16	14.80	9.14	12.00
72" H	Inst	LF	LB	.213	6.43	5.67	---	12.10	15.90	9.97	12.90
Corner posts (2-1/2" o.d.), installed, heavyweight											
36" H	Inst	Ea	HA	.381	26.10	12.50	---	38.60	48.70	32.58	40.70
42" H	Inst	Ea	HA	.400	29.60	13.10	---	42.70	53.60	36.01	44.70
48" H	Inst	Ea	HA	.421	31.40	13.80	---	45.20	56.70	38.40	47.70
60" H	Inst	Ea	HA	.471	37.50	15.50	---	53.00	66.10	45.20	56.00
72" H	Inst	Ea	HA	.533	43.60	17.50	---	61.10	76.10	51.80	64.00
End or gate posts (2-1/2" o.d.), installed, heavyweight											
36" H	Inst	Ea	HA	.381	20.00	12.50	---	32.50	41.70	27.18	34.40
42" H	Inst	Ea	HA	.400	21.80	13.10	---	34.90	44.60	29.01	36.70
48" H	Inst	Ea	HA	.421	24.40	13.80	---	38.20	48.60	32.20	40.60
60" H	Inst	Ea	HA	.471	28.70	15.50	---	44.20	56.10	37.40	47.20
72" H	Inst	Ea	HA	.533	33.10	17.50	---	50.60	64.10	42.50	53.40
Gates, square corner frame, 9 gauge wire, installed											
3' W walkway gates											
36" H	Inst	Ea	HA	.800	73.20	26.30	---	99.50	123.00	85.20	105.00
42" H	Inst	Ea	HA	.800	76.70	26.30	---	103.00	127.00	88.30	108.00
48" H	Inst	Ea	HA	.889	78.40	29.20	---	107.60	134.00	91.50	113.00
60" H	Inst	Ea	HA	.889	94.10	29.20	---	123.30	152.00	105.40	129.00
72" H	Inst	Ea	HA	1.00	110.00	32.80	---	142.80	175.00	121.40	148.00
12' W double driveway gates											
36" H	Inst	Ea	HA	1.60	193.00	52.50	---	245.50	301.00	215.60	263.00
42" H	Inst	Ea	HA	1.60	204.00	52.50	---	256.50	313.00	224.60	273.00
48" H	Inst	Ea	HA	2.00	211.00	65.60	---	276.60	340.00	239.50	293.00
60" H	Inst	Ea	HA	2.00	253.00	65.60	---	318.60	388.00	276.50	336.00
72" H	Inst	Ea	HA	2.67	279.00	87.60	---	366.60	451.00	313.60	382.00

Split rail

Red cedar, 10' L sectional spans

Rails only	Inst	Ea	---	---	11.90	---	---	11.90	11.90	10.50	10.50
Bored 2 rail posts											
5'-6" line or end posts	Inst	Ea	CA	.800	13.90	25.70	---	39.60	54.50	32.10	43.80
5'-6" corner posts	Inst	Ea	CA	.800	15.70	25.70	---	41.40	56.50	33.60	45.60
Bored 3 rail posts											
6'-6" line or end posts	Inst	Ea	CA	.800	17.40	25.70	---	43.10	58.60	35.20	47.40
6'-6" corner posts	Inst	Ea	CA	.800	19.20	25.70	---	44.90	60.60	36.70	49.20

Description	Oper	Unit	Crew Size	Man-Hours Per Unit	Avg Mat'l Unit Cost	Avg Labor Unit Cost	Avg Equip Unit Cost	Avg Total Unit Cost	Avg Price Incl O&P	Avg Total Unit Cost	Avg Price Incl O&P
						Costs Based On Small Volume					**Large Volume**	
Complete fence estimate (does not include gates)											
2 rail, 36" H, 5'-6" post	Inst	LF	CJ	.056	3.94	1.64	---	5.58	7.00	4.73	5.88
3 rail, 48" H, 6'-6" post	Inst	LF	CJ	.065	5.12	1.91	---	7.03	8.75	5.96	7.35
Gate											
2 rails											
3-1/2' W	Inst	Ea	CA	.800	69.70	25.70	---	95.40	119.00	81.60	101.00
5' W	Inst	Ea	CA	1.00	90.60	32.10	---	122.70	152.00	106.10	131.00
3 rails											
3-1/2' W	Inst	Ea	CA	.800	87.10	25.70	---	112.80	139.00	97.10	119.00
5' W	Inst	Ea	CA	1.00	101.00	32.10	---	133.10	164.00	115.40	142.00

				Costs Based On Small Volume					Large Volume	
Description	Oper	Unit	Crew Size	Man-Hours Per Unit	Avg Mat'l Unit Cost	Avg Labor Unit Cost	Avg Equip Unit Cost	Avg Total Unit Cost	Avg Price Incl O&P	Avg Total Unit Cost	Avg Price Incl O&P

Fiberglass panels

Corrugated; 8', 10', or 12' L panels; 2-1/2" W x 1/2" D corrugation

Nailed on wood frame

4 oz., 0.03" T, 26" W	Inst	SF	CA	.053	1.60	1.70	---	3.30	**4.39**	2.75	3.62
Fire retardant	Inst	SF	CA	.053	2.21	1.70	---	3.91	**5.09**	3.32	4.27
5 oz., 0.037" T, 26" W	Inst	SF	CA	.053	1.97	1.70	---	3.67	**4.82**	3.09	4.01
Fire retardant	Inst	SF	CA	.053	2.72	1.70	---	4.42	**5.68**	3.78	4.80
6 oz., 0.045" T, 26" W	Inst	SF	CA	.053	2.34	1.70	---	4.04	**5.24**	3.43	4.40
Fire retardant	Inst	SF	CA	.053	3.23	1.70	---	4.93	**6.27**	4.25	5.34
8 oz., 0.06" T, 26" W	Inst	SF	CA	.053	2.71	1.70	---	4.41	**5.67**	3.77	4.79
Fire retardant	Inst	SF	CA	.053	3.73	1.70	---	5.43	**6.84**	4.72	5.88

Flat panels; 8', 10', or 12' L panels; clear, green, or white

5 oz., 48" W	Inst	SF	CA	.053	1.07	1.70	---	2.77	**3.78**	2.26	3.05
Fire retardant	Inst	SF	CA	.053	1.07	1.70	---	2.77	**3.78**	2.26	3.05
6 oz., 48" W	Inst	SF	CA	.053	2.10	1.70	---	3.80	**4.97**	3.21	4.15
Fire retardant	Inst	SF	CA	.053	3.62	1.70	---	5.32	**6.71**	4.61	5.76
8 oz., 48" W	Inst	SF	CA	.053	2.71	1.70	---	4.41	**5.67**	3.77	4.79
Fire retardant	Inst	SF	CA	.053	4.67	1.70	---	6.37	**7.92**	5.59	6.88

Solar block; 8', 10', 12' L panels; 2-1/2" W x 1/2" D corrugation; nailed on wood frame

5 oz., 26" W	Inst	SF	CA	.053	1.21	1.70	---	2.91	**3.94**	2.39	3.20

Accessories

Wood corrugated

2-1/2" W x 1-1/2" D x 6' L	Inst	Ea	---	---	3.18	---	---	3.18	**3.18**	2.93	2.93
2-1/2" W x 1-1/2" D x 8' L	Inst	Ea	---	---	3.86	---	---	3.86	**3.86**	3.56	3.56
2-1/2" W x 3/4" D x 6' L	Inst	Ea	---	---	1.46	---	---	1.46	**1.46**	1.35	1.35
2-1/2" W x 3/4" D x 8' L	Inst	Ea	---	---	1.88	---	---	1.88	**1.88**	1.73	1.73

Rubber corrugated

1" x 3"	Inst	Ea	---	---	1.46	---	---	1.46	**1.46**	1.35	1.35

Polyfoam corrugated

1" x 3"	Inst	Ea	---	---	1.03	---	---	1.03	**1.03**	.95	.95

Vertical crown molding

Wood

1-1/2" x 6' L	Inst	Ea	---	---	2.65	---	---	2.65	**2.65**	2.45	2.45
1-1/2" x 8' L	Inst	Ea	---	---	3.51	---	---	3.51	**3.51**	3.23	3.23
Polyfoam, 1" x 1" x 3' L	Inst	Ea	---	---	1.28	---	---	1.28	**1.28**	1.18	1.18
Rubber, 1" x 1" x 3' L	Inst	Ea	---	---	2.43	---	---	2.43	**2.43**	2.23	2.23

Fireplaces

Description	Oper	Unit	Crew Size	Man-Hours Per Unit	Avg Mat'l Unit Cost	Avg Labor Unit Cost	Avg Equip Unit Cost	Avg Total Unit Cost	Avg Price Incl O&P	Avg Total Unit Cost	Avg Price Incl O&P
								Costs Based On Small Volume		Large Volume	

Woodburning, prefabricated. No masonry support required, installs directly on floor. Ceramic-backed firebox with black vitreous enamel side panels. No finish plastering or brick hearthwork included. Fire screen, 9 " (i.d.) factory-built insulated chimneys with flue, lining, damper, and flashing with rain cap included. Chimney height from floor to where chimney exits through roof

Description	Oper	Unit	Crew Size	Man-Hours Per Unit	Avg Mat'l Unit Cost	Avg Labor Unit Cost	Avg Equip Unit Cost	Avg Total Unit Cost	Avg Price Incl O&P	Avg Total Unit Cost	Avg Price Incl O&P
36" W fireplace unit with:											
Up to 9'-0" chimney height	Inst	LS	CJ	18.8	980.00	552.00	---	1532.00	**1950.00**	1262.00	**1580.00**
9'-3" to 12'-2" chimney	Inst	LS	CJ	20.5	1030.00	602.00	---	1632.00	**2090.00**	1339.00	**1680.00**
12'-3" to 15'-1" chimney	Inst	LS	CJ	22.2	1080.00	652.00	---	1732.00	**2220.00**	1419.00	**1780.00**
15'-2" to 18'-0" chimney	Inst	LS	CJ	24.6	1130.00	723.00	---	1853.00	**2380.00**	1510.00	**1900.00**
18'-1" to 20'-11" chimney	Inst	LS	CJ	27.1	1180.00	796.00	---	1976.00	**2550.00**	1613.00	**2030.00**
21'-3" to 23'-10" chimney	Inst	LS	CJ	30.8	1230.00	905.00	---	2135.00	**2770.00**	1717.00	**2180.00**
23'-11" to 24'-9" chimney	Inst	LS	CJ	34.8	1280.00	1020.00	---	2300.00	**3010.00**	1853.00	**2370.00**
42" W fireplace unit with:											
Up to 9'-0" chimney height	Inst	LS	CJ	18.8	1110.00	552.00	---	1662.00	**2100.00**	1381.00	**1710.00**
9'-3" to 12'-2" chimney	Inst	LS	CJ	20.5	1170.00	602.00	---	1772.00	**2240.00**	1461.00	**1820.00**
12'-3" to 15'-1" chimney	Inst	LS	CJ	22.2	1220.00	652.00	---	1872.00	**2380.00**	1556.00	**1930.00**
15'-2" to 18'-0" chimney	Inst	LS	CJ	24.6	1280.00	723.00	---	2003.00	**2560.00**	1650.00	**2060.00**
18'-1" to 20'-11" chimney	Inst	LS	CJ	27.1	1340.00	796.00	---	2136.00	**2730.00**	1753.00	**2200.00**
21'-3" to 23'-10" chimney	Inst	LS	CJ	30.8	1400.00	905.00	---	2305.00	**2960.00**	1877.00	**2360.00**
23'-11" to 24'-9" chimney	Inst	LS	CJ	34.8	1450.00	1020.00	---	2470.00	**3210.00**	2013.00	**2550.00**
Accessories											
Log lighter with gas valve (straight or angle pattern)	Inst	Ea	SA	2.05	32.90	74.50	---	107.40	**147.00**	78.60	**106.00**
Log lighter, less gas valve (straight, angle, tee pattern)	Inst	Ea	SA	1.03	12.30	37.40	---	49.70	**69.20**	35.60	**48.70**
Gas valve for log lighter	Inst	Ea	SA	1.03	21.90	37.40	---	59.30	**80.20**	44.50	**58.80**
Spare parts											
Gas valve key	Inst	Ea	---	---	1.51	---	---	1.51	**1.51**	1.39	**1.39**
Stem extender	Inst	Ea	---	---	2.74	---	---	2.74	**2.74**	2.52	**2.52**
Extra long	Inst	Ea	---	---	3.77	---	---	3.77	**3.77**	3.47	**3.47**
Valve floor plate	Inst	Ea	---	---	3.15	---	---	3.15	**3.15**	2.90	**2.90**
Lighter burner tube (12" to 17")	Inst	Ea	---	---	6.30	---	---	6.30	**6.30**	5.80	**5.80**

Fireplace mantels. See Mantels, fireplace, page 174

Flashing. See Sheet metal, page 233

Floor finishes. See individual items.

Floor joists. See Framing, page 124

				Costs Based On Small Volume					Large Volume	
Description	Oper	Unit	Crew Size	Man-Hours Per Unit	Avg Mat'l Unit Cost	Avg Labor Unit Cost	Avg Equip Unit Cost	Avg Total Unit Cost	Avg Price Incl O&P	Avg Total Unit Cost	Avg Price Incl O&P

Food centers

Includes wiring, connection and installation in exposed drainboard only

Built-in models, 4-1/4" x 6-3/4" x 10" rough cut,

Description	Oper	Unit	Crew Size	Man-Hours Per Unit	Avg Mat'l Unit Cost	Avg Labor Unit Cost	Avg Equip Unit Cost	Avg Total Unit Cost	Avg Price Incl O&P	Avg Total Unit Cost	Avg Price Incl O&P
1/4 hp, 110 volts, 6 speed	Inst	Ea	EA	3.81	520.00	131.00	---	651.00	790.00	537.80	647.00
Options											
Blender	Inst	Ea	---	---	55.90	---	---	55.90	55.90	47.90	47.90
Citrus fruit juicer	Inst	Ea	---	---	27.40	---	---	27.40	27.40	23.50	23.50
Food processor	Inst	Ea	---	---	283.00	---	---	283.00	283.00	242.00	242.00
Ice crusher	Inst	Ea	---	---	88.80	---	---	88.80	88.80	76.10	76.10
Knife sharpener	Inst	Ea	---	---	58.30	---	---	58.30	58.30	50.00	50.00
Meat grinder, shredder/slicer with power post	Inst	Ea	---	---	339.00	---	---	339.00	339.00	290.00	290.00
Mixer	Inst	Ea	---	---	159.00	---	---	159.00	159.00	136.00	136.00

Footings. See Concrete, page 69
Formica. See Countertops, page 74
Forming. See Concrete, page 70
Foundations. See Concrete, page 70

Framing

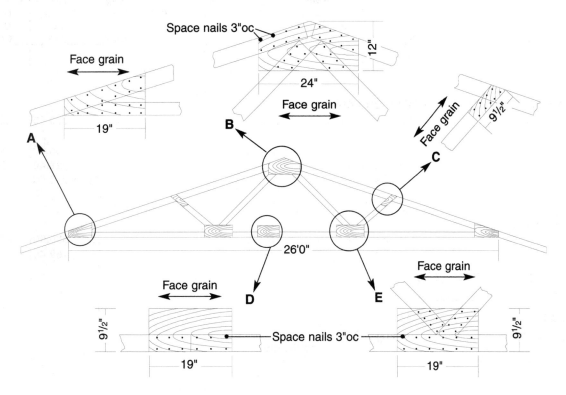

Space nails 3"oc

Face grain

12"

24"

Face grain

Face grain

9½"

19"

A

B

C

26'0"

D

E

Face grain

Face grain

9½"

9½"

Space nails 3"oc

19"

19"

Construction of a 26 foot W truss:

A Bevel-heel gusset
B Peak gusset
C Upper chord intermediate gusset
D Splice of lower chord
E Lower chord intermediate gusset

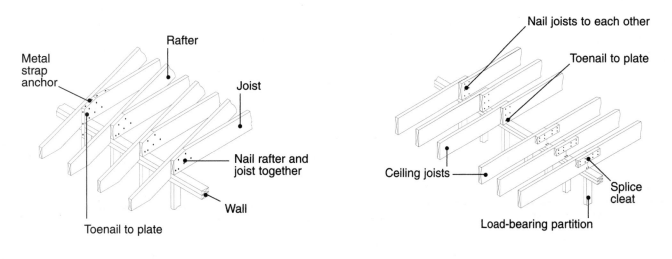

Rafter

Metal
strap
anchor

Joist

Nail joists to each other

Toenail to plate

Nail rafter and
joist together

Ceiling joists

Wall

Splice
cleat

Toenail to plate

Load-bearing partition

Metal joist hanger

Wood hanger

A

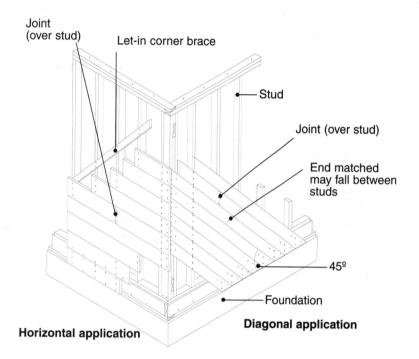

Joint (over stud)

Let-in corner brace

Stud

Joint (over stud)

End matched may fall between studs

45º

Foundation

Horizontal application

Diagonal application

Application of wood sheathing:
- **A** Horizontal and diagonal
- **B** Started at subfloor
- **C** Started at foundation wall

B

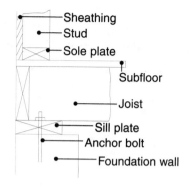

Sheathing
Stud
Sole plate
Subfloor
Joist
Sill plate
Anchor bolt
Foundation wall

C

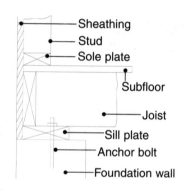

Sheathing
Stud
Sole plate
Subfloor
Joist
Sill plate
Anchor bolt
Foundation wall

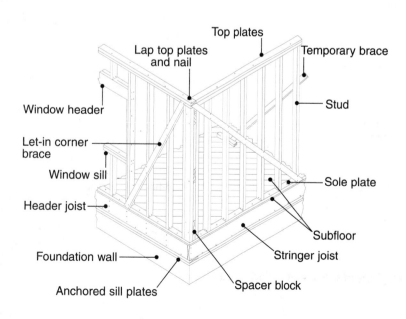

Lap top plates and nail

Top plates

Temporary brace

Window header

Stud

Let-in corner brace

Window sill

Header joist

Sole plate

Foundation wall

Subfloor

Stringer joist

Anchored sill plates

Spacer block

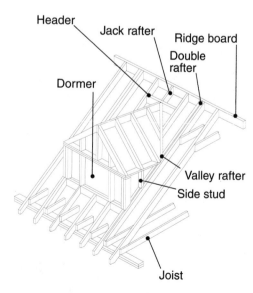

Header

Jack rafter

Ridge board

Double rafter

Dormer

Valley rafter

Side stud

Joist

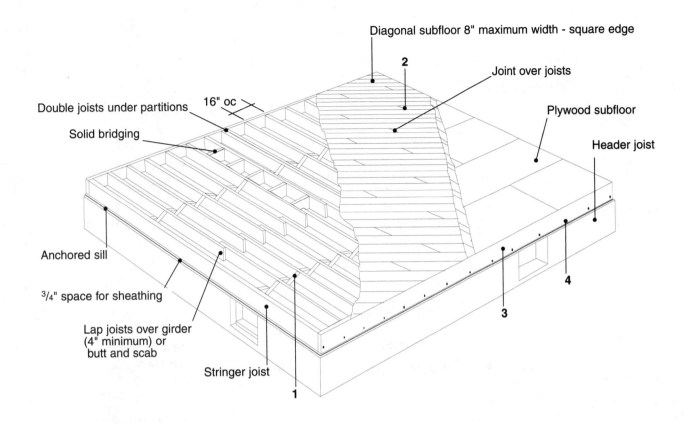

Diagonal subfloor 8" maximum width - square edge

Joint over joists

Plywood subfloor

Header joist

Double joists under partitions

16" oc

Solid bridging

Anchored sill

³/₄" space for sheathing

Lap joists over girder (4" minimum) or butt and scab

Stringer joist

Floor framing:
1. Nailing bridge to joists
2. Nailing board subfloor to joists
3. Nailing header to joists
4. Toenailing header to sill

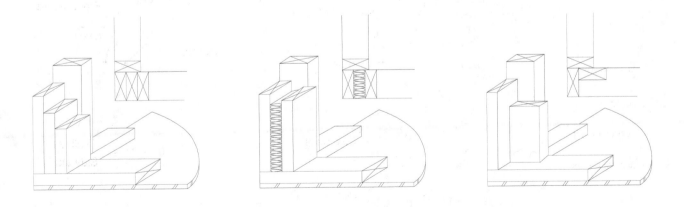

Stud arrangements at exterior corners

Framing, rough carpentry
Dimension lumber
Beams; set on steel columns, not wood columns

Description	Oper	Unit	Crew Size	Man-Hours Per Unit	Avg Mat'l Unit Cost	Avg Labor Unit Cost	Avg Equip Unit Cost	Avg Total Unit Cost	Avg Price Incl O&P	Avg Total Unit Cost	Avg Price Incl O&P
Built-up (2 pieces)											
4" x 6" - 10'	Demo	LF	LB	.028	---	.75	---	.75	1.11	.64	.95
4" x 6" - 10'	Inst	LF	2C	.056	1.20	1.80	.16	3.16	4.24	2.74	3.66
4" x 8" - 10'	Demo	LF	LB	.032	---	.85	---	.85	1.27	.72	1.07
4" x 8" - 10'	Inst	LF	2C	.054	1.67	1.73	.16	3.56	4.68	3.10	4.06
4" x 10" - 10'	Demo	LF	LB	.035	---	.93	---	.93	1.39	.80	1.19
4" x 10" - 10'	Inst	LF	2C	.058	2.37	1.86	.17	4.40	5.69	3.81	4.91
4" x 12" - 12'	Demo	LF	LB	.038	---	1.01	---	1.01	1.51	.85	1.27
4" x 12" - 12'	Inst	LF	2C	.062	2.92	1.99	.18	5.09	6.52	4.45	5.69
6" x 8" - 10'	Demo	LF	LD	.051	---	1.50	---	1.50	2.23	1.26	1.88
6" x 8" - 10'	Inst	LF	CW	.086	4.19	2.53	.13	6.85	8.74	6.07	7.73
6" x 10" - 10'	Demo	LF	LD	.056	---	1.64	---	1.64	2.45	1.38	2.06
6" x 10" - 10'	Inst	LF	CW	.093	5.23	2.73	.14	8.10	10.30	7.17	9.04
6" x 12" - 12'	Demo	LF	LD	.061	---	1.79	---	1.79	2.67	1.53	2.28
6" x 12" - 12'	Inst	LF	CW	.099	6.26	2.91	.14	9.31	11.70	8.25	10.30
Built-up (3 pieces)											
6" x 8" - 10'	Demo	LF	LD	.051	---	1.50	---	1.50	2.23	1.29	1.93
6" x 8" - 10'	Inst	LF	CW	.089	2.54	2.61	.13	5.28	6.97	4.57	6.01
6" x 10" - 10'	Demo	LF	LD	.056	---	1.64	---	1.64	2.45	1.41	2.10
6" x 10" - 10'	Inst	LF	CW	.095	3.58	2.79	.14	6.51	8.44	5.68	7.35
6" x 12" - 12'	Demo	LF	LD	.062	---	1.82	---	1.82	2.71	1.56	2.32
6" x 12" - 12'	Inst	LF	CW	.102	4.41	3.00	.15	7.56	9.71	6.62	8.48
9" x 10" - 12'	Demo	LF	LD	.070	---	2.06	---	2.06	3.06	1.73	2.58
9" x 10" - 12'	Inst	LF	CW	.111	7.87	3.26	.16	11.29	14.10	10.05	12.50
9" x 12" - 12'	Demo	LF	LD	.076	---	2.23	---	2.23	3.33	1.91	2.84
9" x 12" - 12'	Inst	LF	CW	.119	9.42	3.50	.17	13.09	16.20	11.67	14.40
Single member											
2" x 6"	Demo	LF	LB	.020	---	.53	---	.53	.79	.45	.67
2" x 6"	Inst	LF	2C	.036	.69	1.16	.11	1.96	2.64	1.70	2.30
2" x 8"	Demo	LF	LB	.022	---	.59	---	.59	.87	.48	.71
2" x 8"	Inst	LF	2C	.038	.92	1.22	.11	2.25	3.00	1.98	2.63
2" x 10"	Demo	LF	LB	.023	---	.61	---	.61	.91	.53	.79
2" x 10"	Inst	LF	2C	.041	1.27	1.32	.12	2.71	3.55	2.32	3.04
2" x 12"	Demo	LF	LB	.025	---	.67	---	.67	.99	.56	.83
2" x 12"	Inst	LF	2C	.043	1.55	1.38	.12	3.05	3.97	2.65	3.43
3" x 6"	Demo	LF	LB	.023	---	.61	---	.61	.91	.51	.75
3" x 6"	Inst	LF	2C	.040	1.63	1.28	.12	3.03	3.92	2.66	3.43
3" x 8"	Demo	LF	LB	.025	---	.67	---	.67	.99	.56	.83
3" x 8"	Inst	LF	2C	.043	2.18	1.38	.12	3.68	4.70	3.24	4.11
3" x 10"	Demo	LF	LB	.028	---	.75	---	.75	1.11	.64	.95
3" x 10"	Inst	LF	2C	.046	2.70	1.48	.13	4.31	5.45	3.80	4.79

Description	Oper	Unit	Crew Size	Man-Hours Per Unit	Costs Based On Small Volume				Avg Price Incl O&P	Large Volume	
					Avg Mat'l Unit Cost	Avg Labor Unit Cost	Avg Equip Unit Cost	Avg Total Unit Cost		Avg Total Unit Cost	Avg Price Incl O&P
3" x 12"	Demo	LF	LB	.030	---	.80	---	.80	1.19	.69	1.03
3" x 12"	Inst	LF	2C	.049	3.22	1.57	.14	4.93	6.20	4.38	5.49
3" x 14"	Demo	LF	LB	.033	---	.88	---	.88	1.31	.75	1.11
3" x 14"	Inst	LF	2C	.052	3.89	1.67	.15	5.71	7.13	5.06	6.31
4" x 6"	Demo	LF	LB	.025	---	.67	---	.67	.99	.56	.83
4" x 6"	Inst	LF	2C	.043	1.99	1.38	.13	3.50	4.49	3.10	3.96
4" x 8"	Demo	LF	LB	.029	---	.77	---	.77	1.15	.67	.99
4" x 8"	Inst	LF	2C	.048	2.64	1.54	.14	4.32	5.49	3.81	4.82
4" x 10"	Demo	LF	LB	.032	---	.85	---	.85	1.27	.72	1.07
4" x 10"	Inst	LF	2C	.052	3.27	1.67	.15	5.09	6.41	4.46	5.61
4" x 12"	Demo	LF	LB	.036	---	.96	---	.96	1.43	.83	1.23
4" x 12"	Inst	LF	2C	.057	3.90	1.83	.16	5.89	7.39	5.17	6.46
4" x 14"	Demo	LF	LH	.044	---	1.17	---	1.17	1.75	.99	1.47
4" x 14"	Inst	LF	CS	.070	4.74	2.12	.14	7.00	8.77	6.17	7.70
4" x 16"	Demo	LF	LH	.048	---	1.28	---	1.28	1.91	1.07	1.59
4" x 16"	Inst	LF	CS	.075	5.63	2.27	.14	8.04	10.00	7.12	8.83
6" x 8"	Demo	LF	LJ	.048	---	1.28	---	1.28	1.91	1.09	1.63
6" x 8"	Inst	LF	CW	.081	5.09	2.38	.12	7.59	9.54	6.72	8.42
6" x 10"	Demo	LF	LJ	.054	---	1.44	---	1.44	2.14	1.23	1.83
6" x 10"	Inst	LF	CW	.088	6.44	2.58	.13	9.15	11.40	8.13	10.10
6" x 12"	Demo	LF	LJ	.060	---	1.60	---	1.60	2.38	1.36	2.02
6" x 12"	Inst	LF	CW	.095	8.04	2.79	.14	10.97	13.60	9.73	12.00
6" x 14"	Demo	LF	LJ	.066	---	1.76	---	1.76	2.62	1.49	2.22
6" x 14"	Inst	LF	CW	.103	9.36	3.03	.15	12.54	15.50	11.13	13.70
6" x 16"	Demo	LF	LJ	.072	---	1.92	---	1.92	2.86	1.63	2.42
6" x 16"	Inst	LF	CW	.110	11.10	3.23	.16	14.49	17.80	13.00	15.80
8" x 8"	Demo	LF	LJ	.056	---	1.49	---	1.49	2.22	1.28	1.91
8" x 8"	Inst	LF	CW	.091	7.33	2.67	.13	10.13	12.60	8.98	11.10
8" x 10"	Demo	LF	LJ	.068	---	1.81	---	1.81	2.70	1.55	2.30
8" x 10"	Inst	LF	CW	.104	9.36	3.05	.15	12.56	15.50	11.10	13.70
8" x 12"	Demo	LF	LJ	.073	---	1.94	---	1.94	2.90	1.65	2.46
8" x 12"	Inst	LF	CW	.110	11.60	3.23	.16	14.99	18.40	13.40	16.30
8" x 14"	Demo	LF	LJ	.080	---	2.13	---	2.13	3.18	1.81	2.70
8" x 14"	Inst	LF	CW	.119	14.00	3.50	.17	17.67	21.50	15.72	19.10
8" x 16"	Demo	LF	LJ	.089	---	2.37	---	2.37	3.53	2.02	3.02
8" x 16"	Inst	LF	CW	.129	16.30	3.79	.19	20.28	24.60	18.09	21.90

Blocking, horizontal, for studs

Description	Oper	Unit	Crew Size	Man-Hours Per Unit	Avg Mat'l Unit Cost	Avg Labor Unit Cost	Avg Equip Unit Cost	Avg Total Unit Cost	Avg Price Incl O&P	Avg Total Unit Cost	Avg Price Incl O&P
2" x 4" - 12"	Demo	LF	1L	.013	---	.35	---	.35	.52	.29	.44
2" x 4" - 12"	Inst	LF	CA	.042	.42	1.35	.25	2.02	2.75	1.75	2.38
2" x 4" - 16"	Demo	LF	1L	.012	---	.32	---	.32	.48	.29	.44
2" x 4" - 16"	Inst	LF	CA	.034	.42	1.09	.20	1.71	2.32	1.48	2.00
2" x 4" - 24"	Demo	LF	1L	.012	---	.32	---	.32	.48	.27	.40
2" x 4" - 24"	Inst	LF	CA	.027	.42	.87	.16	1.45	1.94	1.25	1.67
2" x 6" - 12"	Demo	LF	1L	.015	---	.40	---	.40	.60	.35	.52
2" x 6" - 12"	Inst	LF	CA	.045	.58	1.44	.27	2.29	3.10	2.01	2.72
2" x 6" - 16"	Demo	LF	1L	.015	---	.40	---	.40	.60	.32	.48
2" x 6" - 16"	Inst	LF	CA	.037	.58	1.19	.22	1.99	2.67	1.75	2.34

Description	Oper	Unit	Crew Size	Man-Hours Per Unit	Costs Based On Small Volume					Large Volume	
					Avg Mat'l Unit Cost	Avg Labor Unit Cost	Avg Equip Unit Cost	Avg Total Unit Cost	Avg Price Incl O&P	Avg Total Unit Cost	Avg Price Incl O&P
2" x 6" - 24"	Demo	LF	1L	.014	---	.37	---	.37	.56	.32	.48
2" x 6" - 24"	Inst	LF	CA	.029	.58	.93	.17	1.68	2.23	1.47	1.95
2" x 8" - 12"	Demo	LF	1L	.018	---	.48	---	.48	.71	.40	.60
2" x 8" - 12"	Inst	LF	CA	.049	.82	1.57	.28	2.67	3.58	2.29	3.05
2" x 8" - 16"	Demo	LF	1L	.017	---	.45	---	.45	.67	.37	.56
2" x 8" - 16"	Inst	LF	CA	.040	.82	1.28	.23	2.33	3.10	2.02	2.68
2" x 8" - 24"	Demo	LF	1L	.016	---	.43	---	.43	.64	.37	.56
2" x 8" - 24"	Inst	LF	CA	.031	.82	.99	.18	1.99	2.62	1.76	2.30

Bracing, diagonal let-ins

Studs, 12" oc

Description	Oper	Unit	Crew Size	Man-Hours Per Unit	Avg Mat'l Unit Cost	Avg Labor Unit Cost	Avg Equip Unit Cost	Avg Total Unit Cost	Avg Price Incl O&P	Avg Total Unit Cost	Avg Price Incl O&P
1" x 6"	Demo	Set	1L	.013	---	.35	---	.35	.52	.29	.44
1" x 6"	Inst	Set	CA	.066	.37	2.12	.38	2.87	3.98	2.46	3.41
Studs, 16" oc											
1" x 6"	Demo	Set	1L	.012	---	.32	---	.32	.48	.27	.40
1" x 6"	Inst	Set	CA	.051	.37	1.64	.30	2.31	3.18	1.96	2.70
Studs, 24" oc											
1" x 6"	Demo	Set	1L	.011	---	.29	---	.29	.44	.27	.40
1" x 6"	Inst	Set	CA	.039	.37	1.25	.23	1.85	2.53	1.58	2.16

Bridging, "X" type

For joists 12" oc

Description	Oper	Unit	Crew Size	Man-Hours Per Unit	Avg Mat'l Unit Cost	Avg Labor Unit Cost	Avg Equip Unit Cost	Avg Total Unit Cost	Avg Price Incl O&P	Avg Total Unit Cost	Avg Price Incl O&P
1" x 3"	Demo	Set	1L	.017	---	.45	---	.45	.67	.40	.60
1" x 3"	Inst	Set	CA	.067	.57	2.15	.39	3.11	4.27	2.68	3.67
2" x 2"	Demo	Set	1L	.018	---	.48	---	.48	.71	.40	.60
2" x 2"	Inst	Set	CA	.068	.86	2.18	.40	3.44	4.66	2.96	4.01
For joists 16" oc											
1" x 3"	Demo	Set	1L	.018	---	.48	---	.48	.71	.40	.60
1" x 3"	Inst	Set	CA	.069	.61	2.21	.40	3.22	4.42	2.75	3.76
2" x 2"	Demo	Set	1L	.018	---	.48	---	.48	.71	.43	.64
2" x 2"	Inst	Set	CA	.069	.91	2.21	.40	3.52	4.77	3.04	4.11
For joists 24" oc											
1" x 3"	Demo	Set	1L	.019	---	.51	---	.51	.75	.43	.64
1" x 3"	Inst	Set	CA	.019	.68	.61	.11	1.40	1.81	1.21	1.56
2" x 2"	Demo	Set	1L	.070	---	1.86	---	1.86	2.78	1.60	2.38
2" x 2"	Inst	Set	CA	.071	1.01	2.28	.41	3.70	4.99	3.18	4.27

Bridging, solid, between joists

Description	Oper	Unit	Crew Size	Man-Hours Per Unit	Avg Mat'l Unit Cost	Avg Labor Unit Cost	Avg Equip Unit Cost	Avg Total Unit Cost	Avg Price Incl O&P	Avg Total Unit Cost	Avg Price Incl O&P
2" x 6" - 12"	Demo	Set	1L	.016	---	.43	---	.43	.64	.37	.56
2" x 6" - 12"	Inst	Set	CA	.047	.58	1.51	.28	2.37	3.21	2.04	2.77
2" x 8" - 12"	Demo	Set	1L	.018	---	.48	---	.48	.71	.40	.60
2" x 8" - 12"	Inst	Set	CA	.051	.82	1.64	.30	2.76	3.70	2.36	3.16
2" x 10" - 12"	Demo	Set	1L	.021	---	.56	---	.56	.83	.48	.71
2" x 10" - 12"	Inst	Set	CA	.054	1.17	1.73	.32	3.22	4.27	2.79	3.68
2" x 12" - 12"	Demo	Set	1L	.023	---	.61	---	.61	.91	.51	.75
2" x 12" - 12"	Inst	Set	CA	.058	1.44	1.86	.34	3.64	4.79	3.14	4.12
2" x 6" - 16"	Demo	Set	1L	.015	---	.40	---	.40	.60	.35	.52
2" x 6" - 16"	Inst	Set	CA	.039	.58	1.25	.23	2.06	2.77	1.78	2.39
2" x 8" - 16"	Demo	Set	1L	.017	---	.45	---	.45	.67	.40	.60
2" x 8" - 16"	Inst	Set	CA	.042	.82	1.35	.24	2.41	3.20	2.06	2.73

Description	Oper	Unit	Crew Size	Man-Hours Per Unit	Costs Based On Small Volume				Large Volume	
					Avg Mat'l Unit Cost	Avg Labor Unit Cost	Avg Equip Unit Cost	Avg Total Unit Cost	Avg Price Incl O&P	Avg Total Unit Cost	Avg Price Incl O&P
2" x 10" - 16"	Demo	Set	1L	.020	---	.53	---	.53	.79	.45	.67
2" x 10" - 16"	Inst	Set	CA	.045	1.17	1.44	.26	2.87	3.77	2.48	3.25
2" x 12" - 16"	Demo	Set	1L	.022	---	.59	---	.59	.87	.48	.71
2" x 12" - 16"	Inst	Set	CA	.048	1.44	1.54	.28	3.26	4.25	2.81	3.65
2" x 6" - 24"	Demo	Set	1L	.014	---	.37	---	.37	.56	.32	.48
2" x 6" - 24"	Inst	Set	CA	.030	.58	.96	.18	1.72	2.29	1.51	2.01
2" x 8" - 24"	Demo	Set	1L	.016	---	.43	---	.43	.64	.37	.56
2" x 8" - 24"	Inst	Set	CA	.033	.82	1.06	.19	2.07	2.72	1.79	2.35
2" x 10" - 24"	Demo	Set	1L	.019	---	.51	---	.51	.75	.43	.64
2" x 10" - 24"	Inst	Set	CA	.035	1.17	1.12	.20	2.49	3.23	2.17	2.81
2" x 12" - 24"	Demo	Set	1L	.021	---	.56	---	.56	.83	.48	.71
2" x 12" - 24"	Inst	Set	CA	.037	1.44	1.19	.22	2.85	3.66	2.51	3.21

Columns or posts, without base or cap, hardware, or chamfer corners

Description	Oper	Unit	Crew Size	Man-Hours Per Unit	Avg Mat'l Unit Cost	Avg Labor Unit Cost	Avg Equip Unit Cost	Avg Total Unit Cost	Avg Price Incl O&P	Avg Total Unit Cost	Avg Price Incl O&P
4" x 4" - 8'	Demo	LF	LB	.032	---	.85	---	.85	1.27	.72	1.07
4" x 4" - 8'	Inst	LF	2C	.060	1.26	1.93	.17	3.36	4.51	2.93	3.92
4" x 6" - 8'	Demo	LF	LB	.036	---	.96	---	.96	1.43	.80	1.19
4" x 6" - 8'	Inst	LF	2C	.064	1.90	2.05	.19	4.14	5.46	3.62	4.75
4" x 8" - 8'	Demo	LF	LB	.039	---	1.04	---	1.04	1.55	.88	1.31
4" x 8" - 8'	Inst	LF	2C	.069	2.56	2.21	.20	4.97	6.47	4.32	5.60
6" x 6" - 8'	Demo	LF	LB	.044	---	1.17	---	1.17	1.75	.99	1.47
6" x 6" - 8'	Inst	LF	2C	.076	3.76	2.44	.22	6.42	8.20	5.68	7.23
6" x 8" - 8'	Demo	LF	LB	.049	---	1.31	---	1.31	1.94	1.12	1.67
6" x 8" - 8'	Inst	LF	2C	.082	5.01	2.63	.24	7.88	9.95	6.97	8.77
6" x 10" - 8'	Demo	LF	LB	.055	---	1.47	---	1.47	2.18	1.23	1.83
6" x 10" - 8'	Inst	LF	2C	.089	6.36	2.86	.26	9.48	11.90	8.38	10.40
8" x 8" - 8'	Demo	LF	LB	.059	---	1.57	---	1.57	2.34	1.33	1.98
8" x 8" - 8'	Inst	LF	2C	.096	7.25	3.08	.28	10.61	13.20	9.41	11.70
8" x 10" - 8'	Demo	LF	LB	.070	---	1.86	---	1.86	2.78	1.57	2.34
8" x 10" - 8'	Inst	LF	2C	.108	9.28	3.47	.32	13.07	16.20	11.54	14.30

Fascia

Description	Oper	Unit	Crew Size	Man-Hours Per Unit	Avg Mat'l Unit Cost	Avg Labor Unit Cost	Avg Equip Unit Cost	Avg Total Unit Cost	Avg Price Incl O&P	Avg Total Unit Cost	Avg Price Incl O&P
1" x 4" - 12'	Demo	LF	LB	.014	---	.37	---	.37	.56	.32	.48
1" x 4" - 12'	Inst	LF	2C	.049	.24	1.57	.14	1.95	2.77	1.66	2.35

Furring strips on ceilings

Description	Oper	Unit	Crew Size	Man-Hours Per Unit	Avg Mat'l Unit Cost	Avg Labor Unit Cost	Avg Equip Unit Cost	Avg Total Unit Cost	Avg Price Incl O&P	Avg Total Unit Cost	Avg Price Incl O&P
1" x 3" on wood	Demo	LF	LB	.030	---	.80	---	.80	1.19	.69	1.03
1" x 3" on wood	Inst	LF	2C	.033	.18	1.06	.10	1.34	1.90	1.14	1.61

Furring strips on walls

Description	Oper	Unit	Crew Size	Man-Hours Per Unit	Avg Mat'l Unit Cost	Avg Labor Unit Cost	Avg Equip Unit Cost	Avg Total Unit Cost	Avg Price Incl O&P	Avg Total Unit Cost	Avg Price Incl O&P
1" x 3" on wood	Demo	LF	LB	.023	---	.61	---	.61	.91	.53	.79
1" x 3" on wood	Inst	LF	2C	.025	.18	.80	.07	1.05	1.48	.93	1.30
1" x 3" on masonry	Demo	LF	LB	.030	---	.80	---	.80	1.19	.69	1.03
1" x 3" on masonry	Inst	LF	2C	.033	.18	1.06	.10	1.34	1.90	1.14	1.61
1" x 3" on concrete	Demo	LF	LB	.041	---	1.09	---	1.09	1.63	.93	1.39
1" x 3" on concrete	Inst	LF	2C	.045	.18	1.44	.13	1.75	2.50	1.49	2.12

Description	Oper	Unit	Crew Size	Man-Hours Per Unit	Avg Mat'l Unit Cost	Avg Labor Unit Cost	Avg Equip Unit Cost	Avg Total Unit Cost	Avg Price Incl O&P	Avg Total Unit Cost	Avg Price Incl O&P
Headers or lintels, over openings											
4 feet wide											
4" x 6"	Demo	LF	1L	.028	---	.75	---	.75	1.11	.64	.95
4" x 6"	Inst	LF	CA	.048	1.89	1.54	.28	3.71	4.76	3.23	4.13
4" x 8"	Demo	LF	1L	.031	---	.83	---	.83	1.23	.69	1.03
4" x 8"	Inst	LF	CA	.051	2.55	1.64	.30	4.49	5.69	3.91	4.94
4" x 12"	Demo	LF	1L	.038	---	1.01	---	1.01	1.51	.85	1.27
4" x 12"	Inst	LF	CA	.058	3.81	1.86	.34	6.01	7.51	5.26	6.56
4" x 14"	Demo	LF	1L	.040	---	1.07	---	1.07	1.59	.91	1.35
4" x 14"	Inst	LF	CA	.060	4.65	1.93	.35	6.93	8.59	6.12	7.56
6 feet wide											
4" x 12"	Demo	LF	LB	.032	---	.85	---	.85	1.27	.72	1.07
4" x 12"	Inst	LF	2C	.048	3.81	1.54	.14	5.49	6.83	4.85	6.02
8 feet wide											
4" x 12"	Demo	LF	LB	.029	---	.77	---	.77	1.15	.64	.95
4" x 12"	Inst	LF	2C	.042	3.81	1.35	.12	5.28	6.52	4.63	5.71
10 feet wide											
4" x 12"	Demo	LF	LB	.025	---	.67	---	.67	.99	.59	.87
4" x 12"	Inst	LF	2C	.035	3.81	1.12	.10	5.03	6.17	4.46	5.46
4" x 14"	Demo	LF	LB	.030	---	.80	---	.80	1.19	.69	1.03
4" x 14"	Inst	LF	2C	.041	4.65	1.32	.12	6.09	7.44	5.40	6.59
12 feet wide											
4" x 14"	Demo	LF	LB	.028	---	.75	---	.75	1.11	.64	.95
4" x 14"	Inst	LF	2C	.038	4.65	1.22	.11	5.98	7.29	5.34	6.50
4" x 16"	Demo	LF	LB	.033	---	.88	---	.88	1.31	.75	1.11
4" x 16"	Inst	LF	2C	.044	5.53	1.41	.13	7.07	8.61	6.30	7.64
14 feet wide											
4" x 16"	Demo	LF	LB	.032	---	.85	---	.85	1.27	.72	1.07
4" x 16"	Inst	LF	2C	.042	5.53	1.35	.12	7.00	8.50	6.22	7.53
16 feet wide											
4" x 16"	Demo	LF	LB	.031	---	.83	---	.83	1.23	.69	1.03
4" x 16"	Inst	LF	2C	.040	5.53	1.28	.12	6.93	8.41	6.19	7.49
18 feet wide											
4" x 16"	Demo	LF	LB	.030	---	.80	---	.80	1.19	.67	.99
4" x 16"	Inst	LF	2C	.038	5.53	1.22	.11	6.86	8.30	6.12	7.38
Joists, ceiling/floor, per LF of stick											
2" x 4" -6'	Demo	LF	LB	.018	---	.48	---	.48	.71	.43	.64
2" x 4" -6'	Inst	LF	2C	.022	.42	.71	.07	1.20	1.61	1.05	1.41
2" x 4" -8'	Demo	LF	LB	.016	---	.43	---	.43	.64	.35	.52
2" x 4" -8'	Inst	LF	2C	.019	.42	.61	.05	1.08	1.45	.94	1.26
2" x 4" -10'	Demo	LF	LB	.014	---	.37	---	.37	.56	.32	.48
2" x 4" -10'	Inst	LF	2C	.017	.42	.55	.05	1.02	1.35	.87	1.15
2" x 4" -12'	Demo	LF	LB	.013	---	.35	---	.35	.52	.29	.44
2" x 4" -12'	Inst	LF	2C	.015	.42	.48	.04	.94	1.25	.84	1.10
2" x 6" - 8'	Demo	LF	LB	.018	---	.48	---	.48	.71	.40	.60
2" x 6" - 8'	Inst	LF	2C	.021	.58	.67	.06	1.31	1.74	1.16	1.53
2" x 6" - 10'	Demo	LF	LB	.016	---	.43	---	.43	.64	.35	.52
2" x 6" - 10'	Inst	LF	2C	.019	.58	.61	.05	1.24	1.63	1.09	1.43

Description	Oper Unit	Crew Size	Man-Hours Per Unit	Avg Mat'l Unit Cost	Avg Labor Unit Cost	Avg Equip Unit Cost	Avg Total Unit Cost	Avg Price Incl O&P	Avg Total Unit Cost	Avg Price Incl O&P
						Costs Based On Small Volume			Large Volume	
2" x 6" - 12'	Demo LF	LB	.014	---	.37	---	.37	.56	.32	.48
2" x 6" - 12'	Inst LF	2C	.017	.58	.55	.05	1.18	1.54	1.05	1.37
2" x 6" - 14'	Demo LF	LB	.013	---	.35	---	.35	.52	.29	.44
2" x 6" - 14'	Inst LF	2C	.016	.58	.51	.05	1.14	1.49	.99	1.28
2" x 8" - 10'	Demo LF	LB	.018	---	.48	---	.48	.71	.40	.60
2" x 8" - 10'	Inst LF	2C	.021	.82	.67	.06	1.55	2.01	1.36	1.76
2" x 8" - 12'	Demo LF	LB	.016	---	.43	---	.43	.64	.37	.56
2" x 8" - 12'	Inst LF	2C	.019	.82	.61	.06	1.49	1.92	1.29	1.66
2" x 8" - 14'	Demo LF	LB	.015	---	.40	---	.40	.60	.35	.52
2" x 8" - 14'	Inst LF	2C	.018	.82	.58	.05	1.45	1.86	1.25	1.60
2" x 8" - 16'	Demo LF	LB	.014	---	.37	---	.37	.56	.32	.48
2" x 8" - 16'	Inst LF	2C	.017	.82	.55	.05	1.42	1.81	1.22	1.55
2" x 10" - 12'	Demo LF	LB	.018	---	.48	---	.48	.71	.40	.60
2" x 10" - 12'	Inst LF	2C	.021	1.17	.67	.06	1.90	2.42	1.67	2.11
2" x 10" - 14'	Demo LF	LB	.017	---	.45	---	.45	.67	.37	.56
2" x 10" - 14'	Inst LF	2C	.020	1.17	.64	.06	1.87	2.37	1.64	2.06
2" x 10" - 16'	Demo LF	LB	.016	---	.43	---	.43	.64	.37	.56
2" x 10" - 16'	Inst LF	2C	.019	1.17	.61	.05	1.83	2.31	1.60	2.02
2" x 10" - 18'	Demo LF	LB	.015	---	.40	---	.40	.60	.35	.52
2" x 10" - 18'	Inst LF	2C	.018	1.17	.58	.05	1.80	2.26	1.56	1.96
2" x 12" - 14'	Demo LF	LB	.019	---	.51	---	.51	.75	.43	.64
2" x 12" - 14'	Inst LF	2C	.022	1.44	.71	.06	2.21	2.78	1.92	2.40
2" x 12" - 16'	Demo LF	LB	.018	---	.48	---	.48	.71	.40	.60
2" x 12" - 16'	Inst LF	2C	.020	1.44	.64	.06	2.14	2.68	1.89	2.35
2" x 12" - 18'	Demo LF	LB	.017	---	.45	---	.45	.67	.37	.56
2" x 12" - 18'	Inst LF	2C	.019	1.44	.61	.06	2.11	2.63	1.89	2.35
2" x 12" - 20'	Demo LF	LB	.016	---	.43	---	.43	.64	.37	.56
2" x 12" - 20'	Inst LF	2C	.019	1.44	.61	.05	2.10	2.62	1.85	2.30
3" x 8" - 12'	Demo LF	LB	.020	---	.53	---	.53	.79	.45	.67
3" x 8" - 12'	Inst LF	2C	.023	2.08	.74	.07	2.89	3.57	2.58	3.18
3" x 8" - 14'	Demo LF	LB	.019	---	.51	---	.51	.75	.43	.64
3" x 8" - 14'	Inst LF	2C	.021	2.08	.67	.06	2.81	3.46	2.51	3.08
3" x 8" - 16'	Demo LF	LB	.018	---	.48	---	.48	.71	.40	.60
3" x 8" - 16'	Inst LF	2C	.020	2.08	.64	.06	2.78	3.41	2.48	3.03
3" x 8" - 18'	Demo LF	LB	.017	---	.45	---	.45	.67	.37	.56
3" x 8" - 18'	Inst LF	2C	.019	2.08	.61	.06	2.75	3.37	2.44	2.98
3" x 10" - 16'	Demo LF	LB	.021	---	.56	---	.56	.83	.45	.67
3" x 10" - 16'	Inst LF	2C	.023	2.60	.74	.07	3.41	4.17	3.05	3.73
3" x 10" - 18'	Demo LF	LB	.020	---	.53	---	.53	.79	.45	.67
3" x 10" - 18'	Inst LF	2C	.022	2.60	.71	.06	3.37	4.11	3.01	3.67
3" x 10" - 20'	Demo LF	LB	.019	---	.51	---	.51	.75	.43	.64
3" x 10" - 20'	Inst LF	2C	.021	2.60	.67	.06	3.33	4.06	2.98	3.62
3" x 12" - 16'	Demo LF	LB	.023	---	.61	---	.61	.91	.51	.75
3" x 12" - 16'	Inst LF	2C	.026	2.60	.83	.08	3.51	4.32	3.12	3.82
3" x 12" - 18'	Demo LF	LB	.022	---	.59	---	.59	.87	.48	.71
3" x 12" - 18'	Inst LF	2C	.024	3.11	.77	.07	3.95	4.80	3.52	4.27

				Costs Based On Small Volume					Large Volume	
Description	Oper	Unit	Crew Size	Man-Hours Per Unit	Avg Mat'l Unit Cost	Avg Labor Unit Cost	Avg Equip Unit Cost	Avg Total Unit Cost	Avg Price Incl O&P	Avg Total Unit Cost	Avg Price Incl O&P
3" x 12" - 20'	Demo	LF	LB	.021	---	.56	---	.56	.83	.48	.71
3" x 12" - 20'	Inst	LF	2C	.023	3.11	.74	.07	3.92	4.75	3.52	4.27
3" x 12" - 22'	Demo	LF	LB	.020	---	.53	---	.53	.79	.45	.67
3" x 12" - 22'	Inst	LF	2C	.023	3.11	.74	.07	3.92	4.75	3.49	4.22

Joists, ceiling/floor, per SF of area

Description	Oper	Unit	Crew Size	Man-Hours Per Unit	Avg Mat'l Unit Cost	Avg Labor Unit Cost	Avg Equip Unit Cost	Avg Total Unit Cost	Avg Price Incl O&P	Avg Total Unit Cost	Avg Price Incl O&P
2" x 4" - 6', 12" oc	Demo	SF	LB	.019	---	.51	---	.51	.75	.43	.64
2" x 4" - 6', 12" oc	Inst	SF	2C	.023	.43	.74	.07	1.24	1.67	1.06	1.42
2" x 4" - 6', 16" oc	Demo	SF	LB	.016	---	.43	---	.43	.64	.35	.52
2" x 4" - 6', 16" oc	Inst	SF	2C	.019	.33	.61	.06	1.00	1.35	.86	1.17
2" x 4" - 6', 24" oc	Demo	SF	LB	.012	---	.32	---	.32	.48	.27	.40
2" x 4" - 6', 24" oc	Inst	SF	2C	.015	.22	.48	.04	.74	1.02	.63	.85
2" x 6" - 8', 12" oc	Demo	SF	LB	.018	---	.48	---	.48	.71	.40	.60
2" x 6" - 8', 12" oc	Inst	SF	2C	.022	.60	.71	.06	1.37	1.81	1.17	1.54
2" x 6" - 8', 16" oc	Demo	SF	LB	.015	---	.40	---	.40	.60	.35	.52
2" x 6" - 8', 16" oc	Inst	SF	2C	.018	.46	.58	.05	1.09	1.45	.94	1.24
2" x 6" - 8', 24" oc	Demo	SF	LB	.012	---	.32	---	.32	.48	.27	.40
2" x 6" - 8', 24" oc	Inst	SF	2C	.014	.31	.45	.04	.80	1.07	.70	.93
2" x 8" - 10', 12" oc	Demo	SF	LB	.018	---	.48	---	.48	.71	.40	.60
2" x 8" - 10', 12" oc	Inst	SF	2C	.021	.84	.67	.06	1.57	2.04	1.38	1.78
2" x 8" - 10', 16" oc	Demo	SF	LB	.015	---	.40	---	.40	.60	.35	.52
2" x 8" - 10', 16" oc	Inst	SF	2C	.018	.63	.58	.05	1.26	1.64	1.09	1.42
2" x 8" - 10', 24" oc	Demo	SF	LB	.012	---	.32	---	.32	.48	.27	.40
2" x 8" - 10', 24" oc	Inst	SF	2C	.014	.44	.45	.04	.93	1.22	.81	1.06
2" x 10" - 12', 12" oc	Demo	SF	LB	.019	---	.51	---	.51	.75	.43	.64
2" x 10" - 12', 12" oc	Inst	SF	2C	.022	1.20	.71	.06	1.97	2.50	1.69	2.14
2" x 10" - 12', 16" oc	Demo	SF	LB	.016	---	.43	---	.43	.64	.35	.52
2" x 10" - 12', 16" oc	Inst	SF	2C	.018	.91	.58	.05	1.54	1.96	1.34	1.70
2" x 10" - 12', 24" oc	Demo	SF	LB	.012	---	.32	---	.32	.48	.27	.40
2" x 10" - 12', 24" oc	Inst	SF	2C	.014	.62	.45	.04	1.11	1.43	.97	1.24
2" x 12" - 14', 12" oc	Demo	SF	LB	.019	---	.51	---	.51	.75	.43	.64
2" x 12" - 14', 12" oc	Inst	SF	2C	.022	1.48	.71	.06	2.25	2.82	1.99	2.49
2" x 12" - 14', 16" oc	Demo	SF	LB	.016	---	.43	---	.43	.64	.37	.56
2" x 12" - 14', 16" oc	Inst	SF	2C	.019	1.12	.61	.05	1.78	2.25	1.56	1.97
2" x 12" - 14', 24" oc	Demo	SF	LB	.012	---	.32	---	.32	.48	.29	.44
2" x 12" - 14', 24" oc	Inst	SF	2C	.014	.76	.45	.04	1.25	1.59	1.11	1.40
3" x 8" - 14', 12" oc	Demo	SF	LB	.019	---	.51	---	.51	.75	.43	.64
3" x 8" - 14', 12" oc	Inst	SF	2C	.022	2.13	.71	.06	2.90	3.57	2.59	3.18
3" x 8" - 14', 16" oc	Demo	SF	LB	.016	---	.43	---	.43	.64	.37	.56
3" x 8" - 14', 16" oc	Inst	SF	2C	.019	1.61	.61	.05	2.27	2.82	2.02	2.50
3" x 8" - 14', 24" oc	Demo	SF	LB	.012	---	.32	---	.32	.48	.27	.40
3" x 8" - 14', 24" oc	Inst	SF	2C	.014	1.10	.45	.04	1.59	1.98	1.41	1.75
3" x 10" - 16', 12" oc	Demo	SF	LB	.021	---	.56	---	.56	.83	.48	.71
3" x 10" - 16', 12" oc	Inst	SF	2C	.024	2.66	.77	.07	3.50	4.28	3.10	3.78

Description	Oper	Unit	Crew Size	Man-Hours Per Unit	Costs Based On Small Volume					Large Volume	
					Avg Mat'l Unit Cost	Avg Labor Unit Cost	Avg Equip Unit Cost	Avg Total Unit Cost	Avg Price Incl O&P	Avg Total Unit Cost	Avg Price Incl O&P
3" x 10" - 16', 16" oc	Demo	SF	LB	.018	---	.48	---	.48	.71	.40	.60
3" x 10" - 16', 16" oc	Inst	SF	2C	.020	2.02	.64	.06	2.72	3.35	2.42	2.96
3" x 10" - 16', 24" oc	Demo	SF	LB	.013	---	.35	---	.35	.52	.29	.44
3" x 10" - 16', 24" oc	Inst	SF	2C	.015	1.37	.48	.04	1.89	2.34	1.70	2.09
3" x 12" - 18', 12" oc	Demo	SF	LB	.022	---	.59	---	.59	.87	.51	.75
3" x 12" - 18', 12" oc	Inst	SF	2C	.025	3.20	.80	.07	4.07	4.95	3.62	4.39
3" x 12" - 18', 16" oc	Demo	SF	LB	.019	---	.51	---	.51	.75	.43	.64
3" x 12" - 18', 16" oc	Inst	SF	2C	.021	2.42	.67	.06	3.15	3.85	2.82	3.44
3" x 12" - 18', 24" oc	Demo	SF	LB	.014	---	.37	---	.37	.56	.32	.48
3" x 12" - 18', 24" oc	Inst	SF	2C	.016	1.65	.51	.05	2.21	2.72	1.95	2.38

Ledgers

Nailed

Description	Oper	Unit	Crew Size	Man-Hours Per Unit	Avg Mat'l Unit Cost	Avg Labor Unit Cost	Avg Equip Unit Cost	Avg Total Unit Cost	Avg Price Incl O&P	Avg Total Unit Cost	Avg Price Incl O&P
2" x 4" - 12'	Demo	LF	LB	.015	---	.40	---	.40	.60	.35	.52
2" x 4" - 12'	Inst	LF	2C	.022	.47	.71	.06	1.24	1.66	1.09	1.46
2" x 6" - 12'	Demo	LF	LB	.017	---	.45	---	.45	.67	.40	.60
2" x 6" - 12'	Inst	LF	2C	.024	.64	.77	.07	1.48	1.96	1.31	1.74
2" x 8" - 12'	Demo	LF	LB	.019	---	.51	---	.51	.75	.43	.64
2" x 8" - 12'	Inst	LF	2C	.027	.87	.87	.08	1.82	2.38	1.59	2.07

Bolted

Labor/material costs are for ledgers with pre-embedded bolts
Labor costs include securing ledgers

Description	Oper	Unit	Crew Size	Man-Hours Per Unit	Avg Mat'l Unit Cost	Avg Labor Unit Cost	Avg Equip Unit Cost	Avg Total Unit Cost	Avg Price Incl O&P	Avg Total Unit Cost	Avg Price Incl O&P
3" x 6" - 12'	Demo	LF	LB	.026	---	.69	---	.69	1.03	.59	.87
3" x 6" - 12'	Inst	LF	2C	.037	1.52	1.19	.11	2.82	3.64	2.45	3.16
3" x 8" - 12'	Demo	LF	LB	.029	---	.77	---	.77	1.15	.64	.95
3" x 8" - 12'	Inst	LF	2C	.040	2.07	1.28	.12	3.47	4.43	3.06	3.89
3" x 10" - 12'	Demo	LF	LB	.032	---	.85	---	.85	1.27	.72	1.07
3" x 10" - 12'	Inst	LF	2C	.043	2.59	1.38	.13	4.10	5.18	3.64	4.58
3" x 12" - 12'	Demo	LF	LB	.034	---	.91	---	.91	1.35	.77	1.15
3" x 12" - 12'	Inst	LF	2C	.046	3.10	1.48	.13	4.71	5.91	4.17	5.22

Patio framing

Wood deck: 4" x 4" rough sawn beams (4'-0" oc) leveled 1/16" to 1/8" and nailed
to pre-set concrete piers with woodblock on top; 2" T x 4" W decking (S4S)
with 1/2" spacing, double-nailed beam junctures

Description	Oper	Unit	Crew Size	Man-Hours Per Unit	Avg Mat'l Unit Cost	Avg Labor Unit Cost	Avg Equip Unit Cost	Avg Total Unit Cost	Avg Price Incl O&P	Avg Total Unit Cost	Avg Price Incl O&P
Fir	Demo	SF	LB	.050	---	1.33	---	1.33	1.98	1.12	1.67
Fir	Inst	SF	CN	.071	4.36	2.20	.17	6.73	8.49	5.97	7.51
Redwood	Demo	SF	LB	.050	---	1.33	---	1.33	1.98	1.12	1.67
Redwood	Inst	SF	CN	.071	4.67	2.20	.17	7.04	8.84	6.25	7.83

Wood awning: 4" x 4" columns (10'-0" oc) nailed to wood; 2" x 6" beams nailed horizontally
either side of columns; 2" x 6" ledger nailed to wall studs; 2" x 6" joists (4'-0" oc) nailed to
to ledger and toe-nailed on top of beams; 2" x 2" (4" oc) nailed to joists for sunscreen

Description	Oper	Unit	Crew Size	Man-Hours Per Unit	Avg Mat'l Unit Cost	Avg Labor Unit Cost	Avg Equip Unit Cost	Avg Total Unit Cost	Avg Price Incl O&P	Avg Total Unit Cost	Avg Price Incl O&P
Redwood, rough sawn	Demo	SF	LB	.065	---	1.73	---	1.73	2.58	1.47	2.18
Redwood, rough sawn	Inst	SF	CS	.113	1.35	3.42	.22	4.99	6.90	4.31	5.94

				Costs Based On Small Volume						Large Volume	
Description	Oper	Unit	Crew Size	Man-Hours Per Unit	Avg Mat'l Unit Cost	Avg Labor Unit Cost	Avg Equip Unit Cost	Avg Total Unit Cost	Avg Price Incl O&P	Avg Total Unit Cost	Avg Price Incl O&P

Plates; joined with studs before setting

Double top, nailed

2" x 4" - 8'	Demo	LF	LB	.023	---	.61	---	.61	.91	.53	.79
2" x 4" - 8'	Inst	LF	2C	.046	.77	1.48	.13	2.38	3.23	2.06	2.79
2" x 6" - 8'	Demo	LF	LB	.023	---	.61	---	.61	.91	.53	.79
2" x 6" - 8'	Inst	LF	2C	.046	1.20	1.48	.13	2.81	3.72	2.44	3.23

Single bottom, nailed

2" x 4" - 8'	Demo	LF	LB	.014	---	.37	---	.37	.56	.32	.48
2" x 4" - 8'	Inst	LF	2C	.023	.43	.74	.07	1.24	1.67	1.09	1.47
2" x 6" - 8'	Demo	LF	LB	.014	---	.37	---	.37	.56	.32	.48
2" x 6" - 8'	Inst	LF	2C	.023	.59	.74	.07	1.40	1.86	1.24	1.64

Sill or bottom, bolted. Labor/material costs are for plates with pre-embedded bolts, labor cost includes securing plates

2" x 4" - 8'	Demo	LF	LB	.027	---	.72	---	.72	1.07	.61	.91
2" x 4" - 8'	Inst	LF	2C	.034	.39	1.09	.10	1.58	2.19	1.36	1.88
2" x 6" - 8'	Demo	LF	LB	.027	---	.72	---	.72	1.07	.61	.91
2" x 6" - 8'	Inst	LF	2C	.034	.55	1.09	.10	1.74	2.37	1.50	2.04

Rafters, per LF of stick

Common, gable or hip, to 1/3 pitch

2" x 4" - 8' Avg.	Demo	LF	LB	.025	---	.67	---	.67	.99	.59	.87
2" x 4" - 8' Avg.	Inst	LF	2C	.028	.42	.90	.08	1.40	1.91	1.19	1.61
2" x 4" - 10' Avg.	Demo	LF	LB	.022	---	.59	---	.59	.87	.51	.75
2" x 4" - 10' Avg.	Inst	LF	2C	.024	.42	.77	.07	1.26	1.71	1.08	1.46
2" x 4" - 12' Avg.	Demo	LF	LB	.020	---	.53	---	.53	.79	.45	.67
2" x 4" - 12' Avg.	Inst	LF	2C	.021	.42	.67	.06	1.15	1.55	1.01	1.35
2" x 4" - 14' Avg.	Demo	LF	LB	.018	---	.48	---	.48	.71	.40	.60
2" x 4" - 14' Avg.	Inst	LF	2C	.019	.42	.61	.06	1.09	1.46	.94	1.26
2" x 4" - 16' Avg.	Demo	LF	LB	.017	---	.45	---	.45	.67	.37	.56
2" x 4" - 16' Avg.	Inst	LF	2C	.018	.42	.58	.05	1.05	1.40	.90	1.20
2" x 6" - 10' Avg.	Demo	LF	LB	.025	---	.67	---	.67	.99	.56	.83
2" x 6" - 10' Avg.	Inst	LF	2C	.027	.58	.87	.08	1.53	2.05	1.34	1.79
2" x 6" - 12' Avg.	Demo	LF	LB	.022	---	.59	---	.59	.87	.51	.75
2" x 6" - 12' Avg.	Inst	LF	2C	.024	.58	.77	.07	1.42	1.89	1.26	1.68
2" x 6" - 14' Avg.	Demo	LF	LB	.020	---	.53	---	.53	.79	.45	.67
2" x 6" - 14' Avg.	Inst	LF	2C	.022	.58	.71	.06	1.35	1.79	1.19	1.57
2" x 6" - 16' Avg.	Demo	LF	LB	.019	---	.51	---	.51	.75	.43	.64
2" x 6" - 16' Avg.	Inst	LF	2C	.020	.58	.64	.06	1.28	1.69	1.13	1.48
2" x 6" - 18' Avg.	Demo	LF	LB	.017	---	.45	---	.45	.67	.40	.60
2" x 6" - 18' Avg.	Inst	LF	2C	.019	.58	.61	.05	1.24	1.63	1.09	1.43
2" x 8" - 12' Avg.	Demo	LF	LB	.026	---	.69	---	.69	1.03	.59	.87
2" x 8" - 12' Avg.	Inst	LF	2C	.028	.82	.90	.08	1.80	2.37	1.54	2.02
2" x 8" - 14' Avg.	Demo	LF	LB	.023	---	.61	---	.61	.91	.53	.79
2" x 8" - 14' Avg.	Inst	LF	2C	.025	.82	.80	.07	1.69	2.22	1.46	1.91
2" x 8" - 16' Avg.	Demo	LF	LB	.022	---	.59	---	.59	.87	.51	.75
2" x 8" - 16' Avg.	Inst	LF	2C	.023	.82	.74	.07	1.63	2.12	1.43	1.86

Description	Oper	Unit	Crew Size	Man-Hours Per Unit	Avg Mat'l Unit Cost	Avg Labor Unit Cost	Avg Equip Unit Cost	Avg Total Unit Cost	Avg Price Incl O&P	Avg Total Unit Cost	Avg Price Incl O&P
2" x 8" - 18' Avg.	Demo	LF	LB	.021	---	.56	---	.56	.83	.45	.67
2" x 8" - 18' Avg.	Inst	LF	2C	.022	.82	.71	.06	1.59	2.06	1.39	1.80
2" x 8" - 20' Avg.	Demo	LF	LB	.019	---	.51	---	.51	.75	.45	.67
2" x 8" - 20' Avg.	Inst	LF	2C	.021	.82	.67	.06	1.55	2.01	1.36	1.76

Common, gable or hip, to 3/8-1/2 pitch

Description	Oper	Unit	Crew Size	Man-Hours Per Unit	Avg Mat'l Unit Cost	Avg Labor Unit Cost	Avg Equip Unit Cost	Avg Total Unit Cost	Avg Price Incl O&P	Avg Total Unit Cost	Avg Price Incl O&P
2" x 4" - 8' Avg.	Demo	LF	LB	.030	---	.80	---	.80	1.19	.67	.99
2" x 4" - 8' Avg.	Inst	LF	2C	.032	.42	1.03	.09	1.54	2.11	1.36	1.86
2" x 4" - 10' Avg.	Demo	LF	LB	.025	---	.67	---	.67	.99	.59	.87
2" x 4" - 10' Avg.	Inst	LF	2C	.028	.42	.90	.08	1.40	1.91	1.22	1.66
2" x 4" - 12' Avg.	Demo	LF	LB	.023	---	.61	---	.61	.91	.51	.75
2" x 4" - 12' Avg.	Inst	LF	2C	.024	.42	.77	.07	1.26	1.71	1.11	1.51
2" x 4" - 14' Avg.	Demo	LF	LB	.020	---	.53	---	.53	.79	.45	.67
2" x 4" - 14' Avg.	Inst	LF	2C	.022	.42	.71	.06	1.19	1.60	1.04	1.40
2" x 4" - 16' Avg.	Demo	LF	LB	.019	---	.51	---	.51	.75	.43	.64
2" x 4" - 16' Avg.	Inst	LF	2C	.020	.42	.64	.06	1.12	1.51	.98	1.31
2" x 6" - 10' Avg.	Demo	LF	LB	.029	---	.77	---	.77	1.15	.64	.95
2" x 6" - 10' Avg.	Inst	LF	2C	.031	.58	.99	.09	1.66	2.25	1.44	1.94
2" x 6" - 12' Avg.	Demo	LF	LB	.025	---	.67	---	.67	.99	.59	.87
2" x 6" - 12' Avg.	Inst	LF	2C	.027	.58	.87	.08	1.53	2.05	1.34	1.79
2" x 6" - 14' Avg.	Demo	LF	LB	.023	---	.61	---	.61	.91	.51	.75
2" x 6" - 14' Avg.	Inst	LF	2C	.025	.58	.80	.07	1.45	1.94	1.26	1.68
2" x 6" - 16' Avg.	Demo	LF	LB	.021	---	.56	---	.56	.83	.48	.71
2" x 6" - 16' Avg.	Inst	LF	2C	.023	.58	.74	.07	1.39	1.84	1.20	1.58
2" x 6" - 18' Avg.	Demo	LF	LB	.019	---	.51	---	.51	.75	.43	.64
2" x 6" - 18' Avg.	Inst	LF	2C	.021	.58	.67	.06	1.31	1.74	1.16	1.53
2" x 8" - 12' Avg.	Demo	LF	LB	.029	---	.77	---	.77	1.15	.64	.95
2" x 8" - 12' Avg.	Inst	LF	2C	.031	.82	.99	.09	1.90	2.53	1.64	2.17
2" x 8" - 14' Avg.	Demo	LF	LB	.026	---	.69	---	.69	1.03	.59	.87
2" x 8" - 14' Avg.	Inst	LF	2C	.028	.82	.90	.08	1.80	2.37	1.57	2.06
2" x 8" - 16' Avg.	Demo	LF	LB	.024	---	.64	---	.64	.95	.53	.79
2" x 8" - 16' Avg.	Inst	LF	2C	.026	.82	.83	.08	1.73	2.27	1.50	1.96
2" x 8" - 18' Avg.	Demo	LF	LB	.023	---	.61	---	.61	.91	.51	.75
2" x 8" - 18' Avg.	Inst	LF	2C	.024	.82	.77	.07	1.66	2.17	1.46	1.91
2" x 8" - 20' Avg.	Demo	LF	LB	.021	---	.56	---	.56	.83	.48	.71
2" x 8" - 20' Avg.	Inst	LF	2C	.023	.82	.74	.07	1.63	2.12	1.40	1.81

Common, cut-up roofs, to 1/3 pitch

Description	Oper	Unit	Crew Size	Man-Hours Per Unit	Avg Mat'l Unit Cost	Avg Labor Unit Cost	Avg Equip Unit Cost	Avg Total Unit Cost	Avg Price Incl O&P	Avg Total Unit Cost	Avg Price Incl O&P
2" x 4" - 8' Avg.	Demo	LF	LB	.032	---	.85	---	.85	1.27	.72	1.07
2" x 4" - 8' Avg.	Inst	LF	2C	.035	.42	1.12	.10	1.64	2.27	1.44	1.98
2" x 4" - 10' Avg.	Demo	LF	LB	.027	---	.72	---	.72	1.07	.61	.91
2" x 4" - 10' Avg.	Inst	LF	2C	.030	.42	.96	.09	1.47	2.02	1.26	1.72
2" x 4" - 12' Avg.	Demo	LF	LB	.024	---	.64	---	.64	.95	.53	.79
2" x 4" - 12' Avg.	Inst	LF	2C	.026	.42	.83	.08	1.33	1.81	1.16	1.57
2" x 4" - 14' Avg.	Demo	LF	LB	.022	---	.59	---	.59	.87	.48	.71
2" x 4" - 14' Avg.	Inst	LF	2C	.024	.42	.77	.07	1.26	1.71	1.09	1.47
2" x 4" - 16' Avg.	Demo	LF	LB	.020	---	.53	---	.53	.79	.45	.67
2" x 4" - 16' Avg.	Inst	LF	2C	.022	.42	.71	.06	1.19	1.60	1.02	1.37

Description	Oper	Unit	Crew Size	Man-Hours Per Unit	Avg Mat'l Unit Cost	Avg Labor Unit Cost	Avg Equip Unit Cost	Avg Total Unit Cost	Avg Price Incl O&P	Avg Total Unit Cost	Avg Price Incl O&P
						Costs Based On Small Volume				Large Volume	
2" x 6" - 10' Avg.	Demo	LF	LB	.031	---	.83	---	.83	1.23	.69	1.0
2" x 6" - 10' Avg.	Inst	LF	2C	.033	.59	1.06	.10	1.75	2.37	1.52	2.0
2" x 6" - 12' Avg.	Demo	LF	LB	.027	---	.72	---	.72	1.07	.61	.9
2" x 6" - 12' Avg.	Inst	LF	2C	.029	.59	.93	.08	1.60	2.15	.87	1.2
2" x 6" - 14' Avg.	Demo	LF	LB	.024	---	.64	---	.64	.95	.53	.7
2" x 6" - 14' Avg.	Inst	LF	2C	.026	.59	.83	.08	1.50	2.01	.77	1.1
2" x 6" - 16' Avg.	Demo	LF	LB	.022	---	.59	---	.59	.87	.51	.7
2" x 6" - 16' Avg.	Inst	LF	2C	.024	.59	.77	.07	1.43	1.90	1.24	1.6
2" x 6" - 18' Avg.	Demo	LF	LB	.020	---	.53	---	.53	.79	.45	.6
2" x 6" - 18' Avg.	Inst	LF	2C	.022	.59	.71	.06	1.36	1.80	1.20	1.5
2" x 8" - 12' Avg.	Demo	LF	LB	.030	---	.80	---	.80	1.19	.69	1.0
2" x 8" - 12' Avg.	Inst	LF	2C	.032	.83	1.03	.09	1.95	2.58	1.72	2.2
2" x 8" - 14' Avg.	Demo	LF	LB	.027	---	.72	---	.72	1.07	.61	.9
2" x 8" - 14' Avg.	Inst	LF	2C	.029	.83	.93	.09	1.85	2.44	1.61	2.1
2" x 8" - 16' Avg.	Demo	LF	LB	.025	---	.67	---	.67	.99	.56	.8
2" x 8" - 16' Avg.	Inst	LF	2C	.027	.83	.87	.08	1.78	2.33	1.55	2.0
2" x 8" - 18' Avg.	Demo	LF	LB	.023	---	.61	---	.61	.91	.53	.7
2" x 8" - 18' Avg.	Inst	LF	2C	.025	.83	.80	.07	1.70	2.23	1.47	1.9
2" x 8" - 20' Avg.	Demo	LF	LB	.022	---	.59	---	.59	.87	.51	.7
2" x 8" - 20' Avg.	Inst	LF	2C	.024	.83	.77	.07	1.67	2.18	1.44	1.8

Common, cut-up roofs, to 3/8-1/2 pitch

Description	Oper	Unit	Crew Size	Man-Hours Per Unit	Avg Mat'l Unit Cost	Avg Labor Unit Cost	Avg Equip Unit Cost	Avg Total Unit Cost	Avg Price Incl O&P	Avg Total Unit Cost	Avg Price Incl O&P
2" x 4" - 8' Avg.	Demo	LF	LB	.036	---	.96	---	.96	1.43	.83	1.2
2" x 4" - 8' Avg.	Inst	LF	2C	.040	.42	1.28	.12	1.82	2.53	1.58	2.1
2" x 4" - 10' Avg.	Demo	LF	LB	.031	---	.83	---	.83	1.23	.69	1.0
2" x 4" - 10' Avg.	Inst	LF	2C	.033	.42	1.06	.10	1.58	2.17	1.37	1.8
2" x 4" - 12' Avg.	Demo	LF	LB	.027	---	.72	---	.72	1.07	.61	.9
2" x 4" - 12' Avg.	Inst	LF	2C	.029	.42	.93	.09	1.44	1.97	1.26	1.7
2" x 4" - 14' Avg.	Demo	LF	LB	.024	---	.64	---	.64	.95	.56	.8
2" x 4" - 14' Avg.	Inst	LF	2C	.026	.42	.83	.08	1.33	1.81	1.17	1.5
2" x 4" - 16' Avg.	Demo	LF	LB	.022	---	.59	---	.59	.87	.51	.7
2" x 4" - 16' Avg.	Inst	LF	2C	.024	.42	.77	.07	1.26	1.71	1.12	1.5
2" x 6" - 10' Avg.	Demo	LF	LB	.034	---	.91	---	.91	1.35	.77	1.1
2" x 6" - 10' Avg.	Inst	LF	2C	.037	.59	1.19	.11	1.89	2.57	1.62	2.2
2" x 6" - 12' Avg.	Demo	LF	LB	.030	---	.80	---	.80	1.19	.67	.9
2" x 6" - 12' Avg.	Inst	LF	2C	.032	.59	1.03	.09	1.71	2.31	1.49	2.0
2" x 6" - 14' Avg.	Demo	LF	LB	.027	---	.72	---	.72	1.07	.61	.9
2" x 6" - 14' Avg.	Inst	LF	2C	.029	.59	.93	.08	1.60	2.15	1.41	1.8
2" x 6" - 16' Avg.	Demo	LF	LB	.024	---	.64	---	.64	.95	.56	.8
2" x 6" - 16' Avg.	Inst	LF	2C	.026	.59	.83	.08	1.50	2.01	1.32	1.7
2" x 6" - 18' Avg.	Demo	LF	LB	.022	---	.59	---	.59	.87	.51	.7
2" x 6" - 18' Avg.	Inst	LF	2C	.024	.59	.77	.07	1.43	1.90	1.27	1.6
2" x 8" - 12' Avg.	Demo	LF	LB	.033	---	.88	---	.88	1.31	.75	1.1
2" x 8" - 12' Avg.	Inst	LF	2C	.036	.83	1.16	.10	2.09	2.79	1.79	2.3
2" x 8" - 14' Avg.	Demo	LF	LB	.030	---	.80	---	.80	1.19	.67	.9
2" x 8" - 14' Avg.	Inst	LF	2C	.032	.83	1.03	.09	1.95	2.58	1.69	2.2
2" x 8" - 16' Avg.	Demo	LF	LB	.027	---	.72	---	.72	1.07	.61	.9
2" x 8" - 16' Avg.	Inst	LF	2C	.030	.83	.96	.09	1.88	2.49	1.61	2.1

Description	Oper Unit	Crew Size	Man-Hours Per Unit	Avg Mat'l Unit Cost	Avg Labor Unit Cost	Avg Equip Unit Cost	Avg Total Unit Cost	Avg Price Incl O&P	Avg Total Unit Cost	Avg Price Incl O&P
							Costs Based On Small Volume		**Large Volume**	
2" x 8" - 18' Avg.	Demo LF	LB	.025	---	.67	---	.67	.99	.59	.87
2" x 8" - 18' Avg.	Inst LF	2C	.027	.83	.87	.08	1.78	2.33	1.55	2.03
2" x 8" - 20' Avg.	Demo LF	LB	.024	---	.64	---	.64	.95	.53	.79
2" x 8" - 20' Avg.	Inst LF	2C	.026	.83	.83	.08	1.74	2.29	1.51	1.97

Rafters, per SF of area

Gable or hip, to 1/3 pitch

Description	Oper Unit	Crew	MH	Mat'l	Labor	Equip	Total	Price	Total	Price
2" x 4" - 12" oc	Demo SF	LB	.022	---	.59	---	.59	.87	.51	.75
2" x 4" - 12" oc	Inst SF	2C	.024	.47	.77	.07	1.31	1.77	1.16	1.57
2" x 4" - 16" oc	Demo SF	LB	.019	---	.51	---	.51	.75	.43	.64
2" x 4" - 16" oc	Inst SF	2C	.020	.37	.64	.06	1.07	1.45	.94	1.26
2" x 4" - 24" oc	Demo SF	LB	.014	---	.37	---	.37	.56	.32	.48
2" x 4" - 24" oc	Inst SF	2C	.016	.27	.51	.05	.83	1.13	.70	.94
2" x 6" - 12" oc	Demo SF	LB	.023	---	.61	---	.61	.91	.51	.75
2" x 6" - 12" oc	Inst SF	2C	.025	.65	.80	.07	1.52	2.02	1.31	1.74
2" x 6" - 16" oc	Demo SF	LB	.019	---	.51	---	.51	.75	.43	.64
2" x 6" - 16" oc	Inst SF	2C	.021	.50	.67	.06	1.23	1.65	1.09	1.45
2" x 6" - 24" oc	Demo SF	LB	.015	---	.40	---	.40	.60	.32	.48
2" x 6" - 24" oc	Inst SF	2C	.016	.36	.51	.05	.92	1.23	.79	1.05
2" x 8" - 12" oc	Demo SF	LB	.024	---	.64	---	.64	.95	.53	.79
2" x 8" - 12" oc	Inst SF	2C	.026	.89	.83	.08	1.80	2.36	1.56	2.03
2" x 8" - 16" oc	Demo SF	LB	.020	---	.53	---	.53	.79	.45	.67
2" x 8" - 16" oc	Inst SF	2C	.022	.68	.71	.06	1.45	1.90	1.24	1.62
2" x 8" - 24" oc	Demo SF	LB	.015	---	.40	---	.40	.60	.35	.52
2" x 8" - 24" oc	Inst SF	2C	.017	.48	.55	.05	1.08	1.42	.92	1.21

Gable or hip, to 3/8-1/2 pitch

Description	Oper Unit	Crew	MH	Mat'l	Labor	Equip	Total	Price	Total	Price
2" x 4" - 12" oc	Demo SF	LB	.026	---	.69	---	.69	1.03	.59	.87
2" x 4" - 12" oc	Inst SF	2C	.028	.47	.90	.08	1.45	1.97	1.27	1.72
2" x 4" - 16" oc	Demo SF	LB	.022	---	.59	---	.59	.87	.51	.75
2" x 4" - 16" oc	Inst SF	2C	.024	.37	.77	.07	1.21	1.65	1.04	1.41
2" x 4" - 24" oc	Demo SF	LB	.017	---	.45	---	.45	.67	.37	.56
2" x 4" - 24" oc	Inst SF	2C	.018	.27	.58	.05	.90	1.23	.77	1.05
2" x 6" - 12" oc	Demo SF	LB	.026	---	.69	---	.69	1.03	.59	.87
2" x 6" - 12" oc	Inst SF	2C	.028	.65	.90	.08	1.63	2.18	1.42	1.89
2" x 6" - 16" oc	Demo SF	LB	.022	---	.59	---	.59	.87	.51	.75
2" x 6" - 16" oc	Inst SF	2C	.024	.50	.77	.07	1.34	1.80	1.16	1.55
2" x 6" - 24" oc	Demo SF	LB	.017	---	.45	---	.45	.67	.37	.56
2" x 6" - 24" oc	Inst SF	2C	.018	.36	.58	.05	.99	1.33	.85	1.14
2" x 8" - 12" oc	Demo SF	LB	.027	---	.72	---	.72	1.07	.61	.91
2" x 8" - 12" oc	Inst SF	2C	.029	.89	.93	.08	1.90	2.50	1.63	2.13
2" x 8" - 16" oc	Demo SF	LB	.022	---	.59	---	.59	.87	.51	.75
2" x 8" - 16" oc	Inst SF	2C	.024	.68	.77	.07	1.52	2.01	1.31	1.72
2" x 8" - 24" oc	Demo SF	LB	.014	---	.37	---	.37	.56	.32	.48
2" x 8" - 24" oc	Inst SF	2C	.015	.48	.48	.04	1.00	1.31	.89	1.16

| | | | Costs Based On Small Volume | | | | | | Large Volume | |
| | | | | | | | | | | |
Description	Oper	Unit	Crew Size	Man-Hours Per Unit	Avg Mat'l Unit Cost	Avg Labor Unit Cost	Avg Equip Unit Cost	Avg Total Unit Cost	Avg Price Incl O&P	Avg Total Unit Cost	Avg Price Incl O&P
Cut-up roofs, to 1/3 pitch											
2" x 4" - 12" oc	Demo	SF	LB	.028	---	.75	---	.75	1.11	.64	.9
2" x 4" - 12" oc	Inst	SF	2C	.030	.48	.96	.09	1.53	2.09	1.35	1.8
2" x 4" - 16" oc	Demo	SF	LB	.023	---	.61	---	.61	.91	.53	.7
2" x 4" - 16" oc	Inst	SF	2C	.026	.38	.83	.07	1.28	1.76	1.11	1.5
2" x 4" - 24" oc	Demo	SF	LB	.018	---	.48	---	.48	.71	.40	.6
2" x 4" - 24" oc	Inst	SF	2C	.019	.27	.61	.06	.94	1.29	.85	1.1
2" x 6" - 12" oc	Demo	SF	LB	.027	---	.72	---	.72	1.07	.61	.9
2" x 6" - 12" oc	Inst	SF	2C	.030	.66	.96	.09	1.71	2.29	1.46	1.9
2" x 6" - 16" oc	Demo	SF	LB	.023	---	.61	---	.61	.91	.53	.7
2" x 6" - 16" oc	Inst	SF	2C	.025	.51	.80	.07	1.38	1.86	1.19	1.6
2" x 6" - 24" oc	Demo	SF	LB	.018	---	.48	---	.48	.71	.40	.6
2" x 6" - 24" oc	Inst	SF	2C	.019	.37	.61	.06	1.04	1.40	.89	1.2
2" x 8" - 12" oc	Demo	SF	LB	.028	---	.75	---	.75	1.11	.64	.9
2" x 8" - 12" oc	Inst	SF	2C	.030	.90	.96	.09	1.95	2.57	1.70	2.2
2" x 8" - 16" oc	Demo	SF	LB	.023	---	.61	---	.61	.91	.53	.7
2" x 8" - 16" oc	Inst	SF	2C	.025	.69	.80	.07	1.56	2.07	1.39	1.8
2" x 8" - 24" oc	Demo	SF	LB	.018	---	.48	---	.48	.71	.40	.6
2" x 8" - 24" oc	Inst	SF	2C	.019	.49	.61	.06	1.16	1.54	1.00	1.3
Cut-up roofs, to 3/8-1/2 pitch											
2" x 4" - 12" oc	Demo	SF	LB	.032	---	.85	---	.85	1.27	.72	1.0
2" x 4" - 12" oc	Inst	SF	2C	.034	.48	1.09	.10	1.67	2.29	1.46	1.9
2" x 4" - 16" oc	Demo	SF	LB	.026	---	.69	---	.69	1.03	.59	.8
2" x 4" - 16" oc	Inst	SF	2C	.029	.38	.93	.08	1.39	1.91	1.21	1.6
2" x 4" - 24" oc	Demo	SF	LB	.020	---	.53	---	.53	.79	.45	.6
2" x 4" - 24" oc	Inst	SF	2C	.022	.27	.71	.06	1.04	1.43	.91	1.2
2" x 6" - 12" oc	Demo	SF	LB	.030	---	.80	---	.80	1.19	.69	1.0
2" x 6" - 12" oc	Inst	SF	2C	.033	.66	1.06	.10	1.82	2.45	1.57	2.1
2" x 6" - 16" oc	Demo	SF	LB	.026	---	.69	---	.69	1.03	.59	.8
2" x 6" - 16" oc	Inst	SF	2C	.028	.51	.90	.08	1.49	2.01	1.30	1.7
2" x 6" - 24" oc	Demo	SF	LB	.020	---	.53	---	.53	.79	.45	.6
2" x 6" - 24" oc	Inst	SF	2C	.021	.37	.67	.06	1.10	1.50	.96	1.3
2" x 8" - 12" oc	Demo	SF	LB	.031	---	.83	---	.83	1.23	.69	1.0
2" x 8" - 12" oc	Inst	SF	2C	.033	.90	1.06	.10	2.06	2.72	1.78	2.3
2" x 8" - 16" oc	Demo	SF	LB	.026	---	.69	---	.69	1.03	.59	.8
2" x 8" - 16" oc	Inst	SF	2C	.028	.69	.90	.08	1.67	2.22	1.46	1.9
2" x 8" - 24" oc	Demo	SF	LB	.020	---	.53	---	.53	.79	.45	.6
2" x 8" - 24" oc	Inst	SF	2C	.021	.49	.67	.06	1.22	1.63	1.07	1.4
Roof decking, solid, T&G, dry for plank-and-beam construction											
2" x 6" - 12'	Demo	SF	LB	.029	---	.77	---	.77	1.15	.67	.9
2" x 6" - 12'	Inst	SF	2C	.039	.12	1.25	.11	1.48	2.13	1.26	1.8
2" x 8" - 12'	Demo	SF	LB	.021	---	.56	---	.56	.83	.48	.7
2" x 8" - 12'	Inst	SF	2C	.028	.08	.90	.08	1.06	1.52	.91	1.3

Description	Oper	Unit	Crew Size	Man-Hours Per Unit	Avg Mat'l Unit Cost	Avg Labor Unit Cost	Avg Equip Unit Cost	Avg Total Unit Cost	Avg Price Incl O&P	Avg Total Unit Cost	Avg Price Incl O&P
										Large Volume	
tuds/plates, per LF of stick											
Walls or partitions											
2" x 4" - 8' Avg.	Demo	LF	LB	.018	---	.48	---	.48	**.71**	.43	**.64**
2" x 4" - 8' Avg.	Inst	LF	2C	.022	.40	.71	.06	1.17	**1.58**	1.03	**1.39**
2" x 4" - 10' Avg.	Demo	LF	LB	.016	---	.43	---	.43	**.64**	.37	**.56**
2" x 4" - 10' Avg.	Inst	LF	2C	.019	.40	.61	.06	1.07	**1.43**	.96	**1.28**
2" x 4" - 12' Avg.	Demo	LF	LB	.014	---	.37	---	.37	**.56**	.32	**.48**
2" x 4" - 12' Avg.	Inst	LF	2C	.017	.40	.55	.05	1.00	**1.33**	.88	**1.18**
2" x 6" - 8' Avg.	Demo	LF	LB	.021	---	.56	---	.56	**.83**	.48	**.71**
2" x 6" - 8' Avg.	Inst	LF	2C	.025	.56	.80	.07	1.43	**1.92**	1.23	**1.65**
2" x 6" - 10' Avg.	Demo	LF	LB	.018	---	.48	---	.48	**.71**	.43	**.64**
2" x 6" - 10' Avg.	Inst	LF	2C	.022	.56	.71	.06	1.33	**1.76**	1.16	**1.54**
2" x 6" - 12' Avg.	Demo	LF	LB	.016	---	.43	---	.43	**.64**	.37	**.56**
2" x 6" - 12' Avg.	Inst	LF	2C	.020	.56	.64	.06	1.26	**1.67**	1.10	**1.44**
2" x 8" - 8' Avg.	Demo	LF	LB	.023	---	.61	---	.61	**.91**	.53	**.79**
2" x 8" - 8' Avg.	Inst	LF	2C	.028	.78	.90	.08	1.76	**2.32**	1.53	**2.02**
2" x 8" - 10' Avg.	Demo	LF	LB	.021	---	.56	---	.56	**.83**	.45	**.67**
2" x 8" - 10' Avg.	Inst	LF	2C	.024	.78	.77	.07	1.62	**2.12**	1.42	**1.86**
2" x 8" - 12' Avg.	Demo	LF	LB	.019	---	.51	---	.51	**.75**	.43	**.64**
2" x 8" - 12' Avg.	Inst	LF	2C	.022	.78	.71	.06	1.55	**2.02**	1.35	**1.76**
Gable ends											
2" x 4" - 3' Avg.	Demo	LF	LB	.029	---	.77	---	.77	**1.15**	.64	**.95**
2" x 4" - 3' Avg.	Inst	LF	2C	.055	.42	1.76	.16	2.34	**3.29**	2.03	**2.84**
2" x 4" - 4' Avg.	Demo	LF	LB	.023	---	.61	---	.61	**.91**	.53	**.79**
2" x 4" - 4' Avg.	Inst	LF	2C	.044	.42	1.41	.13	1.96	**2.73**	1.68	**2.33**
2" x 4" - 5' Avg.	Demo	LF	LB	.020	---	.53	---	.53	**.79**	.45	**.67**
2" x 4" - 5' Avg.	Inst	LF	2C	.037	.42	1.19	.11	1.72	**2.37**	1.46	**2.02**
2" x 4" - 6' Avg.	Demo	LF	LB	.017	---	.45	---	.45	**.67**	.40	**.60**
2" x 4" - 6' Avg.	Inst	LF	2C	.032	.42	1.03	.09	1.54	**2.11**	1.36	**1.86**
2" x 6" - 3' Avg.	Demo	LF	LB	.030	---	.80	---	.80	**1.19**	.67	**.99**
2" x 6" - 3' Avg.	Inst	LF	2C	.056	.58	1.80	.16	2.54	**3.52**	2.18	**3.01**
2" x 6" - 4' Avg.	Demo	LF	LB	.024	---	.64	---	.64	**.95**	.56	**.83**
2" x 6" - 4' Avg.	Inst	LF	2C	.045	.58	1.44	.13	2.15	**2.96**	1.86	**2.55**
2" x 6" - 5' Avg.	Demo	LF	LB	.021	---	.56	---	.56	**.83**	.48	**.71**
2" x 6" - 5' Avg.	Inst	LF	2C	.038	.58	1.22	.11	1.91	**2.61**	1.65	**2.24**
2" x 6" - 6' Avg.	Demo	LF	LB	.019	---	.51	---	.51	**.75**	.43	**.64**
2" x 6" - 6' Avg.	Inst	LF	2C	.034	.58	1.09	.10	1.77	**2.40**	1.54	**2.09**
2" x 8" - 3' Avg.	Demo	LF	LB	.031	---	.83	---	.83	**1.23**	.69	**1.03**
2" x 8" - 3' Avg.	Inst	LF	2C	.057	.82	1.83	.17	2.82	**3.86**	2.41	**3.29**
2" x 8" - 4' Avg.	Demo	LF	LB	.025	---	.67	---	.67	**.99**	.59	**.87**
2" x 8" - 4' Avg.	Inst	LF	2C	.046	.82	1.48	.13	2.43	**3.29**	2.09	**2.83**
2" x 8" - 5' Avg.	Demo	LF	LB	.022	---	.59	---	.59	**.87**	.51	**.75**
2" x 8" - 5' Avg.	Inst	LF	2C	.039	.82	1.25	.11	2.18	**2.93**	1.89	**2.53**
2" x 8" - 6' Avg.	Demo	LF	LB	.020	---	.53	---	.53	**.79**	.45	**.67**
2" x 8" - 6' Avg.	Inst	LF	2C	.035	.82	1.12	.10	2.04	**2.73**	1.78	**2.37**

				Costs Based On Small Volume					Large Volume	
Description	Oper	Unit	Crew Size	Man-Hours Per Unit	Avg Mat'l Unit Cost	Avg Labor Unit Cost	Avg Equip Unit Cost	Avg Total Unit Cost	Avg Price Incl O&P	Avg Total Unit Cost	Avg Price Incl O&P

Studs/plates, per SF of area

Walls or partitions

2" x 4" - 8', 12" oc	Demo	SF	LB	.031	---	.83	---	.83	**1.23**	.72	1.0
2" x 4" - 8', 12" oc	Inst	SF	2C	.038	.69	1.22	.11	2.02	**2.73**	1.75	2.3
2" x 4" - 8', 16" oc	Demo	SF	LB	.028	---	.75	---	.75	**1.11**	.64	.9
2" x 4" - 8', 16" oc	Inst	SF	2C	.034	.57	1.09	.10	1.76	**2.39**	1.52	2.0
2" x 4" - 8', 24" oc	Demo	SF	LB	.024	---	.64	---	.64	**.95**	.53	.7
2" x 4" - 8', 24" oc	Inst	SF	2C	.029	.43	.93	.08	1.44	**1.97**	1.26	1.7
2" x 4" - 10', 12" oc	Demo	SF	LB	.026	---	.69	---	.69	**1.03**	.59	.8
2" x 4" - 10', 12" oc	Inst	SF	2C	.032	.66	1.03	.09	1.78	**2.39**	1.55	2.0
2" x 4" - 10', 16" oc	Demo	SF	LB	.023	---	.61	---	.61	**.91**	.53	.7
2" x 4" - 10', 16" oc	Inst	SF	2C	.028	.54	.90	.08	1.52	**2.05**	1.33	1.7
2" x 4" - 10', 24" oc	Demo	SF	LB	.019	---	.51	---	.51	**.75**	.43	.6
2" x 4" - 10', 24" oc	Inst	SF	2C	.023	.41	.74	.07	1.22	**1.65**	1.07	1.4
2" x 4" - 12', 12" oc	Demo	SF	LB	.023	---	.61	---	.61	**.91**	.51	.7
2" x 4" - 12', 12" oc	Inst	SF	2C	.028	.65	.90	.08	1.63	**2.18**	1.40	1.8
2" x 4" - 12', 16" oc	Demo	SF	LB	.020	---	.53	---	.53	**.79**	.45	.6
2" x 4" - 12', 16" oc	Inst	SF	2C	.024	.51	.77	.07	1.35	**1.81**	1.20	1.6
2" x 4" - 12', 24" oc	Demo	SF	LB	.016	---	.43	---	.43	**.64**	.37	.5
2" x 4" - 12', 24" oc	Inst	SF	2C	.020	.39	.64	.06	1.09	**1.47**	.95	1.2
2" x 6" - 8', 12" oc	Demo	SF	LB	.036	---	.96	---	.96	**1.43**	.80	1.1
2" x 6" - 8', 12" oc	Inst	SF	2C	.043	.97	1.38	.13	2.48	**3.32**	2.15	2.8
2" x 6" - 8', 16" oc	Demo	SF	LB	.032	---	.85	---	.85	**1.27**	.72	1.0
2" x 6" - 8', 16" oc	Inst	SF	2C	.038	.79	1.22	.11	2.12	**2.85**	1.87	2.5
2" x 6" - 8', 24" oc	Demo	SF	LB	.027	---	.72	---	.72	**1.07**	.61	.9
2" x 6" - 8', 24" oc	Inst	SF	2C	.033	.60	1.06	.09	1.75	**2.37**	1.52	2.0
2" x 6" - 10', 12" oc	Demo	SF	LB	.030	---	.80	---	.80	**1.19**	.67	.9
2" x 6" - 10', 12" oc	Inst	SF	2C	.036	.92	1.16	.10	2.18	**2.89**	1.89	2.5
2" x 6" - 10', 16" oc	Demo	SF	LB	.026	---	.69	---	.69	**1.03**	.59	.8
2" x 6" - 10', 16" oc	Inst	SF	2C	.032	.74	1.03	.09	1.86	**2.48**	1.62	2.1
2" x 6" - 10', 24" oc	Demo	SF	LB	.022	---	.59	---	.59	**.87**	.51	.7
2" x 6" - 10', 24" oc	Inst	SF	2C	.026	.56	.83	.08	1.47	**1.98**	1.32	1.7
2" x 6" - 12', 12" oc	Demo	SF	LB	.026	---	.69	---	.69	**1.03**	.59	.8
2" x 6" - 12', 12" oc	Inst	SF	2C	.031	.90	.99	.09	1.98	**2.62**	1.76	2.3
2" x 6" - 12', 16" oc	Demo	SF	LB	.023	---	.61	---	.61	**.91**	.51	.7
2" x 6" - 12', 16" oc	Inst	SF	2C	.027	.72	.87	.08	1.67	**2.21**	1.46	1.9
2" x 6" - 12', 24" oc	Demo	SF	LB	.019	---	.51	---	.51	**.75**	.43	.6
2" x 6" - 12', 24" oc	Inst	SF	2C	.023	.54	.74	.07	1.35	**1.80**	1.16	1.5
2" x 8" - 8', 12" oc	Demo	SF	LB	.040	---	1.07	---	1.07	**1.59**	.91	1.3
2" x 8" - 8', 12" oc	Inst	SF	2C	.048	1.35	1.54	.14	3.03	**4.00**	2.64	3.4
2" x 8" - 8', 16" oc	Demo	SF	LB	.036	---	.96	---	.96	**1.43**	.80	1.1
2" x 8" - 8', 16" oc	Inst	SF	2C	.043	1.09	1.38	.12	2.59	**3.44**	2.24	2.9

Description	Oper	Unit	Crew Size	Man-Hours Per Unit	Avg Mat'l Unit Cost	Avg Labor Unit Cost	Avg Equip Unit Cost	Avg Total Unit Cost	Avg Price Incl O&P	Avg Total Unit Cost	Avg Price Incl O&P
								Costs Based On Small Volume		Large Volume	
2" x 8" - 8', 24" oc	Demo	SF	LB	.030	---	.80	---	.80	**1.19**	.69	**1.03**
2" x 8" - 8', 24" oc	Inst	SF	2C	.036	.84	1.16	.11	2.11	**2.81**	1.83	**2.44**
2" x 8" - 10', 12" oc	Demo	SF	LB	.034	---	.91	---	.91	**1.35**	.77	**1.15**
2" x 8" - 10', 12" oc	Inst	SF	2C	.040	1.29	1.28	.12	2.69	**3.53**	2.34	**3.06**
2" x 8" - 10', 16" oc	Demo	SF	LB	.030	---	.80	---	.80	**1.19**	.67	**.99**
2" x 8" - 10', 16" oc	Inst	SF	2C	.035	1.03	1.12	.10	2.25	**2.97**	1.97	**2.59**
2" x 8" - 10', 24" oc	Demo	SF	LB	.025	---	.67	---	.67	**.99**	.56	**.83**
2" x 8" - 10', 24" oc	Inst	SF	2C	.030	.78	.96	.09	1.83	**2.43**	1.56	**2.07**
2" x 8" - 12', 12" oc	Demo	SF	LB	.030	---	.80	---	.80	**1.19**	.67	**.99**
2" x 8" - 12', 12" oc	Inst	SF	2C	.035	1.25	1.12	.10	2.47	**3.22**	2.17	**2.82**
2" x 8" - 12', 16" oc	Demo	SF	LB	.026	---	.69	---	.69	**1.03**	.59	**.87**
2" x 8" - 12', 16" oc	Inst	SF	2C	.031	1.00	.99	.09	2.08	**2.73**	1.80	**2.36**
2" x 8" - 12', 24" oc	Demo	SF	LB	.021	---	.56	---	.56	**.83**	.48	**.71**
2" x 8" - 12', 24" oc	Inst	SF	2C	.025	.74	.80	.07	1.61	**2.12**	1.39	**1.83**

Gable ends

Description	Oper	Unit	Crew Size	Man-Hours Per Unit	Avg Mat'l Unit Cost	Avg Labor Unit Cost	Avg Equip Unit Cost	Avg Total Unit Cost	Avg Price Incl O&P	Avg Total Unit Cost	Avg Price Incl O&P
2" x 4" - 3' Avg., 12" oc	Demo	SF	LB	.087	---	2.32	---	2.32	**3.45**	1.97	**2.94**
2" x 4" - 3' Avg., 12" oc	Inst	SF	2C	.167	1.27	5.36	.49	7.12	**9.99**	6.12	**8.57**
2" x 4" - 3' Avg., 16" oc	Demo	SF	LB	.082	---	2.18	---	2.18	**3.25**	1.84	**2.74**
2" x 4" - 3' Avg., 16" oc	Inst	SF	2C	.155	1.19	4.97	.45	6.61	**9.28**	5.71	**7.99**
2" x 4" - 3' Avg., 24" oc	Demo	SF	LB	.074	---	1.97	---	1.97	**2.94**	1.68	**2.50**
2" x 4" - 3' Avg., 24" oc	Inst	SF	2C	.143	1.09	4.59	.42	6.10	**8.56**	5.22	**7.31**
2" x 4" - 4' Avg., 12" oc	Demo	SF	LB	.058	---	1.55	---	1.55	**2.30**	1.33	**1.98**
2" x 4" - 4' Avg., 12" oc	Inst	SF	2C	.110	1.06	3.53	.32	4.91	**6.83**	4.25	**5.90**
2" x 4" - 4' Avg., 16" oc	Demo	SF	LB	.054	---	1.44	---	1.44	**2.14**	1.23	**1.83**
2" x 4" - 4' Avg., 16" oc	Inst	SF	2C	.103	.98	3.31	.30	4.59	**6.39**	3.93	**5.46**
2" x 4" - 4' Avg., 24" oc	Demo	SF	LB	.049	---	1.31	---	1.31	**1.94**	1.12	**1.67**
2" x 4" - 4' Avg., 24" oc	Inst	SF	2C	.093	.89	2.98	.27	4.14	**5.77**	3.58	**4.96**
2" x 4" - 5' Avg., 12" oc	Demo	SF	LB	.046	---	1.23	---	1.23	**1.83**	1.04	**1.55**
2" x 4" - 5' Avg., 12" oc	Inst	SF	2C	.086	.97	2.76	.25	3.98	**5.51**	3.43	**4.74**
2" x 4" - 5' Avg., 16" oc	Demo	SF	LB	.041	---	1.09	---	1.09	**1.63**	.93	**1.39**
2" x 4" - 5' Avg., 16" oc	Inst	SF	2C	.078	.89	2.50	.23	3.62	**5.01**	3.12	**4.30**
2" x 4" - 5' Avg., 24" oc	Demo	SF	LB	.037	---	.99	---	.99	**1.47**	.83	**1.23**
2" x 4" - 5' Avg., 24" oc	Inst	SF	2C	.069	.79	2.21	.20	3.20	**4.43**	2.78	**3.84**
2" x 4" - 6' Avg., 12" oc	Demo	SF	LB	.038	---	1.01	---	1.01	**1.51**	.85	**1.27**
2" x 4" - 6' Avg., 12" oc	Inst	SF	2C	.070	.91	2.25	.20	3.36	**4.62**	2.88	**3.95**
2" x 4" - 6' Avg., 16" oc	Demo	SF	LB	.033	---	.88	---	.88	**1.31**	.75	**1.11**
2" x 4" - 6' Avg., 16" oc	Inst	SF	2C	.062	.80	1.99	.18	2.97	**4.08**	2.58	**3.54**
2" x 4" - 6' Avg., 24" oc	Demo	SF	LB	.029	---	.77	---	.77	**1.15**	.67	**.99**
2" x 4" - 6' Avg., 24" oc	Inst	SF	2C	.054	.71	1.73	.16	2.60	**3.58**	2.25	**3.08**
2" x 6" - 3' Avg., 12" oc	Demo	SF	LB	.090	---	2.40	---	2.40	**3.57**	2.05	**3.06**
2" x 6" - 3' Avg., 12" oc	Inst	SF	2C	.170	1.78	5.46	.50	7.74	**10.70**	6.68	**9.25**
2" x 6" - 3' Avg., 16" oc	Demo	SF	LB	.085	---	2.26	---	2.26	**3.37**	1.92	**2.86**
2" x 6" - 3' Avg., 16" oc	Inst	SF	2C	.160	1.68	5.13	.47	7.28	**10.10**	6.27	**8.68**

Description	Oper Unit	Crew Size	Man-Hours Per Unit	Avg Mat'l Unit Cost	Avg Labor Unit Cost	Avg Equip Unit Cost	Avg Total Unit Cost	Avg Price Incl O&P	Avg Total Unit Cost	Avg Price Incl O&P
							Costs Based On Small Volume		**Large Volume**	
2" x 6" - 3' Avg., 24" oc	Demo SF	LB	.077	---	2.05	---	2.05	**3.06**	1.76	**2.62**
2" x 6" - 3' Avg., 24" oc	Inst SF	2C	.145	1.54	4.65	.42	6.61	**9.17**	5.73	**7.93**
2" x 6" - 4' Avg., 12" oc	Demo SF	LB	.061	---	1.63	---	1.63	**2.42**	1.39	**2.06**
2" x 6" - 4' Avg., 12" oc	Inst SF	2C	.113	1.49	3.63	.33	5.45	**7.48**	4.70	**6.44**
2" x 6" - 4' Avg., 16" oc	Demo SF	LB	.057	---	1.52	---	1.52	**2.26**	1.28	**1.91**
2" x 6" - 4' Avg., 16" oc	Inst SF	2C	.105	1.38	3.37	.31	5.06	**6.95**	4.37	**5.98**
2" x 6" - 4' Avg., 24" oc	Demo SF	LB	.051	---	1.36	---	1.36	**2.02**	1.17	**1.75**
2" x 6" - 4' Avg., 24" oc	Inst SF	2C	.096	1.25	3.08	.28	4.61	**6.34**	3.97	**5.44**
2" x 6" - 5' Avg., 12" oc	Demo SF	LB	.048	---	1.28	---	1.28	**1.91**	1.09	**1.63**
2" x 6" - 5' Avg., 12" oc	Inst SF	2C	.089	1.36	2.86	.26	4.48	**6.11**	3.86	**5.25**
2" x 6" - 5' Avg., 16" oc	Demo SF	LB	.044	---	1.17	---	1.17	**1.75**	.99	**1.47**
2" x 6" - 5' Avg., 16" oc	Inst SF	2C	.080	1.25	2.57	.23	4.05	**5.52**	3.51	**4.77**
2" x 6" - 5' Avg., 24" oc	Demo SF	LB	.039	---	1.04	---	1.04	**1.55**	.88	**1.31**
2" x 6" - 5' Avg., 24" oc	Inst SF	2C	.071	1.10	2.28	.21	3.59	**4.89**	3.11	**4.22**
2" x 6" - 6' Avg., 12" oc	Demo SF	LB	.040	---	1.07	---	1.07	**1.59**	.91	**1.35**
2" x 6" - 6' Avg., 12" oc	Inst SF	2C	.073	1.27	2.34	.21	3.82	**5.18**	3.32	**4.49**
2" x 6" - 6' Avg., 16" oc	Demo SF	LB	.036	---	.96	---	.96	**1.43**	.80	**1.19**
2" x 6" - 6' Avg., 16" oc	Inst SF	2C	.065	1.13	2.09	.19	3.41	**4.62**	2.94	**3.98**
2" x 6" - 6' Avg., 24" oc	Demo SF	LB	.031	---	.83	---	.83	**1.23**	.69	**1.03**
2" x 6" - 6' Avg., 24" oc	Inst SF	2C	.056	.98	1.80	.16	2.94	**3.98**	2.57	**3.47**
2" x 8" - 3' Avg., 12" oc	Demo SF	LB	.094	---	2.50	---	2.50	**3.73**	2.13	**3.18**
2" x 8" - 3' Avg., 12" oc	Inst SF	2C	.174	2.50	5.58	.51	8.59	**11.80**	7.40	**10.10**
2" x 8" - 3' Avg., 16" oc	Demo SF	LB	.088	---	2.34	---	2.34	**3.49**	2.00	**2.98**
2" x 8" - 3' Avg., 16" oc	Inst SF	2C	.162	2.35	5.20	.47	8.02	**11.00**	6.92	**9.45**
2" x 8" - 3' Avg., 24" oc	Demo SF	LB	.080	---	2.13	---	2.13	**3.18**	1.81	**2.70**
2" x 8" - 3' Avg., 24" oc	Inst SF	2C	.148	2.15	4.75	.43	7.33	**10.00**	6.32	**8.63**
2" x 8" - 4' Avg., 12" oc	Demo SF	LB	.064	---	1.70	---	1.70	**2.54**	1.47	**2.18**
2" x 8" - 4' Avg., 12" oc	Inst SF	2C	.117	2.08	3.75	.34	6.17	**8.36**	5.32	**7.18**
2" x 8" - 4' Avg., 16" oc	Demo SF	LB	.059	---	1.57	---	1.57	**2.34**	1.36	**2.02**
2" x 8" - 4' Avg., 16" oc	Inst SF	2C	.108	1.93	3.47	.32	5.72	**7.74**	4.94	**6.68**
2" x 8" - 4' Avg., 24" oc	Demo SF	LB	.054	---	1.44	---	1.44	**2.14**	1.23	**1.83**
2" x 8" - 4' Avg., 24" oc	Inst SF	2C	.098	1.74	3.14	.29	5.17	**7.01**	4.45	**6.02**
2" x 8" - 5' Avg., 12" oc	Demo SF	LB	.051	---	1.36	---	1.36	**2.02**	1.15	**1.71**
2" x 8" - 5' Avg., 12" oc	Inst SF	2C	.091	1.90	2.92	.27	5.09	**6.84**	4.43	**5.94**
2" x 8" - 5' Avg., 16" oc	Demo SF	LB	.047	---	1.25	---	1.25	**1.87**	1.07	**1.59**
2" x 8" - 5' Avg., 16" oc	Inst SF	2C	.083	1.70	2.66	.24	4.60	**6.19**	3.98	**5.33**
2" x 8" - 5' Avg., 24" oc	Demo SF	LB	.041	---	1.09	---	1.09	**1.63**	.93	**1.39**
2" x 8" - 5' Avg., 24" oc	Inst SF	2C	.073	1.50	2.34	.21	4.05	**5.45**	3.51	**4.71**
2" x 8" - 6' Avg., 12" oc	Demo SF	LB	.043	---	1.15	---	1.15	**1.71**	.96	**1.43**
2" x 8" - 6' Avg., 12" oc	Inst SF	2C	.075	1.77	2.41	.22	4.40	**5.87**	3.82	**5.09**
2" x 8" - 6' Avg., 16" oc	Demo SF	LB	.038	---	1.01	---	1.01	**1.51**	.85	**1.27**
2" x 8" - 6' Avg., 16" oc	Inst SF	2C	.067	1.57	2.15	.19	3.91	**5.22**	3.40	**4.52**
2" x 8" - 6' Avg., 24" oc	Demo SF	LB	.033	---	.88	---	.88	**1.31**	.75	**1.11**
2" x 8" - 6' Avg., 24" oc	Inst SF	2C	.058	1.37	1.86	.17	3.40	**4.54**	2.93	**3.90**

Description	Oper	Unit	Crew Size	Man-Hours Per Unit	Costs Based On Small Volume				Avg Price Incl O&P	Large Volume	
					Avg Mat'l Unit Cost	Avg Labor Unit Cost	Avg Equip Unit Cost	Avg Total Unit Cost		Avg Total Unit Cost	Avg Price Incl O&P

Studs/plates, per LF of wall or partition

Walls or partitions

Description	Oper	Unit	Crew Size	Man-Hours Per Unit	Avg Mat'l Unit Cost	Avg Labor Unit Cost	Avg Equip Unit Cost	Avg Total Unit Cost	Avg Price Incl O&P	Avg Total Unit Cost	Avg Price Incl O&P
2" x 4" - 8', 12" oc	Demo	LF	LB	.250	---	6.66	---	6.66	9.92	5.67	8.45
2" x 4" - 8', 12" oc	Inst	LF	2C	.302	5.30	9.69	.88	15.87	21.50	13.85	18.70
2" x 4" - 10', 12" oc	Demo	LF	LB	.262	---	6.98	---	6.98	10.40	5.91	8.81
2" x 4" - 10', 12" oc	Inst	LF	2C	.320	6.33	10.30	.93	17.56	23.60	15.24	20.50
2" x 4" - 12', 12" oc	Demo	LF	LB	.271	---	7.22	---	7.22	10.80	6.18	9.21
2" x 4" - 12', 12" oc	Inst	LF	2C	.333	7.38	10.70	.97	19.05	25.50	16.55	22.10
2" x 4" - 8', 16" oc	Demo	LF	LB	.222	---	5.91	---	5.91	8.81	5.01	7.46
2" x 4" - 8', 16" oc	Inst	LF	2C	.267	4.28	8.57	.78	13.63	18.60	11.92	16.20
2" x 4" - 10', 16" oc	Demo	LF	LB	.232	---	6.18	---	6.18	9.21	5.27	7.86
2" x 4" - 10', 16" oc	Inst	LF	2C	.281	5.06	9.02	.82	14.90	20.20	12.97	17.50
2" x 4" - 12', 16" oc	Demo	LF	LB	.242	---	6.45	---	6.45	9.60	5.46	8.14
2" x 4" - 12', 16" oc	Inst	LF	2C	.291	5.79	9.34	.85	15.98	21.50	13.87	18.60
2" x 4" - 8', 24" oc	Demo	LF	LB	.190	---	5.06	---	5.06	7.54	4.32	6.43
2" x 4" - 8', 24" oc	Inst	LF	2C	.232	3.22	7.44	.68	11.34	15.60	9.86	13.50
2" x 4" - 10', 24" oc	Demo	LF	LB	.195	---	5.19	---	5.19	7.74	4.40	6.55
2" x 4" - 10', 24" oc	Inst	LF	2C	.235	3.79	7.54	.69	12.02	16.40	10.44	14.20
2" x 4" - 12', 24" oc	Demo	LF	LB	.198	---	5.27	---	5.27	7.86	4.48	6.67
2" x 4" - 12', 24" oc	Inst	LF	2C	.239	4.26	7.67	.70	12.63	17.10	10.97	14.80
2" x 6" - 8', 12" oc	Demo	LF	LB	.286	---	7.62	---	7.62	11.40	6.45	9.60
2" x 6" - 8', 12" oc	Inst	LF	2C	.340	7.50	10.90	.99	19.39	26.00	16.97	22.70
2" x 6" - 10', 12" oc	Demo	LF	LB	.296	---	7.89	---	7.89	11.80	6.77	10.10
2" x 6" - 10', 12" oc	Inst	LF	2C	.356	8.93	11.40	1.04	21.37	28.50	18.65	24.70
2" x 6" - 12', 12" oc	Demo	LF	LB	.314	---	8.36	---	8.36	12.50	7.11	10.60
2" x 6" - 12', 12" oc	Inst	LF	2C	.372	10.40	11.90	1.09	23.39	31.00	20.62	27.10
2" x 6" - 8', 16" oc	Demo	LF	LB	.254	---	6.77	---	6.77	10.10	5.75	8.57
2" x 6" - 8', 16" oc	Inst	LF	2C	.308	6.06	9.88	.90	16.84	22.70	14.65	19.70
2" x 6" - 10', 16" oc	Demo	LF	LB	.267	---	7.11	---	7.11	10.60	5.99	8.93
2" x 6" - 10', 16" oc	Inst	LF	2C	.320	7.13	10.30	.93	18.36	24.50	15.94	21.30
2" x 6" - 12', 16" oc	Demo	LF	LB	.271	---	7.22	---	7.22	10.80	6.18	9.21
2" x 6" - 12', 16" oc	Inst	LF	2C	.327	8.22	10.50	.95	19.67	26.10	17.09	22.60
2" x 6" - 8', 24" oc	Demo	LF	LB	.216	---	5.75	---	5.75	8.57	4.90	7.30
2" x 6" - 8', 24" oc	Inst	LF	2C	.262	4.58	8.41	.76	13.75	18.60	11.91	16.10
2" x 6" - 10', 24" oc	Demo	LF	LB	.222	---	5.91	---	5.91	8.81	5.01	7.46
2" x 6" - 10', 24" oc	Inst	LF	2C	.267	5.33	8.57	.78	14.68	19.80	12.70	17.00
2" x 6" - 12', 24" oc	Demo	LF	LB	.225	---	5.99	---	5.99	8.93	5.06	7.54
2" x 6" - 12', 24" oc	Inst	LF	2C	.267	6.06	8.57	.78	15.41	20.60	13.50	18.00
2" x 8" - 8', 12" oc	Demo	LF	LB	.320	---	8.52	---	8.52	12.70	7.22	10.80
2" x 8" - 8', 12" oc	Inst	LF	2C	.381	10.60	12.20	1.11	23.91	31.60	20.87	27.50
2" x 8" - 10', 12" oc	Demo	LF	LB	.333	---	8.87	---	8.87	13.20	7.62	11.40
2" x 8" - 10', 12" oc	Inst	LF	2C	.400	12.60	12.80	1.17	26.57	34.90	23.09	30.20
2" x 8" - 12', 12" oc	Demo	LF	LB	.356	---	9.48	---	9.48	14.10	8.05	12.00
2" x 8" - 12', 12" oc	Inst	LF	2C	.421	14.70	13.50	1.23	29.43	38.40	25.54	33.20

				Costs Based On Small Volume						Large Volume	
Description	Oper Unit	Crew Size	Man-Hours Per Unit	Avg Mat'l Unit Cost	Avg Labor Unit Cost	Avg Equip Unit Cost	Avg Total Unit Cost	Avg Price Incl O&P	Avg Total Unit Cost	Avg Price Incl O&P
2" x 8" - 8', 16" oc	Demo LF	LB	.286	---	7.62	---	7.62	11.40	6.45	9.6
2" x 8" - 8', 16" oc	Inst LF	2C	.340	8.49	10.90	.99	20.38	27.10	17.75	23.6
2" x 8" - 10', 16" oc	Demo LF	LB	.296	---	7.89	---	7.89	11.80	6.77	10.1
2" x 8" - 10', 16" oc	Inst LF	2C	.356	10.00	11.40	1.04	22.44	29.70	19.50	25.7
2" x 8" - 12', 16" oc	Demo LF	LB	.308	---	8.21	---	8.21	12.20	6.98	10.4
2" x 8" - 12', 16" oc	Inst LF	2C	.364	11.60	11.70	1.06	24.36	31.90	21.08	27.6
2" x 8" - 8', 24" oc	Demo LF	LB	.242	---	6.45	---	6.45	9.60	5.46	8.1
2" x 8" - 8', 24" oc	Inst LF	2C	.291	6.46	9.34	.85	16.65	22.30	14.36	19.2
2" x 8" - 10', 24" oc	Demo LF	LB	.246	---	6.55	---	6.55	9.76	5.62	8.3
2" x 8" - 10', 24" oc	Inst LF	2C	.296	7.48	9.50	.86	17.84	23.70	15.42	20.4
2" x 8" - 12', 24" oc	Demo LF	LB	.254	---	6.77	---	6.77	10.10	5.75	8.5
2" x 8" - 12', 24" oc	Inst LF	2C	.302	8.49	9.69	.88	19.06	25.20	16.59	21.9

Sheathing, walls

Boards, 1" x 8"

Horizontal	Demo SF	LB	.025	---	.67	---	.67	.99	.56	.83
Horizontal	Inst SF	2C	.027	.78	.87	.08	1.73	2.28	1.51	1.98
Diagonal	Demo SF	LB	.028	---	.75	---	.75	1.11	.64	.95
Diagonal	Inst SF	2C	.030	.86	.96	.09	1.91	2.52	1.68	2.22

Plywood

3/8"	Demo SF	LB	.016	---	.43	---	.43	.64	.37	.56
3/8"	Inst SF	2C	.018	.44	.58	.05	1.07	1.42	.92	1.22
1/2"	Demo SF	LB	.016	---	.43	---	.43	.64	.37	.56
1/2"	Inst SF	2C	.018	.59	.58	.05	1.22	1.60	1.06	1.38
5/8"	Demo SF	LB	.016	---	.43	---	.43	.64	.37	.56
5/8"	Inst SF	2C	.018	.73	.58	.05	1.36	1.76	1.18	1.52

Particleboard

1/2"	Demo SF	LB	.016	---	.43	---	.43	.64	.37	.56
1/2"	Inst SF	2C	.018	.51	.58	.05	1.14	1.50	.98	1.29

OSB strand board

3/8"	Demo SF	LB	.016	---	.43	---	.43	.64	.37	.56
3/8"	Inst SF	2C	.018	.40	.58	.05	1.03	1.38	.88	1.18
1/2"	Demo SF	LB	.016	---	.43	---	.43	.64	.37	.56
1/2"	Inst SF	2C	.018	.45	.58	.05	1.08	1.43	.93	1.23
5/8"	Demo SF	LB	.016	---	.43	---	.43	.64	.37	.56
5/8"	Inst SF	2C	.018	.73	.58	.05	1.36	1.76	1.18	1.52

Sheathing, roof

Boards, 1" x 8"

Horizontal	Demo SF	LB	.021	---	.56	---	.56	.83	.48	.71
Horizontal	Inst SF	2C	.023	.78	.74	.07	1.59	2.07	1.40	1.83
Diagonal	Demo SF	LB	.024	---	.64	---	.64	.95	.56	.83
Diagonal	Inst SF	2C	.026	.86	.83	.08	1.77	2.32	1.55	2.01

Description	Oper	Unit	Crew Size	Man-Hours Per Unit	Avg Mat'l Unit Cost	Avg Labor Unit Cost	Avg Equip Unit Cost	Avg Total Unit Cost	Avg Price Incl O&P	Avg Total Unit Cost	Avg Price Incl O&P
								Costs Based On Small Volume		**Large Volume**	
Plywood											
1/2"	Demo	SF	LB	.015	---	.40	---	.40	.60	.35	.52
1/2"	Inst	SF	2C	.016	.57	.51	.05	1.13	**1.48**	.97	**1.25**
5/8"	Demo	SF	LB	.015	---	.40	---	.40	.60	.35	.52
5/8"	Inst	SF	2C	.016	.69	.51	.05	1.25	**1.61**	1.09	**1.39**
3/4"	Demo	SF	LB	.016	---	.43	---	.43	.64	.35	.52
3/4"	Inst	SF	2C	.017	.81	.55	.05	1.41	**1.80**	1.24	**1.59**
Sheathing, subfloor											
Boards, 1" x 8"											
Horizontal	Demo	SF	LB	.023	---	.61	---	.61	**.91**	.51	**.75**
Horizontal	Inst	SF	2C	.025	.81	.80	.07	1.68	**2.21**	1.45	**1.90**
Diagonal	Demo	SF	LB	.026	---	.69	---	.69	**1.03**	.59	**.87**
Diagonal	Inst	SF	2C	.028	.88	.90	.08	1.86	**2.44**	1.63	**2.13**
Plywood											
5/8"	Demo	SF	LB	.015	---	.40	---	.40	.60	.35	.52
5/8"	Inst	SF	2C	.017	.73	.55	.05	1.33	**1.71**	1.15	**1.47**
3/4"	Demo	SF	LB	.017	---	.45	---	.45	.67	.37	.56
3/4"	Inst	SF	2C	.018	.85	.58	.05	1.48	**1.89**	1.28	**1.64**
1-1/8"	Demo	SF	LB	.022	---	.59	---	.59	.87	.51	.75
1-1/8"	Inst	SF	2C	.024	1.24	.77	.07	2.08	**2.65**	1.82	**2.31**
Particleboard											
5/8"	Demo	SF	LB	.015	---	.40	---	.40	.60	.35	.52
5/8"	Inst	SF	2C	.017	.57	.55	.05	1.17	**1.52**	1.01	**1.31**
3/4"	Demo	SF	LB	.017	---	.45	---	.45	.67	.37	.56
3/4"	Inst	SF	2C	.018	.74	.58	.05	1.37	**1.77**	1.18	**1.52**
Underlayment											
Plywood, 3/8"	Demo	SF	LB	.015	---	.40	---	.40	.60	.35	.52
Plywood, 3/8"	Inst	SF	2C	.017	.48	.55	.05	1.08	**1.42**	.93	**1.22**
Hardboard, 0.215"	Demo	SF	LB	.015	---	.40	---	.40	.60	.35	.52
Hardboard, 0.215"	Inst	SF	2C	.017	.34	.55	.05	.94	**1.26**	.80	**1.07**
Trusses, shop fabricated, wood "W" type											
1/8 pitch											
3" rise in 12" run											
20' span	Demo	Ea	LJ	1.032	---	27.50	---	27.50	**41.00**	23.00	**34.30**
20' span	Inst	Ea	CX	1.10	36.90	35.30	1.61	73.81	**97.00**	65.07	**85.20**
22' span	Demo	Ea	LJ	1.032	---	27.50	---	27.50	**41.00**	23.00	**34.30**
22' span	Inst	Ea	CX	1.10	41.10	35.30	1.61	78.01	**102.00**	68.97	**89.60**
24' span	Demo	Ea	LJ	1.032	---	27.50	---	27.50	**41.00**	23.00	**34.30**
24' span	Inst	Ea	CX	1.10	43.10	35.30	1.61	80.01	**104.00**	70.67	**91.70**
26' span	Demo	Ea	LJ	1.032	---	27.50	---	27.50	**41.00**	23.00	**34.30**
26' span	Inst	Ea	CX	1.10	47.20	35.30	1.61	84.11	**109.00**	74.47	**96.00**
28' span	Demo	Ea	LJ	1.032	---	27.50	---	27.50	**41.00**	23.00	**34.30**
28' span	Inst	Ea	CX	1.10	49.20	35.30	1.61	86.11	**111.00**	76.27	**98.10**

					Costs Based On Small Volume					Large Volume	
Description	Oper	Unit	Crew Size	Man-Hours Per Unit	Avg Mat'l Unit Cost	Avg Labor Unit Cost	Avg Equip Unit Cost	Avg Total Unit Cost	Avg Price Incl O&P	Avg Total Unit Cost	Avg Price Incl O&P
30' span	Demo	Ea	LJ	1.07	---	28.50	---	28.50	42.50	24.40	36.3
30' span	Inst	Ea	CX	1.19	53.40	38.20	1.73	93.33	120.00	82.06	105.0
32' span	Demo	Ea	LJ	1.07	---	28.50	---	28.50	42.50	24.40	36.3
32' span	Inst	Ea	CX	1.19	54.80	38.20	1.73	94.73	122.00	83.36	107.0
34' span	Demo	Ea	LJ	1.07	---	28.50	---	28.50	42.50	24.40	36.3
34' span	Inst	Ea	CX	1.19	57.30	38.20	1.73	97.23	125.00	85.66	109.0
36' span	Demo	Ea	LJ	1.14	---	30.40	---	30.40	45.30	25.80	38.5
36' span	Inst	Ea	CX	1.23	61.50	39.50	1.79	102.79	132.00	90.51	115.0

5/24 pitch
5" rise in 12" run

20' span	Demo	Ea	LJ	1.032	---	27.50	---	27.50	41.00	23.00	34.3
20' span	Inst	Ea	CX	1.10	29.40	35.30	1.61	66.31	88.40	58.27	77.4
22' span	Demo	Ea	LJ	1.032	---	27.50	---	27.50	41.00	23.00	34.3
22' span	Inst	Ea	CX	1.10	33.00	35.30	1.61	69.91	92.50	61.57	81.1
24' span	Demo	Ea	LJ	1.032	---	27.50	---	27.50	41.00	23.00	34.3
24' span	Inst	Ea	CX	1.10	34.80	35.30	1.61	71.71	94.60	63.27	83.1
26' span	Demo	Ea	LJ	1.032	---	27.50	---	27.50	41.00	23.00	34.3
26' span	Inst	Ea	CX	1.10	39.00	35.30	1.61	75.91	99.40	66.97	87.4
28' span	Demo	Ea	LJ	1.032	---	27.50	---	27.50	41.00	23.00	34.3
28' span	Inst	Ea	CX	1.10	46.40	35.30	1.61	83.31	108.00	73.77	95.2
30' span	Demo	Ea	LJ	1.07	---	28.50	---	28.50	42.50	24.40	36.3
30' span	Inst	Ea	CX	1.19	50.00	38.20	1.73	89.93	117.00	78.96	102.0
32' span	Demo	Ea	LJ	1.07	---	28.50	---	28.50	42.50	24.40	36.3
32' span	Inst	Ea	CX	1.19	51.30	38.20	1.73	91.23	118.00	80.16	103.0
34' span	Demo	Ea	LJ	1.07	---	28.50	---	28.50	42.50	24.40	36.3
34' span	Inst	Ea	CX	1.19	56.60	38.20	1.73	96.53	124.00	84.96	109.0
36' span	Demo	Ea	LJ	1.14	---	30.40	---	30.40	45.30	25.80	38.5
36' span	Inst	Ea	CX	1.23	58.40	39.50	1.79	99.69	128.00	87.71	112.0

1/4 pitch
6" rise in 12" run

20' span	Demo	Ea	LJ	1.032	---	27.50	---	27.50	41.00	23.00	34.30
20' span	Inst	Ea	CX	1.10	30.00	35.30	1.61	66.91	89.00	58.77	78.00
22' span	Demo	Ea	LJ	1.032	---	27.50	---	27.50	41.00	23.00	34.30
22' span	Inst	Ea	CX	1.10	33.50	35.30	1.61	70.41	93.10	62.07	81.70
24' span	Demo	Ea	LJ	1.032	---	27.50	---	27.50	41.00	23.00	34.30
24' span	Inst	Ea	CX	1.10	37.60	35.30	1.61	74.51	97.90	65.77	86.00
26' span	Demo	Ea	LJ	1.032	---	27.50	---	27.50	41.00	23.00	34.30
26' span	Inst	Ea	CX	1.10	39.00	35.30	1.61	75.91	99.40	66.97	87.40
28' span	Demo	Ea	LJ	1.032	---	27.50	---	27.50	41.00	23.00	34.30
28' span	Inst	Ea	CX	1.10	47.00	35.30	1.61	83.91	109.00	74.27	95.80
30' span	Demo	Ea	LJ	1.07	---	28.50	---	28.50	42.50	24.40	36.30
30' span	Inst	Ea	CX	1.19	48.90	38.20	1.73	88.83	115.00	77.96	101.00
32' span	Demo	Ea	LJ	1.07	---	28.50	---	28.50	42.50	24.40	36.30
32' span	Inst	Ea	CX	1.19	53.00	38.20	1.73	92.93	120.00	81.76	105.00
34' span	Demo	Ea	LJ	1.07	---	28.50	---	28.50	42.50	24.40	36.30
34' span	Inst	Ea	CX	1.19	57.10	38.20	1.73	97.03	125.00	85.46	109.00
36' span	Demo	Ea	LJ	1.14	---	30.40	---	30.40	45.30	25.80	38.50
36' span	Inst	Ea	CX	1.23	60.10	39.50	1.79	101.39	130.00	89.21	114.00

Garage doors

Description	Oper	Unit	Crew Size	Man-Hours Per Unit	Avg Mat'l Unit Cost	Avg Labor Unit Cost	Avg Equip Unit Cost	Avg Total Unit Cost	Avg Price Incl O&P	Avg Total Unit Cost	Avg Price Incl O&P
					Costs Based On Small Volume					Large Volume	
Detach & reset operations											
Wood, aluminum, or hardboard											
Single	Reset	Ea	LB	3.08	---	82.10	---	82.10	**122.00**	71.10	**106.00**
Double	Reset	Ea	LB	4.10	---	109.00	---	109.00	**163.00**	94.80	**141.00**
Steel											
Single	Reset	Ea	LB	3.52	---	93.80	---	93.80	**140.00**	85.30	**127.00**
Double	Reset	Ea	LB	4.92	---	131.00	---	131.00	**195.00**	107.00	**159.00**
Remove operations											
Wood, aluminum, or hardboard											
Single	Demo	Ea	LB	3.08	---	82.10	---	82.10	**122.00**	53.30	**79.40**
Double	Demo	Ea	LB	4.10	---	109.00	---	109.00	**163.00**	71.10	**106.00**
Steel											
Single	Demo	Ea	LB	3.52	---	93.80	---	93.80	**140.00**	61.00	**90.90**
Double	Demo	Ea	LB	4.92	---	131.00	---	131.00	**195.00**	85.30	**127.00**
Replace operations											
Aluminum frame with plastic skin bonded to polystyrene foam core											
Jamb type with hardware and deluxe lock											
8' x 7', single	Inst	Ea	2C	5.71	332.00	183.00	---	515.00	**657.00**	416.00	**524.00**
8' x 8', single	Inst	Ea	2C	5.71	432.00	183.00	---	615.00	**772.00**	503.00	**624.00**
9' x 7', single	Inst	Ea	2C	5.71	362.00	183.00	---	545.00	**692.00**	443.00	**555.00**
9' x 8', single	Inst	Ea	2C	5.71	466.00	183.00	---	649.00	**811.00**	533.00	**658.00**
16' x 7', double	Inst	Ea	2C	7.62	598.00	245.00	---	843.00	**1050.00**	690.00	**854.00**
16' x 8', double	Inst	Ea	2C	7.62	775.00	245.00	---	1020.00	**1260.00**	844.00	**1030.00**
Track type with hardware and deluxe lock											
8' x 7', single	Inst	Ea	2C	5.71	392.00	183.00	---	575.00	**725.00**	468.00	**584.00**
9' x 7', single	Inst	Ea	2C	5.71	431.00	183.00	---	614.00	**770.00**	502.00	**623.00**
16' x 7', double	Inst	Ea	2C	7.62	700.00	245.00	---	945.00	**1170.00**	779.00	**956.00**
Sectional type with hardware and key lock											
8' x 7', single	Inst	Ea	2C	5.71	560.00	183.00	---	743.00	**919.00**	615.00	**752.00**
9' x 7', single	Inst	Ea	2C	5.71	612.00	183.00	---	795.00	**979.00**	660.00	**805.00**
16' x 7', double	Inst	Ea	2C	7.62	1020.00	245.00	---	1265.00	**1540.00**	1057.00	**1280.00**
Fiberglass											
Jamb type with hardware and deluxe lock											
8' x 7', single	Inst	Ea	2C	5.71	293.00	183.00	---	476.00	**612.00**	382.00	**485.00**
8' x 8', single	Inst	Ea	2C	5.71	411.00	183.00	---	594.00	**748.00**	485.00	**603.00**
9' x 7', single	Inst	Ea	2C	5.71	350.00	183.00	---	533.00	**678.00**	432.00	**542.00**
9' x 8', single	Inst	Ea	2C	5.71	456.00	183.00	---	639.00	**800.00**	524.00	**648.00**
16' x 7', double	Inst	Ea	2C	7.62	621.00	245.00	---	866.00	**1080.00**	711.00	**877.00**
16' x 8', double	Inst	Ea	2C	7.62	826.00	245.00	---	1071.00	**1320.00**	889.00	**1080.00**
Track type with hardware and deluxe lock											
8' x 7', single	Inst	Ea	2C	5.71	392.00	183.00	---	575.00	**725.00**	468.00	**584.00**
9' x 7', single	Inst	Ea	2C	5.71	431.00	183.00	---	614.00	**770.00**	502.00	**623.00**
16' x 7', double	Inst	Ea	2C	7.62	666.00	245.00	---	911.00	**1130.00**	750.00	**922.00**
Sectional type with hardware and key lock											
8' x 7', single	Inst	Ea	2C	5.71	348.00	183.00	---	531.00	**675.00**	430.00	**540.00**
9' x 7', single	Inst	Ea	2C	5.71	378.00	183.00	---	561.00	**710.00**	457.00	**570.00**
16' x 7', double	Inst	Ea	2C	7.62	594.00	245.00	---	839.00	**1050.00**	687.00	**850.00**

					Costs Based On Small Volume					Large Volume	
Description	Oper	Unit	Crew Size	Man-Hours Per Unit	Avg Mat'l Unit Cost	Avg Labor Unit Cost	Avg Equip Unit Cost	Avg Total Unit Cost	Avg Price Incl O&P	Avg Total Unit Cost	Avg Price Incl O&P
Steel											
Jamb type with hardware and deluxe lock											
8' x 7', single	Inst	Ea	2C	5.71	254.00	183.00	---	437.00	**567.00**	348.00	446.00
8' x 8', single	Inst	Ea	2C	5.71	339.00	183.00	---	522.00	**665.00**	423.00	531.00
9' x 7', single	Inst	Ea	2C	5.71	278.00	183.00	---	461.00	**595.00**	370.00	470.00
9' x 8', single	Inst	Ea	2C	5.71	378.00	183.00	---	561.00	**710.00**	457.00	570.00
16' x 7', double	Inst	Ea	2C	7.62	472.00	245.00	---	717.00	**910.00**	581.00	728.00
16' x 8', double	Inst	Ea	2C	7.62	643.00	245.00	---	888.00	**1110.00**	730.00	899.00
Track type with hardware and deluxe lock											
8' x 7', single	Inst	Ea	2C	5.71	287.00	183.00	---	470.00	**605.00**	377.00	479.00
9' x 7', single	Inst	Ea	2C	5.71	320.00	183.00	---	503.00	**642.00**	406.00	512.00
16' x 7', double	Inst	Ea	2C	7.62	512.00	245.00	---	757.00	**956.00**	616.00	769.00
Sectional type with hardware and key lock											
8' x 7', single	Inst	Ea	2C	5.71	279.00	183.00	---	462.00	**596.00**	371.00	472.00
9' x 7', single	Inst	Ea	2C	5.71	292.00	183.00	---	475.00	**610.00**	381.00	484.00
16' x 7', double	Inst	Ea	2C	7.62	561.00	245.00	---	806.00	**1010.00**	659.00	817.00
Wood											
Jamb type with hardware and deluxe lock											
8' x 7', single	Inst	Ea	2C	5.71	349.00	183.00	---	532.00	**676.00**	431.00	541.00
8' x 8', single	Inst	Ea	2C	5.71	362.00	183.00	---	545.00	**692.00**	443.00	555.00
9' x 7', single	Inst	Ea	2C	5.71	381.00	183.00	---	564.00	**713.00**	459.00	573.00
9' x 8', single	Inst	Ea	2C	5.71	471.00	183.00	---	654.00	**816.00**	537.00	663.00
16' x 7', double	Inst	Ea	2C	7.62	617.00	245.00	---	862.00	**1080.00**	707.00	873.00
16' x 8', double	Inst	Ea	2C	7.62	797.00	245.00	---	1042.00	**1280.00**	863.00	1050.00
Track type with hardware and deluxe lock											
8' x 7', single	Inst	Ea	2C	5.71	398.00	183.00	---	581.00	**732.00**	474.00	590.00
9' x 7', single	Inst	Ea	2C	5.71	431.00	183.00	---	614.00	**770.00**	502.00	623.00
16' x 7', double	Inst	Ea	2C	7.62	688.00	245.00	---	933.00	**1160.00**	769.00	944.00
Sectional type with hardware and key lock											
8' x 7', single	Inst	Ea	2C	5.71	387.00	183.00	---	570.00	**720.00**	464.00	579.00
9' x 7', single	Inst	Ea	2C	5.71	415.00	183.00	---	598.00	**752.00**	488.00	607.00
16' x 7', double	Inst	Ea	2C	7.62	771.00	245.00	---	1016.00	**1250.00**	841.00	1030.00

Garage door operators

Radio controlled for single or double doors.
Labor includes wiring, connection and installation

Description	Oper	Unit	Crew Size	Man-Hours Per Unit	Costs Based On Small Volume						Large Volume	
					Avg Mat'l Unit Cost	Avg Labor Unit Cost	Avg Equip Unit Cost	Avg Total Unit Cost	Avg Price Incl O&P		Avg Total Unit Cost	Avg Price Incl O&P
Chain drive, 1/4 hp, with receiver and												
one transmitter	Inst	LS	ED	7.11	200.00	236.00	---	436.00	**575.00**		347.00	**454.00**
Screw-worm drive, 1/3 hp, with receiver and												
one transmitter	Inst	LS	ED	7.11	235.00	236.00	---	471.00	**615.00**		377.00	**488.00**
Deluxe models, 1/2 hp, with receiver, transmitter, and time delay light												
Chain drive, not for												
vault-type garages	Inst	LS	ED	7.11	265.00	236.00	---	501.00	**650.00**		402.00	**518.00**
Screw drive with threaded												
worm screw	Inst	LS	ED	7.11	300.00	236.00	---	536.00	**690.00**		432.00	**552.00**
Additional transmitters, ADD	Inst	Ea	EA	---	42.00	---	---	42.00	**48.30**		35.70	**41.10**
Exterior key switch	Inst	Ea	EA	.889	25.00	30.60	---	55.60	**73.40**		44.20	**57.90**
To remove and replace												
unit and receiver	Inst	LS	2C	3.56	---	114.00	---	114.00	**171.00**		85.70	**129.00**

				Costs Based On Small Volume					Large Volume	
Description	Oper	Unit	Crew Size	Man-Hours Per Unit	Avg Mat'l Unit Cost	Avg Labor Unit Cost	Avg Equip Unit Cost	Avg Total Unit Cost	Avg Price Incl O&P	Avg Total Unit Cost	Avg Price Incl O&P

Garbage disposers

Includes wall switch and labor includes rough-in.
See also Trash compactors, page 278

Frequently encountered applications

Detach & reset operations

Garbage disposer	Reset	Ea	SA	1.25	---	45.40	---	45.40	**66.80**	36.40	**53.40**

Remove operations

Garbage disposer	Demo	Ea	SA	1.00	---	36.40	---	36.40	**53.40**	29.10	**42.70**

Replace operations

Standard, 1/3 HP	Inst	Ea	SC	9.09	112.00	322.00	---	434.00	**602.00**	349.40	**484.00**
Average, 1/2 HP	Inst	Ea	SC	9.09	128.00	322.00	---	450.00	**619.00**	362.00	**499.00**
High, 3/4 HP	Inst	Ea	SC	9.09	196.00	322.00	---	518.00	**698.00**	418.00	**563.00**
Premium, 1 HP	Inst	Ea	SC	9.09	322.00	322.00	---	644.00	**843.00**	522.00	**683.00**

In-Sink-Erator Products

Model "Badger 1," 1/3 HP, continuous feed, 1 year parts protection	Inst	Ea	SC	9.09	112.00	322.00	---	434.00	**602.00**	349.40	**484.00**
Model "Badger V," 1/2 HP, continuous feed, 1 year parts protection	Inst	Ea	SC	9.09	128.00	322.00	---	450.00	**619.00**	362.00	**499.00**
Model 333, 3/4 HP, continuous feed, 4 year parts protection	Inst	Ea	SC	9.09	196.00	322.00	---	518.00	**698.00**	418.00	**563.00**
Model 77, 1 HP, automatic reversing feed, 7 year parts protection, stainless steel construction	Inst	Ea	SC	9.09	322.00	322.00	---	644.00	**843.00**	522.00	**683.00**
Model 777SS 1 HP, continuous feed, 7 year parts protection, stainless steel construction	Inst	Ea	SC	9.09	322.00	322.00	---	644.00	**843.00**	522.00	**683.00**
Model 17, 3/4 HP, batch feed, auto reversing, 5 year parts protection, stainless steel construction	Inst	Ea	SC	9.09	419.00	322.00	---	741.00	**954.00**	602.00	**775.00**
Septic, 3 year parts protection, stainless steel grind elements	Inst	Ea	SC	9.09	265.00	322.00	---	587.00	**777.00**	475.00	**628.00**

Adjustments

To only remove and reset garbage disposer	Reset	Ea	SA	1.25	---	45.40	---	45.40	**66.80**	36.40	**53.40**
To only remove garbage disposer	Demo	Ea	SA	1.00	---	36.40	---	36.40	**53.40**	29.10	**42.70**

Parts and accessories

Stainless steel stopper	Inst	Ea	---	---	9.99	---	---	9.99	**9.99**	8.23	**8.23**
Dishwasher connector kit	Inst	Ea	---	---	8.93	---	---	8.93	**8.93**	7.35	**7.35**
Flexible tail pipe	Inst	Ea	---	---	11.50	---	---	11.50	**11.50**	9.45	**9.45**
Power cord accessory kit	Inst	Ea	---	---	9.78	---	---	9.78	**9.78**	8.05	**8.05**
Service wrench	Inst	Ea	---	---	4.89	---	---	4.89	**4.89**	4.03	**4.03**
Plastic stopper	Inst	Ea	---	---	4.89	---	---	4.89	**4.89**	4.03	**4.03**
Deluxe mounting gasket	Inst	Ea	---	---	7.01	---	---	7.01	**7.01**	5.78	**5.78**

Girders. See Framing, page 120

Description	Oper	Unit	Crew Size	Man-Hours Per Unit	Avg Mat'l Unit Cost	Avg Labor Unit Cost	Avg Equip Unit Cost	Avg Total Unit Cost	Avg Price Incl O&P	Avg Total Unit Cost	Avg Price Incl O&P
Glass and glazing											
3/16" T float with putty in wood sash											
8" x 12"	Demo	SF	GA	.533	---	15.90	---	15.90	**23.50**	9.53	**14.10**
8" x 12"	Inst	SF	GA	.444	6.56	13.20	---	19.76	**26.80**	13.87	**18.30**
12" x 16"	Demo	SF	GA	.296	---	8.81	---	8.81	**13.00**	5.30	**7.84**
12" x 16"	Inst	SF	GA	.242	5.37	7.21	---	12.58	**16.60**	9.17	**11.70**
14" x 20"	Demo	SF	GA	.242	---	7.21	---	7.21	**10.70**	4.32	**6.39**
14" x 20"	Inst	SF	GA	.190	4.90	5.66	---	10.56	**13.80**	7.81	**9.89**
16" x 24"	Demo	SF	GA	.190	---	5.66	---	5.66	**8.37**	3.39	**5.02**
16" x 24"	Inst	SF	GA	.148	4.57	4.41	---	8.98	**11.60**	6.78	**8.47**
24" x 26"	Demo	SF	GA	.140	---	4.17	---	4.17	**6.17**	2.50	**3.70**
24" x 26"	Inst	SF	GA	.111	4.11	3.31	---	7.42	**9.41**	5.71	**7.03**
36" x 24"	Demo	SF	GA	.107	---	3.19	---	3.19	**4.72**	1.91	**2.82**
36" x 24"	Inst	SF	GA	.086	3.91	2.56	---	6.47	**8.09**	5.08	**6.17**
1/8" T float with putty in steel sash											
12" x 16"	Demo	SF	GA	.296	---	8.81	---	8.81	**13.00**	5.30	**7.84**
12" x 16"	Inst	SF	GA	.242	7.56	7.21	---	14.77	**19.00**	11.14	**13.90**
16" x 20"	Demo	SF	GA	.205	---	6.10	---	6.10	**9.03**	3.66	**5.42**
16" x 20"	Inst	SF	GA	.167	6.53	4.97	---	11.50	**14.50**	8.87	**10.90**
16" x 24"	Demo	SF	GA	.190	---	5.66	---	5.66	**8.37**	3.39	**5.02**
16" x 24"	Inst	SF	GA	.148	6.28	4.41	---	10.69	**13.40**	8.32	**10.20**
24" x 24"	Demo	SF	GA	.140	---	4.17	---	4.17	**6.17**	2.50	**3.70**
24" x 24"	Inst	SF	GA	.111	5.54	3.31	---	8.85	**11.00**	7.00	**8.45**
28" x 32"	Demo	SF	GA	.103	---	3.07	---	3.07	**4.54**	1.85	**2.73**
28" x 32"	Inst	SF	GA	.083	5.14	2.47	---	7.61	**9.31**	6.13	**7.31**
36" x 36"	Demo	SF	GA	.083	---	2.47	---	2.47	**3.66**	1.49	**2.20**
36" x 36"	Inst	SF	GA	.065	4.79	1.94	---	6.73	**8.13**	5.48	**6.47**
36" x 48"	Demo	SF	GA	.070	---	2.08	---	2.08	**3.08**	1.25	**1.85**
36" x 48"	Inst	SF	GA	.053	4.56	1.58	---	6.14	**7.35**	5.06	**5.93**
1/4" T float											
With putty and points in wood sash											
72" x 48"	Demo	SF	GA	.072	---	2.14	---	2.14	**3.17**	1.28	**1.90**
72" x 48"	Inst	SF	GB	.087	4.53	2.59	---	7.12	**8.82**	5.64	**6.79**
With aluminum channel and rigid neoprene rubber in aluminum sash											
48" x 96"	Demo	SF	GA	.076	---	2.26	---	2.26	**3.35**	1.37	**2.03**
48" x 96"	Inst	SF	GB	.076	4.02	2.26	---	6.28	**7.77**	5.00	**6.02**
96" x 96"	Demo	SF	GA	.074	---	2.20	---	2.20	**3.26**	1.31	**1.94**
96" x 96"	Inst	SF	GC	.062	3.97	1.85	---	5.82	**7.10**	4.69	**5.58**
1" T insulating glass (2 pieces 1/4" T float with 1/2" air space)											
with putty and points in wood sash											
To 6.0 SF	Demo	SF	GA	.267	---	7.95	---	7.95	**11.80**	4.76	**7.05**
To 6.0 SF	Inst	SF	GA	.222	10.90	6.61	---	17.51	**21.80**	13.80	**16.70**
6.1 SF to 12.0 SF	Demo	SF	GA	.121	---	3.60	---	3.60	**5.33**	2.17	**3.22**
6.1 SF to 12.0 SF	Inst	SF	GA	.099	9.83	2.95	---	12.78	**15.20**	10.63	**12.40**
12.1 SF to 18.0 SF	Demo	SF	GA	.089	---	2.65	---	2.65	**3.92**	1.58	**2.34**
12.1 SF to 18.0 SF	Inst	SF	GA	.074	9.38	2.20	---	11.58	**13.60**	9.78	**11.30**
18.1 SF to 24.0 SF	Demo	SF	GA	.092	---	2.74	---	2.74	**4.05**	1.64	**2.42**
18.1 SF to 24.0 SF	Inst	SF	GB	.107	9.07	3.19	---	12.26	**14.70**	10.10	**11.80**

Description	Oper	Unit	Crew Size	Man-Hours Per Unit	Costs Based On Small Volume						Large Volume	
					Avg Mat'l Unit Cost	Avg Labor Unit Cost	Avg Equip Unit Cost	Avg Total Unit Cost	Avg Price Incl O&P		Avg Total Unit Cost	Avg Price Incl O&P
Aluminum sliding door glass with aluminum channel and rigid neoprene rubber												
3/16" T tempered glass												
34" W x 76" H	Demo	SF	GA	.068	---	2.03	---	2.03	3.00		1.22	1.81
34" W x 76" H	Inst	SF	GA	.054	3.31	1.61	---	4.92	6.02		3.97	4.74
46" W x 76" H	Demo	SF	GA	.054	---	1.61	---	1.61	2.38		.98	1.45
46" W x 76" H	Inst	SF	GB	.081	3.28	2.41	---	5.69	7.18		4.39	5.37
5/8" T insulating glass (2 pieces 3/16" T tempered with 1/4" T air space)												
34" W x 76" H	Demo	SF	GA	.078	---	2.32	---	2.32	3.44		1.40	2.07
34" W x 76" H	Inst	SF	GA	.058	6.64	1.73	---	8.37	9.86		7.03	8.13
46" W x 76" H	Demo	SF	GA	.062	---	1.85	---	1.85	2.73		1.10	1.63
46" W x 76" H	Inst	SF	GB	.086	6.55	2.56	---	9.11	11.00		7.46	8.79

Grading. See Concrete, page 69

Glu-lam products

Beams

3-1/8" thick, SF pricing based on 16' oc

Description	Oper	Unit	Crew Size	Man-Hours Per Unit	Avg Mat'l Unit Cost	Avg Labor Unit Cost	Avg Equip Unit Cost	Avg Total Unit Cost	Avg Price Incl O&P	Avg Total Unit Cost	Avg Price Incl O&P
9" deep, 20' long	Demo	LF	LK	.205	---	6.15	.64	6.79	9.81	4.41	6.37
9" deep, 20' long	Inst	LF	CY	.239	13.90	7.24	.64	21.78	27.50	14.40	18.20
9" deep, 20' long	Demo	BF	LK	.068	---	2.04	.21	2.25	3.25	1.46	2.11
9" deep, 20' long	Inst	BF	CY	.080	4.62	2.42	.21	7.25	9.16	4.80	6.05
9" deep, 20' long	Demo	SF	LK	.011	---	.33	.04	.37	.53	.23	.33
9" deep, 20' long	Inst	SF	CY	.013	.75	.39	.04	1.18	1.49	.79	1.00
10-1/2" deep, 20' long	Demo	LF	LK	.205	---	6.15	.64	6.79	9.81	4.41	6.37
10-1/2" deep, 20' long	Inst	LF	CY	.239	16.10	7.24	.64	23.98	30.00	15.85	19.80
10-1/2" deep, 20' long	Demo	BF	LK	.059	---	1.77	.18	1.95	2.82	1.26	1.82
10-1/2" deep, 20' long	Inst	BF	CY	.068	4.59	2.06	.18	6.83	8.55	4.51	5.64
10-1/2" deep, 20' long	Demo	SF	LK	.011	---	.33	.04	.37	.53	.23	.33
10-1/2" deep, 20' long	Inst	SF	CY	.013	.87	.39	.04	1.30	1.63	.87	1.10
12" deep, 20' long	Demo	LF	LK	.205	---	6.15	.64	6.79	9.81	4.41	6.37
12" deep, 20' long	Inst	LF	CY	.239	17.90	7.24	.64	25.78	32.10	17.05	21.20
12" deep, 20' long	Demo	BF	LK	.051	---	1.53	.16	1.69	2.44	1.09	1.58
12" deep, 20' long	Inst	BF	CY	.060	4.46	1.82	.16	6.44	8.02	4.25	5.29
12" deep, 20' long	Demo	SF	LK	.011	---	.33	.04	.37	.53	.23	.33
12" deep, 20' long	Inst	SF	CY	.013	.98	.39	.04	1.41	1.76	.94	1.18
13-1/2" deep, 20' long	Demo	LF	LK	.205	---	6.15	.64	6.79	9.81	4.41	6.37
13-1/2" deep, 20' long	Inst	LF	CY	.239	19.90	7.24	.64	27.78	34.40	18.45	22.80
13-1/2" deep, 20' long	Demo	BF	LK	.046	---	1.38	.14	1.52	2.20	.99	1.43
13-1/2" deep, 20' long	Inst	BF	CY	.053	4.41	1.61	.14	6.16	7.62	4.09	5.06
13-1/2" deep, 20' long	Demo	SF	LK	.011	---	.33	.04	.37	.53	.23	.33
13-1/2" deep, 20' long	Inst	SF	CY	.013	1.10	.39	.04	1.53	1.90	1.02	1.27
15" deep, 20' long	Demo	LF	LK	.205	---	6.15	.64	6.79	9.81	4.41	6.37
15" deep, 20' long	Inst	LF	CY	.239	21.90	7.24	.64	29.78	36.70	19.75	24.30
15" deep, 20' long	Demo	BF	LK	.041	---	1.23	.13	1.36	1.96	.89	1.29
15" deep, 20' long	Inst	BF	CY	.048	4.38	1.45	.13	5.96	7.35	3.94	4.85
15" deep, 20' long	Demo	SF	LK	.011	---	.33	.04	.37	.53	.23	.33
15" deep, 20' long	Inst	SF	CY	.013	1.20	.39	.04	1.63	2.01	1.09	1.35
16-1/2" deep, 20' long	Demo	LF	LK	.205	---	6.15	.64	6.79	9.81	4.41	6.37
16-1/2" deep, 20' long	Inst	LF	CY	.239	23.80	7.24	.64	31.68	38.90	21.05	25.80
16-1/2" deep, 20' long	Demo	BF	LK	.037	---	1.11	.12	1.23	1.77	.80	1.15
16-1/2" deep, 20' long	Inst	BF	CY	.044	4.34	1.33	.12	5.79	7.11	3.82	4.68
16-1/2" deep, 20' long	Demo	SF	LK	.011	---	.33	.04	.37	.53	.23	.33
16-1/2" deep, 20' long	Inst	SF	CY	.013	1.29	.39	.04	1.72	2.11	1.15	1.42
18" deep, 20' long	Demo	LF	LK	.205	---	6.15	.64	6.79	9.81	4.41	6.37
18" deep, 20' long	Inst	LF	CY	.239	25.90	7.24	.64	33.78	41.20	22.35	27.30
18" deep, 20' long	Demo	BF	LK	.034	---	1.02	.11	1.13	1.63	.73	1.05
18" deep, 20' long	Inst	BF	CY	.040	4.32	1.21	.11	5.64	6.90	3.74	4.56
18" deep, 20' long	Demo	SF	LK	.011	---	.33	.04	.37	.53	.23	.33
18" deep, 20' long	Inst	SF	CY	.013	1.43	.39	.04	1.86	2.28	1.24	1.52

					Costs Based On Small Volume				Large Volume	
Description	Oper	Unit	Crew Size	Man-Hours Per Unit	Avg Mat'l Unit Cost	Avg Labor Unit Cost	Avg Equip Unit Cost	Avg Total Unit Cost	Avg Price Incl O&P	Avg Total Unit Cost	Avg Price Incl O&P
19-1/2" deep, 30' long	Demo	LF	LK	.205	---	6.15	.64	6.79	9.81	4.41	6.37
19-1/2" deep, 30' long	Inst	LF	CY	.239	26.60	7.24	.64	34.48	42.10	22.95	27.90
19-1/2" deep, 30' long	Demo	BF	LK	.032	---	.96	.10	1.06	1.53	.69	1.00
19-1/2" deep, 30' long	Inst	BF	CY	.037	4.10	1.12	.10	5.32	6.50	3.52	4.29
19-1/2" deep, 30' long	Demo	SF	LK	.011	---	.33	.04	.37	.53	.23	.33
19-1/2" deep, 30' long	Inst	SF	CY	.013	1.46	.39	.04	1.89	2.31	1.26	1.54
21" deep, 30' long	Demo	LF	LK	.205	---	6.15	.64	6.79	9.81	4.41	6.37
21" deep, 30' long	Inst	LF	CY	.239	28.60	7.24	.64	36.48	44.40	24.25	29.40
21" deep, 30' long	Demo	BF	LK	.029	---	.87	.09	.96	1.39	.63	.91
21" deep, 30' long	Inst	BF	CY	.034	4.10	1.03	.09	5.22	6.35	3.46	4.20
21" deep, 30' long	Demo	SF	LK	.011	---	.33	.04	.37	.53	.23	.33
21" deep, 30' long	Inst	SF	CY	.013	1.56	.39	.04	1.99	2.43	1.33	1.63
22-1/2" deep, 30' long	Demo	LF	LK	.205	---	6.15	.64	6.79	9.81	4.41	6.37
22-1/2" deep, 30' long	Inst	LF	CY	.239	30.50	7.24	.64	38.38	46.60	25.45	30.90
22-1/2" deep, 30' long	Demo	BF	LK	.027	---	.81	.09	.90	1.30	.60	.86
22-1/2" deep, 30' long	Inst	BF	CY	.032	4.07	.97	.09	5.13	6.23	3.41	4.13
22-1/2" deep, 30' long	Demo	SF	LK	.011	---	.33	.04	.37	.53	.23	.33
22-1/2" deep, 30' long	Inst	SF	CY	.013	1.67	.39	.04	2.10	2.55	1.40	1.71
24" deep, 30' long	Demo	LF	LK	.205	---	6.15	.64	6.79	9.81	4.41	6.37
24" deep, 30' long	Inst	LF	CY	.239	32.50	7.24	.64	40.38	48.80	26.75	32.40
24" deep, 30' long	Demo	BF	LK	.026	---	.78	.08	.86	1.24	.56	.81
24" deep, 30' long	Inst	BF	CY	.030	4.05	.91	.08	5.04	6.10	3.33	4.02
24" deep, 30' long	Demo	SF	LK	.011	---	.33	.04	.37	.53	.23	.33
24" deep, 30' long	Inst	SF	CY	.013	1.76	.39	.04	2.19	2.66	1.46	1.77
25-1/2" deep, 30' long	Demo	LF	LK	.205	---	6.15	.64	6.79	9.81	4.41	6.37
25-1/2" deep, 30' long	Inst	LF	CY	.239	34.50	7.24	.64	42.38	51.20	28.15	34.00
25-1/2" deep, 30' long	Demo	BF	LK	.024	---	.72	.08	.80	1.15	.53	.77
25-1/2" deep, 30' long	Inst	BF	CY	.028	4.05	.85	.08	4.98	6.01	3.30	3.97
25-1/2" deep, 30' long	Demo	SF	LK	.011	---	.33	.04	.37	.53	.23	.33
25-1/2" deep, 30' long	Inst	SF	CY	.013	1.89	.39	.04	2.32	2.80	1.55	1.88
27" deep, 30' long	Demo	LF	LK	.205	---	6.15	.64	6.79	9.81	4.41	6.37
27" deep, 30' long	Inst	LF	CY	.239	36.40	7.24	.64	44.28	53.40	29.45	35.50
27" deep, 30' long	Demo	BF	LK	.023	---	.69	.07	.76	1.10	.50	.72
27" deep, 30' long	Inst	BF	CY	.027	4.05	.82	.07	4.94	5.96	3.27	3.93
27" deep, 30' long	Demo	SF	LK	.011	---	.33	.04	.37	.53	.23	.33
27" deep, 30' long	Inst	SF	CY	.013	1.98	.39	.04	2.41	2.91	1.61	1.95

5-1/8" thick, SF pricing based on 16' oc

Description	Oper	Unit	Crew Size	Man-Hours Per Unit	Avg Mat'l Unit Cost	Avg Labor Unit Cost	Avg Equip Unit Cost	Avg Total Unit Cost	Avg Price Incl O&P	Avg Total Unit Cost	Avg Price Incl O&P
12" deep, 20' long	Demo	LF	LK	.205	---	6.15	.64	6.79	9.81	4.41	6.37
12" deep, 20' long	Inst	LF	CY	.239	21.80	7.24	.64	29.68	36.60	19.75	24.30
12" deep, 20' long	Demo	BF	LK	.034	---	1.02	.11	1.13	1.63	.73	1.05
12" deep, 20' long	Inst	BF	CY	.040	3.65	1.21	.11	4.97	6.13	3.29	4.05
12" deep, 20' long	Demo	SF	LK	.011	---	.33	.04	.37	.53	.23	.33
12" deep, 20' long	Inst	SF	CY	.013	1.20	.39	.04	1.63	2.01	1.09	1.35
13-1/2" deep, 20' long	Demo	LF	LK	.205	---	6.15	.64	6.79	9.81	4.41	6.37
13-1/2" deep, 20' long	Inst	LF	CY	.239	24.20	7.24	.64	32.08	39.40	21.35	26.10
13-1/2" deep, 20' long	Demo	BF	LK	.030	---	.90	.09	.99	1.43	.66	.95
13-1/2" deep, 20' long	Inst	BF	CY	.035	3.59	1.06	.09	4.74	5.81	3.15	3.85
13-1/2" deep, 20' long	Demo	SF	LK	.011	---	.33	.04	.37	.53	.23	.33
13-1/2" deep, 20' long	Inst	SF	CY	.013	1.32	.39	.04	1.75	2.15	1.17	1.44

Description	Oper	Unit	Crew Size	Man-Hours Per Unit	Costs Based On Small Volume						Large Volume	
					Avg Mat'l Unit Cost	Avg Labor Unit Cost	Avg Equip Unit Cost	Avg Total Unit Cost	Avg Price Incl O&P		Avg Total Unit Cost	Avg Price Incl O&P
15" deep, 20' long	Demo	LF	LK	.205	---	6.15	.64	6.79	9.81		4.41	6.37
15" deep, 20' long	Inst	LF	CY	.239	26.70	7.24	.64	34.58	42.20		22.95	28.00
15" deep, 20' long	Demo	BF	LK	.027	---	.81	.09	.90	1.30		.60	.86
15" deep, 20' long	Inst	BF	CY	.032	3.57	.97	.09	4.63	5.65		3.08	3.75
15" deep, 20' long	Demo	SF	LK	.011	---	.33	.04	.37	.53		.23	.33
15" deep, 20' long	Inst	SF	CY	.013	1.46	.39	.04	1.89	2.31		1.26	1.54
16-1/2" deep, 20' long	Demo	LF	LK	.205	---	6.15	.64	6.79	9.81		4.41	6.37
16-1/2" deep, 20' long	Inst	LF	CY	.239	29.10	7.24	.64	36.98	45.00		24.55	29.80
16-1/2" deep, 20' long	Demo	BF	LK	.025	---	.75	.08	.83	1.20		.53	.77
16-1/2" deep, 20' long	Inst	BF	CY	.029	3.53	.88	.08	4.49	5.46		2.98	3.62
16-1/2" deep, 20' long	Demo	SF	LK	.011	---	.33	.04	.37	.53		.23	.33
16-1/2" deep, 20' long	Inst	SF	CY	.013	1.58	.39	.04	2.01	2.45		1.34	1.64
18" deep, 20' long	Demo	LF	LK	.205	---	6.15	.64	6.79	9.81		4.41	6.37
18" deep, 20' long	Inst	LF	CY	.239	31.50	7.24	.64	39.38	47.80		26.15	31.70
18" deep, 20' long	Demo	BF	LK	.023	---	.69	.07	.76	1.10		.50	.72
18" deep, 20' long	Inst	BF	CY	.027	3.51	.82	.07	4.40	5.33		2.91	3.51
18" deep, 20' long	Demo	SF	LK	.011	---	.33	.04	.37	.53		.23	.33
18" deep, 20' long	Inst	SF	CY	.013	1.73	.39	.04	2.16	2.62		1.44	1.75
19-1/2" deep, 20' long	Demo	LF	LK	.205	---	6.15	.64	6.79	9.81		4.41	6.37
19-1/2" deep, 20' long	Inst	LF	CY	.239	33.80	7.24	.64	41.68	50.40		27.75	33.50
19-1/2" deep, 20' long	Demo	BF	LK	.021	---	.63	.07	.70	1.01		.46	.67
19-1/2" deep, 20' long	Inst	BF	CY	.025	3.47	.76	.07	4.30	5.20		2.83	3.42
19-1/2" deep, 20' long	Demo	SF	LK	.011	---	.33	.04	.37	.53		.23	.33
19-1/2" deep, 20' long	Inst	SF	CY	.013	1.85	.39	.04	2.28	2.76		1.52	1.84
21" deep, 30' long	Demo	LF	LK	.205	---	6.15	.64	6.79	9.81		4.41	6.37
21" deep, 30' long	Inst	LF	CY	.239	35.10	7.24	.64	42.98	51.80		28.55	34.40
21" deep, 30' long	Demo	BF	LK	.020	---	.60	.06	.66	.95		.43	.62
21" deep, 30' long	Inst	BF	CY	.023	3.35	.70	.06	4.11	4.96		2.72	3.29
21" deep, 30' long	Demo	SF	LK	.011	---	.33	.04	.37	.53		.23	.33
21" deep, 30' long	Inst	SF	CY	.013	1.92	.39	.04	2.35	2.84		1.57	1.90
22-1/2" deep, 30' long	Demo	LF	LK	.205	---	6.15	.64	6.79	9.81		4.41	6.37
22-1/2" deep, 30' long	Inst	LF	CY	.239	37.40	7.24	.64	45.28	54.60		30.15	36.20
22-1/2" deep, 30' long	Demo	BF	LK	.018	---	.54	.06	.60	.86		.40	.58
22-1/2" deep, 30' long	Inst	BF	CY	.021	3.32	.64	.06	4.02	4.83		2.67	3.22
22-1/2" deep, 30' long	Demo	SF	LK	.011	---	.33	.04	.37	.53		.23	.33
22-1/2" deep, 30' long	Inst	SF	CY	.013	2.04	.39	.04	2.47	2.98		1.65	1.99
24" deep, 30' long	Demo	LF	LK	.205	---	6.15	.64	6.79	9.81		4.41	6.37
24" deep, 30' long	Inst	LF	CY	.239	39.80	7.24	.64	47.68	57.20		31.65	38.00
24" deep, 30' long	Demo	BF	LK	.017	---	.51	.05	.56	.81		.36	.52
24" deep, 30' long	Inst	BF	CY	.020	3.32	.61	.05	3.98	4.78		2.63	3.16
24" deep, 30' long	Demo	SF	LK	.011	---	.33	.04	.37	.53		.23	.33
24" deep, 30' long	Inst	SF	CY	.013	2.16	.39	.04	2.59	3.12		1.73	2.09
25-1/2" deep, 30' long	Demo	LF	LK	.205	---	6.15	.64	6.79	9.81		4.41	6.37
25-1/2" deep, 30' long	Inst	LF	CY	.239	42.10	7.24	.64	49.98	60.00		33.25	39.80
25-1/2" deep, 30' long	Demo	BF	LK	.016	---	.48	.05	.53	.77		.33	.48
25-1/2" deep, 30' long	Inst	BF	CY	.019	3.30	.58	.05	3.93	4.71		2.59	3.11
25-1/2" deep, 30' long	Demo	SF	LK	.011	---	.33	.04	.37	.53		.23	.33
25-1/2" deep, 30' long	Inst	SF	CY	.013	2.31	.39	.04	2.74	3.29		1.83	2.20

Description	Oper	Unit	Crew Size	Man-Hours Per Unit	Avg Mat'l Unit Cost	Avg Labor Unit Cost	Avg Equip Unit Cost	Avg Total Unit Cost	Avg Price Incl O&P	Avg Total Unit Cost	Avg Price Incl O&P
					Costs Based On Small Volume					Large Volume	
27" deep, 40' long	Demo	LF	LK	.205	---	6.15	.64	6.79	9.81	4.41	6.37
27" deep, 40' long	Inst	LF	CY	.239	43.60	7.24	.64	51.48	61.70	34.25	41.00
27" deep, 40' long	Demo	BF	LK	.015	---	.45	.05	.50	.72	.33	.48
27" deep, 40' long	Inst	BF	CY	.018	3.24	.55	.05	3.84	4.59	2.55	3.06
27" deep, 40' long	Demo	SF	LK	.011	---	.33	.04	.37	.53	.23	.33
27" deep, 40' long	Inst	SF	CY	.013	2.39	.39	.04	2.82	3.38	1.88	2.26
28-1/2" deep, 40' long	Demo	LF	LK	.205	---	6.15	.64	6.79	9.81	4.41	6.37
28-1/2" deep, 40' long	Inst	LF	CY	.239	46.00	7.24	.64	53.88	64.40	35.85	42.80
28-1/2" deep, 40' long	Demo	BF	LK	.014	---	.42	.04	.46	.67	.30	.43
28-1/2" deep, 40' long	Inst	BF	CY	.017	3.23	.52	.04	3.79	4.53	2.51	3.00
28-1/2" deep, 40' long	Demo	SF	LK	.011	---	.33	.04	.37	.53	.23	.33
28-1/2" deep, 40' long	Inst	SF	CY	.013	2.51	.39	.04	2.94	3.52	1.96	2.35
30" deep, 40' long	Demo	LF	LK	.205	---	6.15	.64	6.79	9.81	4.41	6.37
30" deep, 40' long	Inst	LF	CY	.239	48.30	7.24	.64	56.18	67.10	37.35	44.60
30" deep, 40' long	Demo	BF	LK	.014	---	.42	.04	.46	.67	.30	.43
30" deep, 40' long	Inst	BF	CY	.016	3.23	.48	.04	3.75	4.48	2.48	2.96
30" deep, 40' long	Demo	SF	LK	.011	---	.33	.04	.37	.53	.23	.33
30" deep, 40' long	Inst	SF	CY	.013	2.64	.39	.04	3.07	3.67	2.05	2.45
31-1/2" deep, 40' long	Demo	LF	LK	.205	---	6.15	.64	6.79	9.81	4.41	6.37
31-1/2" deep, 40' long	Inst	LF	CY	.239	50.60	7.24	.64	58.48	69.70	38.95	46.30
31-1/2" deep, 40' long	Demo	BF	LK	.013	---	.39	.04	.43	.62	.27	.39
31-1/2" deep, 40' long	Inst	BF	CY	.015	3.23	.45	.04	3.72	4.44	2.48	2.96
31-1/2" deep, 40' long	Demo	SF	LK	.011	---	.33	.04	.37	.53	.23	.33
31-1/2" deep, 40' long	Inst	SF	CY	.013	2.78	.39	.04	3.21	3.83	2.14	2.56
33" deep, 40' long	Demo	LF	LK	.205	---	6.15	.64	6.79	9.81	4.41	6.37
33" deep, 40' long	Inst	LF	CY	.239	53.00	7.24	.64	60.88	72.40	40.45	48.10
33" deep, 40' long	Demo	BF	LK	.012	---	.36	.04	.40	.58	.27	.39
33" deep, 40' long	Inst	BF	CY	.015	3.20	.45	.04	3.69	4.40	2.43	2.89
33" deep, 40' long	Demo	SF	LK	.011	---	.33	.04	.37	.53	.23	.33
33" deep, 40' long	Inst	SF	CY	.013	2.90	.39	.04	3.33	3.97	2.22	2.65
34-1/2" deep, 40' long	Demo	LF	LK	.205	---	6.15	.64	6.79	9.81	4.41	6.37
34-1/2" deep, 40' long	Inst	LF	CY	.239	55.30	7.24	.64	63.18	75.10	42.05	49.90
34-1/2" deep, 40' long	Demo	BF	LK	.012	---	.36	.04	.40	.58	.26	.38
34-1/2" deep, 40' long	Inst	BF	CY	.014	3.20	.42	.04	3.66	4.36	2.42	2.88
34-1/2" deep, 40' long	Demo	SF	LK	.011	---	.33	.04	.37	.53	.23	.33
34-1/2" deep, 40' long	Inst	SF	CY	.013	3.02	.39	.04	3.45	4.10	2.30	2.74
36" deep, 50' long	Demo	LF	LK	.205	---	6.15	.64	6.79	9.81	4.41	6.37
36" deep, 50' long	Inst	LF	CY	.239	57.00	7.24	.64	64.88	77.00	43.15	51.20
36" deep, 50' long	Demo	BF	LK	.011	---	.33	.04	.37	.53	.23	.33
36" deep, 50' long	Inst	BF	CY	.013	3.17	.39	.04	3.60	4.28	2.40	2.86
36" deep, 50' long	Demo	SF	LK	.011	---	.33	.04	.37	.53	.23	.33
36" deep, 50' long	Inst	SF	CY	.013	3.12	.39	.04	3.55	4.22	2.37	2.82
37-1/2" deep, 50' long	Demo	LF	LK	.205	---	6.15	.64	6.79	9.81	4.41	6.37
37-1/2" deep, 50' long	Inst	LF	CY	.239	59.30	7.24	.64	67.18	79.80	44.75	53.00
37-1/2" deep, 50' long	Demo	BF	LK	.011	---	.33	.03	.36	.52	.23	.33
37-1/2" deep, 50' long	Inst	BF	CY	.013	3.17	.39	.03	3.59	4.27	2.37	2.81
37-1/2" deep, 50' long	Demo	SF	LK	.011	---	.33	.04	.37	.53	.23	.33
37-1/2" deep, 50' long	Inst	SF	CY	.013	3.24	.39	.04	3.67	4.36	2.45	2.91

Description	Oper	Unit	Crew Size	Man-Hours Per Unit	Avg Mat'l Unit Cost	Avg Labor Unit Cost	Avg Equip Unit Cost	Avg Total Unit Cost	Avg Price Incl O&P	Avg Total Unit Cost	Avg Price Incl O&P
								Costs Based On Small Volume		**Large Volume**	
39" deep, 50' long	Demo	LF	LK	.205	---	6.15	.64	6.79	9.81	4.41	6.37
39" deep, 50' long	Inst	LF	CY	.239	61.70	7.24	.64	69.58	82.50	46.25	54.80
39" deep, 50' long	Demo	BF	LK	.011	---	.33	.03	.36	.52	.23	.33
39" deep, 50' long	Inst	BF	CY	.012	3.17	.36	.03	3.56	4.22	2.37	2.81
39" deep, 50' long	Demo	SF	LK	.011	---	.33	.04	.37	.53	.23	.33
39" deep, 50' long	Inst	SF	CY	.013	3.36	.39	.04	3.79	4.50	2.53	3.01

6-3/4" thick, SF pricing based on 16' oc

Description	Oper	Unit	Crew Size	Man-Hours Per Unit	Avg Mat'l Unit Cost	Avg Labor Unit Cost	Avg Equip Unit Cost	Avg Total Unit Cost	Avg Price Incl O&P	Avg Total Unit Cost	Avg Price Incl O&P
30" deep, 30' long	Demo	LF	LK	.205	---	6.15	.64	6.79	9.81	4.41	6.37
30" deep, 30' long	Inst	LF	CY	.239	58.10	7.24	.64	65.98	78.30	43.85	52.00
30" deep, 30' long	Demo	BF	LK	.010	---	.30	.03	.33	.48	.23	.33
30" deep, 30' long	Inst	BF	CY	.012	2.90	.36	.03	3.29	3.91	2.19	2.60
30" deep, 30' long	Demo	SF	LK	.011	---	.33	.04	.37	.53	.23	.33
30" deep, 30' long	Inst	SF	CY	.013	3.18	.39	.04	3.61	4.29	2.41	2.87
31-1/2" deep, 30' long	Demo	LF	LK	.205	---	6.15	.64	6.79	9.81	4.41	6.37
31-1/2" deep, 30' long	Inst	LF	CY	.239	60.80	7.24	.64	68.68	81.50	45.75	54.20
31-1/2" deep, 30' long	Demo	BF	LK	.010	---	.30	.03	.33	.48	.20	.29
31-1/2" deep, 30' long	Inst	BF	CY	.011	2.90	.33	.03	3.26	3.87	2.16	2.56
31-1/2" deep, 30' long	Demo	SF	LK	.011	---	.33	.04	.37	.53	.23	.33
31-1/2" deep, 30' long	Inst	SF	CY	.013	3.32	.39	.04	3.75	4.45	2.50	2.97
33" deep, 30' long	Demo	LF	LK	.205	---	6.15	.64	6.79	9.81	4.41	6.37
33" deep, 30' long	Inst	LF	CY	.239	63.60	7.24	.64	71.48	84.70	47.55	56.30
33" deep, 30' long	Demo	BF	LK	.009	---	.27	.03	.30	.43	.20	.29
33" deep, 30' long	Inst	BF	CY	.011	2.90	.33	.03	3.26	3.87	2.16	2.56
33" deep, 30' long	Demo	SF	LK	.011	---	.33	.04	.37	.53	.23	.33
33" deep, 30' long	Inst	SF	CY	.013	3.48	.39	.04	3.91	4.63	2.61	3.10
34-1/2" deep, 40' long	Demo	LF	LK	.205	---	6.15	.64	6.79	9.81	4.41	6.37
34-1/2" deep, 40' long	Inst	LF	CY	.239	65.30	7.24	.64	73.18	86.70	48.75	57.60
34-1/2" deep, 40' long	Demo	BF	LK	.009	---	.27	.03	.30	.43	.20	.29
34-1/2" deep, 40' long	Inst	BF	CY	.010	2.84	.30	.03	3.17	3.75	2.12	2.51
34-1/2" deep, 40' long	Demo	SF	LK	.011	---	.33	.04	.37	.53	.23	.33
34-1/2" deep, 40' long	Inst	SF	CY	.013	3.57	.39	.04	4.00	4.74	2.67	3.17
36" deep, 40' long	Demo	LF	LK	.205	---	6.15	.64	6.79	9.81	4.41	6.37
36" deep, 40' long	Inst	LF	CY	.239	68.10	7.24	.64	75.98	89.80	50.55	59.70
36" deep, 40' long	Demo	BF	LK	.009	---	.27	.03	.30	.43	.20	.29
36" deep, 40' long	Inst	BF	CY	.010	2.84	.30	.03	3.17	3.75	2.09	2.47
36" deep, 40' long	Demo	SF	LK	.011	---	.33	.04	.37	.53	.23	.33
36" deep, 40' long	Inst	SF	CY	.013	3.72	.39	.04	4.15	4.91	2.77	3.28
37-1/2" deep, 40' long	Demo	LF	LK	.205	---	6.15	.64	6.79	9.81	4.41	6.37
37-1/2" deep, 40' long	Inst	LF	CY	.239	70.80	7.24	.64	78.68	93.00	52.35	61.80
37-1/2" deep, 40' long	Demo	BF	LK	.008	---	.24	.03	.27	.39	.17	.24
37-1/2" deep, 40' long	Inst	BF	CY	.010	2.84	.30	.03	3.17	3.75	2.09	2.47
37-1/2" deep, 40' long	Demo	SF	LK	.011	---	.33	.04	.37	.53	.23	.33
37-1/2" deep, 40' long	Inst	SF	CY	.013	3.87	.39	.04	4.30	5.08	2.87	3.40
39" deep, 50' long	Demo	LF	LK	.205	---	6.15	.64	6.79	9.81	4.41	6.37
39" deep, 50' long	Inst	LF	CY	.239	72.90	7.24	.64	80.78	95.30	53.75	63.40
39" deep, 50' long	Demo	BF	LK	.008	---	.24	.02	.26	.38	.17	.24
39" deep, 50' long	Inst	BF	CY	.009	2.79	.27	.02	3.08	3.64	2.06	2.43
39" deep, 50' long	Demo	SF	LK	.011	---	.33	.04	.37	.53	.23	.33
39" deep, 50' long	Inst	SF	CY	.013	3.99	.39	.04	4.42	5.22	2.95	3.49

				Costs Based On Small Volume					Large Volume	
Description	Oper	Unit	Crew Size	Man-Hours Per Unit	Avg Mat'l Unit Cost	Avg Labor Unit Cost	Avg Equip Unit Cost	Avg Total Unit Cost	Avg Price Incl O&P	Avg Total Unit Cost	Avg Price Incl O&P
40-1/2" deep, 50' long	Demo	LF	LK	.205	---	6.15	.64	6.79	9.81	4.41	6.37
40-1/2" deep, 50' long	Inst	LF	CY	.239	75.60	7.24	.64	83.48	98.50	55.55	65.50
40-1/2" deep, 50' long	Demo	BF	LK	.008	---	.24	.02	.26	.38	.17	.24
40-1/2" deep, 50' long	Inst	BF	CY	.009	2.79	.27	.02	3.08	3.64	2.06	2.43
40-1/2" deep, 50' long	Demo	SF	LK	.011	---	.33	.04	.37	.53	.23	.33
40-1/2" deep, 50' long	Inst	SF	CY	.013	4.13	.39	.04	4.56	5.38	3.04	3.59
42" deep, 50' long	Demo	LF	LK	.205	---	6.15	.64	6.79	9.81	4.41	6.37
42" deep, 50' long	Inst	LF	CY	.239	78.40	7.24	.64	86.28	102.00	57.35	67.60
42" deep, 50' long	Demo	BF	LK	.007	---	.21	.02	.23	.33	.16	.23
42" deep, 50' long	Inst	BF	CY	.009	2.79	.27	.02	3.08	3.64	2.05	2.42
42" deep, 50' long	Demo	SF	LK	.011	---	.33	.04	.37	.53	.23	.33
42" deep, 50' long	Inst	SF	CY	.013	4.28	.39	.04	4.71	5.55	3.14	3.71
43-1/2" deep, 50' long	Demo	LF	LK	.205	---	6.15	.64	6.79	9.81	4.41	6.37
43-1/2" deep, 50' long	Inst	LF	CY	.239	81.10	7.24	.64	88.98	105.00	59.15	69.70
43-1/2" deep, 50' long	Demo	BF	LK	.007	---	.21	.02	.23	.33	.16	.23
43-1/2" deep, 50' long	Inst	BF	CY	.008	2.79	.24	.02	3.05	3.59	2.02	2.38
43-1/2" deep, 50' long	Demo	SF	LK	.011	---	.33	.04	.37	.53	.23	.33
43-1/2" deep, 50' long	Inst	SF	CY	.013	4.44	.39	.04	4.87	5.74	3.25	3.83
45" deep, 50' long	Demo	LF	LK	.205	---	6.15	.64	6.79	9.81	4.41	6.37
45" deep, 50' long	Inst	LF	CY	.239	83.80	7.24	.64	91.68	108.00	61.05	71.80
45" deep, 50' long	Demo	BF	LK	.007	---	.21	.02	.23	.33	.13	.19
45" deep, 50' long	Inst	BF	CY	.008	2.79	.24	.02	3.05	3.59	2.02	2.38
45" deep, 50' long	Demo	SF	LK	.011	---	.33	.04	.37	.53	.23	.33
45" deep, 50' long	Inst	SF	CY	.013	4.58	.39	.04	5.01	5.90	3.34	3.94

8-3/4" thick, SF pricing based on 16' oc

36" deep, 30' long	Demo	LF	LK	.205	---	6.15	.64	6.79	9.81	4.41	6.37
36" deep, 30' long	Inst	LF	CY	.239	79.40	7.24	.64	87.28	103.00	58.15	68.40
36" deep, 30' long	Demo	BF	LK	.007	---	.21	.02	.23	.33	.13	.19
36" deep, 30' long	Inst	BF	CY	.008	2.66	.24	.02	2.92	3.44	1.93	2.27
36" deep, 30' long	Demo	SF	LK	.011	---	.33	.04	.37	.53	.23	.33
36" deep, 30' long	Inst	SF	CY	.013	4.34	.39	.04	4.77	5.62	3.18	3.75
37-1/2" deep, 30' long	Demo	LF	LK	.205	---	6.15	.64	6.79	9.81	4.41	6.37
37-1/2" deep, 30' long	Inst	LF	CY	.239	82.50	7.24	.64	90.38	106.00	60.15	70.80
37-1/2" deep, 30' long	Demo	BF	LK	.007	---	.21	.02	.23	.33	.13	.19
37-1/2" deep, 30' long	Inst	BF	CY	.008	2.64	.24	.02	2.90	3.42	1.92	2.26
37-1/2" deep, 30' long	Demo	SF	LK	.011	---	.33	.04	.37	.53	.23	.33
37-1/2" deep, 30' long	Inst	SF	CY	.013	4.50	.39	.04	4.93	5.81	3.29	3.88
39" deep, 30' long	Demo	LF	LK	.205	---	6.15	.64	6.79	9.81	4.41	6.37
39" deep, 30' long	Inst	LF	CY	.239	85.90	7.24	.64	93.78	110.00	62.35	73.30
39" deep, 30' long	Demo	BF	LK	.006	---	.18	.02	.20	.29	.13	.19
39" deep, 30' long	Inst	BF	CY	.007	2.64	.21	.02	2.87	3.37	1.92	2.26
39" deep, 30' long	Demo	SF	LK	.011	---	.33	.04	.37	.53	.23	.33
39" deep, 30' long	Inst	SF	CY	.013	4.68	.39	.04	5.11	6.01	3.41	4.02
40-1/2" deep, 30' long	Demo	LF	LK	.205	---	6.15	.64	6.79	9.81	4.41	6.37
40-1/2" deep, 30' long	Inst	LF	CY	.239	89.00	7.24	.64	96.88	114.00	64.55	75.80
40-1/2" deep, 30' long	Demo	BF	LK	.006	---	.18	.02	.20	.29	.13	.19
40-1/2" deep, 30' long	Inst	BF	CY	.007	2.64	.21	.02	2.87	3.37	1.92	2.26
40-1/2" deep, 30' long	Demo	SF	LK	.011	---	.33	.04	.37	.53	.23	.33
40-1/2" deep, 30' long	Inst	SF	CY	.013	4.86	.39	.04	5.29	6.22	3.53	4.16

Description	Oper	Unit	Crew Size	Man-Hours Per Unit	Avg Mat'l Unit Cost	Avg Labor Unit Cost	Avg Equip Unit Cost	Avg Total Unit Cost	Avg Price Incl O&P	Avg Total Unit Cost	Avg Price Incl O&P
								Costs Based On Small Volume		Large Volume	
42" deep, 30' long	Demo	LF	LK	.205	---	6.15	.64	6.79	9.81	4.41	6.37
42" deep, 30' long	Inst	LF	CY	.239	92.20	7.24	.64	100.08	117.00	66.55	78.20
42" deep, 30' long	Demo	BF	LK	.006	---	.18	.02	.20	.29	.13	.19
42" deep, 30' long	Inst	BF	CY	.007	2.64	.21	.02	2.87	3.37	1.89	2.22
42" deep, 30' long	Demo	SF	LK	.011	---	.33	.04	.37	.53	.23	.33
42" deep, 30' long	Inst	SF	CY	.013	5.03	.39	.04	5.46	6.42	3.64	4.28
43-1/2" deep, 40' long	Demo	LF	LK	.205	---	6.15	.64	6.79	9.81	4.41	6.37
43-1/2" deep, 40' long	Inst	LF	CY	.239	94.30	7.24	.64	102.18	120.00	68.05	79.80
43-1/2" deep, 40' long	Demo	BF	LK	.006	---	.18	.02	.20	.29	.13	.19
43-1/2" deep, 40' long	Inst	BF	CY	.007	2.60	.21	.02	2.83	3.33	1.86	2.18
43-1/2" deep, 40' long	Demo	SF	LK	.011	---	.33	.04	.37	.53	.23	.33
43-1/2" deep, 40' long	Inst	SF	CY	.013	5.15	.39	.04	5.58	6.55	3.72	4.37
45" deep, 40' long	Demo	LF	LK	.205	---	6.15	.64	6.79	9.81	4.41	6.37
45" deep, 40' long	Inst	LF	CY	.239	97.50	7.24	.64	105.38	124.00	70.15	82.20
45" deep, 40' long	Demo	BF	LK	.005	---	.15	.02	.17	.24	.13	.19
45" deep, 40' long	Inst	BF	CY	.006	2.60	.18	.02	2.80	3.28	1.86	2.18
45" deep, 40' long	Demo	SF	LK	.011	---	.33	.04	.37	.53	.23	.33
45" deep, 40' long	Inst	SF	CY	.013	5.33	.39	.04	5.76	6.76	3.84	4.51
46-1/2" deep, 40' long	Demo	LF	LK	.205	---	6.15	.64	6.79	9.81	4.41	6.37
46-1/2" deep, 40' long	Inst	LF	CY	.239	101.00	7.24	.64	108.88	127.00	72.25	84.70
46-1/2" deep, 40' long	Demo	BF	LK	.005	---	.15	.02	.17	.24	.10	.14
46-1/2" deep, 40' long	Inst	BF	CY	.006	2.60	.18	.02	2.80	3.28	1.86	2.18
46-1/2" deep, 40' long	Demo	SF	LK	.011	---	.33	.04	.37	.53	.23	.33
46-1/2" deep, 40' long	Inst	SF	CY	.013	5.49	.39	.04	5.92	6.94	3.95	4.64
48" deep, 50' long	Demo	LF	LK	.205	---	6.15	.64	6.79	9.81	4.41	6.37
48" deep, 50' long	Inst	LF	CY	.239	103.00	7.24	.64	110.88	130.00	73.85	86.50
48" deep, 50' long	Demo	BF	LK	.005	---	.15	.02	.17	.24	.10	.14
48" deep, 50' long	Inst	BF	CY	.006	2.57	.18	.02	2.77	3.25	1.84	2.16
48" deep, 50' long	Demo	SF	LK	.011	---	.33	.04	.37	.53	.23	.33
48" deep, 50' long	Inst	SF	CY	.013	5.63	.39	.04	6.06	7.11	4.04	4.74
49-1/2" deep, 50' long	Demo	LF	LK	.205	---	6.15	.64	6.79	9.81	4.41	6.37
49-1/2" deep, 50' long	Inst	LF	CY	.239	106.00	7.24	.64	113.88	134.00	75.95	88.90
49-1/2" deep, 50' long	Demo	BF	LK	.005	---	.15	.02	.17	.24	.10	.14
49-1/2" deep, 50' long	Inst	BF	CY	.006	2.57	.18	.02	2.77	3.25	1.84	2.16
49-1/2" deep, 50' long	Demo	SF	LK	.011	---	.33	.04	.37	.53	.23	.33
49-1/2" deep, 50' long	Inst	SF	CY	.013	5.79	.39	.04	6.22	7.29	4.15	4.87
51" deep, 50' long	Demo	LF	LK	.205	---	6.15	.64	6.79	9.81	4.41	6.37
51" deep, 50' long	Inst	LF	CY	.239	109.00	7.24	.64	116.88	137.00	78.05	91.40
51" deep, 50' long	Demo	BF	LK	.005	---	.15	.02	.17	.24	.10	.14
51" deep, 50' long	Inst	BF	CY	.006	2.57	.18	.02	2.77	3.25	1.84	2.16
51" deep, 50' long	Demo	SF	LK	.011	---	.33	.04	.37	.53	.23	.33
51" deep, 50' long	Inst	SF	CY	.013	5.97	.39	.04	6.40	7.50	4.27	5.01
52-1/2" deep, 50' long	Demo	LF	LK	.205	---	6.15	.64	6.79	9.81	4.41	6.37
52-1/2" deep, 50' long	Inst	LF	CY	.239	113.00	7.24	.64	120.88	141.00	80.15	93.80
52-1/2" deep, 50' long	Demo	BF	LK	.005	---	.15	.01	.16	.23	.10	.14
52-1/2" deep, 50' long	Inst	BF	CY	.005	2.57	.15	.01	2.73	3.19	1.84	2.16
52-1/2" deep, 50' long	Demo	SF	LK	.011	---	.33	.04	.37	.53	.23	.33
52-1/2" deep, 50' long	Inst	SF	CY	.013	6.14	.39	.04	6.57	7.69	4.38	5.13

Description	Oper	Unit	Crew Size	Man-Hours Per Unit	Avg Mat'l Unit Cost	Avg Labor Unit Cost	Avg Equip Unit Cost	Avg Total Unit Cost	Avg Price Incl O&P	Avg Total Unit Cost	Avg Price Incl O&P
54" deep, 50' long	Demo	LF	LK	.205	---	6.15	.64	6.79	9.81	4.41	6.37
54" deep, 50' long	Inst	LF	CY	.239	116.00	7.24	.64	123.88	145.00	82.25	96.20
54" deep, 50' long	Demo	BF	LK	.005	---	.15	.01	.16	.23	.10	.14
54" deep, 50' long	Inst	BF	CY	.005	2.57	.15	.01	2.73	3.19	1.81	2.11
54" deep, 50' long	Demo	SF	LK	.011	---	.33	.04	.37	.53	.23	.33
54" deep, 50' long	Inst	SF	CY	.013	6.30	.39	.04	6.73	7.88	4.49	5.26
55-1/2" deep, 50' long	Demo	LF	LK	.205	---	6.15	.64	6.79	9.81	4.41	6.37
55-1/2" deep, 50' long	Inst	LF	CY	.239	119.00	7.24	.64	126.88	148.00	84.35	98.60
55-1/2" deep, 50' long	Demo	BF	LK	.004	---	.12	.01	.13	.19	.10	.14
55-1/2" deep, 50' long	Inst	BF	CY	.005	2.57	.15	.01	2.73	3.19	1.81	2.11
55-1/2" deep, 50' long	Demo	SF	LK	.011	---	.33	.04	.37	.53	.23	.33
55-1/2" deep, 50' long	Inst	SF	CY	.013	6.48	.39	.04	6.91	8.08	4.61	5.40
57" deep, 50' long	Demo	LF	LK	.205	---	6.15	.64	6.79	9.81	4.41	6.37
57" deep, 50' long	Inst	LF	CY	.239	122.00	7.24	.64	129.88	152.00	86.45	101.00
57" deep, 50' long	Demo	BF	LK	.004	---	.12	.01	.13	.19	.10	.14
57" deep, 50' long	Inst	BF	CY	.005	2.57	.15	.01	2.73	3.19	1.81	2.11
57" deep, 50' long	Demo	SF	LK	.011	---	.33	.04	.37	.53	.23	.33
57" deep, 50' long	Inst	SF	CY	.013	6.66	.39	.04	7.09	8.29	4.73	5.54

10-3/4" thick, SF pricing based on 16' oc

Description	Oper	Unit	Crew Size	Man-Hours Per Unit	Avg Mat'l Unit Cost	Avg Labor Unit Cost	Avg Equip Unit Cost	Avg Total Unit Cost	Avg Price Incl O&P	Avg Total Unit Cost	Avg Price Incl O&P
42" deep, 50' long	Demo	LF	LK	.205	---	6.15	.64	6.79	9.81	4.41	6.37
42" deep, 50' long	Inst	LF	CY	.239	103.00	7.24	.64	110.88	130.00	73.75	86.40
42" deep, 50' long	Demo	BF	LK	.005	---	.15	.02	.17	.24	.10	.14
42" deep, 50' long	Inst	BF	CY	.006	2.45	.18	.02	2.65	3.11	1.76	2.07
42" deep, 50' long	Demo	SF	LK	.011	---	.33	.04	.37	.53	.23	.33
42" deep, 50' long	Inst	SF	CY	.013	5.61	.39	.04	6.04	7.08	4.03	4.73
43-1/2" deep, 50' long	Demo	LF	LK	.205	---	6.15	.64	6.79	9.81	4.41	6.37
43-1/2" deep, 50' long	Inst	LF	CY	.239	106.00	7.24	.64	113.88	134.00	76.15	89.10
43-1/2" deep, 50' long	Demo	BF	LK	.005	---	.15	.01	.16	.23	.10	.14
43-1/2" deep, 50' long	Inst	BF	CY	.006	2.45	.18	.01	2.64	3.10	1.76	2.07
43-1/2" deep, 50' long	Demo	SF	LK	.011	---	.33	.04	.37	.53	.23	.33
43-1/2" deep, 50' long	Inst	SF	CY	.013	5.81	.39	.04	6.24	7.31	4.16	4.88
45" deep, 50' long	Demo	LF	LK	.205	---	6.15	.64	6.79	9.81	4.41	6.37
45" deep, 50' long	Inst	LF	CY	.239	110.00	7.24	.64	117.88	138.00	78.55	91.90
45" deep, 50' long	Demo	BF	LK	.005	---	.15	.01	.16	.23	.10	.14
45" deep, 50' long	Inst	BF	CY	.005	2.45	.15	.01	2.61	3.05	1.73	2.02
45" deep, 50' long	Demo	SF	LK	.011	---	.33	.04	.37	.53	.23	.33
45" deep, 50' long	Inst	SF	CY	.013	6.02	.39	.04	6.45	7.55	4.30	5.04
46-1/2" deep, 50' long	Demo	LF	LK	.205	---	6.15	.64	6.79	9.81	4.41	6.37
46-1/2" deep, 50' long	Inst	LF	CY	.239	114.00	7.24	.64	121.88	142.00	80.95	94.60
46-1/2" deep, 50' long	Demo	BF	LK	.004	---	.12	.01	.13	.19	.10	.14
46-1/2" deep, 50' long	Inst	BF	CY	.005	2.45	.15	.01	2.61	3.05	1.73	2.02
46-1/2" deep, 50' long	Demo	SF	LK	.011	---	.33	.04	.37	.53	.23	.33
46-1/2" deep, 50' long	Inst	SF	CY	.013	6.20	.39	.04	6.63	7.76	4.42	5.18
48" deep, 50' long	Demo	LF	LK	.205	---	6.15	.64	6.79	9.81	4.41	6.37
48" deep, 50' long	Inst	LF	CY	.239	117.00	7.24	.64	124.88	146.00	83.25	97.40
48" deep, 50' long	Demo	BF	LK	.004	---	.12	.01	.13	.19	.10	.14
48" deep, 50' long	Inst	BF	CY	.005	2.45	.15	.01	2.61	3.05	1.73	2.02
48" deep, 50' long	Demo	SF	LK	.011	---	.33	.04	.37	.53	.23	.33
48" deep, 50' long	Inst	SF	CY	.013	6.41	.39	.04	6.84	8.00	4.56	5.34

Description	Oper	Unit	Crew Size	Man-Hours Per Unit	Avg Mat'l Unit Cost	Avg Labor Unit Cost	Avg Equip Unit Cost	Avg Total Unit Cost	Avg Price Incl O&P	Avg Total Unit Cost	Avg Price Incl O&P
					Costs Based On Small Volume					**Large Volume**	
49-1/2" deep, 50' long	Demo	LF	LK	.205	---	6.15	.64	6.79	**9.81**	4.41	**6.37**
49-1/2" deep, 50' long	Inst	LF	CY	.239	121.00	7.24	.64	128.88	**150.00**	85.65	**100.00**
49-1/2" deep, 50' long	Demo	BF	LK	.004	---	.12	.01	.13	**.19**	.10	**.14**
49-1/2" deep, 50' long	Inst	BF	CY	.005	2.43	.15	.01	2.59	**3.03**	1.72	**2.01**
49-1/2" deep, 50' long	Demo	SF	LK	.011	---	.33	.04	.37	**.53**	.23	**.33**
49-1/2" deep, 50' long	Inst	SF	CY	.013	6.60	.39	.04	7.03	**8.22**	4.69	**5.49**
51" deep, 50' long	Demo	LF	LK	.205	---	6.15	.64	6.79	**9.81**	4.41	**6.37**
51" deep, 50' long	Inst	LF	CY	.239	124.00	7.24	.64	131.88	**154.00**	88.05	**103.00**
51" deep, 50' long	Demo	BF	LK	.004	---	.12	.01	.13	**.19**	.10	**.14**
51" deep, 50' long	Inst	BF	CY	.005	2.43	.15	.01	2.59	**3.03**	1.72	**2.01**
51" deep, 50' long	Demo	SF	LK	.011	---	.33	.04	.37	**.53**	.23	**.33**
51" deep, 50' long	Inst	SF	CY	.013	6.78	.39	.04	7.21	**8.43**	4.81	**5.63**
52-1/2" deep, 60' long	Demo	LF	LK	.205	---	6.15	.64	6.79	**9.81**	4.41	**6.37**
52-1/2" deep, 60' long	Inst	LF	CY	.239	128.00	7.24	.64	135.88	**158.00**	90.25	**105.00**
52-1/2" deep, 60' long	Demo	BF	LK	.004	---	.12	.01	.13	**.19**	.10	**.14**
52-1/2" deep, 60' long	Inst	BF	CY	.005	2.43	.15	.01	2.59	**3.03**	1.72	**2.01**
52-1/2" deep, 60' long	Demo	SF	LK	.011	---	.33	.04	.37	**.53**	.23	**.33**
52-1/2" deep, 60' long	Inst	SF	CY	.013	6.96	.39	.04	7.39	**8.64**	4.93	**5.77**
54" deep, 60' long	Demo	LF	LK	.205	---	6.15	.64	6.79	**9.81**	4.41	**6.37**
54" deep, 60' long	Inst	LF	CY	.239	131.00	7.24	.64	138.88	**162.00**	92.65	**108.00**
54" deep, 60' long	Demo	BF	LK	.004	---	.12	.01	.13	**.19**	.07	**.10**
54" deep, 60' long	Inst	BF	CY	.004	2.43	.12	.01	2.56	**2.99**	1.72	**2.01**
54" deep, 60' long	Demo	SF	LK	.011	---	.33	.04	.37	**.53**	.23	**.33**
54" deep, 60' long	Inst	SF	CY	.013	7.17	.39	.04	7.60	**8.88**	5.07	**5.93**
55-1/2" deep, 60' long	Demo	LF	LK	.205	---	6.15	.64	6.79	**9.81**	4.41	**6.37**
55-1/2" deep, 60' long	Inst	LF	CY	.239	135.00	7.24	.64	142.88	**167.00**	95.05	**111.00**
55-1/2" deep, 60' long	Demo	BF	LK	.004	---	.12	.01	.13	**.19**	.07	**.10**
55-1/2" deep, 60' long	Inst	BF	CY	.004	2.43	.12	.01	2.56	**2.99**	1.72	**2.01**
55-1/2" deep, 60' long	Demo	SF	LK	.011	---	.33	.04	.37	**.53**	.23	**.33**
55-1/2" deep, 60' long	Inst	SF	CY	.013	7.35	.39	.04	7.78	**9.08**	5.19	**6.06**
57" deep, 60' long	Demo	LF	LK	.205	---	6.15	.64	6.79	**9.81**	4.41	**6.37**
57" deep, 60' long	Inst	LF	CY	.239	138.00	7.24	.64	145.88	**171.00**	97.45	**114.00**
57" deep, 60' long	Demo	BF	LK	.004	---	.12	.01	.13	**.19**	.07	**.10**
57" deep, 60' long	Inst	BF	CY	.004	2.43	.12	.01	2.56	**2.99**	1.72	**2.01**
57" deep, 60' long	Demo	SF	LK	.011	---	.33	.04	.37	**.53**	.23	**.33**
57" deep, 60' long	Inst	SF	CY	.013	7.56	.39	.04	7.99	**9.33**	5.33	**6.23**
58-1/2" deep, 60' long	Demo	LF	LK	.205	---	6.15	.64	6.79	**9.81**	4.41	**6.37**
58-1/2" deep, 60' long	Inst	LF	CY	.239	142.00	7.24	.64	149.88	**175.00**	99.85	**116.00**
58-1/2" deep, 60' long	Demo	BF	LK	.004	---	.12	.01	.13	**.19**	.07	**.10**
58-1/2" deep, 60' long	Inst	BF	CY	.004	2.43	.12	.01	2.56	**2.99**	1.72	**2.01**
58-1/2" deep, 60' long	Demo	SF	LK	.011	---	.33	.04	.37	**.53**	.23	**.33**
58-1/2" deep, 60' long	Inst	SF	CY	.013	7.76	.39	.04	8.19	**9.56**	5.46	**6.37**
60" deep, 60' long	Demo	LF	LK	.205	---	6.15	.64	6.79	**9.81**	4.41	**6.37**
60" deep, 60' long	Inst	LF	CY	.239	146.00	7.24	.64	153.88	**179.00**	102.15	**119.00**
60" deep, 60' long	Demo	BF	LK	.003	---	.09	.01	.10	**.14**	.07	**.10**
60" deep, 60' long	Inst	BF	CY	.004	2.43	.12	.01	2.56	**2.99**	1.72	**2.01**
60" deep, 60' long	Demo	SF	LK	.011	---	.33	.04	.37	**.53**	.23	**.33**
60" deep, 60' long	Inst	SF	CY	.013	7.94	.39	.04	8.37	**9.76**	5.58	**6.51**

				Costs Based On Small Volume					Large Volume	
Description	Oper Unit	Crew Size	Man-Hours Per Unit	Avg Mat'l Unit Cost	Avg Labor Unit Cost	Avg Equip Unit Cost	Avg Total Unit Cost	Avg Price Incl O&P	Avg Total Unit Cost	Avg Price Incl O&P

Purlins
16' long, STR #1, SF pricing based on 8' oc

Description	Oper Unit	Crew Size	Man-Hours Per Unit	Avg Mat'l Unit Cost	Avg Labor Unit Cost	Avg Equip Unit Cost	Avg Total Unit Cost	Avg Price Incl O&P	Avg Total Unit Cost	Avg Price Incl O&P
2" x 8"	Demo LF	LL	.046	---	1.34	.14	1.48	2.14	.96	1.39
2" x 8"	Inst LF	CZ	.103	3.87	3.10	.21	7.18	9.31	4.74	6.13
2" x 8"	Demo BF	LL	.034	---	.99	.11	1.10	1.58	.71	1.02
2" x 8"	Inst BF	CZ	.077	2.91	2.32	.16	5.39	6.98	3.55	4.59
2" x 8"	Demo SF	LL	.005	---	.15	.02	.17	.24	.10	.14
2" x 8"	Inst SF	CZ	.011	.42	.33	.02	.77	1.00	.51	.66
2" x 10"	Demo LF	LL	.046	---	1.34	.14	1.48	2.14	.96	1.39
2" x 10"	Inst LF	CZ	.103	4.58	3.10	.21	7.89	10.10	5.21	6.67
2" x 10"	Demo BF	LL	.027	---	.79	.09	.88	1.26	.58	.84
2" x 10"	Inst BF	CZ	.061	1.92	1.84	.13	3.89	5.09	2.56	3.36
2" x 10"	Demo SF	LL	.005	---	.15	.02	.17	.24	.10	.14
2" x 10"	Inst SF	CZ	.011	.51	.33	.02	.86	1.10	.57	.73
2" x 12"	Demo LF	LL	.046	---	1.34	.14	1.48	2.14	.96	1.39
2" x 12"	Inst LF	CZ	.103	5.13	3.10	.21	8.44	10.80	5.58	7.10
2" x 12"	Demo BF	LL	.023	---	.67	.07	.74	1.07	.49	.70
2" x 12"	Inst BF	CZ	.051	1.67	1.54	.11	3.32	4.33	2.17	2.84
2" x 12"	Demo SF	LL	.005	---	.15	.02	.17	.24	.10	.14
2" x 12"	Inst SF	CZ	.011	.57	.33	.02	.92	1.17	.61	.77
3" x 8"	Demo LF	LL	.046	---	1.34	.14	1.48	2.14	.96	1.39
3" x 8"	Inst LF	CZ	.103	4.82	3.10	.21	8.13	10.40	5.37	6.86
3" x 8"	Demo BF	LL	.023	---	.67	.07	.74	1.07	.49	.70
3" x 8"	Inst BF	CZ	.051	1.50	1.54	.11	3.15	4.14	2.06	2.71
3" x 8"	Demo SF	LL	.005	---	.15	.02	.17	.24	.10	.14
3" x 8"	Inst SF	CZ	.011	.53	.33	.02	.88	1.13	.58	.74
3" x 10"	Demo LF	LL	.046	---	1.34	.14	1.48	2.14	.96	1.39
3" x 10"	Inst LF	CZ	.103	5.55	3.10	.21	8.86	11.20	5.86	7.42
3" x 10"	Demo BF	LL	.018	---	.52	.06	.58	.84	.39	.56
3" x 10"	Inst BF	CZ	.041	1.32	1.23	.09	2.64	3.46	1.75	2.29
3" x 10"	Demo SF	LL	.005	---	.15	.02	.17	.24	.10	.14
3" x 10"	Inst SF	CZ	.011	.62	.33	.02	.97	1.23	.64	.81
3" x 12"	Demo LF	LL	.046	---	1.34	.14	1.48	2.14	.96	1.39
3" x 12"	Inst LF	CZ	.103	6.36	3.10	.21	9.67	12.20	6.40	8.04
3" x 12"	Demo BF	LL	.015	---	.44	.05	.49	.70	.32	.46
3" x 12"	Inst BF	CZ	.034	1.22	1.02	.07	2.31	3.01	1.52	1.97
3" x 12"	Demo SF	LL	.005	---	.15	.02	.17	.24	.10	.14
3" x 12"	Inst SF	CZ	.011	.69	.33	.02	1.04	1.31	.69	.87
3" x 14"	Demo LF	LL	.046	---	1.34	.14	1.48	2.14	.96	1.39
3" x 14"	Inst LF	CZ	.103	7.58	3.10	.21	10.89	13.60	7.21	8.97
3" x 14"	Demo BF	LL	.013	---	.38	.04	.42	.60	.26	.38
3" x 14"	Inst BF	CZ	.029	2.19	.87	.06	3.12	3.89	2.07	2.58
3" x 14"	Demo SF	LL	.005	---	.15	.02	.17	.24	.10	.14
3" x 14"	Inst SF	CZ	.011	.83	.33	.02	1.18	1.47	.78	.97

				Costs Based On Small Volume					Large Volume	
Description	Oper Unit	Crew Size	Man-Hours Per Unit	Avg Mat'l Unit Cost	Avg Labor Unit Cost	Avg Equip Unit Cost	Avg Total Unit Cost	Avg Price Incl O&P	Avg Total Unit Cost	Avg Price Incl O&P

Sub-purlins
8' long, STR #1, SF pricing based on 2' oc

2" x 4"	Demo LF	LL	.020	---	.58	.06	.64	.93	.42	.60
2" x 4"	Inst LF	CZ	.046	.54	1.38	.10	2.02	2.80	1.32	1.83
2" x 4"	Demo BF	LL	.030	---	.87	.09	.96	1.39	.64	.93
2" x 4"	Inst BF	CZ	.068	.81	2.05	.14	3.00	4.14	1.95	2.70
2" x 4"	Demo SF	LL	.009	---	.26	.03	.29	.42	.19	.28
2" x 4"	Inst SF	CZ	.021	.24	.63	.04	.91	1.26	.58	.80
2" x 6"	Demo LF	LL	.020	---	.58	.06	.64	.93	.42	.60
2" x 6"	Inst LF	CZ	.046	.81	1.38	.10	2.29	3.11	1.50	2.04
2" x 6"	Demo BF	LL	.020	---	.58	.06	.64	.93	.42	.60
2" x 6"	Inst BF	CZ	.046	.81	1.38	.10	2.29	3.11	1.50	2.04
2" x 6"	Demo SF	LL	.009	---	.26	.03	.29	.42	.19	.28
2" x 6"	Inst SF	CZ	.021	.36	.63	.04	1.03	1.40	.66	.89
2" x 8"	Demo LF	LL	.020	---	.58	.06	.64	.93	.42	.60
2" x 8"	Inst LF	CZ	.046	1.22	1.38	.10	2.70	3.58	1.77	2.35
2" x 8"	Demo BF	LL	.015	---	.44	.05	.49	.70	.32	.46
2" x 8"	Inst BF	CZ	.034	.92	1.02	.07	2.01	2.66	1.32	1.74
2" x 8"	Demo SF	LL	.009	---	.26	.03	.29	.42	.19	.28
2" x 8"	Inst SF	CZ	.021	.53	.63	.04	1.20	1.60	.77	1.02
3" x 4"	Demo LF	LL	.020	---	.58	.06	.64	.93	.42	.60
3" x 4"	Inst LF	CZ	.046	.92	1.38	.10	2.40	3.23	1.57	2.12
3" x 4"	Demo BF	LL	.020	---	.58	.06	.64	.93	.42	.60
3" x 4"	Inst BF	CZ	.046	.92	1.38	.10	2.40	3.23	1.57	2.12
3" x 4"	Demo SF	LL	.009	---	.26	.03	.29	.42	.19	.28
3" x 4"	Inst SF	CZ	.021	.41	.63	.04	1.08	1.46	.69	.93
3" x 6"	Demo LF	LL	.020	---	.58	.06	.64	.93	.42	.60
3" x 6"	Inst LF	CZ	.046	1.38	1.38	.10	2.86	3.76	1.88	2.47
3" x 6"	Demo BF	LL	.014	---	.41	.04	.45	.65	.29	.42
3" x 6"	Inst BF	CZ	.031	.92	.93	.06	1.91	2.52	1.25	1.64
3" x 6"	Demo SF	LL	.009	---	.26	.03	.29	.42	.19	.28
3" x 6"	Inst SF	CZ	.021	.62	.63	.04	1.29	1.70	.83	1.09
3" x 8"	Demo LF	LL	.020	---	.58	.06	.64	.93	.42	.60
3" x 8"	Inst LF	CZ	.046	1.83	1.38	.10	3.31	4.28	2.18	2.82
3" x 8"	Demo BF	LL	.010	---	.29	.03	.32	.46	.22	.32
3" x 8"	Inst BF	CZ	.023	.92	.69	.05	1.66	2.15	1.09	1.41
3" x 8"	Demo SF	LL	.009	---	.26	.03	.29	.42	.19	.28
3" x 8"	Inst SF	CZ	.021	.80	.63	.04	1.47	1.91	.95	1.23

Description	Oper	Unit	Crew Size	Man-Hours Per Unit	Costs Based On Small Volume					Large Volume	
					Avg Mat'l Unit Cost	Avg Labor Unit Cost	Avg Equip Unit Cost	Avg Total Unit Cost	Avg Price Incl O&P	Avg Total Unit Cost	Avg Price Incl O&P

Ledgers

Bolts 2' oc

Description	Oper	Unit	Crew Size	Man-Hours	Avg Mat'l	Avg Labor	Avg Equip	Avg Total	Avg Price O&P	Avg Total	Avg Price O&P
3" x 8"	Demo	LF	LC	.057	---	1.62	.36	1.98	2.78	1.28	1.8(
3" x 8"	Inst	LF	CS	.085	3.54	2.57	.53	6.64	8.46	4.41	5.6′
3" x 8"	Demo	BF	LC	.029	---	.83	.18	1.01	1.41	.66	.9:
3" x 8"	Inst	BF	CS	.043	1.79	1.30	.27	3.36	4.28	2.21	2.8′
3" x 8"	Demo	SF	LC	.003	---	.09	.02	.11	.15	.07	.0!
3" x 8"	Inst	SF	CS	.004	.18	.12	.03	.33	.42	.23	.2!
3" x 10"	Demo	LF	LC	.057	---	1.62	.36	1.98	2.78	1.28	1.8(
3" x 10"	Inst	LF	CS	.085	4.05	2.57	.53	7.15	9.05	4.75	6.0(
3" x 10"	Demo	BF	LC	.023	---	.65	.14	.79	1.12	.52	.7:
3" x 10"	Inst	BF	CS	.034	1.62	1.03	.21	2.86	3.62	1.89	2.3!
3" x 10"	Demo	SF	LC	.003	---	.09	.02	.11	.15	.07	.0!
3" x 10"	Inst	SF	CS	.004	.21	.12	.03	.36	.45	.25	.3:
3" x 12"	Demo	LF	LC	.057	---	1.62	.36	1.98	2.78	1.28	1.8(
3" x 12"	Inst	LF	CS	.085	4.56	2.57	.53	7.66	9.63	5.09	6.3!
3" x 12"	Demo	BF	LC	.019	---	.54	.12	.66	.93	.42	.5!
3" x 12"	Inst	BF	CS	.029	1.52	.88	.18	2.58	3.24	1.71	2.14
3" x 12"	Demo	SF	LC	.003	---	.09	.02	.11	.15	.07	.0!
3" x 12"	Inst	SF	CS	.004	.23	.12	.03	.38	.48	.26	.3:
4" x 8"	Demo	LF	LC	.057	---	1.62	.36	1.98	2.78	1.28	1.80
4" x 8"	Inst	LF	CS	.085	4.50	2.57	.53	7.60	9.56	5.05	6.34
4" x 8"	Demo	BF	LC	.021	---	.60	.13	.73	1.02	.49	.6(
4" x 8"	Inst	BF	CS	.032	1.68	.97	.20	2.85	3.59	1.89	2.37
4" x 8"	Demo	SF	LC	.003	---	.09	.02	.11	.15	.07	.0!
4" x 8"	Inst	SF	CS	.004	.23	.12	.03	.38	.48	.26	.3:
4" x 10"	Demo	LF	LC	.057	---	1.62	.36	1.98	2.78	1.28	1.80
4" x 10"	Inst	LF	CS	.085	5.22	2.57	.53	8.32	10.40	5.53	6.89
4" x 10"	Demo	BF	LC	.017	---	.48	.11	.59	.83	.38	.54
4" x 10"	Inst	BF	CS	.026	1.58	.79	.16	2.53	3.16	1.66	2.08
4" x 10"	Demo	SF	LC	.003	---	.09	.02	.11	.15	.07	.0!
4" x 10"	Inst	SF	CS	.004	.27	.12	.03	.42	.52	.29	.3(
4" x 12"	Demo	LF	LC	.057	---	1.62	.36	1.98	2.78	1.28	1.80
4" x 12"	Inst	LF	CS	.085	5.97	2.57	.53	9.07	11.30	6.03	7.47
4" x 12"	Demo	BF	LC	.014	---	.40	.09	.49	.68	.32	.44
4" x 12"	Inst	BF	CS	.021	1.50	.64	.13	2.27	2.81	1.51	1.88
4" x 12"	Demo	SF	LC	.003	---	.09	.02	.11	.15	.07	.0!
4" x 12"	Inst	SF	CS	.004	.29	.12	.03	.44	.55	.30	.37

Description	Oper	Unit	Crew Size	Man-Hours Per Unit	Avg Mat'l Unit Cost	Avg Labor Unit Cost	Avg Equip Unit Cost	Avg Total Unit Cost	Avg Price Incl O&P	Avg Total Unit Cost	Avg Price Incl O&P
						Costs Based On Small Volume					Large Volume	

Gutters and downspouts

Aluminum, baked on painted finish (white or brown)

Gutter, 5" box type

Description	Oper	Unit	Crew Size	Man-Hours Per Unit	Avg Mat'l Unit Cost	Avg Labor Unit Cost	Avg Equip Unit Cost	Avg Total Unit Cost	Avg Price Incl O&P	Avg Total Unit Cost	Avg Price Incl O&P
Heavyweight gauge	Inst	LF	UB	.114	1.22	4.07	---	5.29	**7.42**	3.74	**5.17**
Standard weight gauge	Inst	LF	UB	.114	.97	4.07	---	5.04	**7.13**	3.52	**4.92**
End caps	Inst	Ea	UB	---	.70	---	---	.70	**.70**	.63	**.63**
Drop outlet for downspouts	Inst	Ea	UB	---	3.99	---	---	3.99	**3.99**	3.60	**3.60**
Inside/outside corner	Inst	Ea	UB	---	5.01	---	---	5.01	**5.01**	4.52	**4.52**
Joint connector (4 each package)	Inst	Pkg	UB	---	5.42	---	---	5.42	**5.42**	4.89	**4.89**
Strap hanger (10 each package)	Inst	Pkg	UB	---	7.15	---	---	7.15	**7.15**	6.45	**6.45**
Fascia bracket (4 each package)	Inst	Pkg	UB	---	7.15	---	---	7.15	**7.15**	6.45	**6.45**
Spike and ferrule (10 each package)	Inst	Pkg	UB	---	6.42	---	---	6.42	**6.42**	5.79	**5.79**

Downspout, corrugated square

Description	Oper	Unit	Crew Size	Man-Hours Per Unit	Avg Mat'l Unit Cost	Avg Labor Unit Cost	Avg Equip Unit Cost	Avg Total Unit Cost	Avg Price Incl O&P	Avg Total Unit Cost	Avg Price Incl O&P
Standard weight gauge	Inst	LF	UB	.098	1.00	3.49	---	4.49	**6.32**	3.18	**4.41**
Regular/side elbow	Inst	Ea	UB	---	1.70	---	---	1.70	**1.70**	1.54	**1.54**
Downspout holder (4 each package)	Inst	Pkg	UB	---	4.99	---	---	4.99	**4.99**	4.50	**4.50**

Steel, natural finish

Gutter, 4" box type

Description	Oper	Unit	Crew Size	Man-Hours Per Unit	Avg Mat'l Unit Cost	Avg Labor Unit Cost	Avg Equip Unit Cost	Avg Total Unit Cost	Avg Price Incl O&P	Avg Total Unit Cost	Avg Price Incl O&P
Heavyweight gauge	Inst	LF	UB	.124	.79	4.42	---	5.21	**7.45**	3.35	**4.72**
End caps	Inst	Ea	UB	---	.70	---	---	.70	**.70**	.63	**.63**
Drop outlet for downspouts	Inst	Ea	UB	---	2.56	---	---	2.56	**2.56**	2.31	**2.31**
Inside/outside corner	Inst	Ea	UB	---	3.85	---	---	3.85	**3.85**	3.47	**3.47**
Joint connector (4 each package)	Inst	Pkg	UB	---	3.99	---	---	3.99	**3.99**	3.60	**3.60**
Strap hanger (10 each package)	Inst	Pkg	UB	---	5.42	---	---	5.42	**5.42**	4.89	**4.89**
Fascia bracket (4 each package)	Inst	Pkg	UB	---	5.36	---	---	5.36	**5.36**	4.84	**4.84**
Spike and ferrule (10 each package)	Inst	Pkg	UB	---	4.70	---	---	4.70	**4.70**	4.24	**4.24**

Downspout, 30 gauge galvanized steel

Description	Oper	Unit	Crew Size	Man-Hours Per Unit	Avg Mat'l Unit Cost	Avg Labor Unit Cost	Avg Equip Unit Cost	Avg Total Unit Cost	Avg Price Incl O&P	Avg Total Unit Cost	Avg Price Incl O&P
Standard weight gauge	Inst	LF	UB	.124	1.07	4.42	---	5.49	**7.78**	3.61	**5.02**
Regular/side elbow	Inst	Ea	UB	---	1.70	---	---	1.70	**1.70**	1.54	**1.54**
Downspout holder (4 each package)	Inst	Pkg	UB	---	4.56	---	---	4.56	**4.56**	4.12	**4.12**

					Costs Based On Small Volume					Large Volume	
Description	Oper	Unit	Crew Size	Man-Hours Per Unit	Avg Mat'l Unit Cost	Avg Labor Unit Cost	Avg Equip Unit Cost	Avg Total Unit Cost	Avg Price Incl O&P	Avg Total Unit Cost	Avg Price Incl O&P

Vinyl, extruded 5" PVC

White
Gutter	Inst	LF	UB	.124	1.22	4.42	---	5.64	**7.95**	3.74	5.17
End caps	Inst	Ea	UB	---	1.70	---	---	1.70	**1.70**	1.54	1.54
Drop outlet for downspouts	Inst	Ea	UB	---	6.44	---	---	6.44	**6.44**	5.81	5.81
Inside/outside corner	Inst	Ea	UB	---	6.44	---	---	6.44	**6.44**	5.81	5.81
Joint connector	Inst	Ea	UB	---	2.13	---	---	2.13	**2.13**	1.92	1.92
Fascia bracket (4 each package)	Inst	Pkg	UB	---	8.58	---	---	8.58	**8.58**	7.74	7.74
Downspout	Inst	LF	UB	.107	1.36	3.82	---	5.18	**7.21**	3.51	4.79
Downspout driplet	Inst	Ea	UB	---	8.58	---	---	8.58	**8.58**	7.74	7.74
Downspout joiner	Inst	Ea	UB	---	2.85	---	---	2.85	**2.85**	2.57	2.57
Downspout holder (2 each package)	Inst	Pkg	UB	---	8.58	---	---	8.58	**8.58**	7.74	7.74
Regular elbow	Inst	Ea	UB	---	3.13	---	---	3.13	**3.13**	2.83	2.83
Well cap	Inst	Ea	UB	---	3.99	---	---	3.99	**3.99**	3.60	3.60
Well outlet	Inst	Ea	UB	---	5.42	---	---	5.42	**5.42**	4.89	4.89
Expansion joint connector	Inst	Ea	UB	---	7.87	---	---	7.87	**7.87**	7.10	7.10
Rafter adapter (4 each package)	Inst	Pkg	UB	---	5.71	---	---	5.71	**5.71**	5.15	5.15

Brown
Gutter	Inst	LF	UB	.124	1.14	4.42	---	5.56	**7.86**	3.67	5.09
End caps	Inst	Ea	UB	---	2.13	---	---	2.13	**2.13**	1.92	1.92
Drop outlet for downspouts	Inst	Ea	UB	---	7.87	---	---	7.87	**7.87**	7.10	7.10
Inside/outside corner	Inst	Ea	UB	---	8.58	---	---	8.58	**8.58**	7.74	7.74
Joint connector	Inst	Ea	UB	---	2.85	---	---	2.85	**2.85**	2.57	2.57
Fascia bracket (4 each package)	Inst	Pkg	UB	---	10.70	---	---	10.70	**10.70**	9.66	9.66
Downspout	Inst	LF	UB	.107	1.50	3.82	---	5.32	**7.37**	3.63	4.93
Downspout driplet	Inst	Ea	UB	---	10.00	---	---	10.00	**10.00**	9.03	9.03
Downspout joiner	Inst	Ea	UB	---	3.42	---	---	3.42	**3.42**	3.08	3.08
Downspout holder (2 each package)	Inst	Pkg	UB	---	10.70	---	---	10.70	**10.70**	9.68	9.68
Regular elbow	Inst	Ea	UB	---	3.99	---	---	3.99	**3.99**	3.60	3.60
Well cap	Inst	Ea	UB	---	4.70	---	---	4.70	**4.70**	4.24	4.24
Well outlet	Inst	Ea	UB	---	6.85	---	---	6.85	**6.85**	6.18	6.18
Expansion joint connector	Inst	Ea	UB	---	10.00	---	---	10.00	**10.00**	9.03	9.03
Rafter adapter (4 each package)	Inst	Pkg	UB	---	5.71	---	---	5.71	**5.71**	5.15	5.15

Hardboard. See Paneling, page 212

Hardwood flooring

1. Strip flooring is nailed into place over wood sub-flooring or over wood sleeper strips. Using 3¼" W strips leaves 25% cutting and fitting waste; 2¼" W strips leave 33% waste. Nails and the respective cutting and fitting waste have been included in the unit costs.

2. Block flooring is laid in mastic applied to felt-covered wood subfloor. Mastic, 5% block waste and felt are included in material unit costs for block or parquet flooring.

Two types of wood block flooring

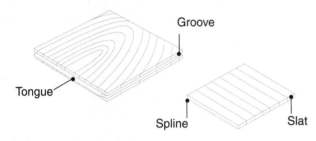

Groove
Tongue
Spline
Slat

Strip flooring:
A Side and end matched
B Side matched
C Square edged

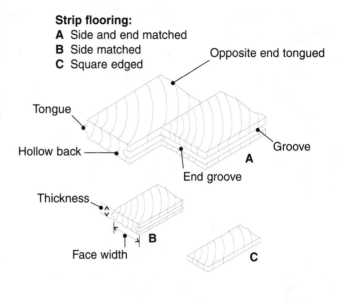

Opposite end tongued
Tongue
Groove
Hollow back
End groove
A
Thickness
B
Face width
C

Installation of first strip of flooring

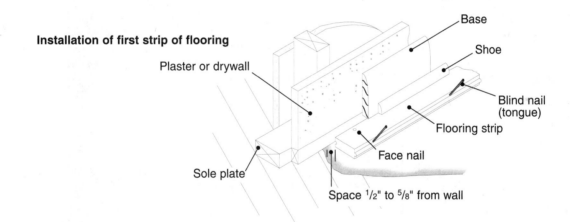

Base
Shoe
Plaster or drywall
Blind nail (tongue)
Flooring strip
Face nail
Sole plate
Space ½" to ⅝" from wall

Nailing of flooring:
A Angle of nailing
B Setting the nail without damage to the flooring

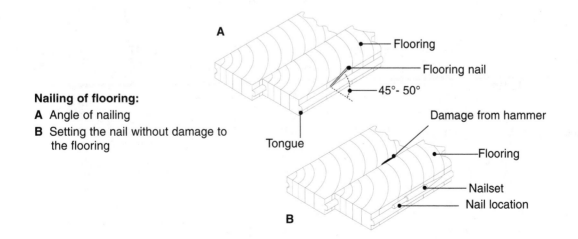

A
Flooring
Flooring nail
45°- 50°
Tongue
Damage from hammer
Flooring
Nailset
Nail location
B

					Costs Based On Small Volume					Large Volume	
Description	Oper	Unit	Crew Size	Man-Hours Per Unit	Avg Mat'l Unit Cost	Avg Labor Unit Cost	Avg Equip Unit Cost	Avg Total Unit Cost	Avg Price Incl O&P	Avg Total Unit Cost	Avg Price Incl O&P

Hardwood flooring
Includes waste and nails

Strip; installed over wood subfloor

Prefinished oak, prime

25/32" x 3-1/4"

Lay floor	Inst	SF	FD	.075	10.70	2.29	---	12.99	**15.70**	10.99	**13.10**
Wax, polish, machine buff	Inst	SF	FC	.016	.04	.50	.04	.58	**.82**	.35	**.49**

25/32" x 2-1/4"

Lay floor	Inst	SF	FD	.100	9.66	3.05	---	12.71	**15.60**	10.48	**12.60**
Wax, polish, machine buff	Inst	SF	FC	.016	.04	.50	.04	.58	**.82**	.35	**.49**

Unfinished

25/32" x 3-1/4", lay floor only
Fir

Vertical grain	Inst	SF	FD	.063	3.32	1.92	---	5.24	**6.64**	4.13	**5.12**
Flat grain	Inst	SF	FD	.063	3.77	1.92	---	5.69	**7.16**	4.53	**5.58**
Yellow pine	Inst	SF	FD	.063	3.23	1.92	---	5.15	**6.54**	4.05	**5.03**

25/32" x 2-1/4", lay floor only

Maple	Inst	SF	FD	.094	4.84	2.87	---	7.71	**9.78**	6.05	**7.50**
Oak	Inst	SF	FD	.094	5.42	2.87	---	8.29	**10.50**	6.56	**8.09**
Yellow pine	Inst	SF	FD	.078	3.29	2.38	---	5.67	**7.28**	4.37	**5.49**

Related materials and operations

Machine sand, fill and finish

New floors	Inst	SF	FC	.031	.12	.96	.09	1.17	**1.64**	.75	**1.04**
Damaged floors	Inst	SF	FC	.047	.16	1.46	.13	1.75	**2.46**	1.09	**1.52**
Wax, polish and machine buff	Inst	SF	FC	.016	.04	.50	.04	.58	**.82**	.35	**.49**

Block
Laid in mastic over wood subfloor covered with felt

Oak, 5/16" x 12" x 12"

Lay floor only

Prefinished	Inst	SF	FD	.058	11.90	1.77	---	13.67	**16.30**	11.67	**13.80**
Unfinished	Inst	SF	FD	.055	6.36	1.68	---	8.04	**9.78**	6.71	**8.04**

Teak, 5/16" x 12" x 12"

Lay floor only

Prefinished	Inst	SF	FD	.058	14.20	1.77	---	15.97	**19.00**	13.87	**16.20**
Unfinished	Inst	SF	FD	.055	8.71	1.68	---	10.39	**12.50**	8.82	**10.50**

Oak, 13/16" x 12" x 12"

Lay floor only

Prefinished	Inst	SF	FD	.058	21.40	1.77	---	23.17	**27.20**	20.17	**23.50**
Unfinished	Inst	SF	FD	.060	15.80	1.83	---	17.63	**20.90**	15.30	**17.90**

Machine sand, fill and finish

New floors	Inst	SF	FC	.031	.12	.96	.09	1.17	**1.64**	.75	**1.04**
Damaged floors	Inst	SF	FC	.047	.16	1.46	.13	1.75	**2.46**	1.09	**1.52**
Wax, polish and machine buff	Inst	SF	FC	.016	.04	.50	.04	.58	**.82**	.35	**.49**

Description	Oper	Unit	Crew Size	Man-Hours Per Unit	Costs Based On Small Volume				Large Volume	
					Avg Mat'l Unit Cost	Avg Labor Unit Cost	Avg Equip Unit Cost	Avg Total Unit Cost	Avg Price Incl O&P	Avg Total Unit Cost	Avg Price Incl O&P

Parquetry, 5/16" x 9" x 9"

Lay floor only

Description	Oper	Unit	Crew Size	Man-Hours Per Unit	Avg Mat'l Unit Cost	Avg Labor Unit Cost	Avg Equip Unit Cost	Avg Total Unit Cost	Avg Price Incl O&P	Avg Total Unit Cost	Avg Price Incl O&P
Oak	Inst	SF	FD	.060	6.36	1.83	---	8.19	**10.00**	6.80	**8.17**
Walnut	Inst	SF	FD	.060	16.40	1.83	---	18.23	**21.50**	15.80	**18.50**
Teak	Inst	SF	FD	.060	8.71	1.83	---	10.54	**12.70**	8.91	**10.60**
Machine sand, fill and finish											
New floors	Inst	SF	FC	.031	.12	.96	.09	1.17	**1.64**	.75	**1.04**
Damaged floors	Inst	SF	FC	.047	.16	1.46	.13	1.75	**2.46**	1.09	**1.52**
Wax, polish and machine buff	Inst	SF	FC	.016	.04	.50	.04	.58	**.82**	.35	**.49**
Acrylic wood parquet blocks											
5/16" x 12" x 12" set in epoxy	Inst	SF	FD	.060	7.71	1.83	---	9.54	**11.60**	8.00	**9.55**
Wax, polish and machine buff	Inst	SF	FC	.016	.04	.50	.04	.58	**.82**	.35	**.49**

					Costs Based On Small Volume					Large Volume	
Description	Oper	Unit	Crew Size	Man-Hours Per Unit	Avg Mat'l Unit Cost	Avg Labor Unit Cost	Avg Equip Unit Cost	Avg Total Unit Cost	Avg Price Incl O&P	Avg Total Unit Cost	Avg Price Incl O&P

Heating

Boilers

Electric fired heaters, includes standard controls and trim, ASME

Hot water

12 KW, 40 MBHP	Inst	Ea	SD	26.7	4460.00	867.00	---	5327.00	6400.00	4669.00	5580.0
24 KW, 82 MBHP	Inst	Ea	SD	27.9	4760.00	906.00	---	5666.00	6800.00	4978.00	5940.0
45 KW, 154 MBHP	Inst	Ea	SD	28.6	5230.00	928.00	---	6158.00	7370.00	5415.00	6450.0
60 KW, 205 MBHP	Inst	Ea	SD	28.9	6200.00	938.00	---	7138.00	8510.00	6308.00	7480.0
240 KW, 820 MBHP	Inst	Ea	SD	32.0	13400	1040	---	14440	16900	12879	1510
480 KW, 1,635 MBHP	Inst	Ea	SD	33.8	23000	1100	---	24100	28100	21621	2510
1,200 KW, 4,095 MBHP	Inst	Ea	SD	35.3	42000	1150	---	43150	50000	38767	4490
2,400 KW, 8,190 MBHP	Inst	Ea	SD	37.5	71600	1220	---	72820	84100	65615	7570

Steam

6 KW, 20 MBHP	Inst	Ea	SD	28.9	10600	938	---	11538	13600	10268	1200
60 KW, 205 MBHP	Inst	Ea	SD	35.3	12800	1150	---	13950	16400	12467	1460
240 KW, 815 MBHP	Inst	Ea	SD	80.0	20200	2600	---	22800	27100	20250	2390
600 KW, 2,047 MBHP	Inst	Ea	SE	213	31200	7120	---	38320	46400	33550	4030
Minimum Job Charge	Inst	Job	SD	17.1	---	555.00	---	555.00	816.00	390.00	573.0

Gas fired, natural or propane, heaters; includes standard controls; MBHP gross output

Cast iron with insulated jacket

Hot water

80 MBHP	Inst	Ea	SF	22.9	1410.00	758.00	---	2168.00	2730.00	1836.00	2290.0
100 MBHP	Inst	Ea	SF	23.3	1610.00	771.00	---	2381.00	2980.00	2029.00	2520.0
122 MBHP	Inst	Ea	SF	23.8	1780.00	788.00	---	2568.00	3200.00	2189.00	2710.0
163 MBHP	Inst	Ea	SF	24.2	2180.00	801.00	---	2981.00	3680.00	2573.00	3150.0
440 MBHP	Inst	Ea	SF	24.5	4250.00	811.00	---	5061.00	6090.00	4453.00	5320.0
2,000 MBHP	Inst	Ea	SF	32.0	13800	1060	---	14860	17400	13295	1550
6,970 MBHP	Inst	Ea	SF	80.0	93800	2650	---	96450	112000	86690	10000

Steam

80 MBHP	Inst	Ea	SF	22.9	1410.00	758.00	---	2168.00	2730.00	1836.00	2290.0
163 MBHP	Inst	Ea	SF	24.2	2180.00	801.00	---	2981.00	3680.00	2573.00	3150.0
440 MBHP	Inst	Ea	SF	24.5	4250.00	811.00	---	5061.00	6090.00	4453.00	5320.0
3,570 MBHP	Inst	Ea	SF	35.3	21700	1170	---	22870	26700	20484	2380
6,970 MBHP	Inst	Ea	SF	80.0	70400	2650	---	73050	84800	65490	7600

Steel with insulated jacket; includes burner and one zone valve

Hot water

50 MBHP	Inst	Ea	SF	14.5	2110.00	480.00	---	2590.00	3130.00	2271.00	2720.0
70 MBHP	Inst	Ea	SF	14.5	2350.00	480.00	---	2830.00	3400.00	2481.00	2970.0
90 MBHP	Inst	Ea	SF	15.2	2380.00	503.00	---	2883.00	3480.00	2527.00	3020.0
105 MBHP	Inst	Ea	SF	16.0	2680.00	530.00	---	3210.00	3860.00	2817.00	3370.0
130 MBHP	Inst	Ea	SF	16.8	3050.00	556.00	---	3606.00	4320.00	3167.00	3780.0
150 MBHP	Inst	Ea	SF	17.8	3550.00	589.00	---	4139.00	4950.00	3650.00	4330.0
185 MBHP	Inst	Ea	SF	19.4	4290.00	642.00	---	4932.00	5880.00	4350.00	5160.0
235 MBHP	Inst	Ea	SF	21.2	5430.00	702.00	---	6132.00	7270.00	5430.00	6410.0
290 MBHP	Inst	Ea	SF	22.9	6130.00	758.00	---	6888.00	8160.00	6106.00	7200.0
480 MBHP	Inst	Ea	SF	35.3	8840	1170	---	10010	11900	8874	1050
640 MBHP	Inst	Ea	SF	45.3	10500	1500	---	12000	14300	10640	1260
800 MBHP	Inst	Ea	SF	54.5	12300	1800	---	14100	16800	12470	1470
960 MBHP	Inst	Ea	SF	58.5	15300	1940	---	17240	20400	15240	1800
Minimum Job Charge	Inst	Job	SF	17.1	---	566.00	---	566.00	832.00	397.00	584.0

Description	Oper	Unit	Crew Size	Man-Hours Per Unit	Costs Based On Small Volume						Large Volume	
					Avg Mat'l Unit Cost	Avg Labor Unit Cost	Avg Equip Unit Cost	Avg Total Unit Cost	Avg Price Incl O&P		Avg Total Unit Cost	Avg Price Incl O&P
Oil fired heaters; includes standard controls; flame retention burner;												
MBHP gross output												
Cast iron with insulated jacket												
Hot water												
110 MBHP	Inst	Ea	SF	32.0	1570.00	1060.00	---	2630.00	3370.00		2215.00	2800.00
200 MBHP	Inst	Ea	SF	45.3	2180.00	1500.00	---	3680.00	4710.00		3110.00	3930.00
1,080 MBHP	Inst	Ea	SF	104	11100	3440	---	14540	17900		12650	15400
1,320 MBHP	Inst	Ea	SF	126	10100	4170	---	14270	17700		12280	15100
2,100 MBHP	Inst	Ea	SF	160	13400	5300	---	18700	23200		16070	19700
4,360 MBHP	Inst	Ea	SF	218	24000	7220	---	31220	38200		27000	32700
6,970 MBHP	Inst	Ea	SF	400	69700	13200	---	82900	99600		72830	87000
Steam												
110 MBHP	Inst	Ea	SF	32.0	1570.00	1060.00	---	2630.00	3370.00		2215.00	2800.00
205 MBHP	Inst	Ea	SF	45.3	2180.00	1500.00	---	3680.00	4710.00		3110.00	3930.00
1,085 MBHP	Inst	Ea	SF	104.3	11100	3450	---	14550	17900		12650	15400
1,360 MBHP	Inst	Ea	SF	126	10100	4170	---	14270	17700		12280	15100
2,175 MBHP	Inst	Ea	SF	160	13400	5300	---	18700	23200		16070	19700
4,360 MBHP	Inst	Ea	SF	218	24000	7220	---	31220	38200		27000	32700
6,970 MBHP	Inst	Ea	SF	400	69700	13200	---	82900	99600		72830	87000
Steel insulated jacket burner												
Hot water												
105 MBHP	Inst	Ea	SF	17.8	2780.00	589.00	---	3369.00	4060.00		2950.00	3530.00
120 MBHP	Inst	Ea	SF	20.0	2810.00	662.00	---	3472.00	4210.00		3037.00	3650.00
140 MBHP	Inst	Ea	SF	22.9	2950.00	758.00	---	3708.00	4500.00		3226.00	3890.00
170 MBHP	Inst	Ea	SF	26.7	3520.00	884.00	---	4404.00	5340.00		3842.00	4630.00
225 MBHP	Inst	Ea	SF	35.3	4190.00	1170.00	---	5360.00	6530.00		4664.00	5650.00
315 MBHP	Inst	Ea	SF	40.0	6160.00	1320.00	---	7480.00	9040.00		6563.00	7860.00
420 MBHP	Inst	Ea	SF	45.3	6930.00	1500.00	---	8430.00	10200.00		7400.00	8870.00
Minimum Job Charge	Inst	Job	SF	17.1	---	566.00	---	566.00	832.00		397.00	584.00
Boiler accessories												
Burners												
Conversion, gas fired, LP or natural												
Residential, gun type, atmospheric input												
72 to 200 MBHP	Inst	Ea	SB	10.7	878.00	337.00	---	1215.00	1500.00		1045.00	1280.00
120 to 360 MBHP	Inst	Ea	SB	12.2	972.00	384.00	---	1356.00	1680.00		1165.00	1430.00
280 to 800 MBHP	Inst	Ea	SB	14.2	1880.00	447.00	---	2327.00	2810.00		2027.00	2440.00
Flame retention, oil fired assembly												
2.0 to 5.0 GPH	Inst	Ea	SB	12.2	670.00	384.00	---	1054.00	1340.00		893.00	1120.00

					Costs Based On Small Volume					Large Volume	
Description	Oper	Unit	Crew Size	Man-Hours Per Unit	Avg Mat'l Unit Cost	Avg Labor Unit Cost	Avg Equip Unit Cost	Avg Total Unit Cost	Avg Price Incl O&P	Avg Total Unit Cost	Avg Price Incl O&P

Forced warm air systems
Duct furnaces
Furnace includes burner, controls, stainless steel heat exchanger
Gas fired with an electric ignition

Outdoor installation, includes vent cap

Description	Oper	Unit	Crew	MHr	Mat'l	Labor	Equip	Total	Price O&P	LV Total	LV Price
225 MBHP output	Inst	Ea	SB	10.7	3750.00	337.00	---	4087.00	4810.00	3642.00	4270.00
375 MBHP output	Inst	Ea	SB	14.2	5430.00	447.00	---	5877.00	6900.00	5237.00	6130.00
450 MBHP output	Inst	Ea	SB	16.3	5700.00	513.00	---	6213.00	7300.00	5527.00	6480.00

Furnaces, hot air heating with blowers and standard controls

Gas or oil lines and couplings not included,
flue piping not included (see page 279)

Electric-fired, UL listed, heat staging, 240 volt run and connection

Description	Oper	Unit	Crew	MHr	Mat'l	Labor	Equip	Total	Price O&P	LV Total	LV Price
30 MBHP	Inst	Ea	UE	7.41	543.00	236.00	---	779.00	973.00	667.00	825.00
75 MBHP	Inst	Ea	UE	7.60	690.00	242.00	---	932.00	1150.00	805.00	985.00
85 MBHP	Inst	Ea	UE	8.33	771.00	265.00	---	1036.00	1280.00	895.00	1090.00
90 MBHP	Inst	Ea	UE	8.89	965.00	283.00	---	1248.00	1530.00	1083.00	1320.00
Minimum Job Charge	Inst	Job	UD	11.4	---	355.00	---	355.00	526.00	249.00	369.00

Gas fired, AGA certified, direct drive models

Description	Oper	Unit	Crew	MHr	Mat'l	Labor	Equip	Total	Price O&P	LV Total	LV Price
40 MBHP	Inst	Ea	UD	5.93	583.00	185.00	---	768.00	944.00	664.00	810.0
65 MBHP	Inst	Ea	UD	6.08	777.00	189.00	---	966.00	1170.00	844.00	1020.0
80 MBHP	Inst	Ea	UD	6.27	824.00	195.00	---	1019.00	1240.00	891.00	1070.0
85 MBHP	Inst	Ea	UD	6.67	851.00	208.00	---	1059.00	1290.00	924.00	1110.0
105 MBHP	Inst	Ea	UD	7.11	878.00	221.00	---	1099.00	1340.00	959.00	1160.0
125 MBHP	Inst	Ea	UD	7.62	1010.00	237.00	---	1247.00	1510.00	1092.00	1310.0
160 MBHP	Inst	Ea	UD	8.21	1040.00	256.00	---	1296.00	1570.00	1130.00	1360.0
200 MBHP	Inst	Ea	UD	8.89	2480.00	277.00	---	2757.00	3260.00	2448.00	2880.0
Minimum Job Charge	Inst	Job	UD	11.4	---	355.00	---	355.00	526.00	249.00	369.00
Gas line with couplings	Inst	LF	SB	.474	---	14.90	---	14.90	22.00	11.20	16.5

Oil fired, UL listed, gun-type burner

Description	Oper	Unit	Crew	MHr	Mat'l	Labor	Equip	Total	Price O&P	LV Total	LV Price
55 MBHP	Inst	Ea	UD	6.08	998.00	189.00	---	1187.00	1430.00	1043.00	1250.0
100 MBHP	Inst	Ea	UD	7.11	1060.00	221.00	---	1281.00	1550.00	1122.00	1350.0
125 MBHP	Inst	Ea	UD	7.62	1470.00	237.00	---	1707.00	2050.00	1508.00	1790.0
150 MBHP	Inst	Ea	UD	8.21	1640.00	256.00	---	1896.00	2270.00	1672.00	1990.0
200 MBHP	Inst	Ea	UD	8.89	2750.00	277.00	---	3027.00	3570.00	2688.00	3160.0
Minimum Job Charge	Inst	Job	UD	11.4	---	355.00	---	355.00	526.00	249.00	369.00
Oil line with couplings	Inst	LF	SB	.267	---	8.41	---	8.41	12.40	6.30	9.2

Combo fired (wood, coal, oil combination), complete with burner

Description	Oper	Unit	Crew	MHr	Mat'l	Labor	Equip	Total	Price O&P	LV Total	LV Price
115 MBHP (based on oil)	Inst	Ea	UD	7.34	2980.00	229.00	---	3209.00	3770.00	2862.00	3350.0
140 MBHP (based on oil)	Inst	Ea	UD	7.88	4790.00	245.00	---	5035.00	5870.00	4515.00	5250.0
150 MBHP (based on oil)	Inst	Ea	UD	8.21	4790.00	256.00	---	5046.00	5890.00	4522.00	5260.0
170 MBHP (based on oil)	Inst	Ea	UD	8.51	5060.00	265.00	---	5325.00	6210.00	4769.00	5550.0
Minimum Job Charge	Inst	Job	UD	11.4	---	355.00	---	355.00	526.00	249.00	369.0
Oil line with couplings	Inst	LF	SB	.267	---	8.41	---	8.41	12.40	6.30	9.2

Description	Oper	Unit	Crew Size	Man-Hours Per Unit	Avg Mat'l Unit Cost	Avg Labor Unit Cost	Avg Equip Unit Cost	Avg Total Unit Cost	Avg Price Incl O&P	Avg Total Unit Cost	Avg Price Incl O&P
										Large Volume	
Space heaters, gas fired											
Unit includes cabinet, grilles, fan, controls, burner and thermostat;											
no flue piping included (see page 279)											
Floor mounted											
60 MBHP	Inst	Ea	SB	2.67	717.00	84.10	---	801.10	948.00	710.00	837.00
180 MBHP	Inst	Ea	SB	5.33	1070.00	168.00	---	1238.00	1470.00	1088.00	1290.00
Suspension mounted, propeller fan											
20 MBHP	Inst	Ea	SB	3.56	536.00	112.00	---	648.00	781.00	568.10	680.00
60 MBHP	Inst	Ea	SB	4.27	663.00	135.00	---	798.00	961.00	700.00	837.00
130 MBHP	Inst	Ea	SB	5.33	978.00	168.00	---	1146.00	1370.00	1009.00	1200.00
320 MBHP	Inst	Ea	SB	10.7	2010.00	337.00	---	2347.00	2810.00	2072.00	2460.00
Powered venter, adapter	ADD	Ea	SB	2.67	340.00	84.10	---	424.10	515.00	370.00	446.00
Wall furnace, self-contained thermostat											
Single capacity, recessed or surface mounted, 1-speed fan											
15 MBHP	Inst	Ea	SB	4.27	590.00	135.00	---	725.00	876.00	633.00	760.00
25 MBHP	Inst	Ea	SB	5.33	610.00	168.00	---	778.00	948.00	677.00	818.00
35 MBHP	Inst	Ea	SB	7.11	817.00	224.00	---	1041.00	1270.00	906.00	1100.00
Dual capacity, recessed or surface mounted, 2-speed blowers											
50 MBHP (direct vent)	Inst	Ea	SB	10.7	817.00	337.00	---	1154.00	1440.00	990.00	1220.00
60 MBHP (up vent)	Inst	Ea	SB	10.7	724.00	337.00	---	1061.00	1330.00	905.00	1120.00
Register kit for circulating heat											
to second room	Inst	Ea	SB	2.67	73.70	84.10	---	157.80	208.00	129.60	169.00
Minimum Job Charge	Inst	Job	SB	11.4	---	359.00	---	359.00	528.00	252.00	370.00
Bathroom heaters, electric fired											
Ceiling Heat-A-Ventlite, includes grille, blower, 4" round duct 13" x 7" dia., 3-way switch											
1,500 watt	Inst	Ea	EA	5.33	318.00	183.00	---	501.00	633.00	425.00	531.00
1,800 watt	Inst	Ea	EA	5.33	352.00	183.00	---	535.00	673.00	456.00	567.00
Ceiling Heat-A-Lite, includes grille, airotor wheel, 13" x 7" dia., 2-way switch											
1,500 watt	Inst	Ea	EA	5.33	235.00	183.00	---	418.00	537.00	350.00	444.00
Ceiling radiant heating using infrared lamps											
Recessed, Heat-A-Lamp											
One bulb, 250 watt lamp	Inst	Ea	EA	4.44	60.30	153.00	---	213.30	292.00	169.50	230.00
Two bulb, 500 watt lamp	Inst	Ea	EA	4.44	103.00	153.00	---	256.00	342.00	208.20	274.00
Three bulb, 750 watt lamp	Inst	Ea	EA	4.44	180.00	153.00	---	333.00	429.00	277.00	354.00
Recessed, Heat-A-Vent											
One bulb, 250 watt lamp	Inst	Ea	EA	4.44	125.00	153.00	---	278.00	366.00	228.00	297.00
Two bulb, 500 watt lamp	Inst	Ea	EA	4.44	145.00	153.00	---	298.00	389.00	246.00	317.00
Wall heaters, recessed											
Fan forced											
1250 watt heating element	Inst	Ea	EA	3.33	166.00	115.00	---	281.00	358.00	236.00	298.00
Radiant heating											
1200 watt heating element	Inst	Ea	EA	3.33	131.00	115.00	---	246.00	318.00	205.00	262.00
1500 watt heating element	Inst	Ea	EA	3.33	138.00	115.00	---	253.00	326.00	211.00	269.00
Wiring, connection, and installation in closed wall or ceiling structure											
ADD	Inst	Ea	EA	3.46	---	119.00	---	119.00	174.00	83.20	122.00

Note: The header spans "Costs Based On Small Volume" over the Man-Hours Per Unit, Avg Mat'l Unit Cost, Avg Labor Unit Cost, Avg Equip Unit Cost, Avg Total Unit Cost, and Avg Price Incl O&P columns; "Large Volume" spans the final two columns (Avg Total Unit Cost, Avg Price Incl O&P).

				Costs Based On Small Volume					Large Volume	
Description	Oper	Unit	Crew Size	Man-Hours Per Unit	Avg Mat'l Unit Cost	Avg Labor Unit Cost	Avg Equip Unit Cost	Avg Total Unit Cost	Avg Price Incl O&P	Avg Total Unit Cost	Avg Price Incl O&P

Insulation

Batt or roll

With wall or ceiling finish already removed

Joists, 16" or 24" oc	Demo	SF	LB	.007	---	.19	---	.19	.28	.13	.2
Rafters, 16" or 24" oc	Demo	SF	LB	.008	---	.21	---	.21	.32	.16	.2
Studs, 16" or 24" oc	Demo	SF	LB	.006	---	.16	---	.16	.24	.13	.2

Place and/or staple, Johns-Manville fiberglass; allowance made for joists, rafters, studs

Joists

Unfaced

3-1/2" T (R-13)

16" oc	Inst	SF	CA	.011	.48	.35	---	.83	1.08	.69	.8

6-1/2" T (R-19)

16" oc	Inst	SF	CA	.011	.56	.35	---	.91	1.17	.76	.9
24" oc	Inst	SF	CA	.007	.56	.22	---	.78	.98	.66	.8

7" T (R-22)

16" oc	Inst	SF	CA	.011	.71	.35	---	1.06	1.35	.90	1.1
24" oc	Inst	SF	CA	.007	.71	.22	---	.93	1.15	.80	.9

9-1/4" T (R-30)

16" oc	Inst	SF	CA	.011	.82	.35	---	1.17	1.47	1.00	1.2
24" oc	Inst	SF	CA	.007	.82	.22	---	1.04	1.28	.90	1.0

Kraft-faced

3-1/2" T (R-11)

16" oc	Inst	SF	CA	.011	.41	.35	---	.76	1.00	.63	.8
24" oc	Inst	SF	CA	.007	.41	.22	---	.63	.81	.53	.6

3-1/2" T (R-13)

16" oc	Inst	SF	CA	.011	.48	.35	---	.83	1.08	.69	.8

6-1/2" T (R-19)

16" oc	Inst	SF	CA	.011	.56	.35	---	.91	1.17	.76	.9
24" oc	Inst	SF	CA	.007	.56	.22	---	.78	.98	.66	.8

7" T (R-22)

16" oc	Inst	SF	CA	.011	.71	.35	---	1.06	1.35	.90	1.1
24" oc	Inst	SF	CA	.007	.71	.22	---	.93	1.15	.80	.9

9-1/4" T (R-30)

16" oc	Inst	SF	CA	.011	.82	.35	---	1.17	1.47	1.00	1.2
24" oc	Inst	SF	CA	.007	.82	.22	---	1.04	1.28	.90	1.0

Foil-faced

4" T (R-11)

16" oc	Inst	SF	CA	.011	.41	.35	---	.76	1.00	.63	.8
24" oc	Inst	SF	CA	.007	.41	.22	---	.63	.81	.53	.6

6-1/2" T (R-19)

16" oc	Inst	SF	CA	.011	.56	.35	---	.91	1.17	.76	.9
24" oc	Inst	SF	CA	.007	.56	.22	---	.78	.98	.66	.8

Description	Oper	Unit	Crew Size	Man-Hours Per Unit	Avg Mat'l Unit Cost	Avg Labor Unit Cost	Avg Equip Unit Cost	Avg Total Unit Cost	Avg Price Incl O&P	Avg Total Unit Cost	Avg Price Incl O&P
					Costs Based On Small Volume					Large Volume	
Rafters											
Unfaced											
3-1/2" T (R-13)											
16" oc	Inst	SF	CA	.016	.48	.51	---	.99	**1.32**	.82	**1.07**
6-1/2" T (R-19)											
16" oc	Inst	SF	CA	.016	.56	.51	---	1.07	**1.41**	.89	**1.15**
24" oc	Inst	SF	CA	.011	.56	.35	---	.91	**1.17**	.76	**.96**
7" T (R-22)											
16" oc	Inst	SF	CA	.016	.71	.51	---	1.22	**1.59**	1.03	**1.31**
24" oc	Inst	SF	CA	.011	.71	.35	---	1.06	**1.35**	.90	**1.12**
9-1/4" T (R-30)											
16" oc	Inst	SF	CA	.016	.82	.51	---	1.33	**1.71**	1.13	**1.43**
24" oc	Inst	SF	CA	.011	.82	.35	---	1.17	**1.47**	1.00	**1.24**
Kraft-faced											
3-1/2" T (R-11)											
16" oc	Inst	SF	CA	.016	.41	.51	---	.92	**1.24**	.76	**1.00**
24" oc	Inst	SF	CA	.011	.41	.35	---	.76	**1.00**	.63	**.81**
3-1/2" T (R-13)											
16" oc	Inst	SF	CA	.016	.48	.51	---	.99	**1.32**	.82	**1.07**
6-1/2" T (R-19)											
16" oc	Inst	SF	CA	.016	.56	.51	---	1.07	**1.41**	.89	**1.15**
24" oc	Inst	SF	CA	.011	.56	.35	---	.91	**1.17**	.76	**.96**
7" T (R-22)											
16" oc	Inst	SF	CA	.016	.71	.51	---	1.22	**1.59**	1.03	**1.31**
24" oc	Inst	SF	CA	.011	.71	.35	---	1.06	**1.35**	.90	**1.12**
9-1/4" T (R-30)											
16" oc	Inst	SF	CA	.016	.82	.51	---	1.33	**1.71**	1.13	**1.43**
24" oc	Inst	SF	CA	.011	.82	.35	---	1.17	**1.47**	1.00	**1.24**
Foil-faced											
4" T (R-11)											
16" oc	Inst	SF	CA	.016	.41	.51	---	.92	**1.24**	.76	**1.00**
24" oc	Inst	SF	CA	.011	.41	.35	---	.76	**1.00**	.63	**.81**
6-1/2" T (R-19)											
16" oc	Inst	SF	CA	.016	.56	.51	---	1.07	**1.41**	.89	**1.15**
24" oc	Inst	SF	CA	.011	.56	.35	---	.91	**1.17**	.76	**.96**
Studs											
Unfaced											
3-1/2" T (R-13)											
16" oc	Inst	SF	CA	.013	.48	.42	---	.90	**1.18**	.75	**.98**
6-1/2" T (R-19)											
16" oc	Inst	SF	CA	.013	.56	.42	---	.98	**1.27**	.82	**1.06**
24" oc	Inst	SF	CA	.009	.56	.29	---	.85	**1.08**	.72	**.91**
7" T (R-22)											
16" oc	Inst	SF	CA	.013	.71	.42	---	1.13	**1.44**	.96	**1.22**
24" oc	Inst	SF	CA	.009	.71	.29	---	1.00	**1.25**	.86	**1.07**
9-1/4" T (R-30)											
16" oc	Inst	SF	CA	.013	.82	.42	---	1.24	**1.57**	1.06	**1.33**
24" oc	Inst	SF	CA	.009	.82	.29	---	1.11	**1.38**	.96	**1.19**

					Costs Based On Small Volume					Large Volume	
Description	Oper	Unit	Crew Size	Man-Hours Per Unit	Avg Mat'l Unit Cost	Avg Labor Unit Cost	Avg Equip Unit Cost	Avg Total Unit Cost	Avg Price Incl O&P	Avg Total Unit Cost	Avg Price Incl O&P
Kraft-faced											
3-1/2" T (R-11)											
16" oc	Inst	SF	CA	.013	.41	.42	---	.83	**1.10**	.69	.9
24" oc	Inst	SF	CA	.009	.41	.29	---	.70	**.90**	.59	.7
3-1/2" T (R-13)											
16" oc	Inst	SF	CA	.013	.48	.42	---	.90	**1.18**	.75	.9
6-1/2" T (R-19)											
16" oc	Inst	SF	CA	.013	.56	.42	---	.98	**1.27**	.82	1.0
24" oc	Inst	SF	CA	.009	.56	.29	---	.85	**1.08**	.72	.9
7" T (R-22)											
16" oc	Inst	SF	CA	.013	.71	.42	---	1.13	**1.44**	.96	1.2
24" oc	Inst	SF	CA	.009	.71	.29	---	1.00	**1.25**	.86	1.0
9-1/4" T (R-30)											
16" oc	Inst	SF	CA	.013	.82	.42	---	1.24	**1.57**	1.06	1.33
24" oc	Inst	SF	CA	.009	.82	.29	---	1.11	**1.38**	.96	1.1
Foil-faced											
4" T (R-11)											
16" oc	Inst	SF	CA	.013	.41	.42	---	.83	**1.10**	.69	.9
24" oc	Inst	SF	CA	.009	.41	.29	---	.70	**.90**	.59	.7
6-1/2" T (R-19)											
16" oc	Inst	SF	CA	.013	.56	.42	---	.98	**1.27**	.82	1.0
24" oc	Inst	SF	CA	.009	.56	.29	---	.85	**1.08**	.72	.9

Loose fill

With ceiling finish already removed

Joists, 16" or 24" oc

4" T	Demo	SF	LB	.005	---	.13	---	.13	**.20**	.11	.16
6" T	Demo	SF	LB	.009	---	.24	---	.24	**.36**	.19	.28

Allowance made for joists, cavities, and cores

Insulating wool, granule or pellet (40 lbs/bag, 4 CF/bag)

Joists, @ 7 lbs/CF density

16" oc

4" T	Inst	SF	CH	.017	.72	.51	---	1.23	**1.60**	1.04	1.34
6" T	Inst	SF	CH	.022	1.08	.67	---	1.75	**2.24**	1.49	1.90

24" oc

4" T	Inst	SF	CH	.017	.74	.51	---	1.25	**1.62**	1.06	1.36
6" T	Inst	SF	CH	.023	1.12	.70	---	1.82	**2.33**	1.52	1.93

Vermiculite/Perlite (approximately 10 lbs/bag, 4 CF/bag)

Joists

16" oc

4" T	Inst	SF	CH	.012	.57	.36	---	.93	**1.20**	.78	1.00
6" T	Inst	SF	CH	.016	.85	.48	---	1.33	**1.70**	1.13	1.43

24" oc

4" T	Inst	SF	CH	.012	.58	.36	---	.94	**1.21**	.80	1.02
6" T	Inst	SF	CH	.017	.88	.51	---	1.39	**1.78**	1.18	1.50

Cavity walls

1" T	Inst	SF	CH	.005	.14	.15	---	.29	**.39**	.24	.32
2" T	Inst	SF	CH	.009	.27	.27	---	.54	**.72**	.46	.61

Block walls (2 cores/block)

8" T block	Inst	SF	CH	.020	.48	.61	---	1.09	**1.46**	.88	1.18
12" T block	Inst	SF	CH	.026	.88	.79	---	1.67	**2.19**	1.41	1.83

Description	Oper	Unit	Crew Size	Man-Hours Per Unit	Costs Based On Small Volume					Large Volume	
					Avg Mat'l Unit Cost	Avg Labor Unit Cost	Avg Equip Unit Cost	Avg Total Unit Cost	Avg Price Incl O&P	Avg Total Unit Cost	Avg Price Incl O&P

rigid
Roofs
1/2" T	Demo	Sq	LB	1.23	---	32.80	---	32.80	**48.80**	25.10	**37.40**
1" T	Demo	Sq	LB	1.45	---	38.60	---	38.60	**57.60**	28.50	**42.50**
Walls, 1/2" T	Demo	SF	LB	.010	---	.27	---	.27	**.40**	.19	**.28**

rigid insulating board
Roofs, over wood decks, 5% waste included
Normal (dry) moisture conditions within building
Nail one ply 15 lb felt, set and nail:

2' x 8' x 1/2" T&G asphalt sheathing	Inst	SF	CN	.026	.66	.81	---	1.47	**1.97**	1.21	**1.61**
4' x 8' x 1/2" asphalt sheathing	Inst	SF	2C	.019	.66	.61	---	1.27	**1.67**	1.04	**1.35**
4' x 8' x 1/2" building block	Inst	SF	2C	.019	.66	.61	---	1.27	**1.67**	1.04	**1.35**

Excessive (humid) moisture conditions within building
Nail and overlap three plies 15 lb felt, mop laps and surface one coat and embed:

2' x 8' x 1/2" T&G asphalt sheathing	Inst	SF	RT	.041	.76	1.37	---	2.13	**3.00**	1.73	**2.40**
4' x 8' x 1/2" asphalt sheathing	Inst	SF	RT	.039	.76	1.30	---	2.06	**2.90**	1.69	**2.35**
4' x 8' x 1/2" building block	Inst	SF	RT	.039	.76	1.30	---	2.06	**2.90**	1.69	**2.35**

Over noncombustible decks, 5% waste included
Normal (dry) moisture conditions within building
Mop one coat and embed:

2' x 8' x 1/2" T&G asphalt sheathing	Inst	SF	RT	.029	.79	.97	---	1.76	**2.41**	1.46	**1.97**
4' x 8' x 1/2" asphalt sheathing	Inst	SF	RT	.028	.79	.94	---	1.73	**2.36**	1.42	**1.92**
4' x 8' x 1/2" building block	Inst	SF	RT	.028	.79	.94	---	1.73	**2.36**	1.42	**1.92**

Excessive (humid) moisture conditions within building
Mop one coat, embed two plies 15 lb felt, mop and embed:

2' x 8' x 1/2" T&G asphalt sheathing	Inst	SF	RT	.051	1.01	1.71	---	2.72	**3.80**	2.18	**3.02**
4' x 8' x 1/2" asphalt sheathing	Inst	SF	RT	.049	1.01	1.64	---	2.65	**3.70**	2.15	**2.96**
4' x 8' x 1/2" building block	Inst	SF	RT	.049	1.01	1.64	---	2.65	**3.70**	2.15	**2.96**

Walls, nailed, 5% waste included
4' x 8' x 1/2" asphalt sheathing

Straight wall	Inst	SF	2C	.015	.60	.48	---	1.08	**1.41**	.89	**1.15**
Cut-up wall	Inst	SF	2C	.018	.60	.58	---	1.18	**1.56**	.96	**1.25**

4' x 8' x 1/2" building board

Straight wall	Inst	SF	2C	.015	.60	.48	---	1.08	**1.41**	.89	**1.15**
Cut-up wall	Inst	SF	2C	.018	.60	.58	---	1.18	**1.56**	.96	**1.25**

Intercom systems. See Electrical, page 107
Jacuzzi whirlpools. See Spas, page 267
Lath & plaster. See Plaster, page 215

Description	Oper	Unit	Crew Size	Man-Hours Per Unit	Avg Mat'l Unit Cost	Avg Labor Unit Cost	Avg Equip Unit Cost	Avg Total Unit Cost	Avg Price Incl O&P	Avg Total Unit Cost	Avg Price Incl O&P
								Costs Based On Small Volume		**Large Volume**	

Lighting fixtures

Labor includes hanging and connecting fixtures

Indoor lighting

Fluorescent

Description	Oper	Unit	Crew Size	Man-Hours Per Unit	Avg Mat'l Unit Cost	Avg Labor Unit Cost	Avg Equip Unit Cost	Avg Total Unit Cost	Avg Price Incl O&P	Avg Total Unit Cost	Avg Price Incl O&P
Pendant mounted worklights											
4' L, two 40 watt RS	Inst	Ea	EA	1.23	57.20	42.30	---	99.50	128.00	78.20	98.50
4' L, two 60 watt RS	Inst	Ea	EA	1.23	93.10	42.30	---	135.40	169.00	110.10	135.00
8' L, two 75 watt RS	Inst	Ea	EA	1.54	108.00	53.00	---	161.00	201.00	130.00	160.00
Recessed mounted light fixture with acrylic diffuser											
1' W x 4' L, two 40 watt RS	Inst	Ea	EA	1.23	61.20	42.30	---	103.50	132.00	81.80	103.00
2' W x 2' L, two U 40 watt RS	Inst	Ea	EA	1.23	66.50	42.30	---	108.80	138.00	86.50	108.00
2' W x 4' L, four 40 watt RS	Inst	Ea	EA	1.54	74.50	53.00	---	127.50	163.00	100.50	126.00
Strip lighting fixtures											
4' L, one 40 watt RS	Inst	Ea	EA	1.23	34.60	42.30	---	76.90	102.00	58.20	75.50
4' L, two 40 watt RS	Inst	Ea	EA	1.23	37.20	42.30	---	79.50	105.00	60.50	78.20
8' L, one 75 watt SL	Inst	Ea	EA	1.54	51.90	53.00	---	104.90	137.00	80.40	103.00
8' L, two 75 watt SL	Inst	Ea	EA	1.54	62.50	53.00	---	115.50	149.00	89.90	114.00
Surface mounted, acrylic diffuser											
1' W x 4' L, two 40 watt RS	Inst	Ea	EA	1.23	90.40	42.30	---	132.70	166.00	107.70	132.00
2' W x 2' L, two U 40 watt RS	Inst	Ea	EA	1.23	113.00	42.30	---	155.30	192.00	127.50	156.00
2' W x 4' L, four 40 watt RS	Inst	Ea	EA	1.54	116.00	53.00	---	169.00	210.00	137.40	168.00
White enameled circline steel ceiling fixtures											
8" W, 22 watt RS	Inst	Ea	EA	1.03	32.40	35.40	---	67.80	89.00	51.60	66.50
12" W, 22 to 32 watt RS	Inst	Ea	EA	1.03	48.60	35.40	---	84.00	108.00	66.00	83.10
16" W, 22 to 40 watt RS	Inst	Ea	EA	1.03	64.80	35.40	---	100.20	126.00	80.40	99.60
Decorative circline fixtures											
12" W, 22 to 32 watt RS	Inst	Ea	EA	1.03	126.00	35.40	---	161.40	197.00	134.90	162.00
16" W, 22 to 40 watt RS	Inst	Ea	EA	1.03	219.00	35.40	---	254.40	303.00	216.90	257.00

Incandescent

Description	Oper	Unit	Crew Size	Man-Hours Per Unit	Avg Mat'l Unit Cost	Avg Labor Unit Cost	Avg Equip Unit Cost	Avg Total Unit Cost	Avg Price Incl O&P	Avg Total Unit Cost	Avg Price Incl O&P
Ceiling fixture, surface mounted											
15" x 5", white bent glass	Inst	Ea	EA	1.03	37.30	35.40	---	72.70	94.60	56.00	71.50
10" x 7", two light, circular	Inst	Ea	EA	1.03	47.00	35.40	---	82.40	106.00	64.60	81.40
8" x 8", one light, screw-in	Inst	Ea	EA	1.03	30.80	35.40	---	66.20	87.10	50.20	64.90
Ceiling fixture, recessed											
Square fixture, drop or flat lens											
8" frame	Inst	Ea	EA	1.54	47.00	53.00	---	100.00	131.00	76.10	98.10
10" frame	Inst	Ea	EA	1.54	53.50	53.00	---	106.50	139.00	81.80	105.00
12" frame	Inst	Ea	EA	1.54	56.70	53.00	---	109.70	143.00	84.70	108.00
Round fixture for concentrated light over small areas											
7" shower fixture frame	Inst	Ea	EA	1.54	47.00	53.00	---	100.00	131.00	76.10	98.10
8" spotlight fixture	Inst	Ea	EA	1.54	45.40	53.00	---	98.40	129.00	74.60	96.50
8" flat lens or stepped baffle frame	Inst	Ea	EA	1.54	47.00	53.00	---	100.00	131.00	76.10	98.10

Description	Oper	Unit	Crew Size	Man-Hours Per Unit	Avg Mat'l Unit Cost	Avg Labor Unit Cost	Avg Equip Unit Cost	Avg Total Unit Cost	Avg Price Incl O&P	Avg Total Unit Cost	Avg Price Incl O&P
								Costs Based On Small Volume		Large Volume	

Track lighting for highlighting effects from a ceiling or wall
Swivel track heads with the ability to slide head along track to a new position

Description	Oper	Unit	Crew	M-H	Mat'l	Labor	Equip	Total	Price O&P	Total	Price O&P
Track heads											
Large cylinder	Inst	Ea	EA	---	55.10	---	---	55.10	**55.10**	48.90	**48.90**
Small cylinder	Inst	Ea	EA	---	45.40	---	---	45.40	**45.40**	40.20	**40.20**
Sphere cylinder	Inst	Ea	EA	---	55.10	---	---	55.10	**55.10**	48.90	**48.90**
Track, 1-7/16" W x 3/4" D											
2' track	Inst	Ea	EA	.513	22.70	17.60	---	40.30	**51.80**	31.60	**39.90**
4' track	Inst	Ea	EA	.769	56.70	26.50	---	83.20	**104.00**	67.50	**83.00**
8' track	Inst	Ea	EA	1.03	84.20	35.40	---	119.60	**149.00**	97.60	**119.00**
Straight connector; joins two track sections end to end											
7-5/8" L	Inst	Ea	EA	---	16.20	---	---	16.20	**16.20**	14.40	**14.40**
L - connector, joins two track sections for 90-degree angle turns											
7-5/8" L	Inst	Ea	EA	---	16.20	---	---	16.20	**16.20**	14.40	**14.40**
Feed in unit, attaches to ceiling or wall outlet box, supplies electrical current to all heads on track(s)	Inst	Ea	EA	1.03	17.80	35.40	---	53.20	**72.20**	38.70	**51.70**
Wall fixtures											
White glass with on/off switch											
1 light, 5" W x 5" H	Inst	Ea	EA	1.03	17.80	35.40	---	53.20	**72.20**	38.70	**51.70**
2 light, 14" W x 5" H	Inst	Ea	EA	1.03	27.50	35.40	---	62.90	**83.40**	47.30	**61.60**
4 light, 24" W x 4" H	Inst	Ea	EA	1.03	40.50	35.40	---	75.90	**98.30**	58.80	**74.80**
Swivel wall fixture											
1 light, 4" W x 8" H	Inst	Ea	EA	1.03	27.90	35.40	---	63.30	**83.80**	47.70	**62.00**
2 light, 9" W x 9" H	Inst	Ea	EA	1.03	48.60	35.40	---	84.00	**108.00**	66.00	**83.10**

Outdoor lighting

Description	Oper	Unit	Crew	M-H	Mat'l	Labor	Equip	Total	Price O&P	Total	Price O&P
Ceiling fixture for porch											
7" W x 13" L, one 75 watt RS	Inst	Ea	EA	1.03	58.30	35.40	---	93.70	**119.00**	74.60	**93.00**
11" W x 4" L, two 60 watt RS	Inst	Ea	EA	1.03	71.30	35.40	---	106.70	**134.00**	86.10	**106.00**
15" W x 4" L, three 60 watt RS	Inst	Ea	EA	1.03	72.90	35.40	---	108.30	**136.00**	87.60	**108.00**
Wall fixture for porch											
6" W x 10" L, one 75 watt RS	Inst	Ea	EA	1.03	63.20	35.40	---	98.60	**124.00**	79.00	**98.00**
6" W x 16" L, one 75 watt RS	Inst	Ea	EA	1.03	79.40	35.40	---	114.80	**143.00**	93.30	**114.00**
Post lantern fixture											
Aluminum cast posts											
84" H post with lantern, set in concrete, includes 50' of conduit, circuit, and cement	Inst	Ea	EA	11.4	373.00	392.00	---	765.00	**1000.00**	606.00	**782.00**
Urethane (in form of simulated redwood) over steel post											
Post with matching lantern, set in concrete, includes 50' of conduit, circuit, and cement	Inst	Ea	EA	11.4	445.00	392.00	---	837.00	**1080.00**	670.00	**856.00**

Linoleum. See Resilient flooring, page 218
Lumber. See Framing, page 120

					Costs Based On Small Volume					Large Volume	
Description	Oper	Unit	Crew Size	Man-Hours Per Unit	Avg Mat'l Unit Cost	Avg Labor Unit Cost	Avg Equip Unit Cost	Avg Total Unit Cost	Avg Price Incl O&P	Avg Total Unit Cost	Avg Price Incl O&P

Mantels, fireplace

Ponderosa pine, kiln-dried, unfinished, assembled

Versailles, ornate, French design											
59" W x 46" H	Inst	Ea	CJ	5.71	2060.00	168.00	---	2228.00	2620.00	1837.00	2150.00
Victorian, ornate, English design											
63" W x 52" H	Inst	Ea	CJ	5.71	1390.00	168.00	---	1558.00	1850.00	1277.00	1510.00
Chelsea, plain, English design											
70" W x 52" H	Inst	Ea	CJ	5.71	1100.00	168.00	---	1268.00	1520.00	1039.00	1240.00
Jamestown, plain, Early American											
68" W x 53" H	Inst	Ea	CJ	5.71	689.00	168.00	---	857.00	1040.00	693.00	838.00

Marlite panels

Description	Oper	Unit	Crew Size	Man-Hours Per Unit	Costs Based On Small Volume					Large Volume	
					Avg Mat'l Unit Cost	Avg Labor Unit Cost	Avg Equip Unit Cost	Avg Total Unit Cost	Avg Price Incl O&P	Avg Total Unit Cost	Avg Price Incl O&P
Panels, 4' x 8', adhesive set	Demo	SF	LB	.012	---	.32	---	.32	.48	.24	.36
Plastic-coated masonite panels; 4' x 8', 5' x 5'; screw applied; channel molding around perimeter; 1/8" T											
Solid colors	Inst	SF	CJ	.067	2.80	1.97	---	4.77	6.17	3.81	4.86
Patterned panels	Inst	SF	CJ	.067	3.05	1.97	---	5.02	6.46	4.03	5.12
Molding, 1/8" panels; corners, divisions, or edging; nailed to framing or sheathing											
Bright anodized	Inst	LF	CJ	.090	.17	2.64	---	2.81	4.16	2.00	2.95
Gold anodized	Inst	LF	CJ	.090	.20	2.64	---	2.84	4.19	2.02	2.97
Colors	Inst	LF	CJ	.090	.24	2.64	---	2.88	4.24	2.06	3.02

Masonry

Brick. All material costs include mortar and waste, 5% on brick and 30% on mortar.

1. **Dimensions**

   a. Standard or regular: 8" L x $2^{1}/_{4}$" H x $3^{3}/_{4}$" W.

   b. Modular: $7^{5}/_{8}$" L x $2^{1}/_{4}$" H x $3^{5}/_{8}$" W.

   c. Norman: $11^{5}/_{8}$" L x $2^{2}/_{3}$" H x $3^{5}/_{8}$" W.

   d. Roman: $11^{5}/_{8}$" L x $1^{5}/_{8}$" H x $3^{5}/_{8}$" W.

2. **Installation**

   a. Mortar joints are $^{3}/_{8}$" thick both horizontally and vertically.

   b. Mortar mix is 1:3, 1 part masonry cement and 3 parts sand.

   c. Galvanized, corrugated wall ties are used on veneers at 1 tie per SF wall area. The tie is $^{7}/_{8}$" x 7" x 16 ga.

   d. Running bond used on veneers and walls.

3. **Notes on Labor.** Output is based on a crew composed of bricklayers and bricktenders at a 1:1 ratio.

4. **Estimating Technique.** Chimneys and columns figured per vertical linear foot with allowances already made for brick waste and mortar waste. Veneers and walls are computed per square foot of wall area.

Concrete (Masonry) Block. All material costs include 3% block waste and 30% mortar waste.

1. **Dimension.** All blocks are two core.

   a. Heavyweight: 8" T blocks weigh approximately 46 lbs/block; 12" T blocks weigh approximately 65 lbs/block.

   b. Lightweight: Also known as haydite blocks. 8" T blocks weigh approximately 30 lbs/block; 12" T blocks weigh approximately 41 lbs/block.

2. **Installation**

   a. Mortar joints are $^{3}/_{8}$" T both horizontally and vertically.

   b. Mortar mix is 1:3, 1 part masonry cement and 3 parts sand.

   c. Reinforcing: Lateral metal is regular truss with 9 gauge sides and ties. Vertical steel is #4 ($^{1}/_{2}$" dia.) rods, at 0.668 lbs/LF.

3. **Notes on Labor.** Output is based on a crew composed of bricklayers and bricktenders.

4. **Estimating Technique.** Figure chimneys and columns per vertical linear foot, with allowances already made for block waste and mortar waste. Veneers and walls are computed per square foot of wall area.

Quarry Tile (on Floor). Includes 5% tile waste.

1. **Dimensions:** 6" square tile is $^{1}/_{2}$" T and 9" square tile is $^{3}/_{4}$" T.

2. **Installation**

   a. Conventional mortar set utilizes portland cement, mortar mix, sand, and water. The mortar dry-cures and bonds to tile.

   b. Dry-set mortar utilizes dry-set portland cement, mortar mix, sand, and water. The mortar dry-cures and bonds to tile.

3. **Notes on Labor.** Output is based on a crew composed of bricklayers and bricktenders.

4. **Estimating Technique.** Compute square feet of floor area.

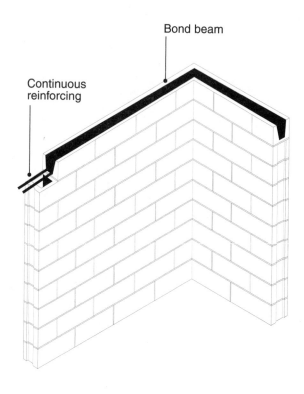

Typical concrete block wall

4" wall thickness

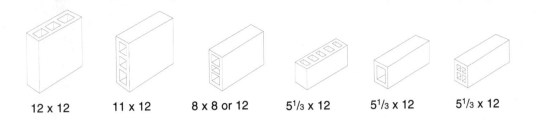

12 x 12   11 x 12   8 x 8 or 12 $5^{1}/_{3}$ x 12 $5^{1}/_{3}$ x 12 $5^{1}/_{3}$ x 12

6" wall thickness

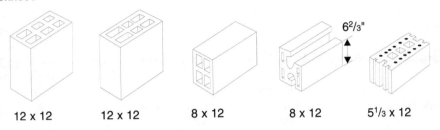

$6^{2}/_{3}$"

12 x 12   12 x 12   8 x 12   8 x 12 $5^{1}/_{3}$ x 12

8" wall thickness

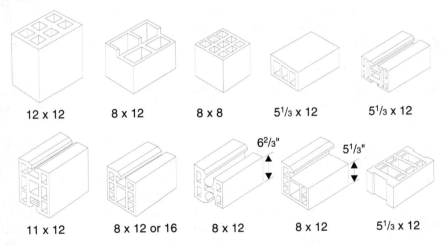

12 x 12   8 x 12   8 x 8 $5^{1}/_{3}$ x 12 $5^{1}/_{3}$ x 12

$6^{2}/_{3}$" $5^{1}/_{3}$"

11 x 12   8 x 12 or 16   8 x 12   8 x 12 $5^{1}/_{3}$ x 12

10" wall thickness

12 x 12   12 x 12 $5^{1}/_{3}$ or 8 x 12 or 16

Common sizes and shapes for clay tile

					Costs Based On Small Volume					Large Volume	
Description	Oper	Unit	Crew Size	Man-Hours Per Unit	Avg Mat'l Unit Cost	Avg Labor Unit Cost	Avg Equip Unit Cost	Avg Total Unit Cost	Avg Price Incl O&P	Avg Total Unit Cost	Avg Price Incl O&P

Masonry

Brick, standard

Running bond, 3/8 " mortar joints

Chimneys

Flue lining included; no scaffolding included

4" T wall with standard brick

Description	Oper	Unit	Crew Size	Man-Hours Per Unit	Avg Mat'l Unit Cost	Avg Labor Unit Cost	Avg Equip Unit Cost	Avg Total Unit Cost	Avg Price Incl O&P	Avg Total Unit Cost	Avg Price Incl O&P
16" x 16" with one 8" x 8" flue	Demo	VLF	LB	1.48	---	39.40	---	39.40	58.70	33.07	47.40
20" x 16" with one 12" x 8" flue	Demo	VLF	LB	1.72	---	45.80	---	45.80	68.30	38.18	54.70
20" x 20" with one 12" x 12" flue	Demo	VLF	LB	1.87	---	49.80	---	49.80	74.20	41.63	59.70
28" x 16" with two 8" x 8" flues	Demo	VLF	LB	2.29	---	61.00	---	61.00	90.90	50.83	72.90
32" x 16" with one 8" x 8" and one 12" x 8" flue	Demo	VLF	LB	2.43	---	64.70	---	64.70	96.50	54.06	77.40
36" x 16" with two 12" x 8" flues	Demo	VLF	LB	2.60	---	69.30	---	69.30	103.00	57.82	82.90
36" x 20" with two 12" x 12" flues	Demo	VLF	LB	2.89	---	77.00	---	77.00	115.00	64.47	92.40
16" x 16" with one 8" x 8" flue	Inst	VLF	BK	1.49	27.10	42.80	---	69.90	94.90	55.90	75.20
20" x 16" with one 12" x 8" flue	Inst	VLF	BK	1.69	35.80	48.60	---	84.40	113.00	67.70	90.30
20" x 20" with one 12" x 12" flue	Inst	VLF	BK	1.89	40.60	54.30	---	94.90	128.00	76.30	102.00
28" x 16" with two 8" x 8" flues	Inst	VLF	BK	2.27	49.40	65.20	---	114.60	154.00	92.00	122.00
32" x 16" with one 8" x 8" and one 12" x 8" flue	Inst	VLF	BK	2.42	57.60	69.60	---	127.20	170.00	102.60	136.00
36" x 16" with two 12" x 8" flues	Inst	VLF	BK	2.60	66.20	74.70	---	140.90	188.00	113.80	150.00
36" x 20" with two 12" x 12" flues	Inst	VLF	BK	2.88	76.50	82.80	---	159.30	211.00	128.80	169.00

8" T wall with standard brick

Description	Oper	Unit	Crew Size	Man-Hours Per Unit	Avg Mat'l Unit Cost	Avg Labor Unit Cost	Avg Equip Unit Cost	Avg Total Unit Cost	Avg Price Incl O&P	Avg Total Unit Cost	Avg Price Incl O&P
24" x 24" with one 8" x 8" flue	Demo	VLF	LB	3.17	---	84.50	---	84.50	126.00	70.59	101.00
28" x 24" with one 12" x 8" flue	Demo	VLF	LB	3.41	---	90.80	---	90.80	135.00	76.01	109.00
28" x 28" with one 12" x 12" flue	Demo	VLF	LB	3.75	---	99.90	---	99.90	149.00	83.25	119.00
36" x 24" with two 8" x 8" flues	Demo	VLF	LB	4.16	---	111.00	---	111.00	165.00	92.45	133.00
40" x 24" with one 8" x 8" and one 12" x 8" flue	Demo	VLF	LB	4.57	---	122.00	---	122.00	181.00	101.80	146.00
44" x 24" with two 12" x 8" flues	Demo	VLF	LB	4.76	---	127.00	---	127.00	189.00	105.90	152.00
44" x 28" with two 12" x 12" flues	Demo	VLF	LB	5.08	---	135.00	---	135.00	202.00	113.20	162.00
24" x 24" with one 8" x 8" flue	Inst	VLF	BK	3.33	51.60	95.70	---	147.30	202.00	117.10	159.00
28" x 24" with one 12" x 8" flue	Inst	VLF	BK	3.68	62.30	106.00	---	168.30	229.00	133.80	181.00
28" x 28" with one 12" x 12" flue	Inst	VLF	BK	3.95	69.80	114.00	---	183.80	249.00	146.10	197.00
36" x 24" with two 8" x 8" flue	Inst	VLF	BK	4.44	80.70	128.00	---	208.70	283.00	166.30	224.00
40" x 24" with one 8" x 8" and one 12" x 8" flue	Inst	VLF	BK	4.95	94.10	142.00	---	236.10	320.00	189.30	254.00
44" x 24" with two 12" x 8" flues	Inst	VLF	BK	5.08	103.00	146.00	---	249.00	336.00	199.80	266.00
44" x 28" with two 12" x 12" flues	Inst	VLF	BK	5.46	116.00	157.00	---	273.00	367.00	219.00	292.00

Columns

Outside dimension; no shoring; solid centers; no scaffolding included

Description	Oper	Unit	Crew Size	Man-Hours Per Unit	Avg Mat'l Unit Cost	Avg Labor Unit Cost	Avg Equip Unit Cost	Avg Total Unit Cost	Avg Price Incl O&P	Avg Total Unit Cost	Avg Price Incl O&P
8" x 8"	Demo	VLF	LB	.473	---	12.60	---	12.60	18.80	10.52	15.10
12" x 8"	Demo	VLF	LB	.780	---	20.80	---	20.80	31.00	17.32	24.90
16" x 8"	Demo	VLF	LB	1.02	---	27.20	---	27.20	40.50	22.64	32.40
20" x 8"	Demo	VLF	LB	1.24	---	33.00	---	33.00	49.20	27.64	39.60
24" x 8"	Demo	VLF	LB	1.46	---	38.90	---	38.90	58.00	32.41	46.50

Description	Oper	Unit	Crew Size	Man-Hours Per Unit	Avg Mat'l Unit Cost	Avg Labor Unit Cost	Avg Equip Unit Cost	Avg Total Unit Cost	Avg Price Incl O&P	Avg Total Unit Cost	Avg Price Incl O&P
						Costs Based On Small Volume					**Large Volume**	
12" x 12"	Demo	VLF	LB	1.13	---	30.10	---	30.10	**44.90**	25.19	**36.10**
16" x 12"	Demo	VLF	LB	1.46	---	38.90	---	38.90	**58.00**	32.41	**46.50**
20" x 12"	Demo	VLF	LB	1.76	---	46.90	---	46.90	**69.90**	39.07	**56.00**
24" x 12"	Demo	VLF	LB	2.04	---	54.40	---	54.40	**81.00**	45.49	**65.10**
28" x 12"	Demo	VLF	LB	2.31	---	61.50	---	61.50	**91.70**	51.48	**73.80**
32" x 12"	Demo	VLF	LB	2.57	---	68.50	---	68.50	**102.00**	57.26	**82.00**
16" x 16"	Demo	VLF	AB	1.86	---	53.30	5.85	59.15	**84.70**	41.29	**59.20**
20" x 16"	Demo	VLF	AB	2.22	---	63.60	6.98	70.58	**101.00**	49.29	**70.60**
24" x 16"	Demo	VLF	AB	2.57	---	73.60	8.08	81.68	**117.00**	57.26	**82.00**
28" x 16"	Demo	VLF	AB	2.86	---	81.90	8.99	90.89	**130.00**	63.59	**91.10**
32" x 16"	Demo	VLF	AB	3.17	---	90.80	9.99	100.79	**144.00**	70.59	**101.00**
36" x 16"	Demo	VLF	AB	3.41	---	97.70	10.70	108.40	**155.00**	76.01	**109.00**
20" x 20"	Demo	VLF	AB	2.63	---	75.30	8.27	83.57	**120.00**	58.49	**83.80**
24" x 20"	Demo	VLF	AB	3.09	---	88.50	9.72	98.22	**141.00**	68.70	**98.40**
28" x 20"	Demo	VLF	AB	3.41	---	97.70	10.70	108.40	**155.00**	76.01	**109.00**
32" x 20"	Demo	VLF	AB	3.75	---	107.00	11.80	118.80	**171.00**	83.25	**119.00**
36" x 20"	Demo	VLF	AB	4.01	---	115.00	12.60	127.60	**183.00**	89.33	**128.00**
24" x 24"	Demo	VLF	AB	3.41	---	97.70	10.70	108.40	**155.00**	76.01	**109.00**
28" x 24"	Demo	VLF	AB	3.81	---	109.00	12.00	121.00	**174.00**	84.89	**122.00**
32" x 24"	Demo	VLF	AB	4.16	---	119.00	13.10	132.10	**189.00**	92.45	**133.00**
36" x 24"	Demo	VLF	AB	4.48	---	128.00	14.10	142.10	**204.00**	99.77	**143.00**
28" x 28"	Demo	VLF	AB	4.23	---	121.00	13.30	134.30	**193.00**	94.12	**135.00**
32" x 28"	Demo	VLF	AB	4.66	---	133.00	14.70	147.70	**212.00**	104.00	**149.00**
36" x 28"	Demo	VLF	AB	5.08	---	145.00	16.00	161.00	**231.00**	113.20	**162.00**
32" x 32"	Demo	VLF	AB	5.08	---	145.00	16.00	161.00	**231.00**	113.20	**162.00**
36" x 32"	Demo	VLF	AB	5.57	---	160.00	17.50	177.50	**254.00**	124.30	**178.00**
36" x 36"	Demo	VLF	AB	6.02	---	172.00	18.90	190.90	**274.00**	134.30	**192.00**
8" x 8" (9.33 Brick/VLF)	Inst	VLF	BK	.553	4.82	15.90	---	20.72	**29.20**	16.13	**22.60**
12" x 8" (14.00 Brick/VLF)	Inst	VLF	BK	.808	7.34	23.20	---	30.54	**43.00**	23.83	**33.30**
16" x 8" (18.67 Brick/VLF)	Inst	VLF	BK	1.05	9.75	30.20	---	39.95	**56.20**	31.06	**43.40**
20" x 8" (23.33 Brick/VLF)	Inst	VLF	BK	1.28	12.20	36.80	---	49.00	**68.80**	38.20	**53.30**
24" x 8" (28.00 Brick/VLF)	Inst	VLF	BK	1.49	14.60	42.80	---	57.40	**80.60**	45.00	**62.70**
12" x 12" (21.00 Brick/VLF)	Inst	VLF	BK	1.17	10.90	33.60	---	44.50	**62.70**	34.67	**48.40**
16" x 12" (28.00 Brick/VLF)	Inst	VLF	BK	1.49	14.60	42.80	---	57.40	**80.60**	45.00	**62.70**
20" x 12" (35.00 Brick/VLF)	Inst	VLF	BK	1.79	18.30	51.40	---	69.70	**97.60**	54.50	**75.80**
24" x 12" (42.00 Brick/VLF)	Inst	VLF	BK	2.09	21.80	60.10	---	81.90	**115.00**	64.20	**89.20**
28" x 12" (49.00 Brick/VLF)	Inst	VLF	BK	2.34	25.50	67.30	---	92.80	**130.00**	73.00	**101.00**
32" x 12" (56.00 Brick/VLF)	Inst	VLF	BK	2.60	29.20	74.70	---	103.90	**145.00**	81.60	**113.00**
16" x 16" (37.33 Brick/VLF)	Inst	VLF	BK	1.89	19.40	54.30	---	73.70	**103.00**	57.80	**80.40**
20" x 16" (46.67 Brick/VLF)	Inst	VLF	BK	2.27	24.30	65.20	---	89.50	**125.00**	70.20	**97.30**
24" x 16" (56.00 Brick/VLF)	Inst	VLF	BK	2.60	29.20	74.70	---	103.90	**145.00**	81.60	**113.00**
28" x 16" (65.33 Brick/VLF)	Inst	VLF	BK	2.88	34.00	82.80	---	116.80	**162.00**	91.90	**127.00**
32" x 16" (74.67 Brick/VLF)	Inst	VLF	BK	3.18	38.90	91.40	---	130.30	**181.00**	102.80	**142.00**
36" x 16" (84.00 Brick/VLF)	Inst	VLF	BK	3.44	43.70	98.90	---	142.60	**198.00**	112.60	**155.00**

				Costs Based On Small Volume						Large Volume	
Description	**Oper**	**Unit**	**Crew Size**	**Man-Hours Per Unit**	**Avg Mat'l Unit Cost**	**Avg Labor Unit Cost**	**Avg Equip Unit Cost**	**Avg Total Unit Cost**	**Avg Price Incl O&P**	**Avg Total Unit Cost**	**Avg Price Incl O&P**
20" x 20" (58.33 Brick/VLF)	Inst	VLF	BK	2.67	30.40	76.70	---	107.10	**149.00**	84.20	116.00
24" x 20" (70.00 Brick/VLF)	Inst	VLF	BK	3.09	36.40	88.80	---	125.20	**174.00**	98.60	136.00
28" x 20" (81.67 Brick/VLF)	Inst	VLF	BK	3.44	42.50	98.90	---	141.40	**196.00**	111.50	153.00
32" x 20" (93.33 Brick/VLF)	Inst	VLF	BK	3.74	48.60	107.00	---	155.60	**216.00**	123.40	169.00
36" x 20" (105.00 Brick/VLF)	Inst	VLF	BK	4.02	54.60	116.00	---	170.60	**235.00**	134.70	184.00
24" x 24" (84.00 Brick/VLF)	Inst	VLF	BK	3.44	43.70	98.90	---	142.60	**198.00**	112.60	155.00
28" x 24" (98.00 Brick/VLF)	Inst	VLF	BK	3.81	51.00	110.00	---	161.00	**222.00**	126.90	174.00
32" x 24" (112.00 Brick/VLF)	Inst	VLF	BK	4.10	58.30	118.00	---	176.30	**243.00**	139.70	191.00
36" x 24" (126.00 Brick/VLF)	Inst	VLF	BK	4.44	65.60	128.00	---	193.60	**266.00**	153.20	209.00
28" x 28" (114.33 Brick/VLF)	Inst	VLF	BK	4.27	59.60	123.00	---	182.60	**251.00**	144.30	197.00
32" x 28" (130.67 Brick/VLF)	Inst	VLF	BK	4.64	68.10	133.00	---	201.10	**277.00**	159.70	218.00
36" x 28" (147.00 Brick/VLF)	Inst	VLF	BK	5.19	76.60	149.00	---	225.60	**310.00**	179.20	244.00
32" x 32" (149.33 Brick/VLF)	Inst	VLF	BK	5.33	77.70	153.00	---	230.70	**318.00**	183.20	250.00
36" x 32" (168.00 Brick/VLF)	Inst	VLF	BK	5.93	87.50	170.00	---	257.50	**355.00**	204.70	278.00
36" x 36" (189.00 Brick/VLF)	Inst	VLF	BK	6.67	98.40	192.00	---	290.40	**399.00**	230.30	313.00
Veneers											
4" T, with air tools	Demo	SF	AB	.070	---	2.00	.22	2.22	**3.19**	1.55	2.23
4" T, with wall ties; no scaffolding included											
Common, 8" x 2-2/3" x 4"	Inst	SF	BO	.182	2.67	5.23	.20	8.10	**11.10**	7.47	10.30
Standard face, 8" x 2-2/3" x 4"	Inst	SF	BO	.194	3.55	5.58	.22	9.35	**12.60**	8.64	11.70
Glazed, 8" x 2-2/3" x 4"	Inst	SF	BO	.203	8.07	5.83	.23	14.13	**18.20**	12.80	16.60
Other brick types/sizes											
8" x 4" x 4"	Inst	SF	BO	.135	2.78	3.88	.15	6.81	**9.13**	6.26	8.45
8" x 3-1/5" x 4"	Inst	SF	BO	.161	4.07	4.63	.18	8.88	**11.80**	8.13	10.90
12" x 4" x 6"	Inst	SF	BO	.097	4.32	2.79	.11	7.22	**9.23**	6.52	8.40
12" x 2-2/3" x 4"	Inst	SF	BO	.131	3.54	3.76	.15	7.45	**9.83**	6.82	9.06
12" x 3-1/5" x 4"	Inst	SF	BO	.114	2.75	3.28	.13	6.16	**8.17**	5.65	7.55
12" x 2" x 4"	Inst	SF	BO	.171	4.80	4.91	.19	9.90	**13.00**	9.03	12.00
12" x 2-2/3" x 6"	Inst	SF	BO	.135	4.13	3.88	.15	8.16	**10.70**	7.43	9.80
12" x 4" x 4"	Inst	SF	BO	.095	3.24	2.73	.11	6.08	**7.90**	5.51	7.21
Walls											
With air tools											
8" T	Demo	SF	AB	.115	---	3.29	.36	3.65	**5.23**	2.73	3.92
12" T	Demo	SF	AB	.164	---	4.70	.52	5.22	**7.47**	3.91	5.60
16" T	Demo	SF	AB	.213	---	6.10	.67	6.77	**9.70**	5.08	7.28
24" T	Demo	SF	AB	.251	---	7.19	.79	7.98	**11.40**	5.97	8.56
With common brick; no scaffolding included											
8" T (13.50 Brick/SF)	Inst	SF	ML	.337	5.24	10.10	.50	15.84	**21.60**	14.58	20.00
12" T (20.25 Brick/SF)	Inst	SF	ML	.457	7.91	13.70	.68	22.29	**30.20**	20.43	27.90
16" T (27.00 Brick/SF)	Inst	SF	ML	.582	10.60	17.50	.86	28.96	**39.10**	26.47	36.10
24" T (40.50 Brick/SF)	Inst	SF	ML	.800	15.80	24.00	1.19	40.99	**55.10**	37.49	50.80

Description	Oper	Unit	Crew Size	Man-Hours Per Unit	Costs Based On Small Volume					Large Volume	
					Avg Mat'l Unit Cost	Avg Labor Unit Cost	Avg Equip Unit Cost	Avg Total Unit Cost	Avg Price Incl O&P	Avg Total Unit Cost	Avg Price Incl O&P

Brick, adobe

Running bond, 3/8 " mortar joints

Walls, with air tools

Description	Oper	Unit	Crew Size	Man-Hrs	Mat'l	Labor	Equip	Total	Price O&P	LV Total	LV Price
4" T	Demo	SF	AB	.059	---	1.69	.18	1.87	**2.68**	1.30	**1.87**
6" T	Demo	SF	AB	.060	---	1.72	.19	1.91	**2.73**	1.33	**1.91**
8" T	Demo	SF	AB	.062	---	1.78	.19	1.97	**2.82**	1.37	**1.96**
12" T	Demo	SF	AB	.065	---	1.86	.21	2.07	**2.97**	1.46	**2.09**

Walls, no scaffolding included

Description	Oper	Unit	Crew Size	Man-Hrs	Mat'l	Labor	Equip	Total	Price O&P	LV Total	LV Price
4" x 4" x 16"	Inst	SF	ML	.139	1.96	4.17	.21	6.34	**8.68**	5.85	**8.07**
6" x 4" x 16"	Inst	SF	ML	.145	1.96	4.35	.22	6.53	**8.96**	6.90	**9.32**
8" x 4" x 16"	Inst	SF	ML	.152	3.80	4.56	.23	8.59	**11.40**	7.84	**10.50**
12" x 4" x 16"	Inst	SF	ML	.168	5.99	5.04	.25	11.28	**14.70**	10.24	**13.40**

Concrete block

Lightweight (haydite) or heavyweight blocks; 2 cores/block, solid face;

includes allowances for lintels, bond beams

Foundations and retaining walls

No excavation included

Without reinforcing or with lateral reinforcing only

With air tools

Description	Oper	Unit	Crew Size	Man-Hrs	Mat'l	Labor	Equip	Total	Price O&P	LV Total	LV Price
8" W x 8" H x 16" L	Demo	SF	AB	.068	---	1.95	.21	2.16	**3.09**	1.52	**2.18**
10" W x 8" H x 16" L	Demo	SF	AB	.074	---	2.12	.23	2.35	**3.37**	1.65	**2.36**
12" W x 8" H x 16" L	Demo	SF	AB	.079	---	2.26	.25	2.51	**3.60**	1.75	**2.50**

Without air tools

Description	Oper	Unit	Crew Size	Man-Hrs	Mat'l	Labor	Equip	Total	Price O&P	LV Total	LV Price
8" W x 8" H x 16" L	Demo	SF	LB	.085	---	2.26	---	2.26	**3.37**	1.57	**2.34**
10" W x 8" H x 16" L	Demo	SF	LB	.091	---	2.42	---	2.42	**3.61**	1.70	**2.54**
12" W x 8" H x 16" L	Demo	SF	LB	.099	---	2.64	---	2.64	**3.93**	1.86	**2.78**

With vertical reinforcing in every other core (2 core blocks) with core concrete filled

With air tools

Description	Oper	Unit	Crew Size	Man-Hrs	Mat'l	Labor	Equip	Total	Price O&P	LV Total	LV Price
8" W x 8" H x 16" L	Demo	SF	AB	.111	---	3.18	.35	3.53	**5.06**	2.48	**3.56**
10" W x 8" H x 16" L	Demo	SF	AB	.120	---	3.44	.38	3.82	**5.47**	2.67	**3.82**
12" W x 8" H x 16" L	Demo	SF	AB	.131	---	3.75	.41	4.16	**5.96**	2.90	**4.15**

Foundations and retaining walls

No reinforcing

Heavyweight blocks

Description	Oper	Unit	Crew Size	Man-Hrs	Mat'l	Labor	Equip	Total	Price O&P	LV Total	LV Price
8" x 8" x 16"	Inst	SF	BO	.135	1.50	3.88	.15	5.53	**7.66**	5.17	**7.20**
10" x 8" x 16"	Inst	SF	BO	.147	1.87	4.22	.16	6.25	**8.61**	5.81	**8.05**
12" x 8" x 16"	Inst	SF	BO	.161	2.27	4.63	.18	7.08	**9.68**	6.55	**9.03**

Lightweight blocks

Description	Oper	Unit	Crew Size	Man-Hrs	Mat'l	Labor	Equip	Total	Price O&P	LV Total	LV Price
8" x 8" x 16"	Inst	SF	BO	.122	1.72	3.51	.14	5.37	**7.34**	4.95	**6.82**
10" x 8" x 16"	Inst	SF	BO	.133	2.08	3.82	.15	6.05	**8.24**	5.61	**7.69**
12" x 8" x 16"	Inst	SF	BO	.145	2.48	4.17	.16	6.81	**9.22**	6.29	**8.58**

Lateral reinforcing every second course

Heavyweight blocks

Description	Oper	Unit	Crew Size	Man-Hrs	Mat'l	Labor	Equip	Total	Price O&P	LV Total	LV Price
8" x 8" x 16"	Inst	SF	BO	.142	1.50	4.08	.16	5.74	**7.97**	5.35	**7.47**
10" x 8" x 16"	Inst	SF	BO	.155	1.87	4.45	.17	6.49	**8.96**	6.03	**8.37**
12" x 8" x 16"	Inst	SF	BO	.171	2.27	4.91	.19	7.37	**10.10**	6.83	**9.44**

Lightweight blocks

Description	Oper	Unit	Crew Size	Man-Hrs	Mat'l	Labor	Equip	Total	Price O&P	LV Total	LV Price
8" x 8" x 16"	Inst	SF	BO	.127	1.72	3.65	.14	5.51	**7.56**	5.11	**7.05**
10" x 8" x 16"	Inst	SF	BO	.140	2.08	4.02	.16	6.26	**8.55**	5.79	**7.96**
12" x 8" x 16"	Inst	SF	BO	.152	2.48	4.37	.17	7.02	**9.53**	6.51	**8.90**

					Costs Based On Small Volume					Large Volume	
Description	**Oper**	**Unit**	**Crew Size**	**Man-Hours Per Unit**	**Avg Mat'l Unit Cost**	**Avg Labor Unit Cost**	**Avg Equip Unit Cost**	**Avg Total Unit Cost**	**Avg Price Incl O&P**	**Avg Total Unit Cost**	**Avg Price Incl O&P**
Vertical reinforcing (No. 4 rod) every second core with core concrete filled											
Heavyweight blocks											
8" x 8" x 16"	Inst	SF	BD	.242	2.39	7.14	.22	9.75	**13.60**	7.61	**10.60**
10" x 8" x 16"	Inst	SF	BD	.267	2.86	7.88	.24	10.98	**15.30**	8.56	**11.80**
12" x 8" x 16"	Inst	SF	BD	.288	3.41	8.50	.26	12.17	**16.80**	9.52	**13.10**
Lightweight blocks											
8" x 8" x 16"	Inst	SF	BD	.222	2.61	6.55	.20	9.36	**13.00**	7.35	**10.10**
10" x 8" x 16"	Inst	SF	BD	.242	3.07	7.14	.22	10.43	**14.40**	8.20	**11.20**
12" x 8" x 16"	Inst	SF	BD	.267	3.62	7.88	.24	11.74	**16.10**	9.23	**12.60**

Exterior walls (above grade)

No bracing or shoring included

Without reinforcing or with lateral reinforcing only

With air tools											
8" W x 8" H x 16" L	Demo	SF	AB	.068	---	1.95	.21	2.16	**3.09**	1.52	**2.18**
10" W x 8" H x 16" L	Demo	SF	AB	.075	---	2.15	.24	2.39	**3.42**	1.66	**2.37**
12" W x 8" H x 16" L	Demo	SF	AB	.083	---	2.38	.26	2.64	**3.78**	1.84	**2.64**
Without air tools											
8" W x 8" H x 16" L	Demo	SF	LB	.086	---	2.29	---	2.29	**3.41**	1.60	**2.38**
10" W x 8" H x 16" L	Demo	SF	LB	.095	---	2.53	---	2.53	**3.77**	1.78	**2.66**
12" W x 8" H x 16" L	Demo	SF	LB	.104	---	2.77	---	2.77	**4.13**	1.94	**2.90**

Exterior walls (above grade)

No reinforcing

Heavyweight blocks											
8" x 8" x 16"	Inst	SF	BO	.145	1.50	4.17	.16	5.83	**8.09**	5.44	**7.61**
10" x 8" x 16"	Inst	SF	BO	.158	1.87	4.54	.18	6.59	**9.10**	6.12	**8.50**
12" x 8" x 16"	Inst	SF	BO	.171	2.27	4.91	.19	7.37	**10.10**	6.83	**9.44**
Lightweight blocks											
8" x 8" x 16"	Inst	SF	BO	.129	1.72	3.71	.14	5.57	**7.64**	5.17	**7.14**
10" x 8" x 16"	Inst	SF	BO	.140	1.82	4.02	.16	6.00	**8.25**	5.79	**7.96**
12" x 8" x 16"	Inst	SF	BO	.152	2.48	4.37	.17	7.02	**9.53**	6.51	**8.90**
Lateral reinforcing every second course											
Heavyweight blocks											
8" x 8" x 16"	Inst	SF	BO	.152	1.50	4.37	.17	6.04	**8.40**	5.66	**7.92**
10" x 8" x 16"	Inst	SF	BO	.164	1.87	4.71	.18	6.76	**9.35**	6.30	**8.77**
12" x 8" x 16"	Inst	SF	BO	.178	2.27	5.12	.20	7.59	**10.40**	7.05	**9.76**
Lightweight blocks											
8" x 8" x 16"	Inst	SF	BO	.135	1.72	3.88	.15	5.75	**7.91**	5.35	**7.41**
10" x 8" x 16"	Inst	SF	BO	.147	1.82	4.22	.16	6.20	**8.55**	6.25	**8.56**
12" x 8" x 16"	Inst	SF	BO	.161	2.48	4.63	.18	7.29	**9.93**	6.74	**9.25**

Partitions (above grade)

No bracing or shoring included

Without reinforcing or with lateral reinforcing only

With air tools											
4" W x 8" H x 16" L	Demo	SF	AB	.069	---	1.98	.22	2.20	**3.14**	1.52	**2.18**
6" W x 8" H x 16" L	Demo	SF	AB	.074	---	2.12	.23	2.35	**3.37**	1.65	**2.36**
8" W x 8" H x 16" L	Demo	SF	AB	.077	---	2.21	.24	2.45	**3.50**	1.72	**2.46**
10" W x 8" H x 16" L	Demo	SF	AB	.086	---	2.46	.27	2.73	**3.92**	1.91	**2.73**
12" W x 8" H x 16" L	Demo	SF	AB	.097	---	2.78	.31	3.09	**4.42**	2.16	**3.09**

Description	Oper	Unit	Crew Size	Man-Hours Per Unit	Costs Based On Small Volume				Large Volume	
					Avg Mat'l Unit Cost	Avg Labor Unit Cost	Avg Equip Unit Cost	Avg Total Unit Cost	Avg Price Incl O&P	Avg Total Unit Cost	Avg Price Incl O&P

Partitions (above grade)

No reinforcing

Heavyweight blocks

Description	Oper	Unit	Crew Size	MH/Unit	Mat'l	Labor	Equip	Total	Price O&P	LV Total	LV Price
4" W x 8" H x 16" L	Inst	SF	BO	.138	1.11	3.97	.15	5.23	7.34	4.90	6.91
6" W x 8" H x 16" L	Inst	SF	BO	.147	1.38	4.22	.16	5.76	8.04	5.39	7.57
8" W x 8" H x 16" L	Inst	SF	BO	.161	1.50	4.63	.18	6.31	8.80	5.89	8.27
10" W x 8" H x 16" L	Inst	SF	BO	.174	1.87	5.00	.19	7.06	9.79	6.58	9.18
12" W x 8" H x 16" L	Inst	SF	BO	.190	2.27	5.46	.21	7.94	11.00	7.38	10.30

Lightweight blocks

Description	Oper	Unit	Crew Size	MH/Unit	Mat'l	Labor	Equip	Total	Price O&P	LV Total	LV Price
4" W x 8" H x 16" L	Inst	SF	BO	.127	1.33	3.65	.14	5.12	7.11	4.77	6.66
6" W x 8" H x 16" L	Inst	SF	BO	.135	1.60	3.88	.15	5.63	7.77	5.25	7.29
8" W x 8" H x 16" L	Inst	SF	BO	.145	1.72	4.17	.16	6.05	8.35	5.62	7.81
10" W x 8" H x 16" L	Inst	SF	BO	.158	1.82	4.54	.18	6.54	9.04	6.31	8.72
12" W x 8" H x 16" L	Inst	SF	BO	.171	2.48	4.91	.19	7.58	10.40	7.02	9.66

Lateral reinforcing every second course

Heavyweight blocks

Description	Oper	Unit	Crew Size	MH/Unit	Mat'l	Labor	Equip	Total	Price O&P	LV Total	LV Price
4" W x 8" H x 16" L	Inst	SF	BO	.145	1.11	4.17	.16	5.44	7.65	5.10	7.21
6" W x 8" H x 16" L	Inst	SF	BO	.155	1.38	4.45	.17	6.00	8.39	5.61	7.88
8" W x 8" H x 16" L	Inst	SF	BO	.171	1.50	4.91	.19	6.60	9.24	6.17	8.68
10" W x 8" H x 16" L	Inst	SF	BO	.186	1.87	5.35	.21	7.43	10.30	6.91	9.67
12" W x 8" H x 16" L	Inst	SF	BO	.203	2.27	5.83	.23	8.33	11.50	7.75	10.80

Lightweight blocks

Description	Oper	Unit	Crew Size	MH/Unit	Mat'l	Labor	Equip	Total	Price O&P	LV Total	LV Price
4" W x 8" H x 16" L	Inst	SF	BO	.133	1.33	3.82	.15	5.30	7.37	4.95	6.93
6" W x 8" H x 16" L	Inst	SF	BO	.142	1.60	4.08	.16	5.84	8.08	5.43	7.57
8" W x 8" H x 16" L	Inst	SF	BO	.152	1.72	4.37	.17	6.26	8.66	5.84	8.13
10" W x 8" H x 16" L	Inst	SF	BO	.164	1.82	4.71	.18	6.71	9.30	6.49	8.99
12" W x 8" H x 16" L	Inst	SF	BO	.178	2.48	5.12	.20	7.80	10.70	7.24	9.98

Fences

Without reinforcing or with lateral reinforcing only

With air tools

Description	Oper	Unit	Crew Size	MH/Unit	Mat'l	Labor	Equip	Total	Price O&P	LV Total	LV Price
6" W x 4" H x 16" L	Demo	SF	AB	.059	---	1.69	.18	1.87	2.68	1.30	1.87
6" W x 6" H x 16" L	Demo	SF	AB	.062	---	1.78	.19	1.97	2.82	1.37	1.96
8" W x 8" H x 16" L	Demo	SF	AB	.065	---	1.86	.21	2.07	2.97	1.46	2.09
10" W x 8" H x 16" L	Demo	SF	AB	.070	---	2.00	.22	2.22	3.19	1.55	2.23
12" W x 8" H x 16" L	Demo	SF	AB	.076	---	2.18	.24	2.42	3.46	1.69	2.42

Without air tools

Description	Oper	Unit	Crew Size	MH/Unit	Mat'l	Labor	Equip	Total	Price O&P	LV Total	LV Price
6" W x 4" H x 16" L	Demo	SF	LB	.074	---	1.97	---	1.97	2.94	1.39	2.06
6" W x 6" H x 16" L	Demo	SF	LB	.077	---	2.05	---	2.05	3.06	1.44	2.14
8" W x 8" H x 16" L	Demo	SF	LB	.082	---	2.18	---	2.18	3.25	1.52	2.26
10" W x 8" H x 16" L	Demo	SF	LB	.104	---	2.77	---	2.77	4.13	1.94	2.90
12" W x 8" H x 16" L	Demo	SF	LB	.095	---	2.53	---	2.53	3.77	1.78	2.66

				Costs Based On Small Volume						Large Volume	
Description	Oper	Unit	Crew Size	Man-Hours Per Unit	Avg Mat'l Unit Cost	Avg Labor Unit Cost	Avg Equip Unit Cost	Avg Total Unit Cost	Avg Price Incl O&P	Avg Total Unit Cost	Avg Price Incl O&P
Fences, lightweight blocks											
No reinforcing											
4" W x 8" H x 16" L	Inst	SF	BO	.115	1.33	3.31	.13	4.77	**6.58**	4.44	6.17
6" W x 4" H x 16" L	Inst	SF	BO	.237	2.73	6.81	.26	9.80	**13.60**	9.12	12.70
6" W x 6" H x 16" L	Inst	SF	BO	.164	2.19	4.71	.18	7.08	**9.72**	6.58	9.09
6" W x 8" H x 16" L	Inst	SF	BO	.122	1.60	3.51	.14	5.25	**7.20**	4.85	6.71
8" W x 8" H x 16" L	Inst	SF	BO	.129	1.72	3.71	.14	5.57	**7.64**	5.17	7.14
10" W x 8" H x 16" L	Inst	SF	BO	.140	1.82	4.02	.16	6.00	**8.25**	5.79	7.96
12" W x 8" H x 16" L	Inst	SF	BO	.152	2.48	4.37	.17	7.02	**9.53**	6.51	8.90
Lateral reinforcing every third course											
4" W x 8" H x 16" L	Inst	SF	BO	.120	1.33	3.45	.13	4.91	**6.80**	4.58	6.39
6" W x 4" H x 16" L	Inst	SF	BO	.244	2.73	7.01	.27	10.01	**13.90**	9.33	13.00
6" W x 6" H x 16" L	Inst	SF	BO	.174	2.19	5.00	.19	7.38	**10.20**	6.86	9.50
6" W x 8" H x 16" L	Inst	SF	BO	.127	1.60	3.65	.14	5.39	**7.42**	5.01	6.94
8" W x 8" H x 16" L	Inst	SF	BO	.135	1.72	3.88	.15	5.75	**7.91**	5.35	7.41
10" W x 8" H x 16" L	Inst	SF	BO	.147	1.82	4.22	.16	6.20	**8.55**	6.00	8.27
12" W x 8" H x 16" L	Inst	SF	BO	.161	2.48	4.63	.18	7.29	**9.93**	6.74	9.25
Concrete slump block											
Running bond, 3/8 " mortar joints											
Walls, with air tools											
4" T	Demo	SF	AB	.062	---	1.78	.19	1.97	**2.82**	1.37	1.96
6" T	Demo	SF	AB	.065	---	1.86	.21	2.07	**2.97**	1.46	2.09
8" T	Demo	SF	AB	.068	---	1.95	.21	2.16	**3.09**	1.52	2.18
12" T	Demo	SF	AB	.083	---	2.38	.26	2.64	**3.78**	1.84	2.64
Walls, no scaffolding included											
4" x 4" x 16"	Inst	SF	ML	.128	1.94	3.84	.19	5.97	**8.14**	5.51	7.56
6" x 4" x 16"	Inst	SF	ML	.133	2.39	3.99	.20	6.58	**8.89**	6.04	8.23
6" x 6" x 16"	Inst	SF	ML	.119	1.77	3.57	.18	5.52	**7.54**	5.09	7.00
8" x 4" x 16"	Inst	SF	ML	.139	2.95	4.17	.21	7.33	**9.82**	6.71	9.06
8" x 6" x 16"	Inst	SF	ML	.123	2.27	3.69	.18	6.14	**8.29**	5.64	7.66
12" x 4" x 16"	Inst	SF	ML	.152	4.86	4.56	.23	9.65	**12.60**	8.77	11.50
12" x 6" x 16"	Inst	SF	ML	.133	3.51	3.99	.20	7.70	**10.20**	7.03	9.36
Concrete slump brick											
Running bond, 3/8 " mortar joints											
Walls, with air tools											
4" T	Demo	SF	AB	.062	---	1.78	.19	1.97	**2.82**	1.37	1.96
Walls, no scaffolding included											
4" x 4" x 8"	Inst	SF	ML	.145	13.20	4.35	.22	17.77	**21.90**	15.83	19.60
4" x 4" x 12"	Inst	SF	ML	.133	6.81	3.99	.20	11.00	**14.00**	9.89	12.70
Concrete textured screen block											
Running bond, 3/8 " mortar joints											
Walls, with air tools											
4" T	Demo	SF	AB	.048	---	1.37	.15	1.52	**2.18**	1.05	1.50
Walls, no scaffolding included											
4" x 6" x 6"	Inst	SF	ML	.200	2.92	6.00	.30	9.22	**12.60**	8.52	11.70
4" x 8" x 8"	Inst	SF	ML	.133	3.98	3.99	.20	8.17	**10.70**	7.43	9.82
4" x 12" x 12"	Inst	SF	ML	.089	2.37	2.67	.13	5.17	**6.83**	4.72	6.28

				Costs Based On Small Volume						Large Volume	
Description	Oper	Unit	Crew Size	Man-Hours Per Unit	Avg Mat'l Unit Cost	Avg Labor Unit Cost	Avg Equip Unit Cost	Avg Total Unit Cost	Avg Price Incl O&P	Avg Total Unit Cost	Avg Price Incl O&P

Glass block
Plain, 4" thick

Description	Oper	Unit	Crew Size	Man-Hours Per Unit	Avg Mat'l Unit Cost	Avg Labor Unit Cost	Avg Equip Unit Cost	Avg Total Unit Cost	Avg Price Incl O&P	Avg Total Unit Cost	Avg Price Incl O&P
6" x 6"	Demo	SF	LB	.091	---	2.42	---	2.42	3.61	1.70	2.54
8" x 8"	Demo	SF	LB	.082	---	2.18	---	2.18	3.25	1.52	2.26
12" x 12"	Demo	SF	LB	.074	---	1.97	---	1.97	2.94	1.39	2.06
6" x 6"	Inst	SF	BK	.284	20.60	8.16	---	28.76	35.90	27.50	34.90
8" x 8"	Inst	SF	BK	.194	12.80	5.58	---	18.38	23.10	17.74	22.60
12" x 12"	Inst	SF	BK	.129	16.30	3.71	---	20.01	24.30	18.55	22.80

Glazed tile
6-T Series. Includes normal allowance for special shapes

Description	Oper	Unit	Crew Size	Man-Hours Per Unit	Avg Mat'l Unit Cost	Avg Labor Unit Cost	Avg Equip Unit Cost	Avg Total Unit Cost	Avg Price Incl O&P	Avg Total Unit Cost	Avg Price Incl O&P
All sizes	Demo	SF	LB	.074	---	1.97	---	1.97	2.94	1.39	2.06
Glazed 1 side											
2" x 5-1/3" x 12"	Inst	SF	BK	.379	3.53	10.90	---	14.43	20.30	11.67	16.30
4" x 5-1/3" x 12"	Inst	SF	BK	.398	5.32	11.40	---	16.72	23.20	13.66	18.80
6" x 5-1/3" x 12"	Inst	SF	BK	.419	6.96	12.00	---	18.96	26.00	15.59	21.20
8" x 5-1/3" x 12"	Inst	SF	BK	.468	8.29	13.50	---	21.79	29.60	17.86	24.10
Glazed 2 sides											
4" x 5-1/3" x 12"	Inst	SF	BK	.498	8.05	14.30	---	22.35	30.60	18.33	24.90
6" x 5-1/3" x 12"	Inst	SF	BK	.531	9.84	15.30	---	25.14	34.10	20.60	27.80
8" x 5-1/3" x 12"	Inst	SF	BK	.612	11.80	17.60	---	29.40	39.80	24.30	32.60

Quarry tile
Floors

Description	Oper	Unit	Crew Size	Man-Hours Per Unit	Avg Mat'l Unit Cost	Avg Labor Unit Cost	Avg Equip Unit Cost	Avg Total Unit Cost	Avg Price Incl O&P	Avg Total Unit Cost	Avg Price Incl O&P
Conventional mortar set	Demo	SF	LB	.043	---	1.15	---	1.15	1.71	.80	1.19
Dry-set mortar	Demo	SF	LB	.037	---	.99	---	.99	1.47	.69	1.03
Conventional mortar set with unmounted tile											
6" x 6" x 1/2" T	Inst	SF	ML	.164	1.31	4.92	.24	6.47	9.08	4.82	6.72
9" x 9" x 3/4" T	Inst	SF	ML	.138	1.84	4.14	.21	6.19	8.50	4.68	6.37
Dry-set mortar with unmounted tile											
6" x 6" x 1/2" T	Inst	SF	ML	.135	4.61	4.05	.20	8.86	11.50	7.05	9.09
9" x 9" x 3/4" T	Inst	SF	ML	.109	5.15	3.27	.16	8.58	11.00	6.92	8.74

Molding

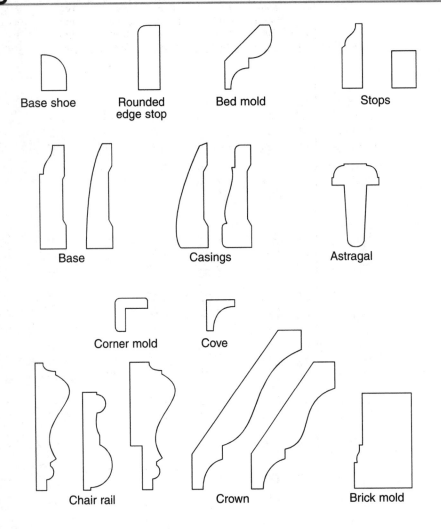

Base shoe

Rounded edge stop

Bed mold

Stops

Base

Casings

Astragal

Corner mold

Cove

Chair rail

Crown

Brick mold

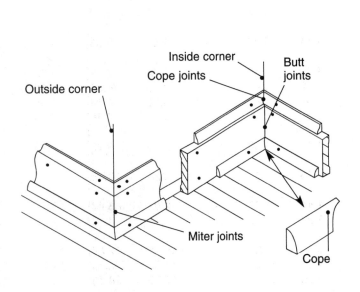

Inside corner

Cope joints

Butt joints

Outside corner

Miter joints

Cope

Base molding installation

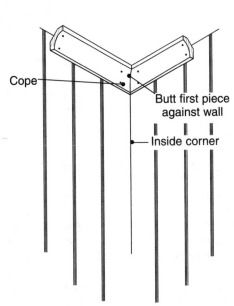

Cope

Butt first piece against wall

Inside corner

Ceiling molding installation

Description	Oper	Unit	Crew Size	Man-Hours Per Unit	Costs Based On Small Volume					Large Volume	
					Avg Mat'l Unit Cost	Avg Labor Unit Cost	Avg Equip Unit Cost	Avg Total Unit Cost	Avg Price Incl O&P	Avg Total Unit Cost	Avg Price Incl O&P

Molding and trim

Removal
At base (floor)	Demo	LF	LB	.023	---	.61	---	.61	.91	.40	.60
At ceiling	Demo	LF	LB	.021	---	.56	---	.56	.83	.35	.52
On wall or cabinets	Demo	LF	LB	.015	---	.40	---	.40	.60	.27	.40

Solid pine
Unfinished

Casing
Flat streamline											
1/2" x 1-5/8"	Inst	LF	CA	.046	1.01	1.48	---	2.49	3.38	1.82	2.43
5/8" x 1-5/8"	Inst	LF	CA	.046	1.09	1.48	---	2.57	3.47	1.90	2.53
5/8" x 2-1/2"	Inst	LF	CA	.046	1.41	1.48	---	2.89	3.84	2.17	2.84
Oval streamline											
1/2" x 1-5/8"	Inst	LF	CA	.046	1.03	1.48	---	2.51	3.40	1.84	2.46
5/8" x 1-5/8"	Inst	LF	CA	.046	1.09	1.48	---	2.57	3.47	1.90	2.53
5/8" x 2-1/2"	Inst	LF	CA	.046	1.61	1.48	---	3.09	4.07	2.34	3.03
711											
5/8" x 1-5/8"	Inst	LF	CA	.046	1.09	1.48	---	2.57	3.47	1.90	2.53
5/8" x 2-1/2"	Inst	LF	CA	.046	1.61	1.48	---	3.09	4.07	2.34	3.03
3 step											
5/8" x 1-5/8"	Inst	LF	CA	.046	.98	1.48	---	2.46	3.34	1.80	2.41
5/8" x 2-1/2"	Inst	LF	CA	.046	1.44	1.48	---	2.92	3.87	2.19	2.86
2 round edge											
5/8" x 2-1/2"	Inst	LF	CA	.046	1.06	1.48	---	2.54	3.43	1.87	2.49
B & B											
1/2" x 1-5/8"	Inst	LF	CA	.046	.98	1.48	---	2.46	3.34	1.80	2.41
4 bead											
1/2" x 1-5/8"	Inst	LF	CA	.046	.92	1.48	---	2.40	3.27	1.75	2.35
1550											
1/2" x 1-5/8"	Inst	LF	CA	.046	.81	1.48	---	2.29	3.15	1.65	2.24
Mullion											
1/4" x 3-1/2"	Inst	LF	CA	.046	1.03	1.48	---	2.51	3.40	1.84	2.46
Universal											
5/8" x 2-1/2"	Inst	LF	CA	.046	1.14	1.48	---	2.62	3.53	1.94	2.57
Colonial											
11/16" x 2-1/2"	Inst	LF	CA	.046	1.30	1.48	---	2.78	3.71	2.08	2.73
11/16" x 2-7/8"	Inst	LF	CA	.046	1.62	1.48	---	3.10	4.08	2.35	3.04
11/16" x 3-1/2"	Inst	LF	CA	.046	1.64	1.48	---	3.12	4.10	2.36	3.05
11/16" x 4-1/2"	Inst	LF	CA	.046	3.01	1.48	---	4.49	5.68	3.54	4.41
356											
11/16" x 2-1/4"	Inst	LF	CA	.046	1.32	1.48	---	2.80	3.73	2.09	2.74
Back band											
11/16" x 1-1/8"	Inst	LF	CA	.046	.75	1.48	---	2.23	3.08	1.60	2.18
Cape Cod											
5/8" x 2-1/2"	Inst	LF	CA	.046	1.44	1.48	---	2.92	3.87	2.19	2.86
11/16" x 3-1/2"	Inst	LF	CA	.046	2.23	1.48	---	3.71	4.78	2.87	3.64
Rosette plinth block w/ edge detail											
3/4" x 2-3/4"	Inst	Ea	CA	.381	3.91	12.20	---	16.11	22.80	11.37	15.90
3/4" x 3-1/2"	Inst	Ea	CA	.381	4.32	12.20	---	16.52	23.30	11.72	16.30
3/4" x 3-3/4"	Inst	Ea	CA	.381	5.00	12.20	---	17.20	24.10	12.30	17.00

					Costs Based On Small Volume					Large Volume	
Description	Oper	Unit	Crew Size	Man-Hours Per Unit	Avg Mat'l Unit Cost	Avg Labor Unit Cost	Avg Equip Unit Cost	Avg Total Unit Cost	Avg Price Incl O&P	Avg Total Unit Cost	Avg Price Incl O&P
Base & base shoe											
Shoe											
3/8" x 3/4"	Inst	LF	CA	.046	.57	1.48	---	2.05	2.87	1.45	2.0
Flat streamline											
7/16" x 1-1/2"	Inst	LF	CA	.046	.66	1.48	---	2.14	2.97	1.53	2.1
Streamline											
3/8" x 2-1/4"	Inst	LF	CA	.046	1.01	1.48	---	2.49	3.38	1.82	2.4
7/16" x 1-5/8"	Inst	LF	CA	.046	.77	1.48	---	2.25	3.10	1.62	2.2
7/16" x 2-1/2"	Inst	LF	CA	.046	1.41	1.48	---	2.89	3.84	2.17	2.8
7/16" x 3-1/2"	Inst	LF	CA	.046	1.70	1.48	---	3.18	4.17	2.42	3.1
1/2" x 1-5/8"	Inst	LF	CA	.046	.83	1.48	---	2.31	3.17	1.67	2.2
1/2" x 2-1/2"	Inst	LF	CA	.046	1.38	1.48	---	2.86	3.80	2.14	2.8
Combination											
1/2" x 1-5/8"	Inst	LF	CA	.046	.77	1.48	---	2.25	3.10	1.62	2.2
1/2" x 2-1/2"	Inst	LF	CA	.046	1.13	1.48	---	2.61	3.51	1.93	2.5
711											
1/2" x 2-1/2"	Inst	LF	CA	.046	1.10	1.48	---	2.58	3.48	1.91	2.54
7/16" x 3-1/2"	Inst	LF	CA	.046	1.21	1.48	---	2.69	3.61	2.00	2.64
3 step											
1/2" x 2-1/2"	Inst	LF	CA	.046	.89	1.48	---	2.37	3.24	1.73	2.3
1/2" x 3-1/2"	Inst	LF	CA	.046	1.29	1.48	---	2.77	3.70	2.07	2.72
1550											
7/16" x 2-1/2"	Inst	LF	CA	.046	1.41	1.48	---	2.89	3.84	2.17	2.84
Cape Cod											
9/16" x 4"	Inst	LF	CA	.046	1.47	1.48	---	2.95	3.90	2.22	2.89
Colonial											
11/16" x 3-1/2"	Inst	LF	CA	.046	1.81	1.48	---	3.29	4.30	2.51	3.23
11/16" x 4-1/2"	Inst	LF	CA	.046	2.05	1.48	---	3.53	4.57	2.72	3.47
Base cap											
11/16" x 1-3/8"	Inst	LF	CA	.046	.85	1.48	---	2.33	3.19	1.69	2.28
Rosette plinth block w/ edge detail											
3/4" x 2-3/4" x 6"	Inst	Ea	CA	.381	3.20	12.20	---	15.40	22.00	10.77	15.20
3/4" x 3-1/2" x 6"	Inst	Ea	CA	.381	3.54	12.20	---	15.74	22.40	11.05	15.50
3/4" x 3-3/4" x 6"	Inst	Ea	CA	.381	4.10	12.20	---	16.30	23.10	11.53	16.10
Colonial plinth block w/ edge detail											
1" x 2-3/4" x 6"	Inst	Ea	CA	.381	3.20	12.20	---	15.40	22.00	10.77	15.20
1" x 3-3/4" x 6"	Inst	Ea	CA	.381	4.10	12.20	---	16.30	23.10	11.53	16.10
Stops											
Round edge											
3/8" x 1/2"	Inst	LF	CA	.046	.59	1.48	---	2.07	2.89	1.46	2.02
3/8" x 3/4"	Inst	LF	CA	.046	.62	1.48	---	2.10	2.93	1.49	2.05
3/8" x 1"	Inst	LF	CA	.046	.77	1.48	---	2.25	3.10	1.62	2.20
3/8" x 1-1/4"	Inst	LF	CA	.046	.91	1.48	---	2.39	3.26	1.74	2.34
3/8" x 1-5/8"	Inst	LF	CA	.046	.93	1.48	---	2.41	3.28	1.76	2.36
1/2" x 3/4"	Inst	LF	CA	.046	.72	1.48	---	2.20	3.04	1.58	2.16
1/2" x 1"	Inst	LF	CA	.046	.75	1.48	---	2.23	3.08	1.60	2.18
1/2" x 1-1/4"	Inst	LF	CA	.046	.90	1.48	---	2.38	3.25	1.73	2.33
1/2" x 1-5/8"	Inst	LF	CA	.046	1.04	1.48	---	2.52	3.41	1.85	2.47
Ogee stop											
7/16" x 1-5/8"	Inst	LF	CA	.046	.65	1.48	---	2.13	2.96	1.52	2.09

Description	Oper	Unit	Crew Size	Man-Hours Per Unit	Avg Mat'l Unit Cost	Avg Labor Unit Cost	Avg Equip Unit Cost	Avg Total Unit Cost	Avg Price Incl O&P	Avg Total Unit Cost	Avg Price Incl O&P
								Costs Based On Small Volume		**Large Volume**	
Crown & cornice											
Crown											
9/16" x 1-5/8"	Inst	LF	CA	.062	1.42	1.99	---	3.41	**4.62**	2.50	3.33
9/16" x 2-1/4"	Inst	LF	CA	.062	1.58	1.99	---	3.57	**4.80**	2.63	3.48
9/16" x 3-1/2"	Inst	LF	CA	.062	1.73	1.99	---	3.72	**4.97**	2.77	3.64
9/16" x 4-1/4"	Inst	LF	CA	.062	1.94	1.99	---	3.93	**5.22**	2.95	3.85
9/16" x 5-1/2"	Inst	LF	CA	.062	2.19	1.99	---	4.18	**5.50**	3.16	4.09
Colonial crown											
9/16" x 3-3/8"	Inst	LF	CA	.062	1.51	1.99	---	3.50	**4.72**	2.58	3.42
9/16" x 4-1/2"	Inst	LF	CA	.062	1.94	1.99	---	3.93	**5.22**	2.95	3.85
Cornice											
1-1/8" x 4-1/2"	Inst	LF	CA	.062	8.05	1.99	---	10.04	**12.20**	8.18	9.86
1-1/4" x 6"	Inst	LF	CA	.062	8.79	1.99	---	10.78	**13.10**	8.81	10.60
Chair rail											
Chair rail											
1/2" x 1-1/2"	Inst	LF	CA	.046	.75	1.48	---	2.23	**3.08**	1.60	2.18
9/16" x 2-1/2"	Inst	LF	CA	.046	1.64	1.48	---	3.12	**4.10**	2.36	3.05
Colonial chair rail											
11/16" x 2-1/2"	Inst	LF	CA	.046	1.26	1.48	---	2.74	**3.66**	2.04	2.69
Outside corner guard											
3/4" x 3/4"	Inst	LF	CA	.046	.76	1.48	---	2.24	**3.09**	1.61	2.19
1" x 1"	Inst	LF	CA	.046	1.41	1.48	---	2.89	**3.84**	2.17	2.84
1-1/8" x 1-1/8"	Inst	LF	CA	.046	2.11	1.48	---	3.59	**4.64**	2.77	3.53
1-3/8" x 1-3/8"	Inst	LF	CA	.046	1.93	1.48	---	3.41	**4.43**	2.62	3.35
Astragal											
Flat astragal											
3/4" x 1-5/8"	Inst	LF	CA	.046	1.87	1.48	---	3.35	**4.36**	2.56	3.28
T astragal											
1-1/4" x 2-1/4"	Inst	LF	CA	.046	3.03	1.48	---	4.51	**5.70**	3.56	4.43
Cap											
Panel cap											
1/2" x 1-1/2"	Inst	LF	CA	.046	1.14	1.48	---	2.62	**3.53**	1.94	2.57
Wainscot cap											
5/8" x 1-1/4"	Inst	LF	CA	.046	1.14	1.48	---	2.62	**3.53**	1.94	2.57
Glass bead & screen											
Insert											
3/16" x 1/2"	Inst	LF	CA	.046	.55	1.48	---	2.03	**2.85**	1.43	1.98
Flat											
1/4" x 3/4"	Inst	LF	CA	.046	.55	1.48	---	2.03	**2.85**	1.43	1.98
Beaded											
1/4" x 3/4"	Inst	LF	CA	.046	.55	1.48	---	2.03	**2.85**	1.43	1.98
Scribe											
3/16" x 3/4"	Inst	LF	CA	.046	.34	1.48	---	1.82	**2.61**	1.25	1.78
Cloverleaf											
3/8" x 3/4"	Inst	LF	CA	.046	.65	1.48	---	2.13	**2.96**	1.52	2.09
Glass bead											
1/4" x 7/16"	Inst	LF	CA	.046	.50	1.48	---	1.98	**2.79**	1.39	1.94
3/8" x 3/8"	Inst	LF	CA	.046	.50	1.48	---	1.98	**2.79**	1.39	1.94

				Costs Based On Small Volume						Large Volume	
Description	Oper	Unit	Crew Size	Man-Hours Per Unit	Avg Mat'l Unit Cost	Avg Labor Unit Cost	Avg Equip Unit Cost	Avg Total Unit Cost	Avg Price Incl O&P	Avg Total Unit Cost	Avg Price Incl O&P
Quarter round											
1/4" x 1/4"	Inst	LF	CA	.046	.37	1.48	---	1.85	2.64	1.28	1.8
3/8" x 3/8"	Inst	LF	CA	.046	.45	1.48	---	1.93	2.73	1.35	1.8
1/2" x 1/2"	Inst	LF	CA	.046	.55	1.48	---	2.03	2.85	1.43	1.9
5/8" x 5/8"	Inst	LF	CA	.046	.59	1.48	---	2.07	2.89	1.46	2.0
3/4" x 3/4"	Inst	LF	CA	.046	.76	1.48	---	2.24	3.09	1.61	2.1
1" x 1"	Inst	LF	CA	.046	1.51	1.48	---	2.99	3.95	2.26	2.9
Cove											
Cove											
3/8" x 3/8"	Inst	LF	CA	.062	.42	1.99	---	2.41	3.47	1.64	2.3
1/2" x 1/2"	Inst	LF	CA	.062	.50	1.99	---	2.49	3.56	1.71	2.4
5/8" x 5/8"	Inst	LF	CA	.062	.76	1.99	---	2.75	3.86	1.93	2.6
3/4" x 3/4"	Inst	LF	CA	.062	.76	1.99	---	2.75	3.86	1.93	2.6
1" x 1"	Inst	LF	CA	.062	1.51	1.99	---	3.50	4.72	2.58	3.4
Sprung cove											
5/8" x 1-5/8"	Inst	LF	CA	.062	1.46	1.99	---	3.45	4.66	2.53	3.3
5/8" x 2-1/4"	Inst	LF	CA	.062	1.79	1.99	---	3.78	5.04	2.81	3.6
Linoleum cove											
7/16" x 1-1/4"	Inst	LF	CA	.062	.86	1.99	---	2.85	3.97	2.02	2.7
Half round											
1/4" x 1/2"	Inst	LF	CA	.046	.42	1.48	---	1.90	2.70	1.32	1.8
1/4" x 5/8"	Inst	LF	CA	.046	.42	1.48	---	1.90	2.70	1.32	1.8
3/8" x 3/4"	Inst	LF	CA	.046	.50	1.48	---	1.98	2.79	1.39	1.9
1/2" x 1"	Inst	LF	CA	.046	.92	1.48	---	2.40	3.27	1.75	2.3
Lattice - batts - S4S											
Lattice											
5/16" x 1-1/4"	Inst	LF	CA	.046	.60	1.48	---	2.08	2.90	1.47	2.0
5/16" x 1-5/8"	Inst	LF	CA	.046	.80	1.48	---	2.28	3.13	1.64	2.2
5/16" x 2-1/2"	Inst	LF	CA	.046	1.14	1.48	---	2.62	3.53	1.94	2.5
Batts											
5/16" x 2-1/2"	Inst	LF	CA	.046	.83	1.48	---	2.31	3.17	1.67	2.2
5/16" x 3-1/2"	Inst	LF	CA	.046	1.22	1.48	---	2.70	3.62	2.00	2.64
S4S											
1/2" x 1/2"	Inst	LF	CA	.046	.39	1.48	---	1.87	2.66	1.29	1.82
1/2" x 3/4"	Inst	LF	CA	.046	.45	1.48	---	1.93	2.73	1.35	1.89
1/2" x 1-1/2"	Inst	LF	CA	.046	.92	1.48	---	2.40	3.27	1.75	2.3
1/2" x 2-1/2"	Inst	LF	CA	.046	1.30	1.48	---	2.78	3.71	2.08	2.73
1/2" x 3-1/2"	Inst	LF	CA	.046	1.77	1.48	---	3.25	4.25	2.48	3.19
1/2" x 5-1/2"	Inst	LF	CA	.046	3.16	1.48	---	4.64	5.85	3.67	4.56
11/16" x 1-1/4"	Inst	LF	CA	.046	1.32	1.48	---	2.80	3.73	2.09	2.74
11/16" x 1-5/8"	Inst	LF	CA	.046	1.64	1.48	---	3.12	4.10	2.36	3.05
11/16" x 2-1/2"	Inst	LF	CA	.046	2.11	1.48	---	3.59	4.64	2.77	3.53
11/16" x 3-1/2"	Inst	LF	CA	.046	2.89	1.48	---	4.37	5.54	3.44	4.30
3/4" x 3/4"	Inst	LF	CA	.046	.83	1.48	---	2.31	3.17	1.67	2.2
1" x 1"	Inst	LF	CA	.046	1.49	1.48	---	2.97	3.93	2.24	2.92
1-9/16" x 1-9/16"	Inst	LF	CA	.046	2.84	1.48	---	4.32	5.48	3.39	4.24
S4S 4EE											
1-1/2" x 1-1/2"	Inst	LF	CA	.046	2.75	1.48	---	4.23	5.38	3.32	4.16

Description	Oper	Unit	Crew Size	Man-Hours Per Unit	Avg Mat'l Unit Cost	Avg Labor Unit Cost	Avg Equip Unit Cost	Avg Total Unit Cost	Avg Price Incl O&P	Avg Total Unit Cost	Avg Price Incl O&P
								Costs Based On Small Volume		Large Volume	
Window stool											
Flat stool											
11/16" x 4-3/4"	Inst	LF	CA	.046	4.73	1.48	---	6.21	7.65	5.01	6.10
11/16" x 5-1/2"	Inst	LF	CA	.046	5.15	1.48	---	6.63	8.14	5.37	6.52
11/16" x 7-1/4"	Inst	LF	CA	.046	7.67	1.48	---	9.15	11.00	7.53	9.00
Rabbeted stool											
7/8" x 2-1/2"	Inst	LF	CA	.046	3.69	1.48	---	5.17	6.46	4.12	5.08
7/8" x 3-1/2"	Inst	LF	CA	.046	5.15	1.48	---	6.63	8.14	5.37	6.52
7/8" x 5-1/2"	Inst	LF	CA	.046	6.23	1.48	---	7.71	9.38	6.30	7.59
Panel & decorative molds											
Panel											
3/8" x 5/8"	Inst	LF	CA	.046	.71	1.48	---	2.19	3.03	1.57	2.15
5/8" x 3/4"	Inst	LF	CA	.046	1.06	1.48	---	2.54	3.43	1.87	2.49
3/4" x 1"	Inst	LF	CA	.046	1.53	1.48	---	3.01	3.97	2.27	2.95
Panel or brick											
1-1/16" x 1-1/2"	Inst	LF	CA	.046	1.98	1.48	---	3.46	4.49	2.66	3.40
Panel											
3/8" x 1-1/8"	Inst	LF	CA	.046	1.06	1.48	---	2.54	3.43	1.87	2.49
9/16" x 1-3/8"	Inst	LF	CA	.046	1.01	1.48	---	2.49	3.38	1.82	2.43
3/4" x 1-5/8"	Inst	LF	CA	.046	2.01	1.48	---	3.49	4.53	2.68	3.42
Beauty M-340											
3/4" x 1-5/8"	Inst	LF	CA	.046	1.51	1.48	---	2.99	3.95	2.26	2.94
Beauty M-400											
3/4" x 1-7/8"	Inst	LF	CA	.046	2.38	1.48	---	3.86	4.95	3.00	3.79
Decorative											
1-3/8" x 3-1/4"	Inst	LF	CA	.046	5.97	1.48	---	7.45	9.08	6.08	7.33
Base cap											
11/16" x 1-3/8"	Inst	LF	CA	.046	1.73	1.48	---	3.21	4.20	2.45	3.16
M340 corner arcs											
3/4" x 1-5/8"	Inst	LF	CA	.381	3.78	12.20	---	15.98	22.70	11.26	15.80
Miscellaneous											
Band mold											
3/4" x 1-5/8"	Inst	LF	CA	.046	1.34	1.48	---	2.82	3.76	2.11	2.77
Parting bead											
3/8" x 3/4"	Inst	LF	CA	.046	1.85	1.48	---	3.33	4.34	2.54	3.26
Apron											
3/8" x 1-1/2"	Inst	LF	CA	.046	.95	1.48	---	2.43	3.31	1.77	2.38
Flat Hoffco											
3/8" x 1"	Inst	LF	CA	.046	.66	1.48	---	2.14	2.97	1.53	2.10
Drip											
3/4" x 1-1/4"	Inst	LF	CA	.046	1.38	1.48	---	2.86	3.80	2.14	2.80
1139 picture											
3/4" x 1-5/8"	Inst	LF	CA	.046	1.47	1.48	---	2.95	3.90	2.22	2.89
Chamfer											
1/2" x 1/2"	Inst	LF	CA	.046	.37	1.48	---	1.85	2.64	1.28	1.81
3/4" x 3/4"	Inst	LF	CA	.046	.47	1.48	---	1.95	2.75	1.37	1.92
1" x 1"	Inst	LF	CA	.046	1.04	1.48	---	2.52	3.41	1.85	2.47
Nose & cove											
1" x 1-5/8"	Inst	LF	CA	.046	2.46	1.48	---	3.94	5.04	3.07	3.87
Cabinet crown											
5/16" x 1-1/4"	Inst	LF	CA	.046	.53	1.48	---	2.01	2.82	1.41	1.96

				Costs Based On Small Volume						Large Volume	
Description	Oper	Unit	Crew Size	Man-Hours Per Unit	Avg Mat'l Unit Cost	Avg Labor Unit Cost	Avg Equip Unit Cost	Avg Total Unit Cost	Avg Price Incl O&P	Avg Total Unit Cost	Avg Price Incl O&P

Finger joint pine
Paint grade
Casing
Prefit
9/16" x 1-1/2"	Inst	LF	CA	.046	.53	1.48	---	2.01	2.82	1.41	1.9

Flat streamline
1/2" x 1-5/8"	Inst	LF	CA	.046	.69	1.48	---	2.17	3.01	1.55	2.
5/8" x 1-5/8"	Inst	LF	CA	.046	.71	1.48	---	2.19	3.03	1.57	2.
11/16" x 2-1/4"	Inst	LF	CA	.046	1.01	1.48	---	2.49	3.38	1.82	2.

711
5/8" x 1-5/8"	Inst	LF	CA	.046	.72	1.48	---	2.20	3.04	1.58	2.
5/8" x 2-1/2"	Inst	LF	CA	.046	1.09	1.48	---	2.57	3.47	1.90	2.

Monterey
5/8" x 2-1/2"	Inst	LF	CA	.046	1.17	1.48	---	2.65	3.56	1.96	2.

3 step
5/8" x 2-1/2"	Inst	LF	CA	.046	1.09	1.48	---	2.57	3.47	1.90	2.

Universal
5/8" x 2-1/2"	Inst	LF	CA	.046	1.10	1.48	---	2.58	3.48	1.91	2.

Colonial
9/16" x 4-1/4"	Inst	LF	CA	.046	1.83	1.48	---	3.31	4.32	2.53	3.
11/16" x 2-7/8"	Inst	LF	CA	.046	1.31	1.48	---	2.79	3.72	2.09	2.7

Clam shell
11/16" x 2"	Inst	LF	CA	.046	.88	1.48	---	2.36	3.23	1.72	2.3

356
11/16" x 2-1/4"	Inst	LF	CA	.046	.97	1.48	---	2.45	3.33	1.79	2.4
9/16" x 2"	Inst	LF	CA	.046	.88	1.48	---	2.36	3.23	1.72	2.3

WM 366
11/16" x 2-1/4"	Inst	LF	CA	.046	.97	1.48	---	2.45	3.33	1.79	2.4

3 fluted
1/2" x 3-1/4"	Inst	LF	CA	.046	1.39	1.48	---	2.87	3.81	2.15	2.8

5 fluted
1/2" x 3-1/2"	Inst	LF	CA	.046	1.48	1.48	---	2.96	3.92	2.23	2.9

Cape Cod
5/8" x 2-1/2"	Inst	LF	CA	.046	1.16	1.48	---	2.64	3.55	1.95	2.5
11/16" x 3-1/2"	Inst	LF	CA	.046	1.92	1.48	---	3.40	4.42	2.61	3.3

Chesapeake
3/4" x 3-1/2"	Inst	LF	CA	.046	1.94	1.48	---	3.42	4.45	2.63	3.3

Coronado
1-1/4" x 2-1/2"	Inst	LF	CA	.046	2.27	1.48	---	3.75	4.82	2.90	3.6

Bell court
1-1/4" x 2-1/2"	Inst	LF	CA	.046	2.27	1.48	---	3.75	4.82	2.90	3.6

Hermosa casing
1-1/4" x 3-1/4"	Inst	LF	CA	.046	3.08	1.48	---	4.56	5.76	3.60	4.4

Cambridge
11/16" x 3-1/4"	Inst	LF	CA	.046	1.89	1.48	---	3.37	4.39	2.58	3.3

275
5/8" x 3"	Inst	LF	CA	.046	1.42	1.48	---	2.90	3.85	2.18	2.8

Description	Oper	Unit	Crew Size	Man-Hours Per Unit	Costs Based On Small Volume					Large Volume	
					Avg Mat'l Unit Cost	Avg Labor Unit Cost	Avg Equip Unit Cost	Avg Total Unit Cost	Avg Price Incl O&P	Avg Total Unit Cost	Avg Price Incl O&P
tops											
2 round edge											
3/8" x 1-1/4"	Inst	LF	CA	.046	.41	1.48	---	1.89	**2.69**	1.31	**1.85**
Round edge											
1/2" x 1-1/4"	Inst	LF	CA	.046	.59	1.48	---	2.07	**2.89**	1.46	**2.02**
1/2" x 1-1/2"	Inst	LF	CA	.046	.76	1.48	---	2.24	**3.09**	1.61	**2.19**
Ogee											
1/2" x 1-3/4"	Inst	LF	CA	.046	.81	1.48	---	2.29	**3.15**	1.65	**2.24**
ase											
Base shoe											
3/8" x 3/4"	Inst	LF	CA	.046	.38	1.48	---	1.86	**2.65**	1.28	**1.81**
Flat											
7/16" x 1-1/2"	Inst	LF	CA	.046	.47	1.48	---	1.95	**2.75**	1.37	**1.92**
Streamline or reversible											
3/8" x 2-1/4"	Inst	LF	CA	.046	.63	1.48	---	2.11	**2.94**	1.50	**2.07**
Streamline											
7/16" x 2-1/2"	Inst	LF	CA	.046	.78	1.48	---	2.26	**3.11**	1.63	**2.21**
7/16" x 3-1/2"	Inst	LF	CA	.046	1.19	1.48	---	2.67	**3.58**	1.98	**2.62**
711											
7/16" x 2-1/2"	Inst	LF	CA	.062	.82	1.99	---	2.81	**3.93**	1.98	**2.73**
7/16" x 3-1/2"	Inst	LF	CA	.062	1.10	1.99	---	3.09	**4.25**	2.23	**3.02**
3 step											
7/16" x 2-1/2"	Inst	LF	CA	.062	.90	1.99	---	2.89	**4.02**	2.05	**2.81**
3 step base											
1/2" x 4-1/2"	Inst	LF	CA	.062	2.04	1.99	---	4.03	**5.33**	3.03	**3.94**
WM 623											
9/16" x 3-1/4"	Inst	LF	CA	.062	1.44	1.99	---	3.43	**4.64**	2.51	**3.34**
WM 618											
9/16" x 5-1/4"	Inst	LF	CA	.062	2.23	1.99	---	4.22	**5.55**	3.19	**4.12**
Monterey											
7/16" x 4-1/4"	Inst	LF	CA	.062	1.47	1.99	---	3.46	**4.68**	2.54	**3.37**
Cape Cod											
9/16" x 4"	Inst	LF	CA	.062	1.85	1.99	---	3.84	**5.11**	2.86	**3.74**
9/16" x 5-1/4"	Inst	LF	CA	.062	2.60	1.99	---	4.59	**5.97**	3.51	**4.49**
"B" Cape Cod base											
9/16" x 5-1/4"	Inst	LF	CA	.062	2.60	1.99	---	4.59	**5.97**	3.51	**4.49**
356 base											
1/2" x 2-1/2"	Inst	LF	CA	.046	1.04	1.48	---	2.52	**3.41**	1.85	**2.47**
1/2" x 3-1/4"	Inst	LF	CA	.046	1.35	1.48	---	2.83	**3.77**	2.12	**2.78**
175 base											
9/16" x 4-1/4"	Inst	LF	CA	.062	1.91	1.99	---	3.90	**5.18**	2.92	**3.81**
Colonial											
9/16" x 4-1/4"	Inst	LF	CA	.062	2.79	1.99	---	4.78	**6.19**	3.67	**4.67**

					Costs Based On Small Volume					Large Volume	
Description	Oper	Unit	Crew Size	Man-Hours Per Unit	Avg Mat'l Unit Cost	Avg Labor Unit Cost	Avg Equip Unit Cost	Avg Total Unit Cost	Avg Price Incl O&P	Avg Total Unit Cost	Avg Price Incl O&P
Crown & cornice											
Colonial crown											
9/16" x 2-1/4"	Inst	LF	CA	.062	1.01	1.99	---	3.00	**4.15**	2.14	2.
9/16" x 3-3/8"	Inst	LF	CA	.062	1.63	1.99	---	3.62	**4.86**	2.68	3.
9/16" x 4-1/2"	Inst	LF	CA	.062	2.23	1.99	---	4.22	**5.55**	3.19	4.
Cornice											
13/16" x 4-5/8"	Inst	LF	CA	.062	4.06	1.99	---	6.05	**7.65**	4.76	5.
1-1/4" x 5-7/8"	Inst	LF	CA	.062	6.26	1.99	---	8.25	**10.20**	6.64	8.
Crown											
3/4" x 6"	Inst	LF	CA	.062	4.25	1.99	---	6.24	**7.87**	4.93	6.
Cornice											
1-1/8" x 3-1/2"	Inst	LF	CA	.062	3.08	1.99	---	5.07	**6.53**	3.92	4.
Georgian crown											
11/16" x 4-1/4"	Inst	LF	CA	.062	2.48	1.99	---	4.47	**5.84**	3.40	4.
Cove											
11/16" x 11/16"	Inst	LF	CA	.046	.49	1.48	---	1.97	**2.78**	1.38	1.
Chair rail											
Chair rail											
9/16" x 2-1/2"	Inst	LF	CA	.046	1.13	1.48	---	2.61	**3.51**	1.93	2.
Colonial chair rail											
3/4" x 3"	Inst	LF	CA	.046	1.72	1.48	---	3.20	**4.19**	2.44	3.
Exterior mold											
Stucco mold											
7/8" x 1-1/4"	Inst	LF	CA	.046	.90	1.48	---	2.38	**3.25**	1.73	2.
WM 180 eastern brick											
1-1/4" x 2"	Inst	LF	CA	.046	1.94	1.48	---	3.42	**4.45**	2.63	3.
Stool & apron											
Flat stool											
11/16" x 4-3/4"	Inst	LF	CA	.046	2.84	1.48	---	4.32	**5.48**	3.39	4.
11/16" x 5-1/2"	Inst	LF	CA	.046	3.78	1.48	---	5.26	**6.56**	4.20	5.
11/16" x 7-1/4"	Inst	LF	CA	.046	4.73	1.48	---	6.21	**7.65**	5.01	6.
Apron											
3/8" x 1-1/2"	Inst	LF	CA	.046	.59	1.48	---	2.07	**2.89**	1.46	2.

Description	Oper	Unit	Crew Size	Man-Hours Per Unit	Avg Mat'l Unit Cost	Avg Labor Unit Cost	Avg Equip Unit Cost	Avg Total Unit Cost	Avg Price Incl O&P	Avg Total Unit Cost	Avg Price Incl O&P
										Large Volume	
Oak											
Hardwood											
Casing											
Streamline casing											
5/8" x 1-5/8"	Inst	LF	CA	.046	1.39	1.48	---	2.87	3.81	2.15	2.81
711 casing											
1/2" x 1-5/8"	Inst	LF	CA	.046	1.31	1.48	---	2.79	3.72	2.09	2.74
5/8" x 2-3/8"	Inst	LF	CA	.046	1.64	1.48	---	3.12	4.10	2.36	3.05
Universal											
5/8" x 2-1/2"	Inst	LF	CA	.046	1.97	1.48	---	3.45	4.48	2.65	3.39
Colonial											
11/16" x 2-7/8"	Inst	LF	CA	.046	1.52	1.48	---	3.00	3.96	2.27	2.95
11/16" x 3-1/2"	Inst	LF	CA	.046	2.13	1.48	---	3.61	4.66	2.79	3.55
11/16" x 4-1/4"	Inst	LF	CA	.046	2.35	1.48	---	3.83	4.92	2.98	3.77
Mull casing											
5/16" x 3-1/4"	Inst	LF	CA	.046	1.81	1.48	---	3.29	4.30	2.51	3.23
356 casing											
9/16" x 2-1/4"	Inst	LF	CA	.046	1.10	1.48	---	2.58	3.48	1.91	2.54
Cape Cod											
5/8" x 2-1/2"	Inst	LF	CA	.046	1.17	1.48	---	2.65	3.56	1.96	2.59
Fluted											
1/2" x 3-1/2"	Inst	LF	CA	.046	1.66	1.48	---	3.14	4.12	2.38	3.08
Victorian											
11/16" x 4-1/4"	Inst	LF	CA	.046	4.57	1.48	---	6.05	7.47	4.88	5.95
Cambridge											
11/16" x 3-1/4"	Inst	LF	CA	.046	2.68	1.48	---	4.16	5.30	3.26	4.09
Rosette plinth block w/ edge detail											
3/4" x 2-3/4"	Inst	Ea	CA	.381	3.94	12.20	---	16.14	22.90	11.40	15.90
3/4" x 3-3/4"	Inst	Ea	CA	.381	4.73	12.20	---	16.93	23.80	12.07	16.70
Base											
Base shoe											
1/2" x 3/4"	Inst	LF	CA	.046	.72	1.48	---	2.20	3.04	1.58	2.16
Colonial											
11/16" x 4-1/4"	Inst	LF	CA	.046	2.78	1.48	---	4.26	5.41	3.35	4.19
Streamline base											
1/2" x 2-1/2"	Inst	LF	CA	.046	1.46	1.48	---	2.94	3.89	2.21	2.88
1/2" x 3-1/2"	Inst	LF	CA	.046	1.66	1.48	---	3.14	4.12	2.38	3.08
711											
1/2" x 2-1/2"	Inst	LF	CA	.046	1.53	1.48	---	3.01	3.97	2.27	2.95
1/2" x 3-1/2"	Inst	LF	CA	.046	2.04	1.48	---	3.52	4.56	2.71	3.46
Cape Cod base											
7/16" x 4-1/4"	Inst	LF	CA	.046	2.58	1.48	---	4.06	5.18	3.17	3.99
Monterey base											
7/16" x 4-1/4"	Inst	LF	CA	.046	2.58	1.48	---	4.06	5.18	3.17	3.99
Rosette plinth block w/ edge detail											
3/4" x 2-3/4"	Inst	Ea	CA	.381	6.06	12.20	---	18.26	25.30	13.21	18.00
3/4" x 3-1/2"	Inst	Ea	CA	.381	7.25	12.20	---	19.45	26.70	14.23	19.20
Colonial plinth block w/ edge detail											
1" x 2-3/4" x 6"	Inst	Ea	CA	.381	4.57	12.20	---	16.77	23.60	11.94	16.50
1" x 3-3/4" x 6"	Inst	Ea	CA	.381	5.59	12.20	---	17.79	24.80	12.81	17.50

				Costs Based On Small Volume						Large Volume	
Description	Oper	Unit	Crew Size	Man-Hours Per Unit	Avg Mat'l Unit Cost	Avg Labor Unit Cost	Avg Equip Unit Cost	Avg Total Unit Cost	Avg Price Incl O&P	Avg Total Unit Cost	Avg Price Incl O&P
Stops											
Round edge											
3/8" x 1-1/4"	Inst	LF	CA	.046	1.02	1.48	---	2.50	3.39	1.83	2.4
Ogee											
1/2" x 1-3/4"	Inst	LF	CA	.046	1.89	1.48	---	3.37	4.39	2.58	3.3
Crown & cornice											
Crown											
1/2" x 1-5/8"	Inst	LF	CA	.062	1.29	1.99	---	3.28	4.47	2.39	3.2
1/2" x 2-1/4"	Inst	LF	CA	.062	1.81	1.99	---	3.80	5.07	2.83	3.7
1/2" x 3-1/2"	Inst	LF	CA	.062	2.59	1.99	---	4.58	5.96	3.50	4.4
Colonial crown											
5/8" x 3-1/2"	Inst	LF	CA	.062	2.68	1.99	---	4.67	6.07	3.58	4.5
5/8" x 4-1/2"	Inst	LF	CA	.062	4.10	1.99	---	6.09	7.70	4.79	5.9
Cornice											
13/16" x 4-5/8"	Inst	LF	CA	.062	4.17	1.99	---	6.16	7.78	4.85	6.0
1-1/4" x 6"	Inst	LF	CA	.062	8.35	1.99	---	10.34	12.60	8.44	10.2
Quarter round											
1/2" x 1/2"	Inst	LF	CA	.046	.66	1.48	---	2.14	2.97	1.53	2.1
3/4 x 3/4"	Inst	LF	CA	.046	1.17	1.48	---	2.65	3.56	1.96	2.5
Cove											
1/2" x 1/2"	Inst	LF	CA	.062	.66	1.99	---	2.65	3.74	1.85	2.5
3/4 x 3/4"	Inst	LF	CA	.062	1.17	1.99	---	3.16	4.33	2.28	3.0
Half round											
3/8" x 3/4"	Inst	LF	CA	.046	.70	1.48	---	2.18	3.02	1.56	2.1
Screen mold											
Flat screen											
3/8" x 3/4"	Inst	LF	CA	.046	.74	1.48	---	2.22	3.07	1.59	2.1
Scribe											
3/16" x 3/4"	Inst	LF	CA	.046	.64	1.48	---	2.12	2.95	1.51	2.0
Cloverleaf											
3/8" x 3/4"	Inst	LF	CA	.046	.88	1.48	---	2.36	3.23	1.72	2.3
Outside corner guard											
3/4" x 3/4"	Inst	LF	CA	.046	1.18	1.48	---	2.66	3.57	1.97	2.6
1" x 1"	Inst	LF	CA	.046	1.35	1.48	---	2.83	3.77	2.12	2.7
1-1/8" x 1-1/8"	Inst	LF	CA	.046	2.02	1.48	---	3.50	4.54	2.69	3.4
Chair rail											
Chair rail											
5/8" x 2-1/2"	Inst	LF	CA	.046	2.12	1.48	---	3.60	4.65	2.78	3.5
Colonial chair rail											
5/8" x 2-1/2"	Inst	LF	CA	.046	2.52	1.48	---	4.00	5.11	3.12	3.9
Victorian											
11/16" x 4-1/4"	Inst	LF	CA	.046	5.64	1.48	---	7.12	8.70	5.79	7.0
Panel cap											
1/2" x 1-1/2"	Inst	LF	CA	.046	1.45	1.48	---	2.93	3.88	2.20	2.8

Description	Oper	Unit	Crew Size	Man-Hours Per Unit	Avg Mat'l Unit Cost	Avg Labor Unit Cost	Avg Equip Unit Cost	Avg Total Unit Cost	Avg Price Incl O&P	Avg Total Unit Cost	Avg Price Incl O&P
								Costs Based On Small Volume		**Large Volume**	
Astragal											
Flat astragal											
3/4" x 1-5/8"	Inst	LF	CA	.046	2.11	1.48	---	3.59	4.64	2.77	3.53
1-3/4 T astragal											
1-1/4" x 2-1/4"	Inst	LF	CA	.046	7.72	1.48	---	9.20	11.10	7.58	9.06
T astragal mahogany											
1-1/4" x 2-1/4" (1-3/8" door)	Inst	LF	CA	.046	3.93	1.48	---	5.41	6.73	4.33	5.32
Nose & cove											
1" x 1-5/8"	Inst	LF	CA	.062	2.59	1.99	---	4.58	5.96	3.50	4.48
Hand rail											
Hand rail											
1-1/2" x 2-1/2"	Inst	LF	CA	.046	4.57	1.48	---	6.05	7.47	4.88	5.95
Plowed											
1-1/2" x 2-1/2"	Inst	LF	CA	.046	5.04	1.48	---	6.52	8.01	5.28	6.41
Panel mold											
Panel mold											
3/8" x 5/8"	Inst	LF	CA	.046	.76	1.48	---	2.24	3.09	1.61	2.19
5/8" x 3/4"	Inst	LF	CA	.046	1.08	1.48	---	2.56	3.46	1.89	2.51
3/4" x 1"	Inst	LF	CA	.046	1.49	1.48	---	2.97	3.93	2.24	2.92
3/4" x 1-5/8"	Inst	LF	CA	.046	2.17	1.48	---	3.65	4.71	2.82	3.58
Beauty mold											
9/16" x 1-3/8"	Inst	LF	CA	.046	1.81	1.48	---	3.29	4.30	2.51	3.23
3/4" x 1-7/8"	Inst	LF	CA	.046	2.77	1.48	---	4.25	5.40	3.34	4.18
Raised panel											
1-1/8" x 1-1/4"	Inst	LF	CA	.046	2.17	1.48	---	3.65	4.71	2.82	3.58
Full round											
1-3/8"	Inst	LF	CA	.046	3.07	1.48	---	4.55	5.74	3.59	4.47
4S											
1" x 4"	Inst	LF	CA	.062	4.17	1.99	---	6.16	7.78	4.85	6.03
1" x 6"	Inst	LF	CA	.062	6.30	1.99	---	8.29	10.20	6.68	8.14
1" x 8"	Inst	LF	CA	.062	8.35	1.99	---	10.34	12.60	8.44	10.20
1" x 10"	Inst	LF	CA	.062	10.80	1.99	---	12.79	15.40	10.52	12.60
1" x 12"	Inst	LF	CA	.062	13.40	1.99	---	15.39	18.40	12.78	15.10
Oak bar nosing											
1-1/4" x 3-1/2"	Inst	LF	CA	.046	7.56	1.48	---	9.04	10.90	7.44	8.90
1-5/8" x 5"	Inst	LF	CA	.046	10.80	1.48	---	12.28	14.60	10.20	12.10
Redwood											
Lattice											
Lattice											
5/16" x 1-1/4"	Inst	LF	CA	.046	.35	1.48	---	1.83	2.62	1.26	1.79
5/16" x 1-5/8"	Inst	LF	CA	.046	.41	1.48	---	1.89	2.69	1.31	1.85
Batts											
5/16" x 2-1/2"	Inst	LF	CA	.046	.64	1.48	---	2.12	2.95	1.51	2.08
5/16" x 3-1/2"	Inst	LF	CA	.046	.89	1.48	---	2.37	3.24	1.73	2.33
5/16" x 5-1/2"	Inst	LF	CA	.046	1.52	1.48	---	3.00	3.96	2.27	2.95

				Costs Based On Small Volume					Large Volume	
Description	Oper	Unit	Crew Size	Man-Hours Per Unit	Avg Mat'l Unit Cost	Avg Labor Unit Cost	Avg Equip Unit Cost	Avg Total Unit Cost	Avg Price Incl O&P	Avg Total Unit Cost	Avg Price Incl O&P

Miscellaneous exterior molds

Bricks											
1-1/2" x 1-1/2"	Inst	LF	CA	.062	1.55	1.99	---	3.54	**4.77**	2.61	3.
Siding											
7/8" x 1-5/8"	Inst	LF	CA	.062	1.14	1.99	---	3.13	**4.30**	2.26	3.
Stucco											
7/8" x 1-1/2"	Inst	LF	CA	.062	.67	1.99	---	2.66	**3.76**	1.86	2.
Watertable without lip											
1-1/2" x 2-3/8"	Inst	LF	CA	.062	2.49	1.99	---	4.48	**5.85**	3.41	4.
Watertable with lip											
1-1/2" x 2-3/8"	Inst	LF	CA	.062	2.49	1.99	---	4.48	**5.85**	3.41	4.
Corrugated											
3/4" x 1-5/8"	Inst	LF	CA	.062	.47	1.99	---	2.46	**3.53**	1.69	2.
Crest											
3/4" x 1-5/8"	Inst	LF	CA	.062	.47	1.99	---	2.46	**3.53**	1.69	2.

Resin flexible molding

Primed paint grade

Diameter casing

711 casing											
5/8" x 2-1/2"	Inst	LF	CA	.046	11.20	1.48	---	12.68	**15.10**	10.54	12.
Universal casing											
5/8" x 2-1/2"	Inst	LF	CA	.046	11.20	1.48	---	12.68	**15.10**	10.54	12.
356 casing											
5/8" x 2-1/2"	Inst	LF	CA	.046	11.20	1.48	---	12.68	**15.10**	10.54	12.
Cape Cod casing											
5/8" x 2-1/2"	Inst	LF	CA	.046	11.20	1.48	---	12.68	**15.10**	10.54	12.

Bases & round corner blocks

Base shoe											
3/8" x 3/4"	Inst	LF	CA	.046	2.11	1.48	---	3.59	**4.64**	2.77	3.
Streamline base											
3/8" x 2-1/4"	Inst	LF	CA	.046	3.44	1.48	---	4.92	**6.17**	3.91	4.
7/16" x 3-1/2"	Inst	LF	CA	.046	5.55	1.48	---	7.03	**8.60**	5.72	6.
711 base											
7/16" x 2-1/2"	Inst	LF	CA	.046	3.32	1.48	---	4.80	**6.03**	3.80	4.
7/16" x 3-1/2"	Inst	LF	CA	.046	5.01	1.48	---	6.49	**7.98**	5.25	6.
WM 623 base											
9/16" x 3-1/4"	Inst	LF	CA	.046	4.59	1.48	---	6.07	**7.49**	4.89	5.
WM 618											
9/16" x 5-1/4"	Inst	LF	CA	.046	10.60	1.48	---	12.08	**14.40**	10.07	11.
Cape Cod base											
9/16" x 5-1/4"	Inst	LF	CA	.046	8.45	1.48	---	9.93	**11.90**	8.21	9.
9/16" x 4"	Inst	LF	CA	.046	6.04	1.48	---	7.52	**9.16**	6.14	7.
"B" Cape Cod base											
9/16" x 5-1/4"	Inst	LF	CA	.046	9.66	1.48	---	11.14	**13.30**	9.24	11.0
Colonial base											
11/16" x 4-1/4"	Inst	LF	CA	.046	7.25	1.48	---	8.73	**10.60**	7.17	8.
Monterey base											
7/16" x 4-1/4"	Inst	LF	CA	.046	7.25	1.48	---	8.73	**10.60**	7.17	8.

Description	Oper	Unit	Crew Size	Man-Hours Per Unit	Costs Based On Small Volume						Large Volume	
					Avg Mat'l Unit Cost	Avg Labor Unit Cost	Avg Equip Unit Cost	Avg Total Unit Cost	Avg Price Incl O&P		Avg Total Unit Cost	Avg Price Incl O&P
356 base												
1-1/2" x 2-1/2"	Inst	LF	CA	.046	3.62	1.48	---	5.10	6.38		4.07	5.02
1-1/2" x 3-1/4"	Inst	LF	CA	.046	5.55	1.48	---	7.03	8.60		5.72	6.92
Step base												
7/16" x 2-1/2"	Inst	LF	CA	.046	3.86	1.48	---	5.34	6.65		4.27	5.25
1/2" x 4-1/2"	Inst	LF	CA	.046	7.61	1.48	---	9.09	11.00		7.49	8.95

Oak grain

Diameter casing

Description	Oper	Unit	Crew Size	Man-Hours Per Unit	Avg Mat'l Unit Cost	Avg Labor Unit Cost	Avg Equip Unit Cost	Avg Total Unit Cost	Avg Price Incl O&P		Avg Total Unit Cost	Avg Price Incl O&P
711 casing												
5/8" x 2-1/2"	Inst	LF	CA	.046	12.10	1.48	---	13.58	16.10		11.36	13.40
356 casing												
11/16" x 2-1/2"	Inst	LF	CA	.046	12.10	1.48	---	13.58	16.10		11.36	13.40

Bases & base shoe

Description	Oper	Unit	Crew Size	Man-Hours Per Unit	Avg Mat'l Unit Cost	Avg Labor Unit Cost	Avg Equip Unit Cost	Avg Total Unit Cost	Avg Price Incl O&P		Avg Total Unit Cost	Avg Price Incl O&P
711 base												
3/8" x 2-1/2"	Inst	LF	CA	.046	4.83	1.48	---	6.31	7.77		5.10	6.21
3/8" x 3-1/2"	Inst	LF	CA	.046	7.25	1.48	---	8.73	10.60		7.17	8.59
Base shoe												
3/8" x 3/4"	Inst	LF	CA	.046	2.42	1.48	---	3.90	5.00		3.03	3.82

Spindles

Western hemlock, clear, kiln dried, turned for decorative applications

Planter design

Description	Oper	Unit	Crew Size	Man-Hours Per Unit	Avg Mat'l Unit Cost	Avg Labor Unit Cost	Avg Equip Unit Cost	Avg Total Unit Cost	Avg Price Incl O&P		Avg Total Unit Cost	Avg Price Incl O&P
1-11/16" x 1-11/16"												
3'-0" H	Inst	Ea	CA	.500	4.47	16.10	---	20.57	29.20		14.53	20.40
4'-0" H	Inst	Ea	CA	.500	5.73	16.10	---	21.83	30.70		15.61	21.70
2-3/8" x 2-3/8"												
3'-0" H	Inst	Ea	CA	.500	9.30	16.10	---	25.40	34.80		18.67	25.20
4'-0" H	Inst	Ea	CA	.500	12.70	16.10	---	28.80	38.70		21.60	28.60
5'-0" H	Inst	Ea	CA	.500	15.50	16.10	---	31.60	41.90		24.00	31.30
6'-0" H	Inst	Ea	CA	.615	19.70	19.70	---	39.40	52.30		29.70	38.70
8'-0" H	Inst	Ea	CA	.615	38.30	19.70	---	58.00	73.60		45.60	57.00
3-1/4" x 3-1/4"												
3'-0" H	Inst	Ea	CA	.500	17.60	16.10	---	33.70	44.30		25.80	33.40
4'-0" H	Inst	Ea	CA	.500	23.60	16.10	---	39.70	51.20		30.90	39.30
5'-0" H	Inst	Ea	CA	.500	29.80	16.10	---	45.90	58.40		36.30	45.40
6'-0" H	Inst	Ea	CA	.615	35.70	19.70	---	55.40	70.60		43.40	54.40
8'-0" H	Inst	Ea	CA	.615	51.50	19.70	---	71.20	88.80		57.00	70.00

Colonial design

Description	Oper	Unit	Crew Size	Man-Hours Per Unit	Avg Mat'l Unit Cost	Avg Labor Unit Cost	Avg Equip Unit Cost	Avg Total Unit Cost	Avg Price Incl O&P		Avg Total Unit Cost	Avg Price Incl O&P
1-11/16" x 1-11/16"												
1'-0" H	Inst	Ea	CA	.444	1.45	14.30	---	15.75	23.00		10.42	15.20
1'-6" H	Inst	Ea	CA	.444	2.59	14.30	---	16.89	24.40		11.40	16.30
2'-0" H	Inst	Ea	CA	.444	3.32	14.30	---	17.62	25.20		12.02	17.00
2'-4" H	Inst	Ea	CA	.500	3.44	16.10	---	19.54	28.00		13.65	19.40
2'-8" H	Inst	Ea	CA	.500	3.86	16.10	---	19.96	28.50		14.01	19.80
3'-0" H	Inst	Ea	CA	.500	4.59	16.10	---	20.69	29.40		14.63	20.60

					Costs Based On Small Volume					Large Volume	
Description	Oper	Unit	Crew Size	Man-Hours Per Unit	Avg Mat'l Unit Cost	Avg Labor Unit Cost	Avg Equip Unit Cost	Avg Total Unit Cost	Avg Price Incl O&P	Avg Total Unit Cost	Avg Price Incl O&P
2-3/8" x 2-3/8"											
1'-0" H	Inst	Ea	CA	.444	2.96	14.30	---	17.26	24.80	11.72	16.?
1'-6" H	Inst	Ea	CA	.444	4.11	14.30	---	18.41	26.10	12.70	17.8
2'-0" H	Inst	Ea	CA	.444	6.22	14.30	---	20.52	28.50	14.51	19.9
2'-4" H	Inst	Ea	CA	.500	6.65	16.10	---	22.75	31.70	16.40	22.6
2'-8" H	Inst	Ea	CA	.500	8.45	16.10	---	24.55	33.80	17.95	24.4
3'-0" H	Inst	Ea	CA	.500	9.30	16.10	---	25.40	34.80	18.67	25.2
3-1/4" x 3-1/4"											
1'-6" H	Inst	Ea	CA	.444	6.94	14.30	---	21.24	29.40	15.13	20.6
2'-0" H	Inst	Ea	CA	.444	9.42	14.30	---	23.72	32.20	17.25	23.1
2'-4" H	Inst	Ea	CA	.500	11.10	16.10	---	27.20	36.80	20.17	26.9
2'-8" H	Inst	Ea	CA	.500	12.50	16.10	---	28.60	38.50	21.40	28.4
3'-0" H	Inst	Ea	CA	.500	17.60	16.10	---	33.70	44.30	25.80	33.4
8'-0" H	Inst	Ea	CA	.615	51.50	19.70	---	71.20	88.80	57.00	70.0
Mediterranean design											
1-11/16" x 1-11/16"											
1'-0" H	Inst	Ea	CA	.444	1.51	14.30	---	15.81	23.10	10.48	15.3
1'-6" H	Inst	Ea	CA	.444	2.48	14.30	---	16.78	24.20	11.30	16.2
2'-0" H	Inst	Ea	CA	.444	2.72	14.30	---	17.02	24.50	11.51	16.5
2'-4" H	Inst	Ea	CA	.500	3.20	16.10	---	19.30	27.80	13.45	19.2
2'-8" H	Inst	Ea	CA	.500	3.56	16.10	---	19.66	28.20	13.75	19.5
3'-0" H	Inst	Ea	CA	.500	4.47	16.10	---	20.57	29.20	14.53	20.4
4'-0" H	Inst	Ea	CA	.500	5.73	16.10	---	21.83	30.70	15.61	21.7
5'-0" H	Inst	Ea	CA	.500	10.30	16.10	---	26.40	35.90	19.55	26.2
2-3/8" x 2-3/8"											
1'-0" H	Inst	Ea	CA	.444	2.72	14.30	---	17.02	24.50	11.51	16.5
1'-6" H	Inst	Ea	CA	.444	3.93	14.30	---	18.23	25.90	12.55	17.6
2'-0" H	Inst	Ea	CA	.444	4.96	14.30	---	19.26	27.10	13.43	18.7
2'-4" H	Inst	Ea	CA	.500	6.34	16.10	---	22.44	31.40	16.14	22.3
2'-8" H	Inst	Ea	CA	.500	7.91	16.10	---	24.01	33.20	17.48	23.8
3'-0" H	Inst	Ea	CA	.500	9.06	16.10	---	25.16	34.50	18.47	25.0
4'-0" H	Inst	Ea	CA	.500	10.80	16.10	---	26.90	36.40	19.92	26.6
5'-0" H	Inst	Ea	CA	.500	14.80	16.10	---	30.90	41.10	23.40	30.6
6'-0" H	Inst	Ea	CA	.615	20.80	19.70	---	40.50	53.50	30.60	39.7
8'-0" H	Inst	Ea	CA	.615	38.30	19.70	---	58.00	73.60	45.60	57.0
3-1/4" x 3-1/4"											
3'-0" H	Inst	Ea	CA	.500	17.10	16.10	---	33.20	43.70	25.30	32.9
4'-0" H	Inst	Ea	CA	.500	20.10	16.10	---	36.20	47.20	27.90	35.9
5'-0" H	Inst	Ea	CA	.500	27.10	16.10	---	43.20	55.20	33.90	42.7
6'-0" H	Inst	Ea	CA	.615	38.00	19.70	---	57.70	73.40	45.40	56.8
8'-0" H	Inst	Ea	CA	.615	60.00	19.70	---	79.70	98.60	64.20	78.4
Spindle rails, 8'-0" H pieces											
For 1-11/16" spindles	Inst	Ea	CA	---	15.20	---	---	15.20	17.50	13.00	15.0
For 2-3/8" spindles	Inst	Ea	CA	---	22.00	---	---	22.00	25.30	18.80	21.7
For 3-1/4" spindles	Inst	Ea	CA	---	25.50	---	---	25.50	29.40	21.90	25.2

Wood bullnose round corners

Description	Oper	Unit	Crew Size	Man-Hours Per Unit	Avg Mat'l Unit Cost	Avg Labor Unit Cost	Avg Equip Unit Cost	Avg Total Unit Cost	Avg Price Incl O&P	Avg Total Unit Cost	Avg Price Incl O&P
										Large Volume	
Pine base R/C											
Colonial base											
11/16" x 4-1/4"	Inst	LF	CA	.046	1.23	1.48	---	2.71	3.63	2.01	2.65
Streamline base											
3/8" x 2-1/4"	Inst	LF	CA	.046	.64	1.48	---	2.12	2.95	1.51	2.08
7/16" x 3-1/2"	Inst	LF	CA	.046	.91	1.48	---	2.39	3.26	1.74	2.34
711 base											
7/16" x 2-1/2"	Inst	LF	CA	.046	.76	1.48	---	2.24	3.09	1.61	2.19
7/16" x 3-1/2"	Inst	LF	CA	.046	1.07	1.48	---	2.55	3.44	1.88	2.50
3 step base											
7/16" x 2-1/2"	Inst	LF	CA	.046	.67	1.48	---	2.15	2.98	1.54	2.11
1/2" x 4-1/2"	Inst	LF	CA	.046	1.98	1.48	---	3.46	4.49	2.66	3.40
WM 623 base											
9/16" x 3-1/4"	Inst	LF	CA	.046	.93	1.48	---	2.41	3.28	1.76	2.36
WM 618 base											
9/16" x 5-1/4"	Inst	LF	CA	.046	1.44	1.48	---	2.92	3.87	2.19	2.86
Monterey base											
7/16" x 4-1/4"	Inst	LF	CA	.046	1.00	1.48	---	2.48	3.36	1.82	2.43
Cape Cod base											
9/16" x 5-1/4"	Inst	LF	CA	.046	1.45	1.48	---	2.93	3.88	2.20	2.87
9/16" x 4"	Inst	LF	CA	.046	1.21	1.48	---	2.69	3.61	2.00	2.64
"B" Cape Cod base											
9/16" x 5-1/4"	Inst	LF	CA	.046	1.45	1.48	---	2.93	3.88	2.20	2.87
356 base											
1/2" x 2-1/4"	Inst	LF	CA	.046	.88	1.48	---	2.36	3.23	1.72	2.32
1/2" x 3-1/4"	Inst	LF	CA	.046	1.11	1.48	---	2.59	3.49	1.91	2.54
Base shoe											
3/8" x 3/4"	Inst	LF	CA	.046	.23	1.48	---	1.71	2.48	1.16	1.67
Oak base R/C											
Base shoe											
3/8" x 3/4"	Inst	LF	CA	.046	.38	1.48	---	1.86	2.65	1.28	1.81
711 base											
3/8" x 2-1/2"	Inst	LF	CA	.046	1.10	1.48	---	2.58	3.48	1.91	2.54
3/8" x 3-1/2"	Inst	LF	CA	.046	1.39	1.48	---	2.87	3.81	2.15	2.81
Pine chair rail R/C											
Chair rail											
9/16" x 2-1/2"	Inst	LF	CA	.046	.98	1.48	---	2.46	3.34	1.80	2.41
Colonial chair rail											
11/16" x 2-1/2"	Inst	LF	CA	.046	.99	1.48	---	2.47	3.35	1.81	2.42
3/4" x 3"	Inst	LF	CA	.046	1.13	1.48	---	2.61	3.51	1.93	2.56
Pine crown & cornice R/C											
Colonial crown											
9/16" x 2-1/4"	Inst	LF	CA	.046	.88	1.48	---	2.36	3.23	1.72	2.32
11/16" x 3-3/8"	Inst	LF	CA	.046	1.19	1.48	---	2.67	3.58	1.98	2.62
11/16" x 4-1/2"	Inst	LF	CA	.046	2.07	1.48	---	3.55	4.59	2.73	3.48
Cornice											
13/16" x 4-5/8"	Inst	LF	CA	.046	2.34	1.48	---	3.82	4.91	2.97	3.76
Georgian crown											
11/16" x 4-1/4"	Inst	LF	CA	.046	2.56	1.48	---	4.04	5.16	3.16	3.97

Painting

Interior and Exterior. There is a paint for almost every type of surface and surface condition. The large variety makes it impractical to consider each paint individually. For this reason, average output and average material cost/unit are based on the paints and prices listed below.

1. **Installation.** Paint can be applied by brush, roller or spray gun. Only application by brush or roller is considered in this section.

2. **Notes on Labor.** Average Manhours per Unit, for both roller and brush, is based on what one painter can do in one day. The output for cleaning is also based on what one painter can do in one day.

3. **Estimating Technique.** Use these techniques to determine quantities for interior and exterior painting before you apply unit costs.

Interior

a. Floors, walls and ceilings. Figure actual area. No deductions for openings.

b. Doors and windows. **Only openings to be painted:** Figure 36 SF or 4 SY for each side of each door and 27 SF for each side of each window. Based on doors 3'-0" x 7'-0" or smaller and windows 3'-0" x 4'-0" or smaller. **Openings to be painted with walls:** Figure wall area plus 27 SF or 3 SY for each side of each door and 18 SF or 2 SY for each side of each window. Based on doors 3'-0" x 7'-0" or smaller and windows 3'-0" x 4'-0" or smaller.

For larger doors and windows, add 1'-0" to height and width and figure area.

c. Base or picture moldings and chair rails. Less than 1'-0" wide, figure one SF/LF. On 1'-0" or larger, figure actual area.

d. Stairs (including treads, risers, cove and stringers). Add 2'-0" width (for treads, risers, etc.) times length plus 2'-0".

e. Balustrades. Add 1'-0" to height, figure two times area to paint two sides.

Exterior

a. Walls. (No deductions for openings.)

Siding. Figure actual area plus 10%.

Shingles. Figure actual area plus 40%.

Characteristics - Interior			Characteristics - Exterior		
Type	Coverage SF/Gal.	Surface	Type	Coverage SF/Gal.	Surface
Latex, flat	450	Plaster/drywall	Oil base*	300	Plain siding & stucco
Latex, enamel	450	Doors, windows, trim	Oil base*	450	Door, windows, trim
Shellac	500	Woodwork	Oil base*	300	Shingle siding
Varnish	500	Woodwork	Stain	200	Shingle siding
Stain	500	Woodwork	Latex, masonry	400	Stucco & masonry

*Certain latex paints may also be used on exterior work.

Brick, stucco, concrete and smooth wood surfaces. Figure actual area.

b. Doors and windows. See Interior, doors and windows.

c. Eaves (including soffit or exposed rafter ends and fascia). **Enclosed.** If sidewalls are to be painted the same color, figure 1.5 times actual area. If sidewalls are to be painted a different color, figure 2.0 times actual area. If sidewalls are not to be painted, figure 2.5 times actual area. **Rafter ends exposed:** If sidewalls are to be painted same color, figure 2.5 times actual area. If sidewalls are to be painted a different color, figure 3.0 times actual area. If sidewalls are not to be painted, figure 4.0 times actual area.

d. Porch rails. See Interior, balustrades (previous page).

e. Gutters and downspouts. Figure 2.0 SF/LF or $^2/_9$ SY/LF.

f. Latticework. Figure 2.0 times actual area for each side.

g. Fences. **Solid fence:** Figure actual area of each side to be painted. **Basketweave:** Figure 1.5 times actual area for each side to be painted.

Calculating Square Foot Coverage

Triangle

To find the number of square feet in any shape triangle or 3 sided surface, multiply the height by the width and divide the total by 2.

Square

Multiply the base measurement in feet times the height in feet.

Rectangle

Multiply the base measurement in feet times the height in feet.

Arch Roof

Multiply length (B) by width (A) and add one-half the total.

Circle

To find the number of square feet in a circle multiply the diameter (distance across) by itself and them multiply this total by .7854.

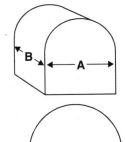

Cylinder

When the circumference (distance around the cylinder) is known, multiply height by circumference. When the diameter (distance across) is known, multiply diameter by 3.1416. This gives circumference. Then multiply by height.

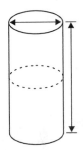

Gambrel Roof

Multiply length (B) by width (A) and add one-third of the total.

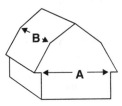

Cone

Determine area of base by multiplying 3.1416 times radius (A) in feet.

Determine the surface area of a cone by multiplying circumference of base (in feet) times one-half of the slant height (B) in feet.

Add the square foot area of the base to the square foot area of the cone for total square foot area.

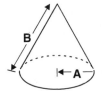

Description	Oper	Unit	Crew Size	Man-Hours Per Unit	Costs Based On Small Volume					Large Volume	
					Avg Mat'l Unit Cost	Avg Labor Unit Cost	Avg Equip Unit Cost	Avg Total Unit Cost	Avg Price Incl O&P	Avg Total Unit Cost	Avg Price Incl O&P

Painting and cleaning

Frequently encountered applications

Interior

Description	Oper	Unit	Crew Size	Man-Hours	Mat'l	Labor	Equip	Total	Price O&P	Total	Price O&P
Wet clean, walls	Inst	SF	NA	.007	.02	.23	---	.25	.36	.18	.2
Wet clean, floors	Inst	SF	NA	.005	.02	.16	---	.18	.26	.14	.2
Wet clean, millwork and trim	Inst	SF	NA	.007	.02	.23	---	.25	.36	.18	.2
Prime or seal, one coat	Inst	SF	NA	.006	.07	.19	---	.26	.36	.19	.2
Paint, one coat	Inst	SF	NA	.007	.10	.23	---	.33	.45	.25	.3
Paint, two coats	Inst	SF	NA	.011	.19	.36	---	.55	.74	.43	.5

Exterior

Description	Oper	Unit	Crew Size	Man-Hours	Mat'l	Labor	Equip	Total	Price O&P	Total	Price O&P
Wet clean, walls	Inst	SF	NA	.008	.02	.26	---	.28	.41	.18	.2
Paint, one coat	Inst	SF	NA	.010	.24	.32	---	.56	.74	.44	.5
Paint, two coats	Inst	SF	NA	.017	.35	.55	---	.90	1.20	.70	.9

Interior

Time calculations include normal materials handling and protection of furniture and other property not to be painted

Preparation

Excluding openings, unless otherwise indicated

Cleaning, wet

Smooth finishes

Description	Oper	Unit	Crew Size	Man-Hours	Mat'l	Labor	Equip	Total	Price O&P	Total	Price O&P
Plaster and drywall	Inst	SF	NA	.007	.02	.23	---	.25	.36	.18	.20
Paneling	Inst	SF	NA	.006	.02	.19	---	.21	.31	.15	.2
Millwork and trim	Inst	SF	NA	.007	.02	.23	---	.25	.36	.18	.20
Floors	Inst	SF	NA	.005	.02	.16	---	.18	.26	.14	.20
Sand finishes	Inst	SF	NA	.010	.02	.32	---	.34	.50	.24	.35

Sheetrock or drywall, repair joints and cracks or pops

Description	Oper	Unit	Crew Size	Man-Hours	Mat'l	Labor	Equip	Total	Price O&P	Total	Price O&P
Tape, fill, and finish	Inst	SF	NA	.013	.04	.42	---	.46	.67	.32	.46

Sheetrock or drywall, thin coat plaster

Description	Oper	Unit	Crew Size	Man-Hours	Mat'l	Labor	Equip	Total	Price O&P	Total	Price O&P
in lieu of taping	Inst	SF	NA	.017	.11	.55	---	.66	.94	.49	.69

Light sanding

Description	Oper	Unit	Crew Size	Man-Hours	Mat'l	Labor	Equip	Total	Price O&P	Total	Price O&P
Before first coat	Inst	SF	NA	.007	.01	.23	---	.24	.35	.17	.25
Before second coat	Inst	SF	NA	.006	.01	.19	---	.20	.30	.17	.25
Before third coat	Inst	SF	NA	.006	.01	.19	---	.20	.30	.14	.20

Liquid removal of paint or varnish

Description	Oper	Unit	Crew Size	Man-Hours	Mat'l	Labor	Equip	Total	Price O&P	Total	Price O&P
Paneling (170 SF/gal)	Inst	SF	NA	.036	.17	1.17	---	1.34	1.91	.97	1.37
Millwork & trim (170 SF/gal)	Inst	SF	NA	.041	.17	1.33	---	1.50	2.15	1.10	1.57
Floors (170 SF/gal)	Inst	SF	NA	.024	.17	.78	---	.95	1.34	.71	.99
Burning off paint	Inst	SF	NA	.057	.03	1.85	---	1.88	2.76	1.32	1.94

One coat application

Excluding openings unless otherwise indicated

Description	Oper	Unit	Costs Based On Small Volume							Large Volume	
			Crew Size	Man-Hours Per Unit	Avg Mat'l Unit Cost	Avg Labor Unit Cost	Avg Equip Unit Cost	Avg Total Unit Cost	Avg Price Incl O&P	Avg Total Unit Cost	Avg Price Incl O&P
Sizing, on sheetrock or plaster											
Smooth finish											
Brush (650 SF/gal)	Inst	SF	NA	.005	.03	.16	---	.19	.27	.12	.17
Roller (625 SF/gal)	Inst	SF	NA	.004	.03	.13	---	.16	.22	.13	.18
Sand finish											
Brush (550 SF/gal)	Inst	SF	NA	.007	.03	.23	---	.26	.37	.19	.27
Roller (525 SF/gal)	Inst	SF	NA	.005	.03	.16	---	.19	.27	.13	.18
Sealer											
Sheetrock or plaster											
Smooth finish											
Brush (300 SF/gal)	Inst	SF	NA	.008	.07	.26	---	.33	.46	.25	.35
Roller (285 SF/gal)	Inst	SF	NA	.006	.07	.19	---	.26	.36	.19	.26
Spray (250 SF/gal)	Inst	SF	NC	.004	.08	.13	---	.21	.29	.17	.22
Sand finish											
Brush (250 SF/gal)	Inst	SF	NA	.011	.08	.36	---	.44	.62	.33	.46
Roller (235 SF/gal)	Inst	SF	NA	.008	.09	.26	---	.35	.48	.27	.38
Spray (210 SF/gal)	Inst	SF	NC	.004	.10	.13	---	.23	.31	.19	.25
Acoustical tile or panels											
Brush (225 SF/gal)	Inst	SF	NA	.009	.09	.29	---	.38	.53	.27	.38
Roller (200 SF/gal)	Inst	SF	NA	.007	.10	.23	---	.33	.45	.25	.34
Spray (160 SF/gal)	Inst	SF	NC	.005	.13	.17	---	.30	.39	.25	.33
Latex											
Drywall or plaster, latex flat											
Smooth finish											
Brush (300 SF/gal)	Inst	SF	NA	.009	.10	.29	---	.39	.54	.28	.39
Roller (285 SF/gal)	Inst	SF	NA	.007	.10	.23	---	.33	.45	.25	.34
Spray (260 SF/gal)	Inst	SF	NC	.005	.11	.17	---	.28	.37	.20	.26
Sand finish											
Brush (250 SF/gal)	Inst	SF	NA	.012	.12	.39	---	.51	.71	.37	.50
Roller (235 SF/gal)	Inst	SF	NA	.010	.13	.32	---	.45	.62	.34	.46
Spray (210 SF/gal)	Inst	SF	NC	.005	.14	.17	---	.31	.40	.23	.29
Texture or stipple applied to drywall, one coat											
Brush (125 SF/gal)	Inst	SF	NA	.010	.24	.32	---	.56	.74	.44	.57
Roller (120 SF/gal)	Inst	SF	NA	.007	.25	.23	---	.48	.61	.38	.48
Paneling, latex enamel											
Brush (300 SF/gal)	Inst	SF	NA	.009	.10	.29	---	.39	.54	.28	.39
Roller (285 SF/gal)	Inst	SF	NA	.007	.10	.23	---	.33	.45	.25	.34
Spray (260 SF/gal)	Inst	SF	NC	.005	.11	.17	---	.28	.37	.20	.26
Acoustical tile or panels, latex flat											
Brush (225 SF/gal)	Inst	SF	NA	.010	.13	.32	---	.45	.62	.35	.47
Roller (210 SF/gal)	Inst	SF	NA	.008	.14	.26	---	.40	.54	.32	.43
Spray (185 SF/gal)	Inst	SF	NC	.006	.16	.20	---	.36	.47	.27	.35

					Costs Based On Small Volume					Large Volume	
Description	Oper	Unit	Crew Size	Man-Hours Per Unit	Avg Mat'l Unit Cost	Avg Labor Unit Cost	Avg Equip Unit Cost	Avg Total Unit Cost	Avg Price Incl O&P	Avg Total Unit Cost	Avg Price Incl O&P
Millwork and trim, latex enamel											
Doors and windows											
Roller and/or brush (360 SF/gal)	Inst	SF	NA	.018	.08	.58	---	.66	.95	.49	
Spray, doors only (325 SF / gal)	Inst	SF	NC	.008	.09	.27	---	.36	.49	.28	
Cabinets											
Roller and/or brush (360 SF/gal)	Inst	SF	NA	.019	.08	.62	---	.70	1.00	.49	
Spray, doors only (325 SF / gal)	Inst	SF	NC	.009	.09	.30	---	.39	.54	.28	
Louvers, spray (300 SF/gal)	Inst	SF	NC	.026	.10	.87	---	.97	1.39	.69	
Picture molding, chair rail, base, ceiling mold etc., less than 6" high											
Note: SF equals LF on trim less than 6" high											
Brush (900 SF/gal)	Inst	SF	NA	.012	.03	.39	---	.42	.61	.29	
Floors, wood											
Brush (405 SF/gal)	Inst	SF	NA	.006	.07	.19	---	.26	.36	.20	
Roller (385 SF/gal)	Inst	SF	NA	.005	.08	.16	---	.24	.33	.17	
Note: The following percentage adjustment for Small Volume also applies to Large											
For custom colors, ADD	Inst	%	---	---	10.0	---	---	---	---	---	
Floor seal											
Brush (450 SF/gal)	Inst	SF	NA	.004	.07	.13	---	.20	.27	.16	
Roller (430 SF/gal)	Inst	SF	NA	.004	.08	.13	---	.21	.28	.17	
Penetrating stainwax (hardwood floors)											
Brush (450 SF/gal)	Inst	SF	NA	.006	.07	.19	---	.26	.36	.19	
Roller (425 SF/gal)	Inst	SF	NA	.005	.08	.16	---	.24	.33	.17	
Natural finishes											
Paneling, brush work unless otherwise indicated											
Stain, brush on - wipe off											
(360 SF/gal)	Inst	SF	NA	.020	.09	.65	---	.74	1.06	.53	
Varnish (380 SF/gal)	Inst	SF	NA	.008	.09	.26	---	.35	.48	.27	
Shellac (630 SF/gal)	Inst	SF	NA	.007	.06	.23	---	.29	.40	.22	
Lacquer											
Brush (450 SF/gal)	Inst	SF	NA	.006	.07	.19	---	.26	.36	.19	
Spray (300 SF/gal)	Inst	SF	NC	.005	.11	.17	---	.28	.37	.20	
Doors and windows, brush work unless otherwise indicated											
Stain, brush on - wipe off											
(450 SF/gal)	Inst	SF	NA	.046	.07	1.49	---	1.56	2.28	1.10	1.6
Varnish (550 SF/gal)	Inst	SF	NA	.034	.07	1.10	---	1.17	1.71	.84	1.2
Shellac (550 SF/gal)	Inst	SF	NA	.031	.07	1.00	---	1.07	1.56	.77	1.1
Lacquer											
Brush (550 SF/gal)	Inst	SF	NA	.028	.06	.91	---	.97	1.41	.70	1.0
Spray doors (300 SF/gal)	Inst	SF	NC	.014	.11	.47	---	.58	.81	.43	.6
Cabinets, brush work unless otherwise indicated											
Stain, brush on - wipe off											
(450 SF/gal)	Inst	SF	NA	.048	.07	1.55	---	1.62	2.38	1.16	1.7
Varnish (550 SF/gal)	Inst	SF	NA	.036	.07	1.17	---	1.24	1.80	.87	1.2
Shellac (550 SF/gal)	Inst	SF	NA	.033	.07	1.07	---	1.14	1.66	.80	1.1
Lacquer											
Brush (550 SF/gal)	Inst	SF	NA	.030	.06	.97	---	1.03	1.50	.73	1.0
Spray (300 SF/gal)	Inst	SF	NC	.015	.11	.50	---	.61	.86	.47	.6
Louvers, lacquer, spray											
(300 SF/gal)	Inst	SF	NC	.024	.11	.80	---	.91	1.30	.67	.9

Description	Oper	Unit	Crew Size	Man-Hours Per Unit	Avg Mat'l Unit Cost	Avg Labor Unit Cost	Avg Equip Unit Cost	Avg Total Unit Cost	Avg Price Incl O&P	Avg Total Unit Cost	Avg Price Incl O&P
								Small Volume		**Large Volume**	
Picture molding, chair rail, base, ceiling mold etc., less than 6" high											
Note: SF equals LF on trim less than 6" high											
Varnish, brush (900 SF/gal)	Inst	SF	NA	.011	.04	.36	---	.40	**.57**	.30	**.43**
Shellac, brush (900 SF/gal)	Inst	SF	NA	.011	.04	.36	---	.40	**.57**	.30	**.43**
Lacquer, spray (700 SF/gal)	Inst	SF	NA	.009	.05	.29	---	.34	**.49**	.23	**.33**
Floors, wood, brush work unless otherwise indicated											
Shellac (450 SF/gal)	Inst	SF	NA	.006	.09	.19	---	.28	**.39**	.21	**.28**
Varnish (500 SF/gal)	Inst	SF	NA	.006	.07	.19	---	.26	**.36**	.19	**.26**
Buffing, by machine	Inst	SF	NA	.004	.02	.13	---	.15	**.21**	.12	**.17**
Waxing and polishing, by hand (1,000 SF/gal)	Inst	SF	NA	.008	.04	.26	---	.30	**.43**	.19	**.27**

wo coat application

Excluding openings, unless otherwise indicated

For sizing or sealer, see One coat application, page 205

Latex

No spray work included, see One coat application, page 205

Description	Oper	Unit	Crew Size	Man-Hours Per Unit	Avg Mat'l Unit Cost	Avg Labor Unit Cost	Avg Equip Unit Cost	Avg Total Unit Cost	Avg Price Incl O&P	Avg Total Unit Cost	Avg Price Incl O&P
Drywall or plaster, latex flat											
Smooth finish											
Brush (170 SF/gal)	Inst	SF	NA	.013	.17	.42	---	.59	**.81**	.45	**.61**
Roller (160 SF/gal)	Inst	SF	NA	.011	.19	.36	---	.55	**.74**	.43	**.57**
Sand finish											
Brush (170 SF/gal)	Inst	SF	NA	.019	.17	.62	---	.79	**1.10**	.58	**.80**
Roller (160 SF/gal)	Inst	SF	NA	.015	.19	.49	---	.68	**.93**	.53	**.71**
Paneling, latex enamel											
Brush (170 SF/gal)	Inst	SF	NA	.013	.17	.42	---	.59	**.81**	.45	**.61**
Roller (160 SF/gal)	Inst	SF	NA	.011	.19	.36	---	.55	**.74**	.43	**.57**
Acoustical tile or panels, latex flat											
Brush (130 SF/gal)	Inst	SF	NA	.017	.23	.55	---	.78	**1.07**	.59	**.80**
Roller (120 SF/gal)	Inst	SF	NA	.013	.25	.42	---	.67	**.90**	.51	**.67**

Millwork and trim, enamel

Description	Oper	Unit	Crew Size	Man-Hours Per Unit	Avg Mat'l Unit Cost	Avg Labor Unit Cost	Avg Equip Unit Cost	Avg Total Unit Cost	Avg Price Incl O&P	Avg Total Unit Cost	Avg Price Incl O&P
Doors and windows											
Roller and/or brush (200 SF/gal)	Inst	SF	NA	.034	.15	1.10	---	1.25	**1.79**	.91	**1.29**
Cabinets											
Roller and/or brush (200 SF/gal)	Inst	SF	NA	.036	.15	1.17	---	1.32	**1.89**	.94	**1.34**

Louvers, see One coat application, page 206

Description	Oper	Unit	Crew Size	Man-Hours Per Unit	Avg Mat'l Unit Cost	Avg Labor Unit Cost	Avg Equip Unit Cost	Avg Total Unit Cost	Avg Price Incl O&P	Avg Total Unit Cost	Avg Price Incl O&P
Picture molding, chair rail, base, ceiling mold etc., less than 6" high											
Note: SF equals LF on trim less than 6" high											
Brush (510 SF/gal)	Inst	SF	NA	.023	.06	.74	---	.80	**1.17**	.57	**.82**
Floors, wood											
Brush (230 SF/gal)	Inst	SF	NA	.010	.13	.32	---	.45	**.62**	.34	**.46**
Roller (220 SF/gal)	Inst	SF	NA	.009	.14	.29	---	.43	**.59**	.31	**.42**
Note: The following percentage adjustment for Small Volume also applies to Large											
For custom colors, ADD	Inst	%	---	---	10.0	---	---	---	**---**	---	**---**

For wood floor seal, penetrating stainwax, or natural finish,

see One coat application, page 206

				Costs Based On Small Volume						Large Volume	
Description	Oper	Unit	Crew Size	Man-Hours Per Unit	Avg Mat'l Unit Cost	Avg Labor Unit Cost	Avg Equip Unit Cost	Avg Total Unit Cost	Avg Price Incl O&P	Avg Total Unit Cost	Avg Price Incl O&P

Exterior

Time calculations include normal materials handling and protection of property not to be painted

Preparation

Excluding openings, unless otherwise indicated

Cleaning, wet

Description	Oper	Unit	Crew Size	Man-Hours Per Unit	Avg Mat'l Unit Cost	Avg Labor Unit Cost	Avg Equip Unit Cost	Avg Total Unit Cost	Avg Price Incl O&P	Avg Total Unit Cost	Avg Price Incl O&P
Plain siding	Inst	SF	NA	.008	.02	.26	---	.28	.41	.18	.2
Exterior doors and trim											
Note: SF equals LF on trim less than 6" high											
	Inst	SF	NA	.008	.02	.26	---	.28	.41	.21	.3
Windows, wash and clean glass	Inst	SF	NA	.009	.02	.29	---	.31	.45	.20	.3
Porch floors and steps	Inst	SF	NA	.005	.02	.16	---	.18	.26	.15	.2
Acid wash											
Gutters and downspouts	Inst	SF	NA	.011	.02	.36	---	.38	.55	.27	.3
Sanding, light											
Porch floors and steps	Inst	SF	NA	.006	.02	.19	---	.21	.31	.17	.2
Sanding and puttying											
Plain siding	Inst	SF	NA	.008	.02	.26	---	.28	.41	.20	.3
Exterior doors and trim											
Note: SF equals LF on trim less than 6" high											
	Inst	SF	NA	.016	.02	.52	---	.54	.79	.37	.5
Puttying sash or reglazing											
Windows (30 SF glass/lb glazing compound)	Inst	SF	NA	.048	.54	1.55	---	2.09	2.89	1.55	2.1

One coat application

Excluding openings, unless otherwise indicated

Latex, flat (unless otherwise indicated)

Description	Oper	Unit	Crew Size	Man-Hours Per Unit	Avg Mat'l Unit Cost	Avg Labor Unit Cost	Avg Equip Unit Cost	Avg Total Unit Cost	Avg Price Incl O&P	Avg Total Unit Cost	Avg Price Incl O&P
Plain siding											
Brush (300 SF/gal)	Inst	SF	NA	.014	.10	.45	---	.55	.78	.41	.5
Roller (275 SF/gal)	Inst	SF	NA	.011	.11	.36	---	.47	.65	.36	.4
Spray (325 SF/gal)	Inst	SF	NC	.007	.09	.23	---	.32	.44	.25	.3
Shingle siding											
Brush (270 SF/gal)	Inst	SF	NA	.013	.11	.42	---	.53	.74	.39	.5
Roller (260 SF/gal)	Inst	SF	NA	.010	.11	.32	---	.43	.60	.33	.4
Spray (300 SF/gal)	Inst	SF	NC	.007	.10	.23	---	.33	.46	.26	.3
Stucco											
Brush (135 SF/gal)	Inst	SF	NA	.014	.22	.45	---	.67	.91	.52	.7
Roller (125 SF/gal)	Inst	SF	NA	.010	.24	.32	---	.56	.74	.44	.5
Spray (150 SF/gal)	Inst	SF	NC	.007	.20	.23	---	.43	.57	.35	.4

Cement walls, see Cement base paint, page 209

Description	Oper	Unit	Crew Size	Man-Hours Per Unit	Avg Mat'l Unit Cost	Avg Labor Unit Cost	Avg Equip Unit Cost	Avg Total Unit Cost	Avg Price Incl O&P	Avg Total Unit Cost	Avg Price Incl O&P
Masonry block, brick, tile; masonry latex											
Brush (180 SF/gal)	Inst	SF	NA	.013	.17	.42	---	.59	.81	.44	.6
Roller (125 SF/gal)	Inst	SF	NA	.008	.24	.26	---	.50	.65	.40	.5
Spray (160 SF/gal)	Inst	SF	NC	.006	.19	.20	---	.39	.50	.30	.3
Doors, exterior side only											
Brush (375 SF/gal)	Inst	SF	NA	.018	.08	.58	---	.66	.95	.49	.7
Roller (375 SF/gal)	Inst	SF	NA	.013	.08	.42	---	.50	.71	.36	.5

Description	Oper	Unit	Crew Size	Man-Hours Per Unit	Avg Mat'l Unit Cost	Avg Labor Unit Cost	Avg Equip Unit Cost	Avg Total Unit Cost	Avg Price Incl O&P	Avg Total Unit Cost	Avg Price Incl O&P
								Costs Based On Small Volume		**Large Volume**	
Windows, exterior side only, brush work											
(450 SF/gal)	Inst	SF	NA	.018	.02	.58	---	.60	.88	.46	.67
Trim, less than 6" high, brush											
Note: SF equals LF on trim less than 6" high											
High gloss (300 SF/gal)	Inst	SF	NA	.013	.10	.42	---	.52	.73	.38	.53
Screens, full; high gloss											
Paint applied to wood only, brush work											
(700 SF/gal)	Inst	SF	NA	.021	.04	.68	---	.72	1.05	.53	.76
Paint applied to wood (brush) and wire (spray)											
(475 SF/gal)	Inst	SF	NC	.025	.06	.83	---	.89	1.30	.66	.95
Storm windows and doors, 2 lites, brush work											
(340 SF/gal)	Inst	SF	NA	.034	.09	1.10	---	1.19	1.73	.86	1.24
Blinds or shutters											
Brush (120 SF/gal)	Inst	SF	NA	.088	.25	2.85	---	3.10	4.49	2.23	3.21
Spray (300 SF/gal)	Inst	SF	NC	.029	.10	.97	---	1.07	1.54	.76	1.09
Gutters and downspouts, brush work											
(225 LF/gal), galvanized	Inst	SF	NA	.019	.13	.62	---	.75	1.05	.54	.75
Porch floors and steps, wood											
Brush (340 SF/gal)	Inst	SF	NA	.008	.09	.26	---	.35	.48	.24	.33
Roller (325 SF/gal)	Inst	SF	NA	.006	.09	.19	---	.28	.39	.21	.28
Shingle roofs											
Brush (135 SF/gal)	Inst	SF	NA	.011	.22	.36	---	.58	.77	.46	.60
Roller (125 SF/gal)	Inst	SF	NA	.008	.24	.26	---	.50	.65	.40	.52
Spray (150 SF/gal)	Inst	SF	NC	.005	.20	.17	---	.37	.47	.31	.40
Note: The following percentage adjustment for Small Volume also applies to Large											
For custom colors, ADD	Inst	%	---	---	10.0	---	---	---	---	---	---
Cement base paint (epoxy concrete enamel)											
Cement walls, smooth finish											
Brush (120 SF/gal)	Inst	SF	NA	.009	.34	.29	---	.63	.81	.49	.62
Roller (110 SF/gal)	Inst	SF	NA	.006	.37	.19	---	.56	.69	.46	.55
Concrete porch floors and steps											
Brush (400 SF/gal)	Inst	SF	NA	.006	.10	.19	---	.29	.40	.22	.29
Roller (375 SF/gal)	Inst	SF	NA	.005	.11	.16	---	.27	.36	.20	.25
Stain											
Shingle siding											
Brush (180 SF/gal)	Inst	SF	NA	.014	.18	.45	---	.63	.87	.48	.66
Roller (170 SF/gal)	Inst	SF	NA	.010	.19	.32	---	.51	.69	.40	.52
Spray (200 SF/gal)	Inst	SF	NC	.008	.16	.27	---	.43	.57	.31	.40
Shingle roofs											
Brush (180 SF/gal)	Inst	SF	NA	.012	.18	.39	---	.57	.77	.42	.56
Roller (170 SF/gal)	Inst	SF	NA	.009	.19	.29	---	.48	.64	.36	.47
Spray (200 SF/gal)	Inst	SF	NC	.006	.16	.20	---	.36	.47	.27	.35

Description	Oper	Unit	Costs Based On Small Volume						Large Volume	
			Crew Size	Man-Hours Per Unit	Avg Mat'l Unit Cost	Avg Labor Unit Cost	Avg Equip Unit Cost	Avg Total Unit Cost	Avg Price Incl O&P	Avg Total Unit Cost	Avg Price Incl O&P

Two coat application

Excluding openings, unless otherwise indicated

Latex, flat (unless otherwise indicated)

Description	Oper	Unit	Crew Size	Man-Hrs/Unit	Avg Mat'l	Avg Labor	Avg Equip	Avg Total	Avg Price O&P	Lg Avg Total	Lg Avg Price O&P
Plain siding											
Brush (170 SF/gal)	Inst	SF	NA	.026	.17	.84	---	1.01	**1.43**	.74	1.04
Roller (155 SF/gal)	Inst	SF	NA	.020	.19	.65	---	.84	**1.17**	.62	.86
Spray (185 SF/gal)	Inst	SF	NC	.012	.16	.40	---	.56	**.77**	.41	.55
Shingle siding											
Brush (150 SF/gal)	Inst	SF	NA	.024	.20	.78	---	.98	**1.37**	.73	1.01
Roller (150 SF/gal)	Inst	SF	NA	.018	.20	.58	---	.78	**1.08**	.60	.82
Spray (170 SF/gal)	Inst	SF	NC	.013	.17	.43	---	.60	**.83**	.46	.62
Stucco											
Brush (90 SF/gal)	Inst	SF	NA	.024	.33	.78	---	1.11	**1.51**	.84	1.13
Roller (85 SF/gal)	Inst	SF	NA	.017	.35	.55	---	.90	**1.20**	.70	.92
Spray (100 SF/gal)	Inst	SF	NC	.013	.30	.43	---	.73	**.97**	.56	.73

Cement wall, see Cement base paint, page 211

Description	Oper	Unit	Crew Size	Man-Hrs/Unit	Avg Mat'l	Avg Labor	Avg Equip	Avg Total	Avg Price O&P	Lg Avg Total	Lg Avg Price O&P
Masonry block, brick, tile; masonry latex											
Brush (120 SF/gal)	Inst	SF	NA	.022	.25	.71	---	.96	**1.33**	.71	.96
Roller (85 SF/gal)	Inst	SF	NA	.015	.35	.49	---	.84	**1.10**	.67	.87
Spray (105 SF/gal)	Inst	SF	NC	.011	.28	.37	---	.65	**.85**	.52	.67
Doors, exterior side only											
Brush (215 SF/gal)	Inst	SF	NA	.034	.14	1.10	---	1.24	**1.78**	.90	1.28
Roller (215 SF/gal)	Inst	SF	NA	.025	.14	.81	---	.95	**1.35**	.70	.99
Windows, exterior side only, brush work											
(255 SF/gal)	Inst	SF	NA	.034	.12	1.10	---	1.22	**1.76**	.88	1.26
Trim, less than 6" high, brush											
Note: SF equals LF on trim less than 6" high											
High gloss (230 SF/gal)	Inst	SF	NA	.024	.13	.78	---	.91	**1.29**	.66	.94
Screens, full; high gloss											
Paint applied to wood only, brush work											
(400 SF/gal)	Inst	SF	NA	.040	.07	1.30	---	1.37	**1.99**	.98	1.42
Paint applied to wood (brush) and wire (spray)											
(270 SF/gal)	Inst	SF	NC	.048	.11	1.60	---	1.71	**2.49**	1.20	1.74
Blinds or shutters											
Brush (65 SF/gal)	Inst	SF	NA	.163	.46	5.28	---	5.74	**8.32**	4.10	5.91
Spray (170 SF/gal)	Inst	SF	NC	.054	.17	1.80	---	1.97	**2.85**	1.43	2.05
Gutters and downspouts, brush work											
(130 LF/gal), galvanized	Inst	SF	NA	.036	.23	1.17	---	1.40	**1.98**	1.01	1.42
Porch floors and steps, wood											
Brush (195 SF/gal)	Inst	SF	NA	.014	.15	.45	---	.60	**.84**	.46	.63
Roller (185 SF/gal)	Inst	SF	NA	.011	.16	.36	---	.52	**.70**	.40	.54
Shingle roofs											
Brush (75 SF/gal)	Inst	SF	NA	.018	.40	.58	---	.98	**1.30**	.77	1.01
Roller (70 SF/gal)	Inst	SF	NA	.014	.42	.45	---	.87	**1.13**	.70	.90
Spray (85 SF/gal)	Inst	SF	NC	.011	.35	.37	---	.72	**.93**	.58	.74
Note: The following percentage adjustment for Small Volume also applies to Large											
For custom colors, ADD	Inst	%	---	---	10.0	---	---	---	---	---	---

Description	Oper	Unit	Crew Size	Man-Hours Per Unit	Avg Mat'l Unit Cost	Avg Labor Unit Cost	Avg Equip Unit Cost	Avg Total Unit Cost	Avg Price Incl O&P	Avg Total Unit Cost	Avg Price Incl O&P
								Costs Based On Small Volume		Large Volume	
Cement base paint (epoxy concrete enamel)											
Cement walls, smooth finish											
Brush (80 SF/gal)	Inst	SF	NA	.016	.51	.52	---	1.03	**1.33**	.81	**1.02**
Roller (75 SF/gal)	Inst	SF	NA	.011	.54	.36	---	.90	**1.12**	.74	**.91**
Concrete porch floors and steps											
Brush (225 SF/gal)	Inst	SF	NA	.011	.18	.36	---	.54	**.73**	.42	**.56**
Roller (210 SF/gal)	Inst	SF	NA	.011	.19	.36	---	.55	**.74**	.43	**.57**
Stain											
Shingle siding											
Brush (105 SF/gal)	Inst	SF	NA	.024	.31	.78	---	1.09	**1.49**	.82	**1.11**
Roller (100 SF/gal)	Inst	SF	NA	.019	.32	.62	---	.94	**1.26**	.71	**.94**
Spray (115 SF/gal)	Inst	SF	NC	.013	.28	.43	---	.71	**.95**	.55	**.72**
Shingle roofs											
Brush (105 SF/gal)	Inst	SF	NA	.020	.31	.65	---	.96	**1.30**	.72	**.97**
Roller (100 SF/gal)	Inst	SF	NA	.015	.32	.49	---	.81	**1.07**	.65	**.85**
Spray (115 SF/gal)	Inst	SF	NC	.010	.28	.33	---	.61	**.80**	.48	**.62**

Paneling

Hardboard and plywood

Description	Oper	Unit	Crew Size	Man-Hours Per Unit	Avg Mat'l Unit Cost	Avg Labor Unit Cost	Avg Equip Unit Cost	Avg Total Unit Cost	Avg Price Incl O&P	Avg Total Unit Cost	Avg Price Incl O&P
								Costs Based On Small Volume		**Large Volume**	

Demolition

Description	Oper	Unit	Crew Size	Man-Hours Per Unit	Avg Mat'l Unit Cost	Avg Labor Unit Cost	Avg Equip Unit Cost	Avg Total Unit Cost	Avg Price Incl O&P	Avg Total Unit Cost	Avg Price Incl O&P
Sheets, plywood or hardboard	Demo	SF	LB	.013	---	.35	---	.35	**.52**	.24	**.36**
Boards, wood	Demo	SF	LB	.015	---	.40	---	.40	**.60**	.27	**.40**

Installation

Waste, nails, and adhesives not included

Economy hardboard

Presdwood, 4' x 8' sheets
Standard

Description	Oper	Unit	Crew Size	Man-Hours Per Unit	Avg Mat'l Unit Cost	Avg Labor Unit Cost	Avg Equip Unit Cost	Avg Total Unit Cost	Avg Price Incl O&P	Avg Total Unit Cost	Avg Price Incl O&P
1/8" T	Inst	SF	2C	.024	.26	.77	---	1.03	**1.45**	.73	**1.02**
1/4" T	Inst	SF	2C	.024	.40	.77	---	1.17	**1.62**	.86	**1.17**
Tempered											
1/8" T	Inst	SF	2C	.024	.45	.77	---	1.22	**1.67**	.90	**1.22**
1/4" T	Inst	SF	2C	.024	.72	.77	---	1.49	**1.98**	1.13	**1.48**
Duolux, 4' x 8' sheets											
Standard											
1/8" T	Inst	SF	2C	.024	.28	.77	---	1.05	**1.48**	.74	**1.03**
1/4" T	Inst	SF	2C	.024	.41	.77	---	1.18	**1.63**	.87	**1.18**
Tempered											
1/8" T	Inst	SF	2C	.024	.49	.77	---	1.26	**1.72**	.93	**1.25**
1/4" T	Inst	SF	2C	.024	.55	.77	---	1.32	**1.79**	.98	**1.31**

Particleboard, 40 lb interior underlayment

Nailed to floors

Description	Oper	Unit	Crew Size	Man-Hours Per Unit	Avg Mat'l Unit Cost	Avg Labor Unit Cost	Avg Equip Unit Cost	Avg Total Unit Cost	Avg Price Incl O&P	Avg Total Unit Cost	Avg Price Incl O&P
3/8" T	Inst	SF	2C	.016	.49	.51	---	1.00	**1.33**	.74	**.96**
1/2" T	Inst	SF	2C	.016	.57	.51	---	1.08	**1.43**	.82	**1.06**
5/8" T	Inst	SF	2C	.016	.64	.51	---	1.15	**1.51**	.88	**1.13**
3/4" T	Inst	SF	2C	.016	.74	.51	---	1.25	**1.62**	.96	**1.22**
Nailed to walls											
3/8" T	Inst	SF	2C	.017	.49	.55	---	1.04	**1.38**	.77	**1.01**
1/2" T	Inst	SF	2C	.017	.57	.55	---	1.12	**1.47**	.85	**1.10**
5/8" T	Inst	SF	2C	.017	.64	.55	---	1.19	**1.55**	.91	**1.17**
3/4" T	Inst	SF	2C	.017	.74	.55	---	1.29	**1.67**	.99	**1.27**

Masonite prefinished 4' x 8' panels

Description	Oper	Unit	Crew Size	Man-Hours Per Unit	Avg Mat'l Unit Cost	Avg Labor Unit Cost	Avg Equip Unit Cost	Avg Total Unit Cost	Avg Price Incl O&P	Avg Total Unit Cost	Avg Price Incl O&P
1/4" T, oak and maple designs	Inst	SF	2C	.031	.66	.99	---	1.65	**2.25**	1.21	**1.62**
1/4" T, nutwood designs	Inst	SF	2C	.031	.77	.99	---	1.76	**2.38**	1.30	**1.72**
1/4" T, weathered white	Inst	SF	2C	.031	.77	.99	---	1.76	**2.38**	1.30	**1.72**
1/4" T, brick or stone designs	Inst	SF	2C	.031	1.08	.99	---	2.07	**2.73**	1.57	**2.03**

Pegboard, 4' x 8' sheets

Presdwood, tempered

Description	Oper	Unit	Crew Size	Man-Hours Per Unit	Avg Mat'l Unit Cost	Avg Labor Unit Cost	Avg Equip Unit Cost	Avg Total Unit Cost	Avg Price Incl O&P	Avg Total Unit Cost	Avg Price Incl O&P
1/8" T	Inst	SF	2C	.024	.57	.77	---	1.34	**1.81**	1.01	**1.35**
1/4" T	Inst	SF	2C	.024	.76	.77	---	1.53	**2.03**	1.17	**1.53**
Duolux, tempered											
1/8" T	Inst	SF	2C	.024	.83	.77	---	1.60	**2.11**	1.24	**1.61**
1/4" T	Inst	SF	2C	.024	1.21	.77	---	1.98	**2.55**	1.56	**1.98**

Description	Oper	Unit	Crew Size	Man-Hours Per Unit	Costs Based On Small Volume						Large Volume	
					Avg Mat'l Unit Cost	Avg Labor Unit Cost	Avg Equip Unit Cost	Avg Total Unit Cost	Avg Price Incl O&P		Avg Total Unit Cost	Avg Price Incl O&P

Unfinished hardwood plywood, applied with nails (nail heads filled)

Ash-sen, flush face

Description	Oper	Unit	Crew Size	MHr	Mat'l	Labor	Equip	Total	Price O&P	LV Total	LV Price
1/8" x 4' x 7', 8'	Inst	SF	CS	.041	1.26	1.24	---	2.50	3.31	1.92	2.49
3/16" x 4' x 8'	Inst	SF	CS	.041	1.53	1.24	---	2.77	3.62	2.15	2.76
1/4" x 4' x 8'	Inst	SF	CS	.041	1.79	1.24	---	3.03	3.92	2.37	3.01
1/4" x 4' x 10'	Inst	SF	CS	.041	2.01	1.24	---	3.25	4.17	2.57	3.24
1/2" x 4' x 10'	Inst	SF	CS	.041	2.28	1.24	---	3.52	4.48	2.81	3.51
3/4" x 4' x 8' vertical core	Inst	SF	CS	.041	2.50	1.24	---	3.74	4.74	3.00	3.73
3/4" x 4' x 8' lumber core	Inst	SF	CS	.041	3.55	1.24	---	4.79	5.94	3.91	4.78

Ash, V-grooved

Description	Oper	Unit	Crew Size	MHr	Mat'l	Labor	Equip	Total	Price O&P	LV Total	LV Price
3/16" x 4' x 8'	Inst	SF	CS	.041	1.38	1.24	---	2.62	3.45	2.02	2.61
1/4" x 4' x 8'	Inst	SF	CS	.041	1.69	1.24	---	2.93	3.81	2.29	2.92
1/4" x 4' x 10'	Inst	SF	CS	.041	1.96	1.24	---	3.20	4.12	2.52	3.18

Birch, natural, "A" grade face

Flush face

Description	Oper	Unit	Crew Size	MHr	Mat'l	Labor	Equip	Total	Price O&P	LV Total	LV Price
1/8" x 4' x 8'	Inst	SF	CS	.041	.98	1.24	---	2.22	2.99	1.67	2.20
3/16" x 4' x 8'	Inst	SF	CS	.041	1.53	1.24	---	2.77	3.62	2.15	2.76
1/4" x 4' x 8'	Inst	SF	CS	.041	1.54	1.24	---	2.78	3.63	2.16	2.77
1/4" x 4' x 10'	Inst	SF	CS	.041	2.01	1.24	---	3.25	4.17	2.57	3.24
3/8" x 4' x 8'	Inst	SF	CS	.041	2.22	1.24	---	3.46	4.41	2.74	3.43
1/2" x 4' x 8'	Inst	SF	CS	.041	2.56	1.24	---	3.80	4.81	3.05	3.79
3/4" x 4' x 8' lumber core	Inst	SF	CS	.041	3.50	1.24	---	4.74	5.89	3.86	4.72

V-grooved

Description	Oper	Unit	Crew Size	MHr	Mat'l	Labor	Equip	Total	Price O&P	LV Total	LV Price
1/4" x 4' x 8', mismatched	Inst	SF	CS	.041	1.92	1.24	---	3.16	4.07	2.48	3.14
1/4" x 4' x 8'	Inst	SF	CS	.041	1.93	1.24	---	3.17	4.08	2.49	3.15
1/4" x 4' x 10'	Inst	SF	CS	.041	2.19	1.24	---	3.43	4.38	2.72	3.41

Birch, select red

Description	Oper	Unit	Crew Size	MHr	Mat'l	Labor	Equip	Total	Price O&P	LV Total	LV Price
1/4" x 4' x 8'	Inst	SF	CS	.041	1.93	1.24	---	3.17	4.08	2.49	3.15

Birch, select white

Description	Oper	Unit	Crew Size	MHr	Mat'l	Labor	Equip	Total	Price O&P	LV Total	LV Price
1/4" x 4' x 8'	Inst	SF	CS	.041	1.93	1.24	---	3.17	4.08	2.49	3.15

Oak, flush face

Description	Oper	Unit	Crew Size	MHr	Mat'l	Labor	Equip	Total	Price O&P	LV Total	LV Price
1/8" x 4' x 8'	Inst	SF	CS	.041	.96	1.24	---	2.20	2.97	1.66	2.19
1/4" x 4' x 8'	Inst	SF	CS	.041	1.22	1.24	---	2.46	3.26	1.88	2.45
1/2" x 4' x 8'	Inst	SF	CS	.041	2.38	1.24	---	3.62	4.60	2.88	3.60

Philippine mahogany

Rotary cut

Description	Oper	Unit	Crew Size	MHr	Mat'l	Labor	Equip	Total	Price O&P	LV Total	LV Price
1/8" x 4' x 8'	Inst	SF	CS	.041	.46	1.24	---	1.70	2.39	1.22	1.69
3/16" x 4' x 8'	Inst	SF	CS	.041	.68	1.24	---	1.92	2.64	1.41	1.90
1/4" x 4' x 8'	Inst	SF	CS	.041	.68	1.24	---	1.92	2.64	1.41	1.90
1/4" x 4' x 10'	Inst	SF	CS	.041	1.18	1.24	---	2.42	3.22	1.84	2.40
1/2" x 4' x 8'	Inst	SF	CS	.041	1.37	1.24	---	2.61	3.44	2.01	2.59
3/4" x 4' x 8' vert. core	Inst	SF	CS	.041	2.08	1.24	---	3.32	4.25	2.63	3.31

V-grooved

Description	Oper	Unit	Crew Size	MHr	Mat'l	Labor	Equip	Total	Price O&P	LV Total	LV Price
3/16" x 4' x 8'	Inst	SF	CS	.041	.72	1.24	---	1.96	2.69	1.44	1.94
1/4" x 4' x 8'	Inst	SF	CS	.041	.80	1.24	---	2.04	2.78	1.52	2.03
1/4" x 4' x 10'	Inst	SF	CS	.041	1.25	1.24	---	2.49	3.30	1.90	2.47

Plaster & Stucco

1. Dimensions (Lath)

a. Gypsum. Plain, perforated and insulating. Normally, each lath is 16" x 48" in $3/8$" or $1/2$" thicknesses; a five-piece bundle covers 3 SY.

b. Wire. Only diamond and riblash are discussed here. Diamond lath is furnished in 27" x 96"-wide sheets covering 16 SY and 20 SY respectively.

c. Wood. Can be fir, pine, redwood, spruce, etc. Bundles may consist of 50 or 100 pieces of $3/8$" x $1^1/2$" x 48" lath covering 3.4 SY and 6.8 SY respectively. There is usually a $3/8$" gap between lath.

2. Dimensions (Plaster)

Only two and three coat gypsum cement plaster are discussed here.

3. Installation (Lath)

Laths are nailed. The types and size of nails vary with the type and thickness of lath. Quantity will vary with oc spacing of studs or joists. Only lath applied to wood will be considered here.

a. Nails for gypsum lath. Common type is 13 gauge blued $19/64$" flathead, $1^1/8$" long, spaced approximately 4" oc, and $1^1/4$" long spaced approximately 5" oc for $3/8$" and $1/2$" lath respectively.

b. Nails for wire lath. For ceiling, common is $1^1/2$" long 11 gauge barbed galvanized with a $7/16$" head diameter.

c. Nails for wood lath. 3d fine common.

d. Gypsum lath may be attached by the use of air-driven staples. Wire lath may be tied to support, usually with 18 gauge wire.

4. Installation (Plaster)

Quantities of materials used to plaster vary with the type of lath and thickness of plaster.

a. Two coat. Brown and finish coat.

b. Three coat. Scratch, brown and finish coat.

For types and quantities of material used, see **Notes on Material Pricing.**

5. Notes on Labor

a. Lath. Average Manhour per Unit is based on what one lather can do in one day.

b. Plaster. Average Manhour per Unit is based on what two plasterers and one laborer can do in one day.

Stucco

1. Dimensions

a. 18 gauge wire

b. 15 lb. felt paper

c. 1" x 18 gauge galvanized netting

d. Mortar of 1:3 mix

2. Installation (Lathing)

a. 18 gauge wire stretched taut horizontally across studs at approximately 8" oc.

b. 15 lb. felt paper placed over wire.

c. 1" x 18 gauge galvanized netting placed over felt.

3. Installation (Mortar)

a. The mortar mix used in this section is a 1:3 mix.

b. One CY of mortar is comprised of 1 CY sand, 9 CF portland cement and 100 lbs. hydrated lime.

c. For mortar requirements for 100 SY of stucco, see table below:

4. Estimating Technique.
Determine area and deduct area of window and door openings. No waste has been included in the following figures unless otherwise noted. For waste, add 10% to total area.

Stucco Thickness	Cubic Yards Per CSY	
	On Masonry	On Netting
1/2"	1.50	1.75
5/8"	1.90	2.20
3/4"	2.20	2.60
1"	2.90	3.40

Plaster and stucco

Description	Oper	Unit	Crew Size	Man-Hours Per Unit	Avg Mat'l Unit Cost	Avg Labor Unit Cost	Avg Equip Unit Cost	Avg Total Unit Cost	Avg Price Incl O&P	Avg Total Unit Cost	Avg Price Incl O&P
								Costs Based On Small Volume		**Large Volume**	

Remove both plaster or stucco and lath or netting to studs or sheathing

Lath (wood or metal) and plaster, walls and ceilings

Description	Oper	Unit	Crew	Man-Hrs	Mat'l	Labor	Equip	Total	Price O&P	L Total	L Price O&P
2 coats	Demo	SY	LB	.188	---	5.01	---	5.01	**7.46**	3.28	**4.88**
3 coats	Demo	SY	LB	.205	---	5.46	---	5.46	**8.14**	3.54	**5.28**

Lath (only), nails included

Gypsum lath, 16" x 48", applied with nails to ceilings or walls, 5% waste included, nails included (.067 lbs per CY)

Description	Oper	Unit	Crew	Man-Hrs	Mat'l	Labor	Equip	Total	Price O&P	L Total	L Price O&P
3/8" T, perforated or plain	Inst	SY	LR	.154	2.50	4.80	---	7.30	**10.00**	5.50	**7.38**
1/2" T, perforated or plain	Inst	SY	LR	.154	2.73	4.80	---	7.53	**10.30**	5.72	**7.63**
3/8" T, insulating, aluminum foil back	Inst	SY	LR	.154	3.10	4.80	---	7.90	**10.70**	6.07	**8.04**
1/2" T, insulating, aluminum foil back	Inst	SY	LR	.154	3.32	4.80	---	8.12	**11.00**	6.28	**8.28**
For installation with staples, DEDUCT	Inst	SY	LR	-.038	---	1.18	---	1.18	**---**	.78	**---**

Metal lath, nailed to ceilings or walls, 5% waste and nails (0.067 lbs per SY) included

Diamond lath (junior mesh), 27" x 96" sheets, nailed to wood members @ 16" oc

Description	Oper	Unit	Crew	Man-Hrs	Mat'l	Labor	Equip	Total	Price O&P	L Total	L Price O&P
3.4 lb black painted	Inst	SY	LR	.103	3.54	3.21	---	6.75	**8.85**	5.46	**6.99**
3.4 lb galvanized	Inst	SY	LR	.103	3.84	3.21	---	7.05	**9.20**	5.75	**7.32**

Riblath, 3/8" high rib, 27" x 96" sheets, nailed to wood members @ 24" oc

Description	Oper	Unit	Crew	Man-Hrs	Mat'l	Labor	Equip	Total	Price O&P	L Total	L Price O&P
3.4 lb painted	Inst	SY	LR	.077	3.95	2.40	---	6.35	**8.12**	5.32	**6.65**
3.4 lb galvanized	Inst	SY	LR	.077	4.65	2.40	---	7.05	**8.92**	5.99	**7.42**

Wood lath, nailed to ceilings or walls, 5% waste and nails included, redwood, "A" grade and better

Description	Oper	Unit	Crew	Man-Hrs	Mat'l	Labor	Equip	Total	Price O&P	L Total	L Price O&P
3/8" x 1-1/2" x 48" @ 3/8" spacing	Inst	SY	LR	.154	5.92	4.80	---	10.72	**14.00**	8.75	**11.10**

Labor adjustments, lath

Description	Oper	Unit	Crew	Man-Hrs	Mat'l	Labor	Equip	Total	Price O&P	L Total	L Price O&P
For gypsum or wood lath above second floor ADD	Inst	SY	LR	.011	---	.34	---	.34	**.51**	.22	**.33**
For metal lath above second floor ADD	Inst	SY	LR	.026	---	.81	---	.81	**1.21**	.53	**.79**

				Man-Hours	Avg Mat'l	Avg Labor	Avg Equip	Avg Total	Avg Price	Avg Total	Avg Price
				Costs Based On Small Volume						Large Volume	
Description	Oper	Unit	Crew Size	Per Unit	Unit Cost	Unit Cost	Unit Cost	Unit Cost	Incl O&P	Unit Cost	Incl O&P

Plaster (only), applied to ceilings and walls, 10% waste included

Material price includes gypsum plaster, sand, hydrated lime and gauging plaster

Two coats gypsum plaster

Description	Oper	Unit	Crew Size	MH/Unit	Mat'l	Labor	Equip	Total	Price O&P	LV Total	LV Price O&P
On gypsum lath	Inst	SY	P3	.320	2.71	9.78	.48	12.97	**18.20**	9.24	**12.8**
On unit masonry, no lath	Inst	SY	P3	.329	2.93	10.10	.49	13.52	**18.80**	9.61	**13.2**

Three coats gypsum plaster

On gypsum lath	Inst	SY	P3	.444	2.91	13.60	.66	17.17	**24.20**	11.98	**16.7**
On unit masonry, no lath	Inst	SY	P3	.444	3.16	13.60	.66	17.42	**24.50**	12.22	**17.0**
On wire lath	Inst	SY	P3	.462	4.64	14.10	.69	19.43	**27.10**	13.97	**19.1**
On wood lath	Inst	SY	P3	.462	3.10	14.10	.69	17.89	**25.30**	12.52	**17.5**

Labor adjustments, plaster

| For plaster above second floor ADD | Inst | SY | P3 | .096 | --- | 2.93 | --- | 2.93 | **4.37** | 1.93 | **2.8** |
| Thin coat plaster over sheetrock (in lieu of taping) | Inst | SY | P3 | .296 | .79 | 9.05 | --- | 9.84 | **14.40** | 6.60 | **9.5** |

Stucco, exterior walls

Netting, galvanized, 1" x 18 ga x 48", with 18 ga wire and 15 lb felt	Inst	SY	LR	.205	3.23	6.39	---	9.62	**13.20**	7.22	**9.7**
Steel-Tex, 49" W x 11-1/2' L rolls, with felt backing	Inst	SY	LR	.154	5.16	4.80	---	9.96	**13.10**	8.04	**10.3**
1 coat work with float finish											
Over masonry	Inst	SY	P3	.320	.63	9.78	.48	10.89	**15.80**	7.30	**10.5**
2 coat work with float finish											
Over masonry	Inst	SY	P3	.615	1.84	18.80	.91	21.55	**31.00**	14.54	**20.8**
Over metal netting	Inst	SY	PE	.727	2.60	21.90	.65	25.15	**36.30**	17.09	**24.4**
3 coat work with float finish											
Over metal netting	Inst	SY	PE	.952	3.51	28.70	.85	33.06	**47.70**	22.48	**32.0**

Plumbing. See individual items

Description	Oper	Unit	Costs Based On Small Volume							Large Volume	
			Crew Size	Man-Hours Per Unit	Avg Mat'l Unit Cost	Avg Labor Unit Cost	Avg Equip Unit Cost	Avg Total Unit Cost	Avg Price Incl O&P	Avg Total Unit Cost	Avg Price Incl O&P

Range hoods

Metal finishes

Labor includes wiring and connection by electrician and installation by carpenter in stud-exposed structure only

Economy model, UL approved; mitered, welded construction; completely assembled and wired; includes fan, motor, washable aluminum filter, and light

Description	Oper	Unit	Crew Size	MHrs	Mat'l	Labor	Equip	Total	Price O&P	LV Total	LV Price O&P
24" wide											
3-1/4" x 10" duct, 160 CFM	Inst	Ea	ED	5.71	82.60	190.00	---	272.60	**372.00**	206.00	**278.00**
Non-ducted, 160 CFM	Inst	Ea	ED	5.71	89.70	190.00	---	279.70	**380.00**	212.20	**285.00**
30" wide											
7" round duct, 240 CFM	Inst	Ea	ED	5.71	82.60	190.00	---	272.60	**372.00**	206.00	**278.00**
3-1/4" x 10" duct, 160 CFM	Inst	Ea	ED	5.71	89.80	190.00	---	279.80	**380.00**	212.30	**285.00**
Non-ducted, 160 CFM	Inst	Ea	ED	5.71	101.00	190.00	---	291.00	**393.00**	222.40	**297.00**
36" wide											
7" round duct, 240 CFM	Inst	Ea	ED	5.71	94.90	190.00	---	284.90	**386.00**	216.90	**291.00**
3-1/4" x 10" duct, 160 CFM	Inst	Ea	ED	5.71	101.00	190.00	---	291.00	**393.00**	222.40	**297.00**
Non-ducted, 160 CFM	Inst	Ea	ED	5.71	120.00	190.00	---	310.00	**415.00**	239.00	**316.00**

Standard model, UL approved; deluxe mitered wrap-around styling; solid state fan control, infinite speed settings; removable filter; built-in damper and dual light assembly

Description	Oper	Unit	Crew Size	MHrs	Mat'l	Labor	Equip	Total	Price O&P	LV Total	LV Price O&P
30" wide x 9" deep											
3-1/4" x 10" duct	Inst	Ea	ED	5.71	161.00	190.00	---	351.00	**462.00**	275.00	**358.00**
Non-ducted	Inst	Ea	ED	5.71	276.00	190.00	---	466.00	**595.00**	377.00	**475.00**
36" wide x 9" deep											
3-1/4" x 10" duct	Inst	Ea	ED	5.71	183.00	190.00	---	373.00	**488.00**	295.00	**380.00**
Non-ducted	Inst	Ea	ED	5.71	196.00	190.00	---	386.00	**503.00**	306.00	**393.00**
42" wide x 9" deep											
3-1/4" x 10" duct	Inst	Ea	ED	7.62	158.00	253.00	---	411.00	**551.00**	317.00	**419.00**
Non-ducted	Inst	Ea	ED	7.62	173.00	253.00	---	426.00	**569.00**	330.00	**435.00**

Decorator/designer model; solid state fan control, infinite speed settings; easy clean aluminum mesh grease filters; enclosed light assembly and switches

Description	Oper	Unit	Crew Size	MHrs	Mat'l	Labor	Equip	Total	Price O&P	LV Total	LV Price O&P
Single faced hood - fan exhausts horizontally or vertically, 24" D canopy x 21" front to back											
30" wide	Inst	Ea	ED	7.62	317.00	253.00	---	570.00	**734.00**	457.00	**581.00**
36" wide	Inst	Ea	ED	7.62	341.00	253.00	---	594.00	**761.00**	478.00	**605.00**
Single faced, contemporary style, duct horizontally or vertically, 9" D canopy											
30" wide											
With 330 cfm power unit	Inst	Ea	ED	7.62	299.00	253.00	---	552.00	**714.00**	441.00	**563.00**
With 410 cfm power unit	Inst	Ea	ED	7.62	341.00	253.00	---	594.00	**761.00**	478.00	**605.00**
36" wide											
With 330 cfm power unit	Inst	Ea	ED	7.62	316.00	253.00	---	569.00	**733.00**	456.00	**580.00**
With 410 cfm power unit	Inst	Ea	ED	7.62	352.00	253.00	---	605.00	**775.00**	488.00	**617.00**
Material adjustments											
Note: The following percentage adjustment for Small Volume also applies to Large											
Stainless steel, ADD	Inst	%		---	---	33.0	---	---	---	---	---

Reinforcing steel. See Concrete, page 69

Resilient flooring

Sheet Products

Linoleum or vinyl sheet are normally installed either over a wood subfloor or a smooth concrete subfloor. When laid over wood, a layer of felt must first be laid in paste. This keeps irregularities in the wood from showing through. The amount of paste required to bond both felt and sheet is approximately 16 gallons (5% waste included) per 100 SY. When laid over smooth concrete subfloor, the sheet products can be bonded directly to the floor. However, the concrete subfloor should not be in direct contact with the ground because of excessive moisture. For bonding to concrete, 8 gallons of paste is required (5% waste included) per 100 SY. After laying the flooring over concrete or wood, wax is applied using 0.5 gallon per 100 SY. Paste, wax, felt (as needed), and 10% sheet waste are included in material costs.

Tile Products

All resilient tile can be bonded the same way as sheet products, either to smooth concrete or to felt over wood subfloor. When bonded to smooth concrete, a concrete primer (0.5 gal/100 SF) is first applied to seal the concrete. The tiles are then bonded to the sealed floor with resilient tile cement (0.6 gal/100 SF). On wood subfloors, felt is first laid in paste (0.9 gal/100 SF) and then the tiles are bonded to the felt with resilient tile emulsion (0.7 gal/100 SF). Bonding materials, felt (as needed), and 10% tile waste are included in material costs.

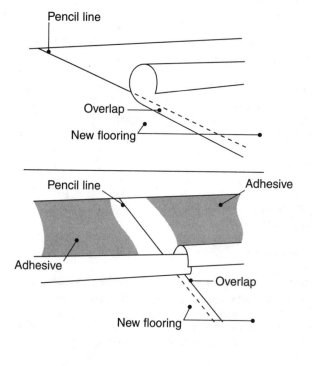

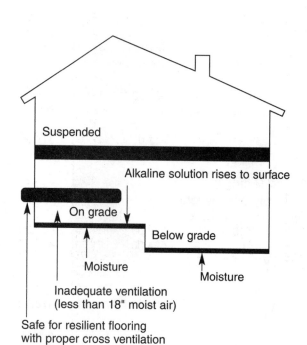

Description	Oper	Unit	Crew Size	Man-Hours Per Unit	Avg Mat'l Unit Cost	Avg Labor Unit Cost	Avg Equip Unit Cost	Avg Total Unit Cost	Avg Price Incl O&P	Avg Total Unit Cost	Avg Price Incl O&P
					Costs Based On Small Volume					Large Volume	

Resilient flooring

Description	Oper	Unit	Crew Size	Man-Hours Per Unit	Avg Mat'l Unit Cost	Avg Labor Unit Cost	Avg Equip Unit Cost	Avg Total Unit Cost	Avg Price Incl O&P	Avg Total Unit Cost	Avg Price Incl O&P
Adhesive set sheet products	Demo	SY	LB	.143	---	3.81	---	3.81	**5.68**	2.66	**3.97**
Adhesive set tile products	Demo	SF	LB	.015	---	.40	---	.40	**.60**	.29	**.44**

Install over smooth concrete subfloor

Tile

Description	Oper	Unit	Crew Size	Man-Hours Per Unit	Avg Mat'l Unit Cost	Avg Labor Unit Cost	Avg Equip Unit Cost	Avg Total Unit Cost	Avg Price Incl O&P	Avg Total Unit Cost	Avg Price Incl O&P
Asphalt, 9" x 9", 1/8" thick, marbleized											
Group B colors, dark	Inst	SF	FB	.040	1.55	1.17	---	2.72	**3.50**	2.21	**2.80**
Group C colors, medium	Inst	SF	FB	.040	1.76	1.17	---	2.93	**3.74**	2.40	**3.02**
Group D colors, light	Inst	SF	FB	.040	1.76	1.17	---	2.93	**3.74**	2.40	**3.02**
Vinyl tile, 12" x 12", including adhesive											
1/16" thick											
Vega II	Inst	SF	FB	.029	1.40	.85	---	2.25	**2.86**	1.84	**2.31**
.080" thick											
Designer slate	Inst	SF	FB	.029	3.02	.85	---	3.87	**4.72**	3.29	**3.98**
1/8" thick											
Embassy oak	Inst	SF	FB	.029	3.72	.85	---	4.57	**5.52**	3.93	**4.71**
Majestic slate	Inst	SF	FB	.029	3.02	.85	---	3.87	**4.72**	3.29	**3.98**
Mediterranean marble	Inst	SF	FB	.029	3.30	.85	---	4.15	**5.04**	3.55	**4.27**
Pecan	Inst	SF	FB	.029	3.86	.85	---	4.71	**5.68**	4.05	**4.85**
Plaza brick	Inst	SF	FB	.029	5.82	.85	---	6.67	**7.94**	5.82	**6.89**
Plymouth plank	Inst	SF	FB	.029	4.21	.85	---	5.06	**6.09**	4.37	**5.22**
Teak	Inst	SF	FB	.029	3.58	.85	---	4.43	**5.36**	3.80	**4.56**
Terrazzo	Inst	SF	FB	.029	2.87	.85	---	3.72	**4.55**	3.17	**3.84**
Vinyl tile, 12" x 12", self-stick											
.080" thick"											
Elite	Inst	SF	FB	.035	2.25	1.02	---	3.27	**4.09**	2.75	**3.40**
Stylglo	Inst	SF	FB	.035	1.69	1.02	---	2.71	**3.45**	2.25	**2.82**
1/16" thick											
Decorator	Inst	SF	FB	.035	1.12	1.02	---	2.14	**2.79**	1.74	**2.24**
Proclaim	Inst	SF	FB	.035	1.47	1.02	---	2.49	**3.19**	2.05	**2.59**
Composition vinyl tile, 12" x 12"											
3/32" thick"											
Designer colors	Inst	SF	FB	.035	1.26	1.02	---	2.28	**2.95**	1.86	**2.37**
Standard colors (marbleized)	Inst	SF	FB	.035	1.61	1.02	---	2.63	**3.35**	2.18	**2.74**
1/8" thick											
Designer colors, pebbled	Inst	SF	FB	.035	2.39	1.02	---	3.41	**4.25**	2.88	**3.55**
Standard colors (marbleized)	Inst	SF	FB	.035	1.40	1.02	---	2.42	**3.11**	1.99	**2.52**
Solid black or white	Inst	SF	FB	.035	2.25	1.02	---	3.27	**4.09**	2.75	**3.40**
Solid colors	Inst	SF	FB	.035	2.46	1.02	---	3.48	**4.33**	2.94	**3.62**

Description	Oper	Unit	Crew Size	Man-Hours Per Unit	Avg Mat'l Unit Cost	Avg Labor Unit Cost	Avg Equip Unit Cost	Avg Total Unit Cost	Avg Price Incl O&P	Avg Total Unit Cost	Avg Price Incl O&P
									Costs Based On Small Volume ← →	**Large Volume**	

Description	Oper	Unit	Crew Size	Man-Hours Per Unit	Avg Mat'l Unit Cost	Avg Labor Unit Cost	Avg Equip Unit Cost	Avg Total Unit Cost	Avg Price Incl O&P	Avg Total Unit Cost	Avg Price Incl O&P
Sheet vinyl											
Armstrong, no-wax, 6' wide											
Designer Solarian (.070")	Inst	SY	FB	.367	39.30	10.70	---	50.00	**61.00**	42.91	**51.7**
Designer Solarian II (.090")	Inst	SY	FB	.367	50.50	10.70	---	61.20	**73.90**	53.01	**63.4**
Imperial Accotone (.065")	Inst	SY	FB	.367	12.60	10.70	---	23.30	**30.30**	18.91	**24.1**
Solarian Supreme (.090")	Inst	SY	FB	.367	53.40	10.70	---	64.10	**77.10**	55.51	**66.3**
Sundial Solarian (.077")	Inst	SY	FB	.367	19.70	10.70	---	30.40	**38.40**	25.21	**31.4**
Mannington, 6' wide											
Vega (.080")	Inst	SY	FB	.367	11.20	10.70	---	21.90	**28.70**	17.61	**22.7**
Vinyl Ease (.073")	Inst	SY	FB	.367	9.83	10.70	---	20.53	**27.10**	16.36	**21.2**
Tarkett, 6' wide											
Preference (.065")	Inst	SY	FB	.367	9.13	10.70	---	19.83	**26.30**	15.72	**20.5**
Softred (.062")	Inst	SY	FB	.367	11.20	10.70	---	21.90	**28.70**	17.61	**22.7**

Install over wood subfloor

Tile

Description	Oper	Unit	Crew Size	Man-Hours Per Unit	Avg Mat'l Unit Cost	Avg Labor Unit Cost	Avg Equip Unit Cost	Avg Total Unit Cost	Avg Price Incl O&P	Avg Total Unit Cost	Avg Price Incl O&P
Asphalt, 9" x 9", 1/8" thick, marbleized											
Group B colors, dark	Inst	SF	FB	.046	1.55	1.34	---	2.89	**3.76**	2.33	**2.9**
Group C colors, medium	Inst	SF	FB	.046	1.76	1.34	---	3.10	**4.00**	2.52	**3.1**
Group D colors, light	Inst	SF	FB	.046	1.76	1.34	---	3.10	**4.00**	2.52	**3.1**
Vinyl tile, 12" x 12", including adhesive											
1/16" thick											
Vega II	Inst	SF	FB	.034	1.40	.99	---	2.39	**3.07**	1.96	**2.48**
.080" thick											
Designer slate	Inst	SF	FB	.034	3.02	.99	---	4.01	**4.93**	3.41	**4.15**
1/8" thick											
Embassy oak	Inst	SF	FB	.034	3.72	.99	---	4.71	**5.74**	4.05	**4.88**
Majestic slate	Inst	SF	FB	.034	3.02	.99	---	4.01	**4.93**	3.41	**4.15**
Mediterranean marble	Inst	SF	FB	.034	3.30	.99	---	4.29	**5.26**	3.67	**4.4**
Pecan	Inst	SF	FB	.034	3.86	.99	---	4.85	**5.90**	4.17	**5.02**
Plaza brick	Inst	SF	FB	.034	5.82	.99	---	6.81	**8.15**	5.94	**7.06**
Plymouth plank	Inst	SF	FB	.034	4.21	.99	---	5.20	**6.30**	4.49	**5.3**
Teak	Inst	SF	FB	.034	3.58	.99	---	4.57	**5.58**	3.92	**4.73**
Terrazzo	Inst	SF	FB	.034	2.87	.99	---	3.86	**4.76**	3.29	**4.01**
Vinyl tile, 12" x 12", self-stick											
.080" thick											
Elite	Inst	SF	FB	.034	2.25	.99	---	3.24	**4.05**	2.72	**3.35**
Stylglo	Inst	SF	FB	.034	1.69	.99	---	2.68	**3.40**	2.22	**2.78**
1/16" thick											
Decorator	Inst	SF	FB	.034	1.12	.99	---	2.11	**2.75**	1.71	**2.19**
Proclaim	Inst	SF	FB	.034	1.47	.99	---	2.46	**3.15**	2.02	**2.55**
Composition vinyl tile, 12" x 12"											
3/32" thick											
Designer colors	Inst	SF	FB	.034	1.26	.99	---	2.25	**2.91**	1.83	**2.33**
Standard colors (marbleized)	Inst	SF	FB	.034	1.61	.99	---	2.60	**3.31**	2.15	**2.70**
1/8" thick											
Designer colors, pebbled	Inst	SF	FB	.034	2.39	.99	---	3.38	**4.21**	2.85	**3.50**
Standard colors (marbleized)	Inst	SF	FB	.034	1.40	.99	---	2.39	**3.07**	1.96	**2.48**
Solid black or white	Inst	SF	FB	.034	2.25	.99	---	3.24	**4.05**	2.72	**3.35**
Solid colors	Inst	SF	FB	.034	2.46	.99	---	3.45	**4.29**	2.91	**3.57**

Description	Oper	Unit	Crew Size	Man-Hours Per Unit	Avg Mat'l Unit Cost	Avg Labor Unit Cost	Avg Equip Unit Cost	Avg Total Unit Cost	Avg Price Incl O&P	Avg Total Unit Cost	Avg Price Incl O&P
						Costs Based On Small Volume				Large Volume	
Sheet vinyl											
Armstrong, no-wax, 6' wide											
Designer Solarian (.070")	Inst	SY	FB	.429	39.30	12.50	---	51.80	**63.60**	44.17	**53.60**
Designer Solarian II (.090")	Inst	SY	FB	.429	50.50	12.50	---	63.00	**76.60**	54.27	**65.20**
Imperial Accotone (.065")	Inst	SY	FB	.429	12.60	12.50	---	25.10	**33.00**	20.17	**26.00**
Solarian Supreme (.090")	Inst	SY	FB	.429	53.40	12.50	---	65.90	**79.80**	56.77	**68.10**
Sundial Solarian (.077")	Inst	SY	FB	.429	19.70	12.50	---	32.20	**41.00**	26.47	**33.20**
Mannington, 6' wide											
Vega (.080")	Inst	SY	FB	.429	11.20	12.50	---	23.70	**31.30**	18.87	**24.50**
Vinyl Ease (.073")	Inst	SY	FB	.429	9.83	12.50	---	22.33	**29.70**	17.62	**23.10**
Tarkett, 6' wide											
Preference (.065")	Inst	SY	FB	.429	9.13	12.50	---	21.63	**28.90**	16.98	**22.30**
Softred (.062")	Inst	SY	FB	.429	11.20	12.50	---	23.70	**31.30**	18.87	**24.50**
Related materials and operations											
Top set base											
Vinyl											
2-1/2" H											
Colors	Inst	LF	FB	.051	.53	1.49	---	2.02	**2.80**	1.53	**2.10**
Wood grain	Inst	LF	FB	.051	.72	1.49	---	2.21	**3.02**	1.69	**2.28**
4" H											
Colors	Inst	LF	FB	.051	.62	1.49	---	2.11	**2.90**	1.61	**2.19**
Wood grain	Inst	LF	FB	.051	.85	1.49	---	2.34	**3.17**	1.81	**2.42**
Linoleum cove, 7/16" x 1-1/4"											
Softwood	Inst	LF	FB	.073	.62	2.13	---	2.75	**3.85**	2.05	**2.83**

Retaining walls, see Concrete, page 70, or Masonry, page 181

Roofing

Fiberglass Shingles, 225 lb., three tab strip.
Three bundle/square; 20 year

1. **Dimensions.** Each shingle is 12" x 36". With a 5" exposure to the weather, 80 shingles are required to cover one square (100 SF).

2. **Installation**

   a. Over wood. After scatter-nailing one ply of 15 lb. felt, the shingles are installed. Four nails per shingle is customary.

   b. Over existing roofing. 15 lb. felt is not required. Shingles are installed the same as over wood, except longer nails are used, i.e., $1^{1}/_{4}$" in lieu of 1".

3. **Estimating Technique.** Determine roof area and add a percentage for starters, ridge, and valleys. The percent to be added varies, but generally:

   a. For plain gable and hip – add 10%.

   b. For gable or hip with dormers or intersecting roof(s) – add 15%.

   c. For gable or hip with dormers and intersecting roof(s) – add 20%.

When ridge or hip shingles are special ordered, reduce the above percentages by approximately 50%. Then apply unit costs to the area calculated (including the allowance above).

Mineral Surfaced Roll Roofing, 90 lb.

1. **Dimensions.** A roll is 36" wide x 36'-0" long, or 108 SF. It covers one square (100 SF), after including 8 SF for head and end laps.

2. **Installation.** Usually, lap cement and $^{7}/_{8}$" galvanized nails are furnished with each roll.

   a. Over wood. Roll roofing is usually applied directly to sheathing with $^{7}/_{8}$" galvanized nails. End and head laps are normally 6" and 2" or 3" respectively.

   b. Over existing roofing. Applied the same as over wood, except nails are not less than $1^{1}/_{4}$" long.

   c. Roll roofing may be installed by the exposed nail or the concealed nail method. In the exposed nail method, nails are visible at laps, and head laps are usually 2". In the concealed nail method, no nails are visible, the head lap is a minimum of 3", and there is a 9"-wide strip of roll installed along rakes and eaves.

3. **Estimating Technique.** Determine the area and add a percentage for hip and/or ridge. The percentage to be added varies, but generally:

   a. For plain gable or hip – add 5%.

   b. For cut-up roof – add 10%.

If metal ridge and/or hip are used, reduce the above percentages by approximately 50%. Then apply unit costs to the area calculated (including the allowance above).

Wood Shingles

1. **Dimensions.** Shingles are available in 16", 18", and 24" lengths and in uniform widths of 4", 5", or 6". But they are commonly furnished in random widths averaging 4" wide.

2. **Installation.** The normal exposure to weather for 16", 18" and 24" shingles on roofs with $^{1}/_{4}$ or steeper pitch is 5", $5^{1}/_{2}$", and $7^{1}/_{2}$" respectively. Where the slope is less than $^{1}/_{4}$, the exposure is usually reduced. Generally, 3d commons are used when shingles are applied to strip sheathing. 5d commons are used when shingles are applied over existing shingles. Two nails per shingle is the general rule, but on some roofs, one nail per shingle is used.

3. **Estimating Technique.** Determine the roof area and add a percentage for waste, starters, ridge, and hip shingles.

   a. Wood shingles. The amount of exposure determines the percentage of 1 square of shingles required to cover 100 SF of roof. Multiply the material cost per square of shingles by the appropriate percent (from the table on the next page) to determine the cost to cover 100 SF of roof with wood shingles. The table does not include cutting waste. NOTE: Nails and the exposure factors in this table have already been calculated into the Average Unit Material Costs.

   b. Nails. The weight of nails required per square varies with the size of the nail and with the shingle exposure. Multiply cost per pound of nails by the appropriate pounds per square. In the table on the next page, pounds of nails per square are based on two nails per shingle. For one nail per shingle, deduct 50%.

   c. The percentage to be added for starters, hip, and ridge shingles varies, but generally:

      1) Plain gable and hip – add 10%.

      2) Gable or hip with dormers or intersecting roof(s) – add 15%.

      3) Gable or hip with dormers and intersecting roof(s) – add 20%.

When ridge and/or hip shingles are special ordered, reduce these percentages by 50%. Then

Shingle Length	Exposure	% of Sq. Required to Cover 100 SF	Lbs of Nails per Square (2 Nails/Shingle)	
			3d	5d
24"	7½"	100	2.3	2.7
24"	7"	107	2.4	2.9
24"	6½"	115	2.6	3.2
24"	6"	125	2.8	3.4
24"	5¾"	130	2.9	3.6
18"	5½"	100	3.1	3.7
18"	5"	110	3.4	4.1
18"	4½"	122	3.8	4.6
16"	5"	100	3.4	4.1
16"	4½"	111	3.8	4.6
16"	4"	125	4.2	5.1
16"	3¾"	133	4.5	5.5

apply unit costs to the area calculated (including the allowance above).

Built-Up Roofing

1. **Dimensions**

   a. 15 lb. felt. Rolls are 36" wide x 144'-0" long or 432 SF – 4 squares per roll. 32 SF (or 8 SF/sq) is for laps.

   b. 30 lb. felt. Rolls are 36" wide x 72'-0" long or 216 SF – 2 squares per roll. 16 SF (or 8 SF/sq) is for laps.

   c. Asphalt. Available in 100 lb. cartons. Average asphalt usage is:

   1) Top coat without gravel – 35 lbs. per square.

   2) Top coat with gravel – 60 lbs. per square.

   3) Each coat (except top coat) – 25 lbs. per square.

   d. Tar. Available in 550 lb. kegs. Average tar usage is:

   1) Top coat without gravel – 40 lbs. per square.

   2) Top coat without gravel – 75 lbs. per square.

3) Each coat (except top coat) – 30 lbs. per square.

   e. Gravel or slag. Normally, 400 lbs. of gravel or 275 lbs. of slag are used to cover one square.

2. **Installation**

   a. Over wood. Normally, one or two plies of felt are applied by scatter nailing before hot mopping is commenced. Subsequent plies of felt and gravel or slag are imbedded in hot asphalt or tar.

   b. Over concrete. Every ply of felt and gravel or slag is imbedded in hot asphalt or tar.

3. **Estimating Technique**

   a. For buildings with parapet walls, use the outside dimensions of the building to determine the area. This area is usually sufficient to include the flashing.

   b. For buildings without parapet walls, determine the area.

   c. Don't deduct any opening less than 10'-0" x 10'-0" (100 SF). Deduct 50% of the opening for openings larger than 100 SF but smaller than 300 SF, and 100% of the openings exceeding 300 SF.

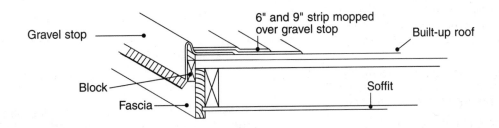

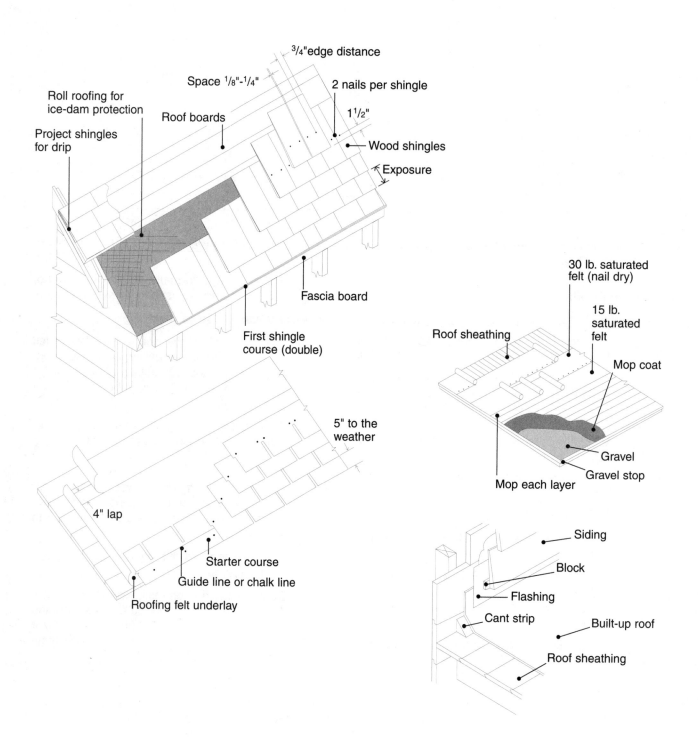

Roll roofing for ice-dam protection

Project shingles for drip

Roof boards

Space 1/8"-1/4"

3/4"edge distance

2 nails per shingle

1 1/2"

Wood shingles

Exposure

Fascia board

First shingle course (double)

5" to the weather

4" lap

Starter course

Guide line or chalk line

Roofing felt underlay

30 lb. saturated felt (nail dry)

15 lb. saturated felt

Mop coat

Roof sheathing

Gravel

Gravel stop

Mop each layer

Siding

Block

Flashing

Cant strip

Built-up roof

Roof sheathing

			Costs Based On Small Volume						Large Volume	
Description	Oper	Unit	Crew Size	Man-Hours Per Unit	Avg Mat'l Unit Cost	Avg Labor Unit Cost	Avg Equip Unit Cost	Avg Total Unit Cost	Avg Price Incl O&P	Avg Total Unit Cost	Avg Price Incl O&P

oofing

Aluminum, nailed to wood

Corrugated (2-1/2"), 26" W, with 3-3/4" side lap and

6" end lap	Demo	Sq	LB	2.00	---	53.30	---	53.30	**79.40**	42.60	**63.50**

Corrugated (2-1/2"), 26" W, with 3-3/4" side lap and 6" end lap

0.0175" thick

Natural	Inst	Sq	UC	2.67	154.00	83.20	---	237.20	**300.00**	212.40	**266.00**
Painted	Inst	Sq	UC	2.67	198.00	83.20	---	281.20	**351.00**	254.40	**315.00**

0.019" thick

Natural	Inst	Sq	UC	2.67	154.00	83.20	---	237.20	**300.00**	212.40	**266.00**
Painted	Inst	Sq	UC	2.67	198.00	83.20	---	281.20	**351.00**	254.40	**315.00**

Composition shingle roofing, 3 tab strip,
12" x 36", nailed, seal down

Demo Shingle, to deck

5" exposure	Demo	Sq	LB	1.25	---	33.30	---	33.30	**49.60**	26.60	**39.70**

Install 25-year shingles, 240 lb/Sq

Over existing roofing

Gable, plain	Inst	Sq	RL	1.67	38.80	56.70	---	95.50	**133.00**	82.10	**112.00**
Gable with dormers	Inst	Sq	RL	1.85	39.20	62.80	---	102.00	**142.00**	87.50	**121.00**
Gable with intersecting roofs	Inst	Sq	RL	1.85	39.50	62.80	---	102.30	**143.00**	87.80	**121.00**
Gable w/dormers & intersections	Inst	Sq	RL	2.08	39.90	70.60	---	110.50	**155.00**	94.70	**132.00**
Hip, plain	Inst	Sq	RL	1.75	39.20	59.40	---	98.60	**137.00**	84.80	**117.00**
Hip with dormers	Inst	Sq	RL	1.95	39.50	66.20	---	105.70	**148.00**	90.60	**125.00**
Hip with intersecting roofs	Inst	Sq	RL	1.95	39.90	66.20	---	106.10	**148.00**	91.00	**126.00**
Hip with dormers & intersections	Inst	Sq	RL	2.21	40.20	75.00	---	115.20	**163.00**	98.40	**137.00**

Over wood decks, includes felt

Gable, plain	Inst	Sq	RL	1.79	43.20	60.80	---	104.00	**144.00**	89.60	**123.00**
Gable with dormers	Inst	Sq	RL	2.00	43.50	67.90	---	111.40	**155.00**	95.80	**132.00**
Gable with intersecting roofs	Inst	Sq	RL	2.00	43.90	67.90	---	111.80	**156.00**	96.10	**132.00**
Gable w/dormers & intersections	Inst	Sq	RL	2.27	44.20	77.00	---	121.20	**170.00**	103.90	**144.00**
Hip, plain	Inst	Sq	RL	1.88	43.50	63.80	---	107.30	**149.00**	92.40	**127.00**
Hip with dormers	Inst	Sq	RL	2.12	43.90	72.00	---	115.90	**162.00**	99.20	**137.00**
Hip with intersecting roofs	Inst	Sq	RL	2.12	44.20	72.00	---	116.20	**162.00**	99.50	**137.00**
Hip with dormers & intersections	Inst	Sq	RL	2.43	44.60	82.50	---	127.10	**179.00**	108.30	**151.00**

Install 30-year laminated shingles

Over existing roofing

Gable, plain	Inst	Sq	RL	1.67	54.80	56.70	---	111.50	**151.00**	97.30	**130.00**
Gable with dormers	Inst	Sq	RL	1.85	55.30	62.80	---	118.10	**161.00**	102.90	**138.00**
Gable with intersecting roofs	Inst	Sq	RL	1.85	55.80	62.80	---	118.60	**162.00**	103.40	**139.00**
Gable w/dormers & intersections	Inst	Sq	RL	2.08	56.30	70.60	---	126.90	**174.00**	110.30	**150.00**
Hip, plain	Inst	Sq	RL	1.75	55.30	59.40	---	114.70	**156.00**	100.20	**134.00**
Hip with dormers	Inst	Sq	RL	1.95	55.80	66.20	---	122.00	**167.00**	106.20	**143.00**
Hip with intersecting roofs	Inst	Sq	RL	1.95	56.30	66.20	---	122.50	**167.00**	106.60	**144.00**
Hip with dormers & intersections	Inst	Sq	RL	2.21	56.80	75.00	---	131.80	**182.00**	114.20	**155.00**

				Costs Based On Small Volume						Large Volume	
Description	Oper	Unit	Crew Size	Man-Hours Per Unit	Avg Mat'l Unit Cost	Avg Labor Unit Cost	Avg Equip Unit Cost	Avg Total Unit Cost	Avg Price Incl O&P	Avg Total Unit Cost	Avg Price Incl O&P
Over wood decks, includes felt											
Gable, plain	Inst	Sq	RL	1.79	59.20	60.80	---	120.00	162.00	104.90	140.
Gable with dormers	Inst	Sq	RL	2.00	59.70	67.90	---	127.60	174.00	111.10	150.
Gable with intersecting roofs	Inst	Sq	RL	2.00	60.20	67.90	---	128.10	174.00	111.60	150.
Gable w/dormers & intersections	Inst	Sq	RL	2.27	60.70	77.00	---	137.70	189.00	119.60	162.
Hip, plain	Inst	Sq	RL	1.88	59.70	63.80	---	123.50	168.00	107.70	144.
Hip with dormers	Inst	Sq	RL	2.12	60.20	72.00	---	132.20	181.00	114.70	155.
Hip with intersecting roofs	Inst	Sq	RL	2.12	60.70	72.00	---	132.70	181.00	115.20	155.
Hip with dormers & intersections	Inst	Sq	RL	2.43	61.20	82.50	---	143.70	198.00	124.10	169.
Install 40-year laminated shingles											
Over existing roofing											
Gable, plain	Inst	Sq	RL	1.67	68.80	56.70	---	125.50	167.00	110.70	145.
Gable with dormers	Inst	Sq	RL	1.85	69.50	62.80	---	132.30	177.00	116.40	154.
Gable with intersecting roofs	Inst	Sq	RL	1.85	70.10	62.80	---	132.90	178.00	117.00	155.
Gable w/dormers & intersections	Inst	Sq	RL	2.08	70.70	70.60	---	141.30	191.00	124.10	165.
Hip, plain	Inst	Sq	RL	1.75	69.50	59.40	---	128.90	172.00	113.70	150.
Hip with dormers	Inst	Sq	RL	1.95	70.10	66.20	---	136.30	183.00	119.80	159.
Hip with intersecting roofs	Inst	Sq	RL	1.95	70.70	66.20	---	136.90	184.00	120.40	160.
Hip with dormers & intersections	Inst	Sq	RL	2.21	71.40	75.00	---	146.40	198.00	128.10	171.
Over wood decks, includes felt											
Gable, plain	Inst	Sq	RL	1.79	73.20	60.80	---	134.00	178.00	118.20	155.
Gable with dormers	Inst	Sq	RL	2.00	73.80	67.90	---	141.70	190.00	124.60	165.
Gable with intersecting roofs	Inst	Sq	RL	2.00	74.50	67.90	---	142.40	191.00	125.20	166.
Gable w/dormers & intersections	Inst	Sq	RL	2.27	75.10	77.00	---	152.10	206.00	133.30	178.
Hip, plain	Inst	Sq	RL	1.88	73.80	63.80	---	137.60	184.00	121.20	160.
Hip with dormers	Inst	Sq	RL	2.12	74.50	72.00	---	146.50	197.00	128.30	170.
Hip with intersecting roofs	Inst	Sq	RL	2.12	75.10	72.00	---	147.10	198.00	128.90	171.
Hip with dormers & intersections	Inst	Sq	RL	2.43	75.70	82.50	---	158.20	215.00	137.90	185.
Install 50-year lifetime laminated shingles											
Over existing roofing											
Gable, plain	Inst	Sq	RL	1.67	79.90	56.70	---	136.60	180.00	121.20	157.
Gable with dormers	Inst	Sq	RL	1.85	80.60	62.80	---	143.40	190.00	127.00	166.
Gable with intersecting roofs	Inst	Sq	RL	1.85	81.30	62.80	---	144.10	191.00	127.70	167.
Gable w/dormers & intersections	Inst	Sq	RL	2.08	82.10	70.60	---	152.70	204.00	134.90	178.
Hip, plain	Inst	Sq	RL	1.75	80.60	59.40	---	140.00	185.00	124.30	162.
Hip with dormers	Inst	Sq	RL	1.95	81.30	66.20	---	147.50	196.00	130.50	171.
Hip with intersecting roofs	Inst	Sq	RL	1.95	82.10	66.20	---	148.30	197.00	131.20	172.
Hip with dormers & intersections	Inst	Sq	RL	2.21	82.80	75.00	---	157.80	212.00	139.00	184.
Over wood decks, includes felt											
Gable, plain	Inst	Sq	RL	1.79	84.20	60.80	---	145.00	191.00	128.70	167.
Gable with dormers	Inst	Sq	RL	2.00	85.00	67.90	---	152.90	203.00	135.20	177.
Gable with intersecting roofs	Inst	Sq	RL	2.00	85.70	67.90	---	153.60	204.00	135.90	178.
Gable w/dormers & intersections	Inst	Sq	RL	2.27	86.40	77.00	---	163.40	219.00	144.10	190.
Hip, plain	Inst	Sq	RL	1.88	85.00	63.80	---	148.80	197.00	131.80	172.
Hip with dormers	Inst	Sq	RL	2.12	85.70	72.00	---	157.70	210.00	139.00	183.
Hip with intersecting roofs	Inst	Sq	RL	2.12	86.40	72.00	---	158.40	211.00	139.70	184.
Hip with dormers & intersections	Inst	Sq	RL	2.43	87.20	82.50	---	169.70	228.00	148.80	198.

Description	Oper	Unit	Crew Size	Man-Hours Per Unit	Avg Mat'l Unit Cost	Avg Labor Unit Cost	Avg Equip Unit Cost	Avg Total Unit Cost	Avg Price Incl O&P	Avg Total Unit Cost	Avg Price Incl O&P
						Costs Based On Small Volume					Large Volume	
Related materials and operations											
Ridge or hip roll											
90 lb mineral surfaced	Inst	LF	2R	.025	---	.89	---	.89	**1.39**	.72	**1.11**
Valley roll											
90 lb mineral surfaced	Inst	LF	2R	.025	.24	.89	---	1.13	**1.66**	.95	**1.37**
Ridge / hip units, 9" x 12" with 5" exp.; 80 pcs / bdle @ 33.3 LF/bdle											
Standard											
Over existing roofing	Inst	LF	2R	.029	1.07	1.04	---	2.11	**2.84**	1.84	**2.45**
Over wood decks	Inst	LF	2R	.030	1.07	1.07	---	2.14	**2.89**	1.88	**2.50**
Architectural											
Over existing roofing	Inst	LF	2R	.029	1.19	1.04	---	2.23	**2.98**	1.95	**2.57**
Over wood decks	Inst	LF	2R	.030	1.19	1.07	---	2.26	**3.03**	1.99	**2.63**
Built-up or membrane roofing											
Install over existing roofing											
One mop coat over smooth surface											
With asphalt	Inst	Sq	RS	.444	10.80	15.00	---	25.80	**35.70**	22.50	**30.60**
With tar	Inst	Sq	RS	.444	19.40	15.00	---	34.40	**45.60**	30.30	**39.70**
Remove gravel, mop one coat, redistribute old gravel											
With asphalt	Inst	Sq	RS	1.13	16.80	38.30	---	55.10	**78.60**	46.80	**66.10**
With tar	Inst	Sq	RS	1.13	29.90	38.30	---	68.20	**93.70**	58.90	**79.90**
Mop in one 30 lb cap sheet, plus smooth top mop coat											
With asphalt	Inst	Sq	RS	1.05	26.20	35.60	---	61.80	**85.30**	53.60	**73.00**
With tar	Inst	Sq	RS	1.05	39.70	35.60	---	75.30	**101.00**	66.00	**87.30**
Remove gravel, mop in one 30 lb cap sheet, mop in and redistribute old gravel											
With asphalt	Inst	Sq	RS	1.82	32.20	61.60	---	93.80	**133.00**	79.90	**112.00**
With tar	Inst	Sq	RS	1.82	50.20	61.60	---	111.80	**153.00**	96.50	**131.00**
Mop in two 15 lb felt plies, plus smooth top mop coat											
With asphalt	Inst	Sq	RS	1.67	32.20	56.60	---	88.80	**125.00**	75.80	**105.00**
With tar	Inst	Sq	RS	1.67	50.60	56.60	---	107.20	**146.00**	92.70	**125.00**
Remove gravel, mop in two 15 lb felt plies, mop in and redistribute old gravel											
With asphalt	Inst	Sq	RS	2.40	38.20	81.30	---	119.50	**170.00**	101.60	**143.00**
With tar	Inst	Sq	RS	2.40	61.10	81.30	---	142.40	**196.00**	122.60	**167.00**
Demo over smooth wood or concrete deck											
3 ply											
With gravel	Demo	Sq	LB	1.82	---	48.50	---	48.50	**72.20**	38.60	**57.60**
Without gravel	Demo	Sq	LB	1.43	---	38.10	---	38.10	**56.80**	30.40	**45.30**
5 ply											
With gravel	Demo	Sq	LB	2.00	---	53.30	---	53.30	**79.40**	42.60	**63.50**
Without gravel	Demo	Sq	LB	1.54	---	41.00	---	41.00	**61.10**	32.80	**48.80**

Description	Oper	Unit	Crew Size	Man-Hours Per Unit	Avg Mat'l Unit Cost	Avg Labor Unit Cost	Avg Equip Unit Cost	Avg Total Unit Cost	Avg Price Incl O&P	Avg Total Unit Cost	Avg Price Incl O&P
Install over smooth wood decks											
Nail one 15 lb felt ply, plus mop in one 90 lb mineral surfaced ply											
With asphalt	Inst	Sq	RS	.750	34.60	25.40	---	60.00	**79.10**	53.30	69.4
With tar	Inst	Sq	RS	.750	40.30	25.40	---	65.70	**85.70**	58.50	75.4
Nail one and mop in two 15 lb felt plies, plus smooth top mop coat											
With asphalt	Inst	Sq	RS	1.82	38.40	61.60	---	100.00	**140.00**	85.80	118.0
With tar	Inst	Sq	RS	1.82	56.70	61.60	---	118.30	**161.00**	102.60	138.0
Nail one and mop in two 15 lb felt plies, plus mop in and distribute gravel											
With asphalt	Inst	Sq	RS	2.40	64.40	81.30	---	145.70	**200.00**	127.40	173.0
With tar	Inst	Sq	RS	2.40	87.30	81.30	---	168.60	**226.00**	148.50	197.0
Nail one and mop in three 15 lb felt plies, plus mop in and distribute gravel											
With asphalt	Inst	Sq	RS	3.00	75.10	102.00	---	177.10	**244.00**	154.00	210.0
With tar	Inst	Sq	RS	3.00	103.00	102.00	---	205.00	**276.00**	179.50	239.0
Nail one and mop in four 15 lb plies, plus mop in and distribute gravel											
With asphalt	Inst	Sq	RS	3.75	85.80	127.00	---	212.80	**295.00**	184.90	253.0
With tar	Inst	Sq	RS	3.75	118.00	127.00	---	245.00	**333.00**	215.00	287.0
Install over smooth concrete decks											
Nail one 15 lb felt ply, plus mop in one 90 lb mineral surfaced ply											
With asphalt	Inst	Sq	RS	.968	40.60	32.80	---	73.40	**97.40**	64.90	85.1
With tar	Inst	Sq	RS	.968	50.80	32.80	---	83.60	**109.00**	74.30	96.6
Nail one and mop in two 15 lb felt plies, plus smooth top mop coat											
With asphalt	Inst	Sq	RS	2.22	43.00	75.20	---	118.20	**166.00**	101.40	141.0
With tar	Inst	Sq	RS	2.22	66.20	75.20	---	141.40	**193.00**	122.70	165.0
Nail one and mop in two 15 lb felt plies, plus mop in and distribute gravel											
With asphalt	Inst	Sq	RS	2.86	69.00	96.80	---	165.80	**229.00**	144.30	197.0
With tar	Inst	Sq	RS	2.86	96.70	96.80	---	193.50	**261.00**	169.90	226.0
Nail one and mop in three 15 lb felt plies, plus mop in and distribute gravel											
With asphalt	Inst	Sq	RS	3.53	79.70	120.00	---	199.70	**277.00**	172.60	237.0
With tar	Inst	Sq	RS	3.53	112.00	120.00	---	232.00	**314.00**	202.50	271.0
Nail one and mop in four 15 lb plies, plus mop in and distribute gravel											
With asphalt	Inst	Sq	RS	4.00	90.40	135.00	---	225.40	**314.00**	195.30	268.0
With tar	Inst	Sq	RS	4.00	128.00	135.00	---	263.00	**357.00**	230.00	308.0

Tile, clay or concrete

Demo tile over wood

Description	Oper	Unit	Crew Size	Man-Hours Per Unit	Avg Mat'l Unit Cost	Avg Labor Unit Cost	Avg Equip Unit Cost	Avg Total Unit Cost	Avg Price Incl O&P	Avg Total Unit Cost	Avg Price Incl O&P
2 piece (interlocking)	Demo	Sq	LB	2.00	---	53.30	---	53.30	**79.40**	42.60	63.5
1 piece	Demo	Sq	LB	1.82	---	48.50	---	48.50	**72.20**	38.60	57.6

Install tile over wood; includes felt

Description	Oper	Unit	Crew Size	Man-Hours Per Unit	Avg Mat'l Unit Cost	Avg Labor Unit Cost	Avg Equip Unit Cost	Avg Total Unit Cost	Avg Price Incl O&P	Avg Total Unit Cost	Avg Price Incl O&P
2 piece (interlocking)											
Red	Inst	Sq	RG	5.56	201.00	182.00	---	383.00	**514.00**	337.00	446.0
Other colors	Inst	Sq	RG	5.56	248.00	182.00	---	430.00	**568.00**	382.00	497.0
Tile shingles											
Mission	Inst	Sq	RG	10.0	110.00	327.00	---	437.00	**634.00**	367.00	526.0
Spanish	Inst	Sq	RG	10.0	134.00	327.00	---	461.00	**661.00**	389.00	552.0

Description	Oper	Unit	Crew Size	Man-Hours Per Unit	Avg Mat'l Unit Cost	Avg Labor Unit Cost	Avg Equip Unit Cost	Avg Total Unit Cost	Avg Price Incl O&P	Large Volume Avg Total Unit Cost	Large Volume Avg Price Incl O&P
ineral surfaced roll											
Single coverage 90 lb/Sq roll, with 6" end lap and 2" head lap	Demo	Sq	LB	.625	---	16.70	---	16.70	**24.80**	13.30	**19.90**
ngle coverage roll on											
lain gable over											
Existing roofing with											
Nails concealed	Inst	Sq	2R	.870	26.20	31.10	---	57.30	**78.30**	49.80	**67.20**
Nails exposed	Inst	Sq	2R	.800	23.80	28.60	---	52.40	**71.70**	45.50	**61.50**
Wood deck with											
Nails concealed	Inst	Sq	2R	.909	26.90	32.50	---	59.40	**81.30**	51.60	**69.80**
Nails exposed	Inst	Sq	2R	.833	24.50	29.80	---	54.30	**74.30**	47.20	**63.80**
lain hip over											
Existing roofing with											
Nails concealed	Inst	Sq	2R	.909	26.90	32.50	---	59.40	**81.30**	51.60	**69.80**
Nails exposed	Inst	Sq	2R	.833	24.50	29.80	---	54.30	**74.30**	47.20	**63.80**
Wood deck with											
Nails concealed	Inst	Sq	2R	1.00	27.40	35.80	---	63.20	**86.90**	54.70	**74.40**
Nails exposed	Inst	Sq	2R	.870	24.90	31.10	---	56.00	**76.90**	48.60	**65.90**
ouble coverage selvage roll, with 19" lap and 17" exposure	Demo	Sq	LB	.909	---	24.20	---	24.20	**36.10**	19.40	**28.90**
ouble coverage selvage roll, with 19" lap and 17" exposure (2 rolls/Sq)											
lain gable over											
Wood deck with nails exposed	Inst	Sq	2R	1.43	54.70	51.20	---	105.90	**142.00**	92.90	**123.00**
elated materials and operations											
Starter strips (36' L rolls) along eaves and up rakes and gable ends on new roofing over wood decks											
9" W	Inst	LF	---	---	.45	---	---	.45	**.52**	.43	**.49**
12" W	Inst	LF	---	---	.57	---	---	.57	**.66**	.54	**.62**
18" W	Inst	LF	---	---	.74	---	---	.74	**.85**	.70	**.81**
24" W	Inst	LF	---	---	.91	---	---	.91	**1.05**	.86	**.99**

Description	Oper	Unit	Crew Size	Man-Hours Per Unit	Avg Mat'l Unit Cost	Avg Labor Unit Cost	Avg Equip Unit Cost	Avg Total Unit Cost	Avg Price Incl O&P	Avg Total Unit Cost	Avg Price Incl O&P
					Costs Based On Small Volume					Large Volume	

Wood

Shakes
24" L with 10" exposure

Description	Oper	Unit	Crew Size	Man-Hours Per Unit	Avg Mat'l Unit Cost	Avg Labor Unit Cost	Avg Equip Unit Cost	Avg Total Unit Cost	Avg Price Incl O&P	Avg Total Unit Cost	Avg Price Incl O&P
1/2" to 3/4" T	Demo	Sq	LB	.800	---	21.30	---	21.30	31.80	17.10	25.
3/4" to 5/4" T	Demo	Sq	LB	.889	---	23.70	---	23.70	35.30	18.90	28.

Shakes, over wood deck; 2 nails/shake, 24" L with 10" exp., 6" W (avg.), red cedar, sawn one side

Description	Oper	Unit	Crew Size	Man-Hours Per Unit	Avg Mat'l Unit Cost	Avg Labor Unit Cost	Avg Equip Unit Cost	Avg Total Unit Cost	Avg Price Incl O&P	Avg Total Unit Cost	Avg Price Incl O&P
Gable											
1/2" to 3/4" T	Inst	Sq	RQ	2.28	184.00	78.00	---	262.00	332.00	237.60	298.
3/4" to 5/4" T	Inst	Sq	RQ	2.54	265.00	86.90	---	351.90	440.00	322.80	399.
Gable with dormers											
1/2" to 3/4" T	Inst	Sq	RQ	2.40	187.00	82.20	---	269.20	342.00	243.70	307.
3/4" to 5/4" T	Inst	Sq	RQ	2.69	270.00	92.10	---	362.10	454.00	330.60	410.
Gable with valleys											
1/2" to 3/4" T	Inst	Sq	RQ	2.40	187.00	82.20	---	269.20	342.00	243.70	307.
3/4" to 5/4" T	Inst	Sq	RQ	2.69	270.00	92.10	---	362.10	454.00	330.60	410.
Hip											
1/2" to 3/4" T	Inst	Sq	RQ	2.40	187.00	82.20	---	269.20	342.00	243.70	307.
3/4" to 5/4" T	Inst	Sq	RQ	2.69	270.00	92.10	---	362.10	454.00	330.60	410.
Hip with valleys											
1/2" to 3/4" T	Inst	Sq	RQ	2.54	191.00	86.90	---	277.90	354.00	250.80	317.
3/4" to 5/4" T	Inst	Sq	RQ	2.82	275.00	96.50	---	371.50	466.00	339.00	421.
Related materials and operations											
Ridge/hip units, 10" W with 10" exp.,											
20 pieces / bundle	Inst	LF	RJ	.022	3.43	.83	---	4.26	5.23	3.94	4.
Roll valley, galvanized, unpainted, 28 gauge, 50' L rolls											
14" W	Inst	LF	UA	.083	1.84	2.96	---	4.80	6.50	4.14	5.
20" W	Inst	LF	UA	.083	1.98	2.96	---	4.94	6.66	4.28	5.
Rosin sized sheathing paper 36" W, 500 SF/roll; nailed											
Over open sheathing	Inst	Sq	RJ	.211	2.17	7.92	---	10.09	14.80	8.38	12.
Over solid sheathing	Inst	Sq	RJ	.143	2.17	5.37	---	7.54	10.80	6.35	9.

Shingles

Description	Oper	Unit	Crew Size	Man-Hours Per Unit	Avg Mat'l Unit Cost	Avg Labor Unit Cost	Avg Equip Unit Cost	Avg Total Unit Cost	Avg Price Incl O&P	Avg Total Unit Cost	Avg Price Incl O&P
16" L with 5" exposure	Demo	Sq	LB	1.67	---	44.50	---	44.50	66.30	35.40	52.
18" L with 5-1/2" exposure	Demo	Sq	LB	1.59	---	42.40	---	42.40	63.10	33.80	50.
24" L with 7-1/2" exposure	Demo	Sq	LB	1.11	---	29.60	---	29.60	44.10	23.70	35.

Shingles, red cedar, No. 1 perfect, 4" W (avg.), 2 nails per shingle

Description	Oper	Unit	Crew Size	Man-Hours Per Unit	Avg Mat'l Unit Cost	Avg Labor Unit Cost	Avg Equip Unit Cost	Avg Total Unit Cost	Avg Price Incl O&P	Avg Total Unit Cost	Avg Price Incl O&P
Over existing roofing materials on											
Gable, plain											
16" L with 5" exposure	Inst	Sq	RM	4.46	271.00	158.00	---	429.00	557.00	384.00	493.
18" L with 5-1/2" exposure	Inst	Sq	RM	4.03	302.00	143.00	---	445.00	568.00	402.00	508.
24" L with 7-1/2" exposure	Inst	Sq	RM	2.91	282.00	103.00	---	385.00	484.00	351.40	437.
Gable with dormers											
16" L with 5" exposure	Inst	Sq	RM	4.63	274.00	164.00	---	438.00	569.00	392.00	503.
18" L with 5-1/2" exposure	Inst	Sq	RM	4.17	305.00	147.00	---	452.00	579.00	408.00	516.
24" L with 7-1/2" exposure	Inst	Sq	RM	3.97	285.00	140.00	---	425.00	545.00	383.00	486.
Gable with intersecting roofs											
16" L with 5" exposure	Inst	Sq	RM	4.63	274.00	164.00	---	438.00	569.00	392.00	503.
18" L with 5-1/2" exposure	Inst	Sq	RM	4.17	305.00	147.00	---	452.00	579.00	408.00	516.
24" L with 7-1/2" exposure	Inst	Sq	RM	3.97	285.00	140.00	---	425.00	545.00	383.00	486.

Description	Oper	Unit	Crew Size	Man-Hours Per Unit	Avg Mat'l Unit Cost	Avg Labor Unit Cost	Avg Equip Unit Cost	Avg Total Unit Cost	Avg Price Incl O&P	Avg Total Unit Cost	Avg Price Incl O&P
								Costs Based On Small Volume		**Large Volume**	
Gable with dormers & intersecting roofs											
16" L with 5" exposure	Inst	Sq	RM	4.81	279.00	170.00	---	449.00	**584.00**	401.00	**516.00**
18" L with 5-1/2" exposure	Inst	Sq	RM	4.31	310.00	152.00	---	462.00	**593.00**	417.00	**529.00**
24" L with 7-1/2" exposure	Inst	Sq	RM	3.09	290.00	109.00	---	399.00	**502.00**	363.40	**453.00**
Hip, plain											
16" L with 5" exposure	Inst	Sq	RM	4.63	274.00	164.00	---	438.00	**569.00**	392.00	**503.00**
18" L with 5-1/2" exposure	Inst	Sq	RM	4.17	305.00	147.00	---	452.00	**579.00**	408.00	**516.00**
24" L with 7-1/2" exposure	Inst	Sq	RM	3.01	285.00	106.00	---	391.00	**492.00**	356.20	**444.00**
Hip with dormers											
16" L with 5" exposure	Inst	Sq	RM	4.81	276.00	170.00	---	446.00	**581.00**	399.00	**513.00**
18" L with 5-1/2" exposure	Inst	Sq	RM	4.39	307.00	155.00	---	462.00	**594.00**	417.00	**529.00**
24" L with 7-1/2" exposure	Inst	Sq	RM	3.16	287.00	112.00	---	399.00	**503.00**	362.50	**453.00**
Hip with intersecting roofs											
16" L with 5" exposure	Inst	Sq	RM	4.81	276.00	170.00	---	446.00	**581.00**	399.00	**513.00**
18" L with 5-1/2" exposure	Inst	Sq	RM	4.39	307.00	155.00	---	462.00	**594.00**	417.00	**529.00**
24" L with 7-1/2" exposure	Inst	Sq	RM	3.16	287.00	112.00	---	399.00	**503.00**	362.50	**453.00**
Hip with dormers & intersecting roofs											
16" L with 5" exposure	Inst	Sq	RM	5.00	281.00	177.00	---	458.00	**597.00**	409.00	**527.00**
18" L with 5-1/2" exposure	Inst	Sq	RM	4.55	313.00	161.00	---	474.00	**609.00**	427.00	**542.00**
24" L with 7-1/2" exposure	Inst	Sq	RM	3.25	292.00	115.00	---	407.00	**514.00**	370.00	**462.00**
Over wood decks on											
Gable, plain											
16" L with 5" exposure	Inst	Sq	RM	4.17	267.00	147.00	---	414.00	**536.00**	372.00	**475.00**
18" L with 5-1/2" exposure	Inst	Sq	RM	3.79	298.00	134.00	---	432.00	**550.00**	391.00	**492.00**
24" L with 7-1/2" exposure	Inst	Sq	RM	2.78	279.00	98.30	---	377.30	**473.00**	343.50	**427.00**
Gable with dormers											
16" L with 5" exposure	Inst	Sq	RM	4.31	269.00	152.00	---	421.00	**546.00**	379.00	**484.00**
18" L with 5-1/2" exposure	Inst	Sq	RM	3.91	301.00	138.00	---	439.00	**560.00**	397.00	**501.00**
24" L with 7-1/2" exposure	Inst	Sq	RM	2.87	281.00	102.00	---	383.00	**481.00**	349.40	**434.00**
Gable with intersecting roofs											
16" L with 5" exposure	Inst	Sq	RM	4.31	269.00	152.00	---	421.00	**546.00**	379.00	**484.00**
18" L with 5-1/2" exposure	Inst	Sq	RM	3.91	301.00	138.00	---	439.00	**560.00**	397.00	**501.00**
24" L with 7-1/2" exposure	Inst	Sq	RM	2.87	281.00	102.00	---	383.00	**481.00**	349.40	**434.00**
Gable with dormers & intersecting roofs											
16" L with 5" exposure	Inst	Sq	RM	4.46	274.00	158.00	---	432.00	**560.00**	387.00	**496.00**
18" L with 5-1/2" exposure	Inst	Sq	RM	4.03	306.00	143.00	---	449.00	**573.00**	405.00	**512.00**
24" L with 7-1/2" exposure	Inst	Sq	RM	2.94	286.00	104.00	---	390.00	**490.00**	355.10	**442.00**
Hip, plain											
16" L with 5" exposure	Inst	Sq	RM	4.31	269.00	152.00	---	421.00	**546.00**	379.00	**484.00**
18" L with 5-1/2" exposure	Inst	Sq	RM	3.91	301.00	138.00	---	439.00	**560.00**	397.00	**501.00**
24" L with 7-1/2" exposure	Inst	Sq	RM	2.87	281.00	102.00	---	383.00	**481.00**	349.40	**434.00**
Hip with dormers											
16" L with 5" exposure	Inst	Sq	RM	4.46	272.00	158.00	---	430.00	**557.00**	385.00	**493.00**
18" L with 5-1/2" exposure	Inst	Sq	RM	4.10	303.00	145.00	---	448.00	**574.00**	405.00	**512.00**
24" L with 7-1/2" exposure	Inst	Sq	RM	3.01	284.00	106.00	---	390.00	**491.00**	355.20	**443.00**
Hip with intersecting roofs											
16" L with 5" exposure	Inst	Sq	RM	4.46	272.00	158.00	---	430.00	**557.00**	385.00	**493.00**
18" L with 5-1/2" exposure	Inst	Sq	RM	4.10	303.00	145.00	---	448.00	**574.00**	405.00	**512.00**
24" L with 7-1/2" exposure	Inst	Sq	RM	3.01	284.00	106.00	---	390.00	**491.00**	355.20	**443.00**
Hip with dormers & intersecting roofs											
16" L with 5" exposure	Inst	Sq	RM	4.63	277.00	164.00	---	441.00	**572.00**	394.00	**506.00**
18" L with 5-1/2" exposure	Inst	Sq	RM	4.24	309.00	150.00	---	459.00	**587.00**	414.00	**524.00**
24" L with 7-1/2" exposure	Inst	Sq	RM	3.09	289.00	109.00	---	398.00	**501.00**	362.40	**451.00**

					Costs Based On Small Volume					Large Volume	
Description	Oper	Unit	Crew Size	Man-Hours Per Unit	Avg Mat'l Unit Cost	Avg Labor Unit Cost	Avg Equip Unit Cost	Avg Total Unit Cost	Avg Price Incl O&P	Avg Total Unit Cost	Avg Price Incl O&P
Related materials and operations											
Ridge/hip units, 40 pieces / bundle											
Over existing roofing											
5" exposure	Inst	LF	RJ	.036	2.89	1.35	---	4.24	**5.42**	3.85	4.8
5-1/2" exposure	Inst	LF	RJ	.034	2.89	1.28	---	4.17	**5.30**	3.77	4.7
7-1/2" exposure	Inst	LF	RJ	.029	1.94	1.09	---	3.03	**3.92**	2.71	3.4
Over wood decks											
5" exposure	Inst	LF	RJ	.033	2.89	1.24	---	4.13	**5.24**	3.77	4.7
5-1/2" exposure	Inst	LF	RJ	.032	2.89	1.20	---	4.09	**5.19**	3.70	4.6
7-1/2" exposure	Inst	LF	RJ	.027	1.94	1.01	---	2.95	**3.80**	2.64	3.3
Roll valley, galvanized, 28 gauge 50' L rolls, unpainted											
18" W	Inst	LF	UA	.083	1.84	2.96	---	4.80	**6.50**	4.14	5.5
24" W	Inst	LF	UA	.083	1.98	2.96	---	4.94	**6.66**	4.28	5.7
Rosin sized sheathing paper, 36" W, 500 SF/roll; nailed											
Over open sheathing	Inst	Sq	RJ	.211	2.17	7.92	---	10.09	**14.80**	8.38	12.2
Over solid sheathing	Inst	Sq	RJ	.143	2.17	5.37	---	7.54	**10.80**	6.35	9.0
Note: The following percentage adjustment for Small Volume also applies to Large											
For No. 2 (red label) grade											
DEDUCT	Inst	%		---	---	-10.0	---	---	---	**---**	---	

Sheathing. See Framing, page 138

Sheet metal

Flashing, general; galvanized

Description	Oper	Unit	Crew Size	Man-Hours Per Unit	Avg Mat'l Unit Cost	Avg Labor Unit Cost	Avg Equip Unit Cost	Avg Total Unit Cost	Avg Price Incl O&P	Avg Total Unit Cost	Avg Price Incl O&P
								Costs Based On Small Volume		Large Volume	
Vertical chimney flashing	Inst	LF	UA	.084	.46	3.00	---	3.46	**4.96**	2.50	**3.57**
"Z" bar flashing											
Standard	Inst	LF	UA	.027	.46	.96	---	1.42	**1.95**	1.08	**1.46**
Old style	Inst	LF	UA	.027	.46	.96	---	1.42	**1.95**	1.08	**1.46**
For plywood siding	Inst	LF	UA	.027	.34	.96	---	1.30	**1.82**	.98	**1.35**
Rain diverter, 1" x 3"											
4' long	Inst	Ea	UA	.229	2.59	8.17	---	10.76	**15.10**	8.00	**11.10**
5' long	Inst	Ea	UA	.229	2.91	8.17	---	11.08	**15.40**	8.28	**11.40**
10' long	Inst	Ea	UA	.320	4.90	11.40	---	16.30	**22.50**	12.50	**17.10**
Roof flashing											
Nosing, roof edging at 90-degree angle or open at 105-degree; galvanized											
3/4" x 3/4"	Inst	LF	UA	.023	.21	.82	---	1.03	**1.46**	.75	**1.05**
1" x 1"	Inst	LF	UA	.023	.21	.82	---	1.03	**1.46**	.75	**1.05**
1" x 2"	Inst	LF	UA	.023	.25	.82	---	1.07	**1.50**	.79	**1.10**
1-1/2" x 1-1/2"	Inst	LF	UA	.023	.25	.82	---	1.07	**1.50**	.79	**1.10**
2" x 2"	Inst	LF	UA	.023	.31	.82	---	1.13	**1.57**	.84	**1.15**
2" x 3"	Inst	LF	UA	.023	.38	.82	---	1.20	**1.65**	.91	**1.24**
2" x 4"	Inst	LF	UA	.023	.53	.82	---	1.35	**1.82**	1.04	**1.38**
3" x 3"	Inst	LF	UA	.029	.53	1.03	---	1.56	**2.14**	1.18	**1.60**
3" x 4"	Inst	LF	UA	.029	.62	1.03	---	1.65	**2.24**	1.26	**1.69**
3" x 5"	Inst	LF	UA	.038	.74	1.36	---	2.10	**2.86**	1.61	**2.17**
4" x 4"	Inst	LF	UA	.038	.74	1.36	---	2.10	**2.86**	1.61	**2.17**
4" x 6"	Inst	LF	UA	.038	.93	1.36	---	2.29	**3.08**	1.78	**2.37**
5" x 5"	Inst	LF	UA	.038	.93	1.36	---	2.29	**3.08**	1.78	**2.37**
6" x 6"	Inst	LF	UA	.038	1.09	1.36	---	2.45	**3.26**	1.92	**2.53**
Roll valley, 50' rolls											
Galvanized, 28 gauge, unpainted											
8" wide	Inst	LF	UA	.038	.60	1.36	---	1.96	**2.70**	1.49	**2.03**
10" wide	Inst	LF	UA	.038	.98	1.36	---	2.34	**3.13**	1.83	**2.43**
20" wide	Inst	LF	UA	.038	1.38	1.36	---	2.74	**3.59**	2.18	**2.83**
Tin seamless, painted											
8" wide	Inst	LF	UA	.038	.47	1.36	---	1.83	**2.55**	1.38	**1.91**
10" wide	Inst	LF	UA	.038	.74	1.36	---	2.10	**2.86**	1.61	**2.17**
20" wide	Inst	LF	UA	.038	1.03	1.36	---	2.39	**3.19**	1.87	**2.47**
Aluminum seamless, .016 gauge											
8" wide	Inst	LF	UA	.038	.56	1.36	---	1.92	**2.65**	1.45	**1.99**
10" wide	Inst	LF	UA	.038	.94	1.36	---	2.30	**3.09**	1.79	**2.38**
20" wide	Inst	LF	UA	.038	1.34	1.36	---	2.70	**3.55**	2.14	**2.78**
"W" valley											
Galvanized hemmed tin											
18" girth	Inst	LF	UA	.038	1.46	1.36	---	2.82	**3.68**	2.25	**2.91**
24" girth	Inst	LF	UA	.038	1.93	1.36	---	3.29	**4.23**	2.66	**3.38**

				Costs Based On Small Volume						Large Volume	
Description	Oper	Unit	Crew Size	Man-Hours Per Unit	Avg Mat'l Unit Cost	Avg Labor Unit Cost	Avg Equip Unit Cost	Avg Total Unit Cost	Avg Price Incl O&P	Avg Total Unit Cost	Avg Price Incl O&P

Gravel stop

4-1/2" galvanized	Inst	LF	UA	.029	.34	1.03	---	1.37	1.92	1.01	1.4
6" galvanized	Inst	LF	UA	.033	.44	1.18	---	1.62	2.25	1.21	1.6
7-1/4" galvanized	Inst	LF	UA	.038	.49	1.36	---	1.85	2.57	1.39	1.9
4-1/2" bonderized	Inst	LF	UA	.029	.43	1.03	---	1.46	2.03	1.09	1.4
6" bonderized	Inst	LF	UA	.033	.57	1.18	---	1.75	2.40	1.33	1.8
7-1/4" bonderized	Inst	LF	UA	.038	.63	1.36	---	1.99	2.73	1.52	2.0

Gutters and downspouts. See page 159

Roof edging (drip edge) with 1/4" kickout, galvanized

1" x 1-1/2"	Inst	LF	UA	.023	.26	.82	---	1.08	1.51	.80	1.1
1-1/2" x 1-1/2"	Inst	LF	UA	.023	.26	.82	---	1.08	1.51	.80	1.1
2" x 2"	Inst	LF	UA	.023	.32	.82	---	1.14	1.58	.86	1.1

Vents

Clothes dryer vent set, aluminum, hood, duct, inside plate

3" diameter	Inst	Set	UA	.727	3.60	25.90	---	29.50	42.50	20.99	30.1
4" diameter	Inst	Set	UA	.727	3.97	25.90	---	29.87	42.90	21.31	30.4

Dormer louvers, half round, 1/8" or 1/4" mesh

18" x 9"

Galvanized, 5-12 pitch	Inst	Ea	UA	1.00	22.40	35.70	---	58.10	78.60	43.60	58.0
Painted, 3-12 pitch	Inst	Ea	UA	1.00	26.50	35.70	---	62.20	83.20	47.20	62.1

24" x 12"

Galvanized, 5-12 pitch	Inst	Ea	UA	1.00	28.10	35.70	---	63.80	85.10	48.70	63.8
Painted, 3-12 pitch	Inst	Ea	UA	1.00	31.90	35.70	---	67.60	89.40	52.00	67.6

Foundation vents, galvanized, 1/4" mesh

6" x 14", stucco	Inst	Ea	UA	.471	1.72	16.80	---	18.52	26.80	13.42	19.3
8" x 14", stucco	Inst	Ea	UA	.471	2.06	16.80	---	18.86	27.20	13.72	19.7
6" x 14", two-way	Inst	Ea	UA	.471	2.06	16.80	---	18.86	27.20	13.72	19.7
6" x 14", for siding (flat type)	Inst	Ea	UA	.471	1.72	16.80	---	18.52	26.80	13.42	19.3
6" x 14", louver type	Inst	Ea	UA	.471	1.98	16.80	---	18.78	27.10	13.66	19.6
6" x 14", foundation insert	Inst	Ea	UA	.471	1.72	16.80	---	18.52	26.80	13.42	19.3

Louver vents, round, 1/8" or 1/4" mesh, galvanized

12" diameter	Inst	Ea	UA	.727	22.90	25.90	---	48.80	64.70	38.10	49.7
14" diameter	Inst	Ea	UA	.727	24.30	25.90	---	50.20	66.40	39.30	51.2
16" diameter	Inst	Ea	UA	.727	26.40	25.90	---	52.30	68.70	41.20	53.2
18" diameter	Inst	Ea	UA	.727	30.30	25.90	---	56.20	73.20	44.60	57.2
24" diameter	Inst	Ea	UA	.727	51.50	25.90	---	77.40	97.50	63.30	78.7

Access doors

Multi-purpose, galvanized

24" x 18", nail-on type	Inst	Ea	UA	.727	14.60	25.90	---	40.50	55.10	30.70	41.2
24" x 24", nail-on type	Inst	Ea	UA	.727	16.40	25.90	---	42.30	57.20	32.30	43.1
24" x 18" x 6" deep box type	Inst	Ea	UA	.727	19.90	25.90	---	45.80	61.20	35.40	46.6

Attic access doors, galvanized

22" x 22"	Inst	Ea	UA	1.00	20.60	35.70	---	56.30	76.50	42.00	56.1
22" x 30"	Inst	Ea	UA	1.00	23.80	35.70	---	59.50	80.20	44.90	59.4
30" x 30"	Inst	Ea	UA	1.00	25.10	35.70	---	60.80	81.70	46.00	60.8

Tub access doors, galvanized

14" x 12"	Inst	Ea	UA	.727	10.20	25.90	---	36.10	50.10	26.84	36.8
14" x 15"	Inst	Ea	UA	.727	10.60	25.90	---	36.50	50.50	27.16	37.2

Description	Oper	Unit	Crew Size	Man-Hours Per Unit	Avg Mat'l Unit Cost	Avg Labor Unit Cost	Avg Equip Unit Cost	Avg Total Unit Cost	Avg Price Incl O&P	Avg Total Unit Cost	Avg Price Incl O&P
						Costs Based On Small Volume					Large Volume	
Utility louvers, aluminum or galvanized											
4" x 6"	Inst	Ea	UA	.286	1.22	10.20	---	11.42	**16.50**	8.21	**11.80**
8" x 6"	Inst	Ea	UA	.286	1.72	10.20	---	11.92	**17.10**	8.65	**12.30**
8" x 8"	Inst	Ea	UA	.286	1.98	10.20	---	12.18	**17.40**	8.89	**12.60**
8" x 12"	Inst	Ea	UA	.286	2.32	10.20	---	12.52	**17.80**	9.18	**12.90**
8" x 18"	Inst	Ea	UA	.286	2.78	10.20	---	12.98	**18.30**	9.59	**13.40**
12" x 12"	Inst	Ea	UA	.286	2.78	10.20	---	12.98	**18.30**	9.59	**13.40**
14" x 14"	Inst	Ea	UA	.286	3.51	10.20	---	13.71	**19.10**	10.24	**14.10**

Shower and tub doors

Shower doors

Hinged shower doors, anodized aluminum frame, tempered hammered glass, hardware included

Description	Oper	Unit	Crew Size	Man-Hours Per Unit	Avg Mat'l Unit Cost	Avg Labor Unit Cost	Avg Equip Unit Cost	Avg Total Unit Cost	Avg Price Incl O&P	Avg Total Unit Cost	Avg Price Incl O&P
					Costs Based On Small Volume					Large Volume	
67" H door, adjustable											
29-3/4" to 31-1/2"											
Obscure glass											
Chrome	Inst	Ea	CA	1.78	311.00	57.10	---	368.10	443.00	298.70	359.0
Polished brass	Inst	Ea	CA	1.78	349.00	57.10	---	406.10	486.00	329.70	394.0
White	Inst	Ea	CA	1.78	445.00	57.10	---	502.10	597.00	408.70	485.0
Bone	Inst	Ea	CA	1.78	445.00	57.10	---	502.10	597.00	408.70	485.0
Clear glass											
Chrome	Inst	Ea	CA	1.78	356.00	57.10	---	413.10	495.00	335.70	401.0
Polished brass	Inst	Ea	CA	1.78	389.00	57.10	---	446.10	533.00	363.70	433.0
White	Inst	Ea	CA	1.78	485.00	57.10	---	542.10	643.00	441.70	523.0
Bone	Inst	Ea	CA	1.78	485.00	57.10	---	542.10	643.00	441.70	523.0
Etched glass											
Chrome	Inst	Ea	CA	1.78	381.00	57.10	---	438.10	524.00	356.70	425.0
Polished brass	Inst	Ea	CA	1.78	417.00	57.10	---	474.10	565.00	385.70	458.0
White	Inst	Ea	CA	1.78	513.00	57.10	---	570.10	676.00	465.70	550.0
Bone	Inst	Ea	CA	1.78	513.00	57.10	---	570.10	676.00	465.70	550.0
Fluted											
Chrome	Inst	Ea	CA	1.78	487.00	57.10	---	544.10	646.00	443.70	525.0
Polished brass	Inst	Ea	CA	1.78	525.00	57.10	---	582.10	690.00	475.70	562.0
White	Inst	Ea	CA	1.78	620.00	57.10	---	677.10	798.00	552.70	651.0
Bone	Inst	Ea	CA	1.78	620.00	57.10	---	677.10	798.00	552.70	651.0
33-1/8" to 34-7/8"											
Obscure glass											
Chrome	Inst	Ea	CA	1.78	311.00	57.10	---	368.10	443.00	298.70	359.0
Polished brass	Inst	Ea	CA	1.78	349.00	57.10	---	406.10	486.00	329.70	394.0
White	Inst	Ea	CA	1.78	445.00	57.10	---	502.10	597.00	408.70	485.0
Bone	Inst	Ea	CA	1.78	485.00	57.10	---	542.10	643.00	441.70	523.0
Clear glass											
Chrome	Inst	Ea	CA	1.78	356.00	57.10	---	413.10	495.00	335.70	401.0
Polished brass	Inst	Ea	CA	1.78	397.00	57.10	---	454.10	542.00	369.70	440.0
White	Inst	Ea	CA	1.78	485.00	57.10	---	542.10	643.00	441.70	523.0
Bone	Inst	Ea	CA	1.78	485.00	57.10	---	542.10	643.00	441.70	523.0
Etched glass											
Chrome	Inst	Ea	CA	1.78	381.00	57.10	---	438.10	524.00	356.70	425.0
Polished brass	Inst	Ea	CA	1.78	417.00	57.10	---	474.10	565.00	385.70	458.0
White	Inst	Ea	CA	1.78	513.00	57.10	---	570.10	676.00	465.70	550.0
Bone	Inst	Ea	CA	1.78	513.00	57.10	---	570.10	676.00	465.70	550.0
Fluted											
Chrome	Inst	Ea	CA	1.78	487.00	57.10	---	544.10	646.00	443.70	525.0
Polished brass	Inst	Ea	CA	1.78	525.00	57.10	---	582.10	690.00	475.70	562.0
White	Inst	Ea	CA	1.78	620.00	57.10	---	677.10	798.00	552.70	651.0
Bone	Inst	Ea	CA	1.78	620.00	57.10	---	677.10	798.00	552.70	651.0

Description	Oper	Unit	Crew Size	Man-Hours Per Unit	Avg Mat'l Unit Cost	Avg Labor Unit Cost	Avg Equip Unit Cost	Avg Total Unit Cost	Avg Price Incl O&P	Avg Total Unit Cost	Avg Price Incl O&P
								Costs Based On Small Volume		**Large Volume**	
23" x 24-1/2" x 23" x 72" neo angle											
Obscure glass											
Chrome	Inst	Ea	CA	1.78	645.00	57.10	---	702.10	828.00	573.70	675.00
Polished brass	Inst	Ea	CA	1.78	791.00	57.10	---	848.10	995.00	693.70	813.00
White	Inst	Ea	CA	1.78	978.00	57.10	---	1035.10	1210.00	847.70	990.00
Bone	Inst	Ea	CA	1.78	978.00	57.10	---	1035.10	1210.00	847.70	990.00
Clear glass											
Chrome	Inst	Ea	CA	1.78	715.00	57.10	---	772.10	908.00	631.70	741.00
Polished brass	Inst	Ea	CA	1.78	861.00	57.10	---	918.10	1080.00	751.70	879.00
White	Inst	Ea	CA	1.78	1030.00	57.10	---	1087.10	1270.00	892.70	1040.00
Bone	Inst	Ea	CA	1.78	1030.00	57.10	---	1087.10	1270.00	892.70	1040.00
Etched glass											
Chrome	Inst	Ea	CA	1.78	728.00	57.10	---	785.10	923.00	642.70	754.00
Polished brass	Inst	Ea	CA	1.78	873.00	57.10	---	930.10	1090.00	761.70	891.00
White	Inst	Ea	CA	1.78	1060.00	57.10	---	1117.10	1300.00	915.70	1070.00
Bone	Inst	Ea	CA	1.78	1060.00	57.10	---	1117.10	1300.00	915.70	1070.00
Fluted											
Chrome	Inst	Ea	CA	1.78	1070.00	57.10	---	1127.10	1320.00	923.70	1080.00
Polished brass	Inst	Ea	CA	1.78	1210.00	57.10	---	1267.10	1480.00	1042.70	1210.00
White	Inst	Ea	CA	1.78	1400.00	57.10	---	1457.10	1700.00	1202.70	1390.00
Bone	Inst	Ea	CA	1.78	1400.00	57.10	---	1457.10	1700.00	1202.70	1390.00

Sliding shower doors, 2 bypassing panels, anodized aluminum frame, tempered hammered glass, hardware included

71-1/2" H doors

Description	Oper	Unit	Crew Size	Man-Hours Per Unit	Avg Mat'l Unit Cost	Avg Labor Unit Cost	Avg Equip Unit Cost	Avg Total Unit Cost	Avg Price Incl O&P	Avg Total Unit Cost	Avg Price Incl O&P
40" to 42" wide opening											
Obscure glass											
Chrome	Inst	Ea	CA	1.94	365.00	62.30	---	427.30	513.00	346.50	415.00
Polished brass	Inst	Ea	CA	1.94	411.00	62.30	---	473.30	566.00	384.50	459.00
White	Inst	Ea	CA	1.94	551.00	62.30	---	613.30	727.00	500.50	591.00
Bone	Inst	Ea	CA	1.94	551.00	62.30	---	613.30	727.00	500.50	591.00
Clear glass											
Chrome	Inst	Ea	CA	1.94	422.00	62.30	---	484.30	578.00	393.50	469.00
Polished brass	Inst	Ea	CA	1.94	470.00	62.30	---	532.30	634.00	433.50	515.00
White	Inst	Ea	CA	1.94	609.00	62.30	---	671.30	793.00	547.50	646.00
Bone	Inst	Ea	CA	1.94	609.00	62.30	---	671.30	793.00	547.50	646.00
Etched glass											
Chrome	Inst	Ea	CA	1.94	441.00	62.30	---	503.30	601.00	409.50	488.00
Polished brass	Inst	Ea	CA	1.94	485.00	62.30	---	547.30	652.00	446.50	529.00
White	Inst	Ea	CA	1.94	632.00	62.30	---	694.30	820.00	566.50	668.00
Bone	Inst	Ea	CA	1.94	632.00	62.30	---	694.30	820.00	566.50	668.00
Fluted											
Chrome	Inst	Ea	CA	1.94	648.00	62.30	---	710.30	838.00	579.50	683.00
Polished brass	Inst	Ea	CA	1.94	693.00	62.30	---	755.30	890.00	617.50	726.00
White	Inst	Ea	CA	1.94	839.00	62.30	---	901.30	1060.00	737.50	864.00
Bone	Inst	Ea	CA	1.94	839.00	62.30	---	901.30	1060.00	737.50	864.00
44" to 48" wide opening											
Obscure glass											
Chrome	Inst	Ea	CA	1.94	365.00	62.30	---	427.30	513.00	346.50	415.00
Polished brass	Inst	Ea	CA	1.94	411.00	62.30	---	473.30	566.00	384.50	459.00
White	Inst	Ea	CA	1.94	551.00	62.30	---	613.30	727.00	500.50	591.00
Bone	Inst	Ea	CA	1.94	551.00	62.30	---	613.30	727.00	500.50	591.00

				Costs Based On Small Volume						Large Volume	
Description	Oper	Unit	Crew Size	Man-Hours Per Unit	Avg Mat'l Unit Cost	Avg Labor Unit Cost	Avg Equip Unit Cost	Avg Total Unit Cost	Avg Price Incl O&P	Avg Total Unit Cost	Avg Price Incl O&P
Clear glass											
Chrome	Inst	Ea	CA	1.94	422.00	62.30	---	484.30	578.00	393.50	469.0
Polished brass	Inst	Ea	CA	1.94	470.00	62.30	---	532.30	634.00	433.50	515.0
White	Inst	Ea	CA	1.94	609.00	62.30	---	671.30	793.00	547.50	646.0
Bone	Inst	Ea	CA	1.94	609.00	62.30	---	671.30	793.00	547.50	646.0
Etched glass											
Chrome	Inst	Ea	CA	1.94	441.00	62.30	---	503.30	601.00	409.50	488.0
Polished brass	Inst	Ea	CA	1.94	485.00	62.30	---	547.30	652.00	446.50	529.0
White	Inst	Ea	CA	1.94	632.00	62.30	---	694.30	820.00	566.50	668.0
Bone	Inst	Ea	CA	1.94	632.00	62.30	---	694.30	820.00	566.50	668.0
Fluted											
Chrome	Inst	Ea	CA	1.94	648.00	62.30	---	710.30	838.00	579.50	683.0
Polished brass	Inst	Ea	CA	1.94	693.00	62.30	---	755.30	890.00	617.50	726.0
White	Inst	Ea	CA	1.94	839.00	62.30	---	901.30	1060.00	737.50	864.0
Bone	Inst	Ea	CA	1.94	839.00	62.30	---	901.30	1060.00	737.50	864.0
56" to 60" wide opening											
Obscure glass											
Chrome	Inst	Ea	CA	1.94	427.00	62.30	---	489.30	584.00	397.50	474.0
Polished brass	Inst	Ea	CA	1.94	471.00	62.30	---	533.30	635.00	434.50	516.0
White	Inst	Ea	CA	1.94	573.00	62.30	---	635.30	752.00	518.50	612.0
Bone	Inst	Ea	CA	1.94	573.00	62.30	---	635.30	752.00	518.50	612.0
Clear glass											
Chrome	Inst	Ea	CA	1.94	485.00	62.30	---	547.30	652.00	446.50	529.0
Polished brass	Inst	Ea	CA	1.94	530.00	62.30	---	592.30	703.00	483.50	572.0
White	Inst	Ea	CA	1.94	667.00	62.30	---	729.30	861.00	596.50	702.0
Bone	Inst	Ea	CA	1.94	667.00	62.30	---	729.30	861.00	596.50	702.0
Etched glass											
Chrome	Inst	Ea	CA	1.94	502.00	62.30	---	564.30	670.00	459.50	545.0
Polished brass	Inst	Ea	CA	1.94	546.00	62.30	---	608.30	721.00	495.50	587.0
White	Inst	Ea	CA	1.94	694.00	62.30	---	756.30	891.00	617.50	727.0
Bone	Inst	Ea	CA	1.94	694.00	62.30	---	756.30	891.00	617.50	727.0
Fluted											
Chrome	Inst	Ea	CA	1.94	691.00	62.30	---	753.30	888.00	615.50	724.0
Polished brass	Inst	Ea	CA	1.94	776.00	62.30	---	838.30	986.00	685.50	805.0
White	Inst	Ea	CA	1.94	898.00	62.30	---	960.30	1130.00	786.50	921.0
Bone	Inst	Ea	CA	1.94	898.00	62.30	---	960.30	1130.00	786.50	921.0

Tub doors

Sliding tub doors, 2 bypassing panels

60" W x 58-1/4" H

Description	Oper	Unit	Crew Size	Man-Hours Per Unit	Avg Mat'l Unit Cost	Avg Labor Unit Cost	Avg Equip Unit Cost	Avg Total Unit Cost	Avg Price Incl O&P	Avg Total Unit Cost	Avg Price Incl O&P
Obscure glass											
Chrome	Inst	Ea	CA	1.94	344.00	62.30	---	406.30	489.00	330.50	396.0
Polished brass	Inst	Ea	CA	1.94	381.00	62.30	---	443.30	531.00	360.50	430.0
Clear glass											
Chrome	Inst	Ea	CA	1.94	393.00	62.30	---	455.30	545.00	369.50	442.0
Polished brass	Inst	Ea	CA	1.94	429.00	62.30	---	491.30	587.00	400.50	476.0

Shower bases or receptors

American Standard Products,
Americast with drain strainer

Description	Oper	Unit	Crew Size	Man-Hours Per Unit	Avg Mat'l Unit Cost	Avg Labor Unit Cost	Avg Equip Unit Cost	Avg Total Unit Cost	Avg Price Incl O&P	Avg Total Unit Cost	Avg Price Incl O&P
								Costs Based On Small Volume		Large Volume	
Detach & reset operations											
Any size	Reset	Ea	SB	3.81	33.20	120.00	---	153.20	**215.00**	111.40	**155.00**
Remove operations											
Any size	Demo	Ea	SB	2.29	---	72.10	---	72.10	**106.00**	50.40	**74.10**
Install rough-in											
Shower base, any size	Inst	Ea	SB	19.0	72.30	599.00	---	671.30	**963.00**	478.50	**684.00**
Replace operations											
32" x 32" x 6-5/8" H											
White	Reset	Ea	SB	5.71	395.00	180.00	---	575.00	**719.00**	452.00	**560.00**
Colors	Reset	Ea	SB	5.71	428.00	180.00	---	608.00	**757.00**	479.00	**591.00**
Premium colors	Reset	Ea	SB	5.71	454.00	180.00	---	634.00	**786.00**	500.00	**615.00**
36" x 36-3/16" x 6-5/8" H											
White	Reset	Ea	SB	5.71	406.00	180.00	---	586.00	**732.00**	461.00	**570.00**
Colors	Reset	Ea	SB	5.71	440.00	180.00	---	620.00	**771.00**	489.00	**602.00**
Premium colors	Reset	Ea	SB	5.71	467.00	180.00	---	647.00	**801.00**	510.00	**627.00**
42-1/8" x 42-1/8" x 6-5/8" H											
White	Reset	Ea	SB	5.71	644.00	180.00	---	824.00	**1010.00**	657.00	**795.00**
Colors	Reset	Ea	SB	5.71	685.00	180.00	---	865.00	**1050.00**	690.00	**834.00**
Premium colors	Reset	Ea	SB	5.71	710.00	180.00	---	890.00	**1080.00**	711.00	**857.00**
48-1/8" x 34-1/4" x 6-5/8" H											
White	Reset	Ea	SB	5.71	547.00	180.00	---	727.00	**893.00**	576.00	**703.00**
Colors	Reset	Ea	SB	5.71	586.00	180.00	---	766.00	**938.00**	608.00	**740.00**
Premium colors	Reset	Ea	SB	5.71	609.00	180.00	---	789.00	**965.00**	628.00	**762.00**
60-1/8" x 34-1/8" x 6-5/8" H											
White	Reset	Ea	SB	5.71	676.00	180.00	---	856.00	**1040.00**	683.00	**825.00**
Colors	Reset	Ea	SB	5.71	717.00	180.00	---	897.00	**1090.00**	716.00	**864.00**
Premium colors	Reset	Ea	SB	5.71	743.00	180.00	---	923.00	**1120.00**	738.00	**889.00**
36-1/4" x 36-1/8" x 6-5/8" H, neo angle											
White	Reset	Ea	SB	5.71	422.00	180.00	---	602.00	**749.00**	473.00	**584.00**
Colors	Reset	Ea	SB	5.71	457.00	180.00	---	637.00	**790.00**	503.00	**618.00**
Premium colors	Reset	Ea	SB	5.71	483.00	180.00	---	663.00	**820.00**	524.00	**642.00**
38-1/8" x 38-1/16" x 6-5/8" H, neo angle											
White	Reset	Ea	SB	5.71	442.00	180.00	---	622.00	**773.00**	490.00	**604.00**
Colors	Reset	Ea	SB	5.71	477.00	180.00	---	657.00	**813.00**	519.00	**637.00**
Premium colors	Reset	Ea	SB	5.71	502.00	180.00	---	682.00	**842.00**	540.00	**661.00**
42-1/4" x 42-1/8" x 6-5/8" H, neo angle											
White	Reset	Ea	SB	5.71	676.00	180.00	---	856.00	**1040.00**	683.00	**825.00**
Colors	Reset	Ea	SB	5.71	717.00	180.00	---	897.00	**1090.00**	716.00	**864.00**
Premium colors	Reset	Ea	SB	5.71	743.00	180.00	---	923.00	**1120.00**	738.00	**889.00**

Shower stalls

Shower stall units with slip-resistant fiberglass floors, and reinforced plastic integral wall surrounds (Aqua Glass), available in white or color, with good quality fittings, single control faucets, and shower head sprayer

Description	Oper	Unit	Crew Size	Man-Hours Per Unit	Avg Mat'l Unit Cost	Avg Labor Unit Cost	Avg Equip Unit Cost	Avg Total Unit Cost	Avg Price Incl O&P	Avg Total Unit Cost	Avg Price Incl O&P
										Large Volume	
Detach & reset operations											
Single integral piece	Reset	Ea	SB	7.62	46.80	240.00	---	286.80	**407.00**	206.50	**291.0**
Multi-piece, nonintegral	Reset	Ea	SB	5.71	46.80	180.00	---	226.80	**318.00**	164.50	**229.0**
Remove operations											
Single integral piece	Demo	Ea	SB	3.81	25.50	120.00	---	145.50	**206.00**	105.10	**148.0**
Multi-piece, nonintegral	Demo	Ea	SB	2.86	25.50	90.10	---	115.60	**162.00**	84.00	**117.0**
Install rough-in											
Shower stall, any size	Inst	Ea	SB	14.3	72.30	450.00	---	522.30	**745.00**	374.50	**531.0**
Replace operations											
Three-wall stall, one integral piece, no door, with plastic drain											
Lasco Compact, Paloma-I 32" x 32" x 72" H											
Stock color	Inst	Ea	SB	11.4	536.00	359.00	---	895.00	**1140.00**	693.00	**878.0**
Lasco Compact, Paloma-II 34" x 34" x 72" H											
Stock color	Inst	Ea	SB	11.4	598.00	359.00	---	957.00	**1220.00**	745.00	**937.0**
Lasco Compact, Paloma-III 36" x 36" x 72" H											
Stock color	Inst	Ea	SB	11.4	573.00	359.00	---	932.00	**1190.00**	724.00	**913.0**
Lasco Compact, Paloma-IV 42" x 34" x 72" H											
Stock color	Inst	Ea	SB	11.4	604.00	359.00	---	963.00	**1220.00**	750.00	**943.0**
Lasco Deluxe, Larissa, 1 seat 47-3/4" x 33-1/2" x 72" H											
Stock color	Inst	Ea	SB	11.4	653.00	359.00	---	1012.00	**1280.00**	790.00	**989.0**
Lasco Deluxe, Camila-I, 2 seat 48" x 35" x 72" H											
Stock color	Inst	Ea	SB	11.4	626.00	359.00	---	985.00	**1250.00**	767.00	**963.0**
Lasco Deluxe, Camila-II, 2 seat 54" x 35" x 72" H											
Stock color	Inst	Ea	SB	12.7	647.00	400.00	---	1047.00	**1330.00**	813.00	**1020.0**
Lasco Deluxe, Camila-III, 2 seat 60" x 35" x 72" H											
Stock color	Inst	Ea	SB	12.7	653.00	400.00	---	1053.00	**1340.00**	818.00	**1030.0**

				Costs Based On Small Volume						Large Volume	
Description	Oper	Unit	Crew Size	Man-Hours Per Unit	Avg Mat'l Unit Cost	Avg Labor Unit Cost	Avg Equip Unit Cost	Avg Total Unit Cost	Avg Price Incl O&P	Avg Total Unit Cost	Avg Price Incl O&P
Two-wall stall, one integral piece, no door, with plastic drain											
Lasco Varina, space saver 36" x 36" x 72" H Stock color	Inst	Ea	SB	11.4	600.00	359.00	---	959.00	**1220.00**	746.00	**939.00**
Lasco Galina-I, space saver, neo angle 38" x 38" x 72" H Stock color	Inst	Ea	SB	11.4	573.00	359.00	---	932.00	**1190.00**	724.00	**913.00**
Lasco Galina-II, space saver, neo angle 38" x 38" x 80" H Stock color	Inst	Ea	SB	11.4	607.00	359.00	---	966.00	**1230.00**	752.00	**945.00**
Lasco Galina-III, space saver, neo angle 36" x 36" x 77" H Stock color	Inst	Ea	SB	11.4	803.00	359.00	---	1162.00	**1450.00**	914.00	**1130.00**
Lasco Neopolitan-I, space saver, neo angle 38" x 38" x 72" H Stock color	Inst	Ea	SB	11.4	643.00	359.00	---	1002.00	**1270.00**	781.00	**979.00**
Multi-piece, nonintegral, nonassembled units with drain											
Lasco Caledon-I, 2-piece 32" x 32" x 72-3/4" H Stock color	Inst	Ea	SB	11.4	580.00	359.00	---	939.00	**1190.00**	729.00	**919.00**
Lasco Caledon-II, 3-piece 36" x 36" x 72-3/4" H Stock color	Inst	Ea	SB	11.4	653.00	359.00	---	1012.00	**1280.00**	790.00	**989.00**
Lasco Caledon-III, 2-piece 48" x 34" x 72" H Stock color	Inst	Ea	SB	12.7	651.00	400.00	---	1051.00	**1340.00**	816.00	**1030.00**
Lasco Caledon-III, 3-piece 48" x 34" x 72-3/4" H Stock color	Inst	Ea	SB	12.7	703.00	400.00	---	1103.00	**1400.00**	859.00	**1080.00**
Lasco Talia-II, 2-piece 36" x 36" x 72" H Stock color	Inst	Ea	SB	11.4	615.00	359.00	---	974.00	**1240.00**	759.00	**953.00**

				Costs Based On Small Volume					Large Volume	
Description	Oper Unit	Crew Size	Man-Hours Per Unit	Avg Mat'l Unit Cost	Avg Labor Unit Cost	Avg Equip Unit Cost	Avg Total Unit Cost	Avg Price Incl O&P	Avg Total Unit Cost	Avg Price Incl O&P

Shower tub units

Shower tub three-wall stall combinations are fiberglass reinforced plastic with integral bath/shower and wall surrounds; includes good quality fittings, single control faucets, and shower head sprayer

Detach & reset operations

Single integral piece	Reset Ea	SB	7.27	46.80	229.00	.00	275.80	**390.00**	206.50	291.00
Multi-piece, non-integral	Reset Ea	SB	7.62	46.80	240.00	.00	286.80	**407.00**	164.50	229.00

Remove operations

Single integral piece	Demo Ea	SB	3.48	25.50	110.00	.00	135.50	**190.00**	105.10	148.00
Multi-piece, non-integral	Demo Ea	SB	3.81	25.50	120.00	.00	145.50	**206.00**	84.00	117.00

Install rough-in

Shower tub	Inst Ea	SB	14.5	72.30	457.00	.00	529.30	**754.00**	418.50	596.00

Replace operations

One-piece combination, (unitized), no door

Lasco Kassia, with grab bar 54-1/2" x 29" x 72" H										
Stock color	Inst Ea	SB	12.3	652.00	387.00	.00	1039.00	**1320.00**	789.00	988.00
Lasco Capriel-I, with grab bar 60" x 30" x 72" H										
Stock color	Inst Ea	SB	12.3	641.00	387.00	.00	1028.00	**1310.00**	780.00	977.00
Lasco Capriel-II, with grab bar 60" x 32" x 74" H										
Stock color	Inst Ea	SB	12.3	648.00	387.00	.00	1035.00	**1310.00**	785.00	984.00
Lasco Capriel-III, with grab bar 60" x 32" x 78" H										
Stock color	Inst Ea	SB	12.3	674.00	387.00	.00	1061.00	**1340.00**	807.00	1010.00

Multi-piece combination, no door

Lasco Mali, 3-piece 60" x 30" x 72" H										
Stock color	Inst Ea	SB	12.3	655.00	387.00	.00	1042.00	**1320.00**	791.00	990.00
Lasco Oriana, 2-piece 60" x 30" x 72" H										
Stock color	Inst Ea	SB	12.3	637.00	387.00	.00	1024.00	**1300.00**	776.00	973.00

Description	Oper	Unit	Crew Size	Man-Hours Per Unit	Avg Mat'l Unit Cost	Avg Labor Unit Cost	Avg Equip Unit Cost	Avg Total Unit Cost	Avg Price Incl O&P	Large Volume Avg Total Unit Cost	Large Volume Avg Price Incl O&P
Shutters											
Aluminum, louvered, 14" W											
3'-0" long	Inst	Pair	CA	1.33	54.00	42.70	---	96.70	126.00	79.40	102.00
6'-8" long	Inst	Pair	CA	1.33	90.00	42.70	---	132.70	168.00	110.90	139.00
Pine or fir, provincial raised panel, primed, in stock, 1-1/16" T											
12" W											
2'-1" long	Inst	Pair	CA	1.33	33.50	42.70	---	76.20	103.00	61.40	81.80
3'-1" long	Inst	Pair	CA	1.33	35.40	42.70	---	78.10	105.00	63.10	83.80
4'-1" long	Inst	Pair	CA	1.33	40.60	42.70	---	83.30	111.00	67.60	89.00
5'-1" long	Inst	Pair	CA	1.33	49.20	42.70	---	91.90	121.00	75.20	97.70
6'-1" long	Inst	Pair	CA	1.33	58.10	42.70	---	100.80	131.00	83.00	107.00
18" W											
2'-1" long	Inst	Pair	CA	1.33	39.80	42.70	---	82.50	110.00	66.90	88.20
3'-1" long	Inst	Pair	CA	1.33	41.90	42.70	---	84.60	112.00	68.80	90.30
4'-1" long	Inst	Pair	CA	1.33	49.80	42.70	---	92.50	121.00	75.70	98.30
5'-1" long	Inst	Pair	CA	1.33	58.40	42.70	---	101.10	131.00	83.20	107.00
6'-1" long	Inst	Pair	CA	1.33	70.10	42.70	---	112.80	145.00	93.50	119.00
Door blinds, 6'-9" long											
1'-3" wide	Inst	Pair	CA	1.33	72.00	42.70	---	114.70	147.00	95.10	121.00
1'-6" wide	Inst	Pair	CA	1.33	90.00	42.70	---	132.70	168.00	110.90	139.00
Birch, special order stationary slat blinds 1-1/16" T											
12" W											
2'-1" long	Inst	Pair	CA	1.33	65.20	42.70	---	107.90	139.00	89.20	114.00
3'-1" long	Inst	Pair	CA	1.33	77.60	42.70	---	120.30	153.00	100.00	126.00
4'-1" long	Inst	Pair	CA	1.33	90.90	42.70	---	133.60	169.00	111.60	140.00
5'-1" long	Inst	Pair	CA	1.33	105.00	42.70	---	147.70	184.00	123.70	153.00
6'-1" long	Inst	Pair	CA	1.33	120.00	42.70	---	162.70	202.00	137.10	169.00
18" W											
2'-1" long	Inst	Pair	CA	1.33	70.90	42.70	---	113.60	146.00	94.10	119.00
3'-1" long	Inst	Pair	CA	1.33	87.30	42.70	---	130.00	164.00	108.50	136.00
4'-1" long	Inst	Pair	CA	1.33	104.00	42.70	---	146.70	183.00	122.70	152.00
5'-1" long	Inst	Pair	CA	1.33	122.00	42.70	---	164.70	204.00	139.10	171.00
6'-1" long	Inst	Pair	CA	1.33	137.00	42.70	---	179.70	222.00	152.10	186.00
Western hemlock, colonial style, primed, 1-1/8" T											
14" wide											
5'-7" long	Inst	Pair	CA	1.33	63.60	42.70	---	106.30	137.00	87.80	112.00
16" wide											
3'-0" long	Inst	Pair	CA	1.33	39.60	42.70	---	82.30	110.00	66.80	88.00
4'-3" long	Inst	Pair	CA	1.33	48.00	42.70	---	90.70	119.00	74.10	96.40
5'-0" long	Inst	Pair	CA	1.33	55.20	42.70	---	97.90	128.00	80.40	104.00
6'-0" long	Inst	Pair	CA	1.33	61.20	42.70	---	103.90	134.00	85.70	110.00
Door blinds, 6'-9" long											
1'-3" wide	Inst	Pair	CA	1.33	79.20	42.70	---	121.90	155.00	101.40	128.00
1'-6" wide	Inst	Pair	CA	1.33	81.60	42.70	---	124.30	158.00	103.50	130.00

Description	Oper	Unit	Crew Size	Man-Hours Per Unit	Avg Mat'l Unit Cost	Avg Labor Unit Cost	Avg Equip Unit Cost	Avg Total Unit Cost	Avg Price Incl O&P	Avg Total Unit Cost	Avg Price Incl O&P
									Costs Based On Small Volume	**Large Volume**	
Cellwood shutters, molded structural foam polystyrene											
Prefinished louver, 16" wide											
3'-3" long	Inst	Pair	CA	1.33	38.80	42.70	---	81.50	**109.00**	66.10	87.20
4'-7" long	Inst	Pair	CA	1.33	49.00	42.70	---	91.70	**120.00**	74.90	97.40
5'-3" long	Inst	Pair	CA	1.33	55.30	42.70	---	98.00	**128.00**	80.50	104.00
6'-3" long	Inst	Pair	CA	1.33	74.90	42.70	---	117.60	**150.00**	97.60	123.00
Door blinds											
16" wide x 6'-9" long	Inst	Pair	CA	1.33	77.30	42.70	---	120.00	**153.00**	99.80	126.00
Prefinished panel, 16" wide											
3'-3" long	Inst	Pair	CA	1.33	40.30	42.70	---	83.00	**110.00**	67.40	88.70
4'-7" long	Inst	Pair	CA	1.33	51.50	42.70	---	94.20	**123.00**	77.20	100.00
5'-3" long	Inst	Pair	CA	1.33	57.40	42.70	---	100.10	**130.00**	82.30	106.00
6'-3" long	Inst	Pair	CA	1.33	74.90	42.70	---	117.60	**150.00**	97.60	123.00
Door blinds											
16" wide x 6'-9" long	Inst	Pair	CA	1.33	81.50	42.70	---	124.20	**158.00**	103.50	130.00

Siding

Aluminum Siding

1. **Dimensions and Descriptions.** The types of aluminum siding discussed are: (1) painted (2) available in various colors, (3) either 0.024" gauge or 0.019" gauge. The types of aluminum siding discussed in this section are:

   a. Clapboard: 12'-6" long; 8" exposure; $^5/_8$" butt; 0.024" gauge.

   b. Board and batten: 10'-0" long; 10" exposure; 1/2" butt; 0.024" gauge.

   c. "Double-five-inch" clapboard: 12'-0" long; 10" exposure; $^1/_2$" butt; 0.024" gauge.

   d. Sculpture clapboard: 12'-6" long; 8" exposure; $^1/_2$" butt; 0.024" gauge.

   e. Shingle siding: 24" long; 12" exposure; 1$^1/_8$" butt; 0.019" gauge.

   Most types of aluminum siding may be available in a smooth or rough texture.

2. **Installation.** Applied horizontally with nails. See chart below.

3. **Estimating Technique.** Determine area and deduct area of window and door openings then add for cutting and fitting waste. In the figures on aluminum siding, no waste has been included.

Wood Siding

The types of wood exterior finishes covered in this section are:

   a. Bevel siding

   b. Drop siding

   c. Vertical siding

   d. Batten siding

   e. Plywood siding with battens

   f. Wood shingle siding

Bevel Siding

1. **Dimension.** Boards are 4", 6", 8", or 12" wide with a thickness of the top edge of $^3/_{16}$". The thickness of the bottom edge is $^9/_{16}$" for 4" and 6" widths and $^{11}/_{16}$" for 8", 10", 12" widths.

2. **Installation.** Applied horizontally with 6d or 8d common nails. See exposure and waste table below.

3. **Estimating Technique.** Determine area and deduct area of window and door openings, then add for lap, exposure, and cutting and fitting waste. In the figures on bevel siding, waste has been included, unless otherwise noted.

Type	Exposure	Butt	Normal Range of Cut & Fit Waste*
Clapboard	8"	$^5/_8$"	6% - 10%
Board & Batten	10"	$^1/_2$"	4% - 8%
Double-Five-Inch	10"	$^1/_2$"	6% - 10%
Sculptured	8"	$^1/_2$"	6% - 10%
Shingle Siding	12"	1$^1/_8$"	4% - 6%

*The amount of waste will vary with the methods of installing siding and can be affected by dimension of the area to be covered.

Board Size	Actual Width	Lap	Exposure	Milling & Lap Waste	Normal Range of Cut & Fit Waste*
$^1/_2$" x 4"	3$^1/_4$"	$^1/_2$"	2$^3/_4$"	31.25%	19% - 23%
$^1/_2$" x 6"	5$^1/_4$"	$^1/_2$"	4$^3/_4$"	21.00%	10% - 13%
$^1/_2$" x 8"	7$^1/_4$"	$^1/_2$"	6$^3/_4$"	15.75%	7% - 10%
$^5/_8$" x 4"	3$^1/_4$"	$^1/_2$"	2$^3/_4$"	31.25%	19% - 23%
$^5/_8$" x 6"	5$^1/_4$"	$^1/_2$"	4$^3/_4$"	21.00%	10% - 13%
$^5/_8$" x 8"	7$^1/_4$"	$^1/_2$"	6$^3/_4$"	15.75%	7% - 10%
$^5/_8$" x 10"	9$^1/_4$"	$^1/_2$"	8$^3/_4$"	12.50%	6% - 10%
$^3/_4$" x 6"	5$^1/_4$"	$^1/_2$"	4$^3/_4$"	21.00%	10% - 13%
$^3/_4$" x 8"	7$^1/_4$"	$^1/_2$"	6$^3/_4$"	15.75%	7% - 10%
$^3/_4$" x 10"	9$^1/_4$"	$^1/_2$"	8$^3/_4$"	12.50%	6% - 10%
$^3/_4$" x 12"	11$^1/_4$"	$^1/_2$"	10$^3/_4$"	10.50%	6% - 10%

*The amount of waste will vary with the methods of installing siding and can be affected by dimension of the area to be covered.

Drop Siding, Tongue and Grooved

1. **Dimension.** The boards are 4", 6", 8", 10", or 12" wide, and all boards are $3/4$" thick at both edges.

2. **Installation.** Applied horizontally, usually with 8d common nails. See exposure and waste table below:

Board Size	Actual Width	Lap +1/16"	Exposure	Milling & Lap Waste	Normal Range of Cut & Waste*
1" x 4"	$3^1/_4$"	$^3/_{16}$"	$3^1/_{16}$"	23.50%	7% - 10%
1" x 6"	$5^1/_4$"	$^3/_{16}$"	$5^1/_{16}$"	15.75%	6% - 8%
1" x 8"	$7^1/_4$"	$^3/_{16}$"	$7^1/_{16}$"	11.75%	5% - 6%
1" x 10"	$9^1/_4$"	$^3/_{16}$"	$9^1/_{16}$"	9.50%	4% - 5%
1" x 12"	$11^1/_4$"	$^3/_{16}$"	$11^1/_{16}$"	8.00%	4% - 5%

*The amount of waste will vary with the methods of installing siding and can be affected by dimensions of the area to be covered.

3. **Estimating Technique.** Determine area and deduct area of window and door openings; then add for lap, exposure, butting and fitting waste. In the figures on drop siding, waste has been included, unless otherwise noted.

Vertical Siding, Tongue and Grooved

1. **Dimensions.** Boards are 8", 10", or 12" wide, and all boards are $3/4$" thick at both top and bottom edges.

2. **Installation.** Applied vertically over horizontal furring strips usually with 8d common nails. See exposure and waste table below:

Board Size	Actual Width	Lap $^1/_{16}$"	Exposure	Milling & Lap Waste	Normal Range of Cut & Fit Waste
1" x 8"	$7^1/_4$"	$^3/_{16}$"	$7^1/_{16}$"	11.75%	5% - 7%
1" x 10"	$9^1/_4$"	$^3/_{16}$"	$9^1/_{16}$"	9.50%	4% - 5%
1" x 12"	$11^1/_4$"	$^3/_{16}$"	$11^1/_{16}$"	8.00%	4% - 5%

3. **Estimating Technique.** Determine area and deduct area of window and door openings; then add for lap, exposure, cutting and fitting waste. In the figures on vertical siding, waste has been included, unless otherwise noted.

Batten Siding

1. **Dimension.** The boards are 8", 10", or 12" wide. If the boards are rough, they are 1" thick; if the boards are dressed, then they are $^{25}/_{32}$" thick. The battens are 1" thick and 2" wide.

2. **Installation.** The 8", 10", or 12"-wide boards are installed vertically over 1" x 3" or 1" x 4" horizontal furring strips. The battens are then installed over the seams of the vertical boards. See exposure and waste table below:

Board Size & Type		Actual Width & Exposure	Milling Waste	Normal Range of Cut & Fit Waste*
1" x 8"	rough sawn	8"	—	3% - 5%
	dressed	$7^1/_2$"	6.25%	3% - 5%
1" x 10"	rough sawn	10"	—	3% - 5%
	dressed	$9^1/_2$"	5.00%	3% - 5%
1" x 12"	rough sawn	12"	—	3% - 5%
	dressed	$11^1/_2$"	4.25%	3% - 5%

*The amount of waste will vary with the methods of installing siding and can be affected by dimension of the area to be covered.

3. **Estimating Technique.** Determine area and deduct area of window and door openings; then add for exposure, cutting and fitting waste. In the figures on batten siding, waste has been included, unless otherwise noted.

Quantity of Battens - No Cut and Fit Waste Included

Board Size and Type		LF per SF or BF
1" x 6"	rough sawn	2.00
	dressed	2.25
1" x 8"	rough sawn	1.55
	dressed	1.60
1" x 10"	rough sawn	1.20
	dressed	1.30
1" x 12"	rough sawn	1.00
	dressed	1.05

Plywood Siding with Battens

1. **Dimensions.** The plywood panels are 4' wide and 8', 9', or 10' long. The panels are $^3/_8$", $^1/_2$" or $^5/_8$" thick. The battens are 1" thick and 2" wide.

2. **Installation.** The panels are applied directly to the studs with the 4'-0" width parallel to the plate. Nails used are 6d commons. The waste will normally range from 3% to 5%.

Quantity of Battens - No Cut and Fit Waste Included	
Spacing	LF per SF or BF
16" oc	.75
24" oc	.50

3. **Estimating Technique.** Determine area and deduct area of window and door openings; then add for cutting and fitting waste. In the figures on plywood siding with battens, waste has been included, unless otherwise noted.

Wood Shingle Siding

1. **Dimension.** The wood shingles have an average width of 4" and length of 16", 18" or 24". The thickness may vary in all cases.

2. **Installation.** On roofs with wood shingles, there should be three thicknesses of wood shingle at every point on the roof. But on exterior walls with wood shingles, only two thicknesses of wood siding shingles are needed. Nail to 1" x 3" or 1" x 4" furring strips using 3d rust-resistant nails. Use 5d rust-resistant nails when applying new shingles over old shingles. See table below:

Shingle Length	Exposure	Shingles per Square, No Waste Included	Lbs. of Nails at Two Nails per Shingle	
			3d	5d
16"	$7^1/_2$"	480	2.3	2.7
18"	$8^1/_2$"	424	2.0	2.4
24"	$11^1/_2$"	313	1.5	1.8

3. **Estimating Technique.** Determine the area and deduct the area of window and door openings. Then add approximately 10% for waste. In the figures on wood shingle siding, waste has been included, unless otherwise noted. In the above table, pounds of nails per square are based on two nails per shingle. For one nail per shingle, deduct 50%.

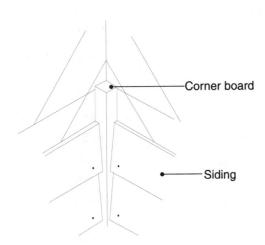

Corner board for application of horizontal siding at interior corner

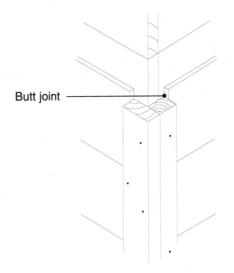

Corner boards for application of horizontal siding at exterior corner

				Costs Based On Small Volume					Large Volume	
Description	Oper Unit	Crew Size	Man-Hours Per Unit	Avg Mat'l Unit Cost	Avg Labor Unit Cost	Avg Equip Unit Cost	Avg Total Unit Cost	Avg Price Incl O&P	Avg Total Unit Cost	Avg Price Incl O&P

Siding

Aluminum, nailed to wood

Clapboard (i.e. lap, drop)

8" exposure	Demo SF	LB	.022	---	.59	---	.59	.87	.43	.64
10" exposure	Demo SF	LB	.018	---	.48	---	.48	.71	.35	.52

Clapboard, horizontal, 1/2" butt

Straight wall

8" exposure	Inst SF	2C	.037	2.53	1.19	---	3.72	4.69	2.98	3.72
10" exposure	Inst SF	2C	.030	2.10	.96	---	3.06	3.86	2.46	3.07

Cut-up wall

8" exposure	Inst SF	2C	.044	3.32	1.41	---	4.73	5.94	3.82	4.75
10" exposure	Inst SF	2C	.036	2.88	1.16	---	4.04	5.05	3.25	4.02

Corrugated (2-1/2"), 26" W, 2-1/2" side lap, 4" end lap

	Demo SF	LB	.019	---	.51	---	.51	.75	.35	.52

0.0175" thick

Natural	Inst SF	UC	.028	2.90	.87	---	3.77	4.63	3.06	3.72
Painted	Inst SF	UC	.028	3.15	.87	---	4.02	4.91	3.27	3.98

0.019" thick

Natural	Inst SF	UC	.028	2.96	.87	---	3.83	4.69	3.11	3.77
Painted	Inst SF	UC	.028	3.21	.87	---	4.08	4.98	3.32	4.02

Panels, 4' x 8' board and batten (12" oc)

	Demo SF	LB	.010	---	.27	---	.27	.40	.19	.28

Straight wall

Smooth	Inst SF	2C	.019	2.60	.61	---	3.21	3.90	2.64	3.18
Rough sawn	Inst SF	2C	.019	2.86	.61	---	3.47	4.20	2.85	3.42

Cut-up wall

Smooth	Inst SF	2C	.022	3.40	.71	---	4.11	4.97	3.40	4.09
Rough sawn	Inst SF	2C	.022	3.66	.71	---	4.37	5.27	3.62	4.35

Shingle, 24" L with 12" exposure

	Demo SF	LB	.016	---	.43	---	.43	.64	.29	.44
Straight wall	Inst SF	2C	.026	.06	.83	---	.89	1.32	.63	.92
Cut-up wall	Inst SF	2C	.032	.06	1.03	---	1.09	1.61	.79	1.16

Hardboard siding, Masonite

Lap, 1/2" T x 12" W x 16' L, with 11" exposure

	Demo SF	LB	.017	---	.45	---	.45	.67	.32	.48

Straight wall

Smooth finish	Inst SF	2C	.033	1.65	1.06	---	2.71	3.49	2.14	2.72
Textured finish	Inst SF	2C	.033	1.95	1.06	---	3.01	3.83	2.39	3.00
Wood-like finish	Inst SF	2C	.033	2.08	1.06	---	3.14	3.98	2.51	3.14

Cut-up wall

Smooth finish	Inst SF	2C	.039	1.69	1.25	---	2.94	3.82	2.31	2.96
Textured finish	Inst SF	2C	.039	2.00	1.25	---	3.25	4.18	2.57	3.25
Wood-like finish	Inst SF	2C	.039	2.13	1.25	---	3.38	4.33	2.68	3.38

Description	Oper	Unit	Crew Size	Man-Hours Per Unit	Avg Mat'l Unit Cost	Avg Labor Unit Cost	Avg Equip Unit Cost	Avg Total Unit Cost	Avg Price Incl O&P	Avg Total Unit Cost	Avg Price Incl O&P
								Costs Based On Small Volume		Large Volume	
Panels, 7/16" T x 4' W x 8' H											
	Demo	SF	LB	.009	---	.24	---	.24	**.36**	.16	**.24**
Grooved/planked panels, textured or wood-like finish											
Primecote	Inst	SF	2C	.019	1.55	.61	---	2.16	**2.70**	1.74	**2.14**
Unprimed	Inst	SF	2C	.019	1.68	.61	---	2.29	**2.85**	1.85	**2.27**
Panel with batten preattached											
Unprimed	Inst	SF	2C	.019	1.77	.61	---	2.38	**2.95**	1.93	**2.36**
Stucco textured panel											
Primecote	Inst	SF	2C	.019	1.82	.61	---	2.43	**3.01**	1.97	**2.41**
Vinyl											
Solid vinyl panels .024 gauge, woodgrained or colors											
10" or double 5" panels	Demo	SF	LB	.017	---	.45	---	.45	**.67**	.32	**.48**
10" or double 5" panels, standard	Inst	SF	2C	.046	.88	1.48	---	2.36	**3.23**	1.78	**2.40**
10" or double 5" panels, insulated	Inst	SF	2C	.046	1.23	1.48	---	2.71	**3.63**	2.07	**2.74**
Starter strip	Inst	LF	2C	.057	.32	1.83	---	2.15	**3.11**	1.55	**2.24**
J channel for corners	Inst	LF	2C	.051	.74	1.64	---	2.38	**3.31**	1.79	**2.46**
Outside corner and corner post	Inst	LF	2C	.057	1.09	1.83	---	2.92	**4.00**	2.20	**2.98**
Casing and trim	Inst	LF	2C	.065	.57	2.09	---	2.66	**3.78**	1.97	**2.78**
Soffit and facia	Inst	LF	2C	.057	.81	1.83	---	2.64	**3.68**	1.97	**2.72**
Wood											
Bevel											
1/2" x 8" with 6-3/4" exp.	Demo	SF	LB	.025	---	.67	---	.67	**.99**	.48	**.71**
5/8" x 10" with 8-3/4" exp.	Demo	SF	LB	.019	---	.51	---	.51	**.75**	.37	**.56**
3/4" x 12" with 10-3/4" exp.	Demo	SF	LB	.016	---	.43	---	.43	**.64**	.29	**.44**
Bevel (horizontal), 1/2" lap, 10' L											
Redwood, kiln-dried, clear VG, pattern											
1/2" x 8", with 6-3/4" exp.											
Rough ends	Inst	SF	2C	.052	3.08	1.67	---	4.75	**6.05**	3.80	**4.78**
Fitted ends	Inst	SF	2C	.068	3.15	2.18	---	5.33	**6.90**	4.21	**5.38**
Mitered ends	Inst	SF	2C	.080	3.23	2.57	---	5.80	**7.57**	4.53	**5.84**
5/8" x 10", with 8-3/4" exp.											
Rough ends	Inst	SF	2C	.041	3.01	1.32	---	4.33	**5.44**	3.48	**4.33**
Fitted ends	Inst	SF	2C	.053	3.08	1.70	---	4.78	**6.09**	3.81	**4.79**
Mitered ends	Inst	SF	2C	.062	3.16	1.99	---	5.15	**6.62**	4.06	**5.15**
3/4" x 12", with 10-3/4" exp.											
Rough ends	Inst	SF	2C	.033	5.21	1.06	---	6.27	**7.58**	5.16	**6.19**
Fitted ends	Inst	SF	2C	.043	5.34	1.38	---	6.72	**8.21**	5.49	**6.65**
Mitered ends	Inst	SF	2C	.050	5.48	1.60	---	7.08	**8.71**	5.76	**7.02**
Red cedar, clear, kiln-dried											
1/2" x 8", with 6-3/4" exp.											
Rough ends	Inst	SF	2C	.052	2.73	1.67	---	4.40	**5.64**	3.51	**4.45**
Fitted ends	Inst	SF	2C	.068	2.79	2.18	---	4.97	**6.48**	3.91	**5.04**
Mitered ends	Inst	SF	2C	.080	2.85	2.57	---	5.42	**7.13**	4.22	**5.48**
5/8" x 10", with 8-3/4" exp.											
Rough ends	Inst	SF	2C	.041	2.86	1.32	---	4.18	**5.26**	3.36	**4.19**
Fitted ends	Inst	SF	2C	.053	2.93	1.70	---	4.63	**5.92**	3.68	**4.64**
Mitered ends	Inst	SF	2C	.062	3.00	1.99	---	4.99	**6.43**	3.93	**5.00**
3/4" x 12", with 10-3/4" exp.											
Rough ends	Inst	SF	2C	.033	4.72	1.06	---	5.78	**7.02**	4.75	**5.72**
Fitted ends	Inst	SF	2C	.043	4.84	1.38	---	6.22	**7.64**	5.07	**6.17**
Mitered ends	Inst	SF	2C	.050	4.96	1.60	---	6.56	**8.11**	5.33	**6.53**

				Costs Based On Small Volume						Large Volume	
Description	Oper	Unit	Crew Size	Man-Hours Per Unit	Avg Mat'l Unit Cost	Avg Labor Unit Cost	Avg Equip Unit Cost	Avg Total Unit Cost	Avg Price Incl O&P	Avg Total Unit Cost	Avg Price Incl O&P
Drop (horizontal) T&G, 1/4" lap											
1" x 8" with 7" exp.	Demo	SF	LB	.024	---	.64	---	.64	.95	.45	.6
1" x 10" with 9" exp.	Demo	SF	LB	.019	---	.51	---	.51	.75	.35	.5
Drop (horizontal) T&G, 1/4" lap, 10' L (avg.)											
Redwood, kiln-dried, clear VG, patterns											
1" x 8", with 7" exp.											
Rough ends	Inst	SF	2C	.049	5.05	1.57	---	6.62	8.17	5.38	6.5
Fitted ends	Inst	SF	2C	.063	5.17	2.02	---	7.19	8.98	5.80	7.1
Mitered ends	Inst	SF	2C	.072	5.29	2.31	---	7.60	9.55	6.13	7.6
1" x 8", with 7-3/4" exp.											
Rough ends	Inst	SF	2C	.044	4.58	1.41	---	5.99	7.39	4.88	5.9
Fitted ends	Inst	SF	2C	.057	4.71	1.83	---	6.54	8.16	5.27	6.5
Mitered ends	Inst	SF	2C	.065	4.83	2.09	---	6.92	8.68	5.58	6.9
1" x 10", with 9" exp.											
Rough ends	Inst	SF	2C	.038	5.30	1.22	---	6.52	7.92	5.36	6.4
Fitted ends	Inst	SF	2C	.049	5.43	1.57	---	7.00	8.60	5.69	6.9
Mitered ends	Inst	SF	2C	.056	5.56	1.80	---	7.36	9.09	5.99	7.3
1" x 10", with 9-3/4" exp.											
Rough ends	Inst	SF	2C	.035	4.92	1.12	---	6.04	7.34	4.97	6.0
Fitted ends	Inst	SF	2C	.045	5.05	1.44	---	6.49	7.97	5.31	6.4
Mitered ends	Inst	SF	2C	.052	5.18	1.67	---	6.85	8.46	5.55	6.7
Red cedar, clear, kiln-dried											
1" x 8", with 7" exp.											
Rough ends	Inst	SF	2C	.049	4.61	1.57	---	6.18	7.66	5.01	6.1
Fitted ends	Inst	SF	2C	.063	4.72	2.02	---	6.74	8.46	5.42	6.7
Mitered ends	Inst	SF	2C	.072	4.83	2.31	---	7.14	9.02	5.75	7.1
1" x 10", with 9" exp.											
Rough ends	Inst	SF	2C	.038	4.77	1.22	---	5.99	7.31	4.92	5.9
Fitted ends	Inst	SF	2C	.049	4.89	1.57	---	6.46	7.98	5.24	6.4
Mitered ends	Inst	SF	2C	.056	5.01	1.80	---	6.81	8.46	5.53	6.8
Pine, kiln-dried, select											
Rough sawn											
1" x 8", with 7-3/4" exp.											
Rough ends	Inst	SF	2C	.044	3.22	1.41	---	4.63	5.82	3.72	4.63
Fitted ends	Inst	SF	2C	.057	3.30	1.83	---	5.13	6.54	4.08	5.15
Mitered ends	Inst	SF	2C	.065	3.39	2.09	---	5.48	7.03	4.35	5.51
1" x 10", with 9-3/4" exp.											
Rough ends	Inst	SF	2C	.035	4.06	1.12	---	5.18	6.35	4.26	5.18
Fitted ends	Inst	SF	2C	.045	4.17	1.44	---	5.61	6.96	4.58	5.62
Mitered ends	Inst	SF	2C	.052	4.28	1.67	---	5.95	7.43	4.80	5.92
Dressed/milled											
1" x 8", with 7" exp.											
Rough ends	Inst	SF	2C	.049	3.84	1.57	---	5.41	6.77	4.36	5.40
Fitted ends	Inst	SF	2C	.063	3.94	2.02	---	5.96	7.56	4.75	5.96
Mitered ends	Inst	SF	2C	.072	4.03	2.31	---	6.34	8.10	5.06	6.39
1" x 10", with 9" exp.											
Rough ends	Inst	SF	2C	.038	4.80	1.22	---	6.02	7.35	4.94	5.98
Fitted ends	Inst	SF	2C	.049	4.92	1.57	---	6.49	8.02	5.26	6.43
Mitered ends	Inst	SF	2C	.056	5.03	1.80	---	6.83	8.48	5.56	6.85

Description	Oper	Unit	Crew Size	Man-Hours Per Unit	Avg Mat'l Unit Cost	Avg Labor Unit Cost	Avg Equip Unit Cost	Avg Total Unit Cost	Avg Price Incl O&P	Avg Total Unit Cost	Avg Price Incl O&P
						Costs Based On Small Volume					**Large Volume**	
Board (1" x 12") and batten (1" x 2", 12" oc)											
Horizontal	Demo	SF	LB	.018	---	.48	---	.48	.71	.35	.52
Vertical											
Standard	Demo	SF	LB	.022	---	.59	---	.59	.87	.43	.64
Reverse	Demo	SF	LB	.021	---	.56	---	.56	.83	.40	.60
Board (1" x 12") and batten (1" x 2", 12" oc); rough sawn, common, standard and better, green											
Horizontal application, @ 10' L											
Redwood	Inst	SF	2C	.041	2.37	1.32	---	3.69	4.70	2.94	3.71
Red cedar	Inst	SF	2C	.041	1.75	1.32	---	3.07	3.99	2.36	3.04
Pine	Inst	SF	2C	.041	2.35	1.32	---	3.67	4.68	2.93	3.70
Vertical, std. (batten on board), @ 8' L											
Redwood	Inst	SF	2C	.051	2.37	1.64	---	4.01	5.18	3.17	4.04
Red cedar	Inst	SF	2C	.051	1.75	1.64	---	3.39	4.47	2.59	3.38
Pine	Inst	SF	2C	.051	2.35	1.64	---	3.99	5.16	3.16	4.03
Vertical, reverse (board on batten), @ 8' L											
Redwood	Inst	SF	2C	.052	2.37	1.67	---	4.04	5.23	3.17	4.04
Red cedar	Inst	SF	2C	.052	1.75	1.67	---	3.42	4.52	2.59	3.38
Pine	Inst	SF	2C	.052	2.35	1.67	---	4.02	5.21	3.16	4.03
Board on board (1" x 12" with 1-1/2" overlap),											
Vertical	Demo	SF	LB	.024	---	.64	---	.64	.95	.45	.67
Board on board (1" x 12" with 1-1/2" overlap) rough sawn common, standard & better green, 8' L; vertical											
Redwood	Inst	SF	2C	.039	2.28	1.25	---	3.53	4.50	2.80	3.52
Red cedar	Inst	SF	2C	.039	1.70	1.25	---	2.95	3.83	2.26	2.90
Pine	Inst	SF	2C	.039	2.40	1.25	---	3.65	4.64	2.91	3.65
Plywood (1/2" T) with battens (1" x 2")											
16" oc battens	Demo	SF	LB	.010	---	.27	---	.27	.40	.19	.28
24" oc battens	Demo	SF	LB	.010	---	.27	---	.27	.40	.19	.28
Plywood (1/2" T) sanded exterior, AB grade with battens (1" x 2") S4S, common, standard & better green											
16" oc											
Fir	Inst	SF	2C	.031	1.69	.99	---	2.68	3.44	2.16	2.73
Pine	Inst	SF	2C	.031	1.74	.99	---	2.73	3.49	2.20	2.77
24" oc											
Fir	Inst	SF	2C	.026	1.64	.83	---	2.47	3.14	1.98	2.48
Pine	Inst	SF	2C	.026	1.67	.83	---	2.50	3.17	2.01	2.51
Shakes only, no sheathing included											
24" L with 11-1/2" exposure											
1/2" to 3/4" T	Demo	SF	LB	.014	---	.37	---	.37	.56	.27	.40
3/4" to 5/4" T	Demo	SF	LB	.016	---	.43	---	.43	.64	.29	.44

Description	Oper	Unit	Crew Size	Man-Hours Per Unit	Avg Mat'l Unit Cost	Avg Labor Unit Cost	Avg Equip Unit Cost	Avg Total Unit Cost	Avg Price Incl O&P	Avg Total Unit Cost	Avg Price Incl O&P
								Costs Based On Small Volume		Large Volume	
Shakes only, over sheathing, 2 nails/shake											
24" L with 11-1/2" exp., @ 6" W, red cedar, sawn one side											
Straight wall											
1/2" to 3/4" T	Inst	SF	CN	.030	1.79	.93	---	2.72	3.45	2.17	2.7?
3/4" to 5/4" T	Inst	SF	CN	.034	2.38	1.05	---	3.43	4.32	2.76	3.4?
Cut-up wall											
1/2" to 3/4" T	Inst	SF	CN	.038	1.83	1.18	---	3.01	3.87	2.39	3.0?
3/4" to 5/4" T	Inst	SF	CN	.042	2.42	1.30	---	3.72	4.74	2.95	3.7?
Related materials and operations											
Rosin sized sheathing paper, 36" W 500 SF/roll; nailed											
Over solid sheathing	Inst	SF	2C	.002	---	.06	---	.06	.10	.03	.0?
Shingles only, no sheathing included											
16" L with 7-1/2" exposure	Demo	SF	LB	.023	---	.61	---	.61	.91	.43	.6?
18" L with 8-1/2" exposure	Demo	SF	LB	.020	---	.53	---	.53	.79	.37	.5?
24" L with 11-1/2" exposure	Demo	SF	LB	.015	---	.40	---	.40	.60	.27	.4?
Shingles only, red cedar, No 1. perfect, @ 4" W, 2 nails/shake, over sheathing											
Straight wall											
16" L with 7-1/2" exposure	Inst	SF	2C	.039	2.09	1.25	---	3.34	4.28	2.65	3.3?
18" L with 8-1/2" exposure	Inst	SF	2C	.035	1.60	1.12	---	2.72	3.52	2.13	2.7?
24" L with 11-1/2" exposure	Inst	SF	2C	.026	2.05	.83	---	2.88	3.61	2.32	2.8?
Cut-up wall											
16" L with 7-1/2" exposure	Inst	SF	2C	.049	2.13	1.57	---	3.70	4.81	2.90	3.7?
18" L with 8-1/2" exposure	Inst	SF	2C	.043	1.63	1.38	---	3.01	3.94	2.35	3.0?
24" L with 11-1/2" exposure	Inst	SF	2C	.032	2.09	1.03	---	3.12	3.94	2.49	3.11
Related materials and operations											
Rosin sized sheathing paper, 36" W 500 SF/roll; nailed											
Over solid sheathing	Inst	SF	2C	.002	---	.06	---	.06	.10	.03	.0?

Description	Oper	Unit	Crew Size	Man-Hours Per Unit	Costs Based On Small Volume						Large Volume	
					Avg Mat'l Unit Cost	Avg Labor Unit Cost	Avg Equip Unit Cost	Avg Total Unit Cost	Avg Price Incl O&P		Avg Total Unit Cost	Avg Price Incl O&P

Sinks

Bathroom

Lavatories (bathroom sinks), with good quality fittings and faucets, concealed overflow

Vitreous china

Countertop units, with self rim (American Standard Products)

Description	Oper	Unit	Crew Size	Man-Hours Per Unit	Avg Mat'l Unit Cost	Avg Labor Unit Cost	Avg Equip Unit Cost	Avg Total Unit Cost	Avg Price Incl O&P	Avg Total Unit Cost	Avg Price Incl O&P
Antiquity											
24-1/2" x 18" with 8" or 4" cc											
Color	Inst	Ea	SA	3.81	421.00	138.00	---	559.00	688.00	468.10	569.00
White	Inst	Ea	SA	3.81	377.00	138.00	---	515.00	638.00	429.10	525.00
Aqualyn											
20" x 17" with 8" or 4" cc											
Color	Inst	Ea	SA	3.81	349.00	138.00	---	487.00	605.00	404.10	496.00
White	Inst	Ea	SA	3.81	316.00	138.00	---	454.00	567.00	375.10	463.00
Ellisse Petite											
24" x 19" with 8" or 4" cc											
Color	Inst	Ea	SA	3.81	514.00	138.00	---	652.00	795.00	549.10	663.00
White	Inst	Ea	SA	3.81	451.00	138.00	---	589.00	722.00	494.10	599.00
Heritage											
26" x 20" with 8" or 4" cc											
Color	Inst	Ea	SA	3.81	906.00	138.00	---	1044.00	1250.00	895.10	1060.00
White	Inst	Ea	SA	3.81	766.00	138.00	---	904.00	1080.00	771.10	918.00
Hexalyn											
22" x 19" with 8" or 4" cc											
Color	Inst	Ea	SA	3.81	463.00	138.00	---	601.00	736.00	505.10	612.00
White	Inst	Ea	SA	3.81	410.00	138.00	---	548.00	675.00	458.10	558.00
Lexington											
22" x 19" with 8" or 4" cc											
Color	Inst	Ea	SA	3.81	590.00	138.00	---	728.00	882.00	616.10	740.00
White	Inst	Ea	SA	3.81	512.00	138.00	---	650.00	792.00	547.10	661.00
Pallas Petite											
24" x 18" with 8" or 4" cc											
Color	Inst	Ea	SA	3.81	628.00	138.00	---	766.00	926.00	650.10	778.00
White	Inst	Ea	SA	3.81	544.00	138.00	---	682.00	829.00	575.10	693.00
Renaissance											
19" diameter with 8" or 4" cc											
Color	Inst	Ea	SA	3.81	345.00	138.00	---	483.00	600.00	400.10	491.00
White	Inst	Ea	SA	3.81	313.00	138.00	---	451.00	564.00	372.10	459.00
Roma											
26" x 20" with 8" or 4" cc w/ sealant											
Color	Inst	Ea	SA	3.81	860.00	138.00	---	998.00	1190.00	854.10	1010.00
White	Inst	Ea	SA	3.81	729.00	138.00	---	867.00	1040.00	738.10	880.00
Rondalyn											
19" diameter with 8" or 4" cc											
Color	Inst	Ea	SA	3.81	333.00	138.00	---	471.00	587.00	390.10	480.00
White	Inst	Ea	SA	3.81	305.00	138.00	---	443.00	554.00	365.10	451.00

					Costs Based On Small Volume					Large Volume	
Description	Oper	Unit	Crew Size	Man-Hours Per Unit	Avg Mat'l Unit Cost	Avg Labor Unit Cost	Avg Equip Unit Cost	Avg Total Unit Cost	Avg Price Incl O&P	Avg Total Unit Cost	Avg Price Incl O&P

Counter and undercounter units (Kohler Products, Artists Edition)

Ankara
Vintage countertop, 17" x 14"

| White | Inst | Ea | SA | 3.81 | 1430.00 | 138.00 | --- | 1568.00 | **1850.00** | 1357.10 | 1590.0 |

Vintage undercounter, 17" x 14"

| White | Inst | Ea | SA | 3.81 | 1240.00 | 138.00 | --- | 1378.00 | **1630.00** | 1187.10 | 1400.0 |

Briar Rose
Vintage countertop, 17" x 14"

| White | Inst | Ea | SA | 3.81 | 1360.00 | 138.00 | --- | 1498.00 | **1770.00** | 1297.10 | 1520.0 |

Vintage undercounter, 17" x 14"

| White | Inst | Ea | SA | 3.81 | 1240.00 | 138.00 | --- | 1378.00 | **1630.00** | 1187.10 | 1400.0 |

Crimson Topaz
Vintage countertop, 17" x 14"

| White | Inst | Ea | SA | 3.81 | 1210.00 | 138.00 | --- | 1348.00 | **1590.00** | 1157.10 | 1360.0 |

Vintage undercounter, 17" x 14"

| White | Inst | Ea | SA | 3.81 | 1210.00 | 138.00 | --- | 1348.00 | **1590.00** | 1157.10 | 1360.0 |

Fables and Flowers
Vintage countertop, 17" x 14"

| White | Inst | Ea | SA | 3.81 | 1860.00 | 138.00 | --- | 1998.00 | **2340.00** | 1727.10 | 2020.0 |

Vintage undercounter, 17" x 14"

| White | Inst | Ea | SA | 3.81 | 1400.00 | 138.00 | --- | 1538.00 | **1820.00** | 1327.10 | 1560.0 |

Peonies and Ivy
Undercounter

| White | Inst | Ea | SA | 3.81 | 1240.00 | 138.00 | --- | 1378.00 | **1630.00** | 1187.10 | 1400.0 |

Provencial
Vintage countertop, 17" x 14"

| White | Inst | Ea | SA | 3.81 | 1430.00 | 138.00 | --- | 1568.00 | **1850.00** | 1357.10 | 1590.0 |

Vintage undercounter, 17" x 14"

| White | Inst | Ea | SA | 3.81 | 1240.00 | 138.00 | --- | 1378.00 | **1630.00** | 1187.10 | 1400.0 |

Pedestal sinks (American Standard Products)

Antiquity w/ pedestal
24-1/2" x 19" with 8" or 4" cc

| Color | Inst | Ea | SA | 3.81 | 394.00 | 138.00 | --- | 532.00 | **657.00** | 444.10 | 542.0 |
| White | Inst | Ea | SA | 3.81 | 514.00 | 138.00 | --- | 652.00 | **795.00** | 549.10 | 663.0 |

Lavatory slab only

| Color | Inst | Ea | SA | 2.00 | 238.00 | 72.70 | --- | 310.70 | **381.00** | 268.20 | 327.0 |
| White | Inst | Ea | SA | 2.00 | 189.00 | 72.70 | --- | 261.70 | **324.00** | 224.20 | 276.0 |

Pedestal only

| Color | Inst | Ea | SA | 2.00 | 156.00 | 72.70 | --- | 228.70 | **286.00** | 195.20 | 243.0 |
| White | Inst | Ea | SA | 2.00 | 125.00 | 72.70 | --- | 197.70 | **251.00** | 168.20 | 212.0 |

Cadet w/ pedestal
24" x 20" with 8" or 4" cc

| Color | Inst | Ea | SA | 3.81 | 536.00 | 138.00 | --- | 674.00 | **820.00** | 568.10 | 685.00 |
| White | Inst | Ea | SA | 3.81 | 472.00 | 138.00 | --- | 610.00 | **747.00** | 513.10 | 621.00 |

Lavatory slab only

| Color | Inst | Ea | SA | 2.00 | 245.00 | 72.70 | --- | 317.70 | **389.00** | 274.20 | 334.00 |
| White | Inst | Ea | SA | 2.00 | 199.00 | 72.70 | --- | 271.70 | **336.00** | 233.20 | 287.00 |

Pedestal only

| Color | Inst | Ea | SA | 2.00 | 97.20 | 72.70 | --- | 169.90 | **219.00** | 143.70 | 184.00 |
| White | Inst | Ea | SA | 2.00 | 79.10 | 72.70 | --- | 151.80 | **198.00** | 127.80 | 166.00 |

Description	Oper	Unit	Crew Size	Man-Hours Per Unit	Avg Mat'l Unit Cost	Avg Labor Unit Cost	Avg Equip Unit Cost	Avg Total Unit Cost	Avg Price Incl O&P	Avg Total Unit Cost	Avg Price Incl O&P
					Costs Based On Small Volume					**Large Volume**	
Ellisse Petite w/ pedestal											
25" x 21" with 8" or 4" cc											
Color	Inst	Ea	SA	3.81	945.00	138.00	---	1083.00	1290.00	928.10	1100.00
White	Inst	Ea	SA	3.81	796.00	138.00	---	934.00	1120.00	797.10	948.00
Lavatory slab only											
Color	Inst	Ea	SA	2.00	444.00	72.70	---	516.70	618.00	449.20	535.00
White	Inst	Ea	SA	2.00	356.00	72.70	---	428.70	516.00	371.20	446.00
Pedestal only											
Color	Inst	Ea	SA	2.00	301.00	72.70	---	373.70	453.00	323.20	390.00
White	Inst	Ea	SA	2.00	244.00	72.70	---	316.70	388.00	273.20	333.00
Lexington w/ pedestal											
24" x 18" with 8" or 4" cc											
Color	Inst	Ea	SA	3.81	793.00	138.00	---	931.00	1120.00	795.10	945.00
White	Inst	Ea	SA	3.81	677.00	138.00	---	815.00	982.00	693.10	828.00
Lavatory slab only											
Color	Inst	Ea	SA	2.00	406.00	72.70	---	478.70	573.00	415.20	496.00
White	Inst	Ea	SA	2.00	331.00	72.70	---	403.70	488.00	349.20	421.00
Pedestal only											
Color	Inst	Ea	SA	2.00	188.00	72.70	---	260.70	323.00	223.20	275.00
White	Inst	Ea	SA	2.00	146.00	72.70	---	218.70	275.00	186.20	233.00
Repertoire w/ pedestal											
23" x 18-1/4" with 8" or 4" cc											
Color	Inst	Ea	SA	3.81	632.00	138.00	---	770.00	930.00	653.10	782.00
White	Inst	Ea	SA	3.81	558.00	138.00	---	696.00	846.00	588.10	708.00
Lavatory slab only											
Color	Inst	Ea	SA	2.00	314.00	72.70	---	386.70	468.00	334.20	403.00
White	Inst	Ea	SA	2.00	260.00	72.70	---	332.70	406.00	287.20	349.00
Pedestal only											
Color	Inst	Ea	SA	2.00	118.00	72.70	---	190.70	242.00	161.20	204.00
White	Inst	Ea	SA	2.00	94.90	72.70	---	167.60	216.00	141.70	182.00

Pedestal sinks (Kohler Products)

Description	Oper	Unit	Crew Size	Man-Hours Per Unit	Avg Mat'l Unit Cost	Avg Labor Unit Cost	Avg Equip Unit Cost	Avg Total Unit Cost	Avg Price Incl O&P	Avg Total Unit Cost	Avg Price Incl O&P
Anatole											
21" x 18" with 8" or 4" cc											
Color	Inst	Ea	SA	3.81	1070.00	138.00	---	1208.00	1430.00	1039.10	1230.00
Premium color	Inst	Ea	SA	3.81	1200.00	138.00	---	1338.00	1580.00	1157.10	1360.00
White	Inst	Ea	SA	3.81	870.00	138.00	---	1008.00	1200.00	862.10	1020.00
Cabernet											
22" x 18" x 33" with 8" or 4" cc											
Color	Inst	Ea	SA	3.81	775.00	138.00	---	913.00	1090.00	779.10	927.00
Premium color	Inst	Ea	SA	3.81	861.00	138.00	---	999.00	1190.00	854.10	1010.00
White	Inst	Ea	SA	3.81	642.00	138.00	---	780.00	942.00	662.10	793.00
27" x 20" x 33" with 8" or 4" cc											
Color	Inst	Ea	SA	3.81	1110.00	138.00	---	1248.00	1480.00	1072.10	1260.00
Premium color	Inst	Ea	SA	3.81	1240.00	138.00	---	1378.00	1630.00	1187.10	1400.00
White	Inst	Ea	SA	3.81	898.00	138.00	---	1036.00	1240.00	887.10	1050.00
Chablis											
24" x 19" with 8" or 4" cc											
Color	Inst	Ea	SA	3.81	876.00	138.00	---	1014.00	1210.00	868.10	1030.00
Premium color	Inst	Ea	SA	3.81	977.00	138.00	---	1115.00	1330.00	957.10	1130.00
White	Inst	Ea	SA	3.81	720.00	138.00	---	858.00	1030.00	730.10	871.00

					Costs Based On Small Volume					Large Volume	
Description	Oper	Unit	Crew Size	Man-Hours Per Unit	Avg Mat'l Unit Cost	Avg Labor Unit Cost	Avg Equip Unit Cost	Avg Total Unit Cost	Avg Price Incl O&P	Avg Total Unit Cost	Avg Price Incl O&P
Chardonnay											
28" x 21-1/2" with 8" or 4" cc											
Color	Inst	Ea	SA	3.81	1520.00	138.00	---	1658.00	1950.00	1427.10	1680.0
Premium color	Inst	Ea	SA	3.81	1710.00	138.00	---	1848.00	2170.00	1607.10	1880.0
White	Inst	Ea	SA	3.81	1210.00	138.00	---	1348.00	1600.00	1167.10	1370.0
Fleur											
28" x 21-1/2" with 8" or 4" cc											
Color	Inst	Ea	SA	3.81	1380.00	138.00	---	1518.00	1790.00	1307.10	1540.0
White	Inst	Ea	SA	3.81	1110.00	138.00	---	1248.00	1480.00	1070.10	1260.0
Pillow Talk											
27-1/2" x 20-1/2" x 33" with 8" or 4" cc											
Color	Inst	Ea	SA	3.81	1810.00	138.00	---	1948.00	2290.00	1697.10	1980.0
Premium color	Inst	Ea	SA	3.81	2060.00	138.00	---	2198.00	2570.00	1907.10	2220.0
White	Inst	Ea	SA	3.81	1440.00	138.00	---	1578.00	1860.00	1367.10	1600.0
Pinoir											
22" x 18" x 33" with 8" or 4" cc											
Color	Inst	Ea	SA	3.81	575.00	138.00	---	713.00	864.00	603.10	724.0
Premium color	Inst	Ea	SA	3.81	631.00	138.00	---	769.00	929.00	652.10	781.0
White	Inst	Ea	SA	3.81	488.00	138.00	---	626.00	765.00	527.10	637.0
Revival											
23-7/8" x 17-7/8" x 34" with 8" or 4" cc											
Color	Inst	Ea	SA	3.81	1030.00	138.00	---	1168.00	1390.00	1003.10	1180.0
Premium color	Inst	Ea	SA	3.81	1150.00	138.00	---	1288.00	1530.00	1117.10	1310.0
White	Inst	Ea	SA	3.81	838.00	138.00	---	976.00	1170.00	834.10	991.0
Wall hung units (American Standard Products)											
Corner Minette											
11" x 16" with 8" or 4" cc											
Color	Inst	Ea	SA	2.86	572.00	104.00	---	676.00	810.00	575.70	685.0
White	Inst	Ea	SA	2.86	478.00	104.00	---	582.00	702.00	493.70	591.0
Declyn											
19" x 17" with 8" or 4" cc											
Color	Inst	Ea	SA	3.81	334.00	138.00	---	472.00	588.00	391.10	481.0
White	Inst	Ea	SA	3.81	310.00	138.00	---	448.00	560.00	369.10	456.0
Penlyn											
18" x 16" with 8" or 4" cc											
White	Inst	Ea	SA	3.81	428.00	138.00	---	566.00	696.00	474.10	576.0
Roxalyn											
20" x 18" with 8" or 4" cc											
Color	Inst	Ea	SA	3.81	516.00	138.00	---	654.00	797.00	551.10	665.0
White	Inst	Ea	SA	3.81	432.00	138.00	---	570.00	700.00	477.10	579.0
Wheelchair, meets ADA requirements											
27" x 20" with 8" or 4" cc											
Color	Inst	Ea	SA	3.81	876.00	138.00	---	1014.00	1210.00	868.10	1030.0
White	Inst	Ea	SA	3.81	688.00	138.00	---	826.00	995.00	703.10	839.0

Description	Oper	Unit	Crew Size	Man-Hours Per Unit	Avg Mat'l Unit Cost	Avg Labor Unit Cost	Avg Equip Unit Cost	Avg Total Unit Cost	Avg Price Incl O&P	Avg Total Unit Cost	Avg Price Incl O&P
										Large Volume	
Wall hung units (Kohler Products)											
Cabernet											
27" x 20-1/2" with 8" or 4" cc											
Color	Inst	Ea	SA	3.81	1120.00	138.00	---	1258.00	**1490.00**	1081.10	1270.00
White	Inst	Ea	SA	3.81	906.00	138.00	---	1044.00	**1250.00**	895.10	1060.00
Chablis											
24" x 19" with 8" or 4" cc											
Color	Inst	Ea	SA	3.81	856.00	138.00	---	994.00	**1190.00**	851.10	1010.00
White	Inst	Ea	SA	3.81	705.00	138.00	---	843.00	**1010.00**	717.10	856.00
Chesapeake											
19" x 17" with 4" cc											
Color	Inst	Ea	SA	3.81	351.00	138.00	---	489.00	**607.00**	405.10	497.00
White	Inst	Ea	SA	3.81	316.00	138.00	---	454.00	**567.00**	375.10	462.00
20" x 18" with 4" cc											
Color	Inst	Ea	SA	3.81	397.00	138.00	---	535.00	**660.00**	446.10	544.00
White	Inst	Ea	SA	3.81	352.00	138.00	---	490.00	**608.00**	406.10	498.00
Fleur											
19-7/8" x 14-3/8" with single hole											
Color	Inst	Ea	SA	3.81	1040.00	138.00	---	1178.00	**1400.00**	1011.10	1190.00
White	Inst	Ea	SA	3.81	845.00	138.00	---	983.00	**1180.00**	840.10	998.00
Greenwich											
18" x 16" with 8" or 4" cc											
White	Inst	Ea	SA	3.81	360.00	138.00	---	498.00	**618.00**	414.10	507.00
20" x 18" with 8" or 4" cc											
Color	Inst	Ea	SA	3.81	370.00	138.00	---	508.00	**629.00**	423.10	517.00
White	Inst	Ea	SA	3.81	331.00	138.00	---	469.00	**584.00**	388.10	478.00
Pinoir											
22" x 18" with 8" or 4" cc											
Color	Inst	Ea	SA	3.81	566.00	138.00	---	704.00	**854.00**	595.10	715.00
White	Inst	Ea	SA	3.81	482.00	138.00	---	620.00	**757.00**	521.10	630.00
Portrait											
19-3/4" x 15-1/8"											
White	Inst	Ea	SA	3.81	617.00	138.00	---	755.00	**913.00**	640.10	767.00
27" x 19-3/8" with 8" cc											
Color	Inst	Ea	SA	3.81	1010.00	138.00	---	1148.00	**1360.00**	985.10	1160.00
White	Inst	Ea	SA	3.81	822.00	138.00	---	960.00	**1150.00**	821.10	975.00
Adjustments											
To only remove and reset lavatory											
	Reset	Ea	SA	2.86	---	104.00	---	104.00	**153.00**	72.70	107.00
To install rough-in	Inst	Ea	SA	10.0	---	364.00	---	364.00	**534.00**	242.00	356.00
te: The following percentage adjustment for Small Volume also applies to Large											
For colors, ADD	Inst	%	---	---	30.0	---	---	---	**---**	---	---
For two 15" L towel bars, ADD	Inst	Pr	---	---	39.60	---	---	39.60	**45.50**	39.60	45.50

257

					Costs Based On Small Volume					Large Volume	
Description	Oper	Unit	Crew Size	Man-Hours Per Unit	Avg Mat'l Unit Cost	Avg Labor Unit Cost	Avg Equip Unit Cost	Avg Total Unit Cost	Avg Price Incl O&P	Avg Total Unit Cost	Avg Price Incl O&P

Enameled cast iron

Countertop units (Kohler Products)

Farmington
19" x 16" with 8" or 4" cc

Color	Inst	Ea	SA	3.81	349.00	138.00	---	487.00	605.00	404.10	496.
White	Inst	Ea	SA	3.81	314.00	138.00	---	452.00	565.00	374.10	461.

Radiant
19" diameter with 4" cc

Color	Inst	Ea	SA	3.81	343.00	138.00	---	481.00	598.00	399.10	490.
White	Inst	Ea	SA	3.81	310.00	138.00	---	448.00	560.00	370.10	456.

Tahoe
20" x 19" with 8" or 4" cc

Color	Inst	Ea	SA	3.81	457.00	138.00	---	595.00	729.00	499.10	605.
White	Inst	Ea	SA	3.81	398.00	138.00	---	536.00	661.00	447.10	545.

Countertop units, with self rim (Kohler Products)

Ellington
19-1/4" x 16-1/4" with 8" or 4" cc

Color	Inst	Ea	SA	3.81	359.00	138.00	---	497.00	616.00	413.10	506.
White	Inst	Ea	SA	3.81	322.00	138.00	---	460.00	574.00	380.10	469.

Ellipse
33" x 19" with 8" cc

Color	Inst	Ea	SA	3.81	876.00	138.00	---	1014.00	1210.00	868.10	1030.
White	Inst	Ea	SA	3.81	720.00	138.00	---	858.00	1030.00	731.10	871.

Farmington
19-1/4" x 16-1/4" with 8" or 4" cc

Color	Inst	Ea	SA	3.81	349.00	138.00	---	487.00	605.00	404.10	496.
White	Inst	Ea	SA	3.81	314.00	138.00	---	452.00	565.00	374.10	461.

Hexsign
22-1/2" x 19" with 8" or 4" cc

Color	Inst	Ea	SA	3.81	480.00	138.00	---	618.00	756.00	519.10	628.
White	Inst	Ea	SA	3.81	416.00	138.00	---	554.00	681.00	463.10	563.

Man's
28" x 19" with 8" cc

Color	Inst	Ea	SA	3.81	602.00	138.00	---	740.00	896.00	627.10	752.
White	Inst	Ea	SA	3.81	509.00	138.00	---	647.00	789.00	545.10	658.

Terragon
19-3/4" x 15-3/4"

Color	Inst	Ea	SA	3.81	585.00	138.00	---	723.00	877.00	612.10	735.
White	Inst	Ea	SA	3.81	496.00	138.00	---	634.00	774.00	534.10	645.

Wall hung units, with shelf back (Kohler Products)

Hampton
19" x 17"

White	Inst	Ea	SA	3.81	493.00	138.00	---	631.00	770.00	531.10	641.

22" x 19"

White	Inst	Ea	SA	3.81	545.00	138.00	---	683.00	831.00	577.10	695.

Marston, corner unit
16" x 16"

Color	Inst	Ea	SA	3.81	733.00	138.00	---	871.00	1050.00	742.10	884.
White	Inst	Ea	SA	3.81	610.00	138.00	---	748.00	905.00	634.10	760.

Description	Oper	Unit	Crew Size	Man-Hours Per Unit	Avg Mat'l Unit Cost	Avg Labor Unit Cost	Avg Equip Unit Cost	Avg Total Unit Cost	Avg Price Incl O&P	Avg Total Unit Cost	Avg Price Incl O&P
										Large Volume	
Taunton											
16" x 14"											
White	Inst	Ea	SA	3.81	575.00	138.00	---	713.00	**865.00**	603.10	**725.00**
Trailer											
13" x 13"											
White	Inst	Ea	SA	4.00	580.00	145.00	---	725.00	**881.00**	583.70	**694.00**
Adjustments											
To only remove and reset lavatory											
	Reset	Ea	SA	2.86	---	104.00	---	104.00	**153.00**	72.70	**107.00**
To install rough-in	Inst	Ea	SA	10.0	---	364.00	---	364.00	**534.00**	242.00	**356.00**
Note: The following percentage adjustment for Small Volume also applies to Large											
For colors, ADD	Inst	%	---	---	25.0	---	---	---	---	---	---
For two 15" L towel bars, ADD	Inst	Pr	---	---	39.60	---	---	39.60	**45.50**	39.60	**45.50**

Americast - Porcelain enameled finish over steel composite material

Countertop units, with self rim (American Standard Products)

Description	Oper	Unit	Crew Size	Man-Hours Per Unit	Avg Mat'l Unit Cost	Avg Labor Unit Cost	Avg Equip Unit Cost	Avg Total Unit Cost	Avg Price Incl O&P	Avg Total Unit Cost	Avg Price Incl O&P
Acclivity											
19" diameter with 8" or 4" cc											
Color	Inst	Ea	SA	3.81	145.00	138.00	---	283.00	**371.00**	225.10	**290.00**
White	Inst	Ea	SA	3.81	300.00	138.00	---	438.00	**549.00**	361.10	**446.00**
Affinity											
20" x 17" with 8" or 4" cc											
Color	Inst	Ea	SA	3.81	365.00	138.00	---	503.00	**623.00**	418.10	**512.00**
White	Inst	Ea	SA	3.81	320.00	138.00	---	458.00	**572.00**	379.10	**467.00**
Adjustments											
To only remove and reset lavatory	Reset	Ea	SA	2.86	---	104.00	---	104.00	**153.00**	72.70	**107.00**
To install rough-in	Inst	Ea	SA	10.0	---	364.00	---	364.00	**534.00**	242.00	**356.00**
Note: The following percentage adjustment for Small Volume also applies to Large											
For colors, ADD	Inst	%	---	---	15.0	---	---	---	---	---	---

Kitchen

(Kitchen and utility), with good quality fittings, faucets, and sprayer

Americast - Porcelain enameled finish over steel composite material (American Standard)

Description	Oper	Unit	Crew Size	Man-Hours Per Unit	Avg Mat'l Unit Cost	Avg Labor Unit Cost	Avg Equip Unit Cost	Avg Total Unit Cost	Avg Price Incl O&P	Avg Total Unit Cost	Avg Price Incl O&P
Silhouette, double bowl with self rim											
33" x 22"											
Color	Inst	Ea	SA	7.27	615.00	264.00	---	879.00	**1100.00**	723.00	**890.00**
Premium color	Inst	Ea	SA	7.27	676.00	264.00	---	940.00	**1170.00**	777.00	**951.00**
White	Inst	Ea	SA	7.27	529.00	264.00	---	793.00	**997.00**	648.00	**803.00**
Silhouette, double bowl with tile edge											
33" x 22"											
Color	Inst	Ea	SA	7.27	517.00	264.00	---	781.00	**983.00**	637.00	**790.00**
Premium color	Inst	Ea	SA	7.27	563.00	264.00	---	827.00	**1040.00**	677.00	**837.00**
White	Inst	Ea	SA	7.27	447.00	264.00	---	711.00	**902.00**	575.00	**719.00**
Silhouette, dual level double bowl with self rim											
38" x 22"											
Color	Inst	Ea	SA	7.27	798.00	264.00	---	1062.00	**1310.00**	884.00	**1070.00**
Premium color	Inst	Ea	SA	7.27	880.00	264.00	---	1144.00	**1400.00**	956.00	**1160.00**
White	Inst	Ea	SA	7.27	677.00	264.00	---	941.00	**1170.00**	778.00	**953.00**
Silhouette, dual level double bowl with tile edge or color matched rim											
38" x 22"											
Color	Inst	Ea	SA	7.27	822.00	264.00	---	1086.00	**1330.00**	905.00	**1100.00**
Premium color	Inst	Ea	SA	7.27	908.00	264.00	---	1172.00	**1430.00**	981.00	**1190.00**
White	Inst	Ea	SA	7.27	695.00	264.00	---	959.00	**1190.00**	794.00	**971.00**

Description	Oper	Unit	Crew Size	Man-Hours Per Unit	Avg Mat'l Unit Cost	Avg Labor Unit Cost	Avg Equip Unit Cost	Avg Total Unit Cost	Avg Price Incl O&P	Avg Total Unit Cost	Avg Price Incl O&P
								Costs Based On Small Volume		**Large Volume**	
Silhouette, dual level double bowl with self rim											
33" x 22"											
Color	Inst	Ea	SA	7.27	646.00	264.00	---	910.00	**1130.00**	750.00	920.
Premium color	Inst	Ea	SA	7.27	707.00	264.00	---	971.00	**1200.00**	804.00	982.
White	Inst	Ea	SA	7.27	551.00	264.00	---	815.00	**1020.00**	667.00	824.
Silhouette, dual level double bowl with tile edge or color matched rim											
33" x 22"											
Color	Inst	Ea	SA	7.27	670.00	264.00	---	934.00	**1160.00**	772.00	946.
Premium color	Inst	Ea	SA	7.27	737.00	264.00	---	1001.00	**1240.00**	831.00	1010.
White	Inst	Ea	SA	7.27	572.00	264.00	---	836.00	**1050.00**	685.00	846.
Silhouette, single bowl with self rim											
25" x 22"											
Color	Inst	Ea	SA	5.71	474.00	208.00	---	682.00	**850.00**	562.00	693.
Premium color	Inst	Ea	SA	5.71	515.00	208.00	---	723.00	**897.00**	598.00	734.
White	Inst	Ea	SA	5.71	413.00	208.00	---	621.00	**780.00**	508.00	631.
Silhouette, single bowl with tile edge											
25" x 22"											
Color	Inst	Ea	SA	5.71	501.00	208.00	---	709.00	**881.00**	586.00	721.
Premium color	Inst	Ea	SA	5.71	544.00	208.00	---	752.00	**931.00**	624.00	764.
White	Inst	Ea	SA	5.71	434.00	208.00	---	642.00	**805.00**	527.00	653.
Silhouette, single bowl with self rim											
18" x 18"											
Color	Inst	Ea	SA	3.81	426.00	138.00	---	564.00	**694.00**	472.10	574.
Premium color	Inst	Ea	SA	3.81	460.00	138.00	---	598.00	**733.00**	502.10	608.
White	Inst	Ea	SA	3.81	376.00	138.00	---	514.00	**635.00**	427.10	523.
Silhouette, single bowl with tile edge											
18" x 18"											
Color	Inst	Ea	SA	3.81	452.00	138.00	---	590.00	**724.00**	495.10	600.
Premium color	Inst	Ea	SA	3.81	491.00	138.00	---	629.00	**768.00**	529.10	639.
White	Inst	Ea	SA	3.81	394.00	138.00	---	532.00	**656.00**	443.10	541.
Enameled cast iron (Kohler Products)											
Bakersfield, single bowl											
31" x 22"											
Color	Inst	Ea	SA	7.27	582.00	264.00	---	846.00	**1060.00**	694.00	856.
Premium color	Inst	Ea	SA	7.27	648.00	264.00	---	912.00	**1130.00**	752.00	922.
White	Inst	Ea	SA	7.27	495.00	264.00	---	759.00	**957.00**	617.00	768.
Bon Vivant, triple bowl, 2 large, 1 small with tile or self rim											
48" x 22"											
Color	Inst	Ea	SA	7.27	1190.00	264.00	---	1454.00	**1760.00**	1232.00	1470.
Premium color	Inst	Ea	SA	7.27	1350.00	264.00	---	1614.00	**1940.00**	1362.00	1630.
White	Inst	Ea	SA	7.27	981.00	264.00	---	1245.00	**1520.00**	1045.00	1260.
Cantina, double with tile or self rim, 2 large											
43" x 22"											
Color	Inst	Ea	SA	7.27	872.00	264.00	---	1136.00	**1390.00**	950.00	1150.
Premium color	Inst	Ea	SA	7.27	981.00	264.00	---	1245.00	**1520.00**	1045.00	1260.
White	Inst	Ea	SA	7.27	727.00	264.00	---	991.00	**1220.00**	822.00	1000.
Ecocycle, double bowl with regular rim											
43" x 22"											
Color	Inst	Ea	SA	7.27	2070.00	264.00	---	2334.00	**2760.00**	2002.00	2360.
Premium color	Inst	Ea	SA	7.27	2350.00	264.00	---	2614.00	**3100.00**	2252.00	2650.
White	Inst	Ea	SA	7.27	1680.00	264.00	---	1944.00	**2320.00**	1662.00	1970.

Description	Oper	Unit	Crew Size	Man-Hours Per Unit	Avg Mat'l Unit Cost	Avg Labor Unit Cost	Avg Equip Unit Cost	Avg Total Unit Cost	Avg Price Incl O&P	Avg Total Unit Cost	Avg Price Incl O&P
								Costs Based On Small Volume		**Large Volume**	
Epicurean, double bowl with cutting board and drainboard, 1 large, 1 small											
43" x 22"											
Color	Inst	Ea	SA	7.27	1040.00	264.00	---	1304.00	**1590.00**	1098.00	**1320.00**
White	Inst	Ea	SA	7.27	872.00	264.00	---	1136.00	**1390.00**	949.00	**1150.00**
Executive Chef, double bowl with tile or self rim, 1 large, 1 medium											
33" x 22"											
Color	Inst	Ea	SA	7.27	783.00	264.00	---	1047.00	**1290.00**	871.00	**1060.00**
Premium color	Inst	Ea	SA	7.27	879.00	264.00	---	1143.00	**1400.00**	956.00	**1160.00**
White	Inst	Ea	SA	7.27	656.00	264.00	---	920.00	**1140.00**	759.00	**931.00**
Lakefield double bowl with tile or self rim, 1 large, 1 small											
33" x 22"											
Color	Inst	Ea	SA	7.27	651.00	264.00	---	915.00	**1140.00**	755.00	**926.00**
Premium color	Inst	Ea	SA	7.27	727.00	264.00	---	991.00	**1220.00**	821.00	**1000.00**
White	Inst	Ea	SA	7.27	550.00	264.00	---	814.00	**1020.00**	666.00	**823.00**
Marsala Hi-Low with self or tile rim											
33" x 22"											
Color	Inst	Ea	SA	7.27	783.00	264.00	---	1047.00	**1290.00**	871.00	**1060.00**
Premium color	Inst	Ea	SA	7.27	879.00	264.00	---	1143.00	**1400.00**	956.00	**1160.00**
White	Inst	Ea	SA	7.27	656.00	264.00	---	920.00	**1140.00**	759.00	**931.00**
Mayfield, single for frame mount with self or tile rim											
24" x 21"											
Color	Inst	Ea	SA	5.71	423.00	208.00	---	631.00	**791.00**	517.00	**642.00**
White	Inst	Ea	SA	5.71	373.00	208.00	---	581.00	**734.00**	473.00	**591.00**
Porcelain on steel											
Double bowl, with self rim											
33" x 22"											
Color	Inst	Ea	SA	7.27	274.00	264.00	---	538.00	**703.00**	423.00	**544.00**
White	Inst	Ea	SA	7.27	259.00	264.00	---	523.00	**686.00**	410.00	**529.00**
Double bowl, dual level with self rim											
33" x 22"											
Color	Inst	Ea	SA	7.27	310.00	264.00	---	574.00	**745.00**	455.00	**581.00**
White	Inst	Ea	SA	7.27	295.00	264.00	---	559.00	**728.00**	442.00	**566.00**
Single bowl											
25" x 22"											
Color	Inst	Ea	SA	5.71	261.00	208.00	---	469.00	**606.00**	375.00	**478.00**
White	Inst	Ea	SA	5.71	248.00	208.00	---	456.00	**590.00**	363.00	**465.00**
Stainless steel (Kohler)											
Triple bowl with self rim											
Ravinia											
43" x 22"	Inst	Ea	SA	7.27	1570.00	264.00	---	1834.00	**2200.00**	1562.00	**1860.00**
Double bowl with self rim											
Ravinia											
42" x 22"	Inst	Ea	SA	7.27	1170.00	264.00	---	1434.00	**1730.00**	1212.00	**1450.00**
33" x 22"	Inst	Ea	SA	7.27	943.00	264.00	---	1207.00	**1470.00**	1012.00	**1220.00**
Single bowl with self rim											
Ravinia											
25" x 22"	Inst	Ea	SA	5.71	661.00	208.00	---	869.00	**1070.00**	727.00	**883.00**
Ballad											
43" x 22"	Inst	Ea	SA	5.71	1360.00	208.00	---	1568.00	**1860.00**	1335.00	**1590.00**
33" x 22"	Inst	Ea	SA	5.71	858.00	208.00	---	1066.00	**1290.00**	900.00	**1080.00**
25" x 22"	Inst	Ea	SA	5.71	538.00	208.00	---	746.00	**924.00**	618.00	**758.00**

					Costs Based On Small Volume					Large Volume	
Description	Oper	Unit	Crew Size	Man-Hours Per Unit	Avg Mat'l Unit Cost	Avg Labor Unit Cost	Avg Equip Unit Cost	Avg Total Unit Cost	Avg Price Incl O&P	Avg Total Unit Cost	Avg Price Incl O&P

Bar

Enameled cast iron (Kohler)

Addison
13" x 13"

Color	Inst	Ea	SA	3.81	380.00	138.00	---	518.00	**640.00**	431.10	527.
Premium color	Inst	Ea	SA	3.81	415.00	138.00	---	553.00	**681.00**	462.10	562.
White	Inst	Ea	SA	3.81	333.00	138.00	---	471.00	**586.00**	390.10	479.

Apertif with tile or self rim
16" x19"

Color	Inst	Ea	SA	3.81	455.00	138.00	---	593.00	**726.00**	497.10	603.
Premium color	Inst	Ea	SA	3.81	501.00	138.00	---	639.00	**780.00**	538.10	650.
White	Inst	Ea	SA	3.81	393.00	138.00	---	531.00	**655.00**	442.10	540.

Entertainer with semicircular self rim or undercounter self rim
22" x18"

Color	Inst	Ea	SA	5.71	402.00	208.00	---	610.00	**768.00**	499.00	621.
Premium color	Inst	Ea	SA	5.71	441.00	208.00	---	649.00	**812.00**	533.00	660.
White	Inst	Ea	SA	5.71	351.00	208.00	---	559.00	**708.00**	454.00	569.

Sorbet with tile or undercounter rim or self rim
15" x 15"

Color	Inst	Ea	SA	3.81	415.00	138.00	---	553.00	**681.00**	462.10	563.
Premium color	Inst	Ea	SA	3.81	456.00	138.00	---	594.00	**728.00**	498.10	604.
White	Inst	Ea	SA	3.81	361.00	138.00	---	499.00	**619.00**	415.10	508.

Stainless steel (Kohler)

Ravinia
22" x 15"

	Inst	Ea	SA	3.81	734.00	138.00	---	872.00	**1050.00**	743.10	885.

Lyric
15" x 15"

	Inst	Ea	SA	3.81	284.00	138.00	---	422.00	**530.00**	347.10	430.

Utility/service

Enameled cast iron (Kohler)

Sutton high back
24" x 18", 8" high back

	Inst	Ea	SA	5.71	834.00	208.00	---	1042.00	**1260.00**	879.00	1060.

Tech
24" x 18", bubbler mound

	Inst	Ea	SA	5.71	462.00	208.00	---	670.00	**836.00**	551.00	681.

Whitby
28" x 28" corner

	Inst	Ea	SA	5.71	930.00	208.00	---	1138.00	**1370.00**	963.00	1150.

Glen Falls, laundry sink
25" x 22"

Color	Inst	Ea	SA	5.71	674.00	208.00	---	882.00	**1080.00**	738.00	895.
White	Inst	Ea	SA	5.71	564.00	208.00	---	772.00	**954.00**	641.00	784.

River Falls, laundry sink
25" x 22"

Color	Inst	Ea	SA	5.71	740.00	208.00	---	948.00	**1160.00**	796.00	962.
White	Inst	Ea	SA	5.71	617.00	208.00	---	825.00	**1010.00**	688.00	838.

Westover, double deep, laundry sink
42" x 21"

Color	Inst	Ea	SA	7.27	1080.00	264.00	---	1344.00	**1630.00**	1133.00	1360.
White	Inst	Ea	SA	7.27	890.00	264.00	---	1154.00	**1410.00**	965.00	1170.

Description	Oper	Unit	Crew Size	Man-Hours Per Unit	Costs Based On Small Volume					Large Volume	
					Avg Mat'l Unit Cost	Avg Labor Unit Cost	Avg Equip Unit Cost	Avg Total Unit Cost	Avg Price Incl O&P	Avg Total Unit Cost	Avg Price Incl O&P
Vitreous china (Kohler)											
Hollister											
28" x 22"	Inst	Ea	SA	5.71	1100.00	208.00	---	1308.00	**1580.00**	1117.00	**1330.00**
Sudbury											
22" x 20"	Inst	Ea	SA	5.71	666.00	208.00	---	874.00	**1070.00**	731.00	**888.00**
Tyrrell											
20" x 20" w/ siphon jet	Inst	Ea	SA	7.27	1180.00	264.00	---	1444.00	**1740.00**	1222.00	**1460.00**
Adjustments											
To only remove and reset sink											
Single bowl	Reset	Ea	SA	2.86	---	104.00	---	104.00	**153.00**	72.70	**107.00**
Double bowl	Reset	Ea	SA	2.96	---	108.00	---	108.00	**158.00**	76.70	**113.00**
Triple bowl	Reset	Ea	SA	3.20	---	116.00	---	116.00	**171.00**	83.20	**122.00**
To install rough-in	Inst	Ea	SA	10.0	---	364.00	---	364.00	**534.00**	242.00	**356.00**
Note: The following percentage adjustment for Small Volume also applies to Large											
For 18" gauge steel, ADD	Inst	%		---	40.0	---	---	---	**---**	---	**---**

Description	Oper	Unit	Crew Size	Man-Hours Per Unit	Avg Mat'l Unit Cost	Avg Labor Unit Cost	Avg Equip Unit Cost	Avg Total Unit Cost	Avg Price Incl O&P	Avg Total Unit Cost	Avg Price Incl O&P
								Costs Based On Small Volume		Large Volume	

Skylights, skywindows, roof windows

Labor costs are for installation of skylight, flashing and roller shade only;
no carpentry or roofing work included

Polycarbonate dome, clear transparent or tinted, roof opening sizes, curb or flush mount, self-flashing

Description	Oper	Unit	Crew Size	Man-Hours Per Unit	Avg Mat'l Unit Cost	Avg Labor Unit Cost	Avg Equip Unit Cost	Avg Total Unit Cost	Avg Price Incl O&P	Avg Total Unit Cost	Avg Price Incl O&P
Single dome											
22" x 22"	Inst	Ea	CA	2.61	47.20	83.80	---	131.00	**180.00**	92.90	124.0
22" x 46"	Inst	Ea	CA	4.30	94.30	138.00	---	232.30	**315.00**	167.70	222.0
30" x 30"	Inst	Ea	CA	4.30	78.60	138.00	---	216.60	**297.00**	153.50	206.0
30" x 46"	Inst	Ea	CA	4.94	137.00	159.00	---	296.00	**395.00**	218.00	284.0
46" x 46"	Inst	Ea	CA	5.80	236.00	186.00	---	422.00	**551.00**	325.00	412.0
Double dome											
22" x 22"	Inst	Ea	CA	2.61	62.80	83.80	---	146.60	**198.00**	106.90	141.0
22" x 46"	Inst	Ea	CA	4.30	137.00	138.00	---	275.00	**364.00**	205.80	266.0
30" x 30"	Inst	Ea	CA	4.30	120.00	138.00	---	258.00	**345.00**	190.80	248.0
30" x 46"	Inst	Ea	CA	4.94	179.00	159.00	---	338.00	**443.00**	256.00	327.0
46" x 46"	Inst	Ea	CA	5.80	288.00	186.00	---	474.00	**610.00**	371.00	465.0
Triple dome											
22" x 22"	Inst	Ea	CA	2.61	83.80	83.80	---	167.60	**222.00**	125.80	162.0
22" x 46"	Inst	Ea	CA	4.30	189.00	138.00	---	327.00	**424.00**	252.80	320.0
30" x 30"	Inst	Ea	CA	4.30	158.00	138.00	---	296.00	**388.00**	224.80	287.0
30" x 46"	Inst	Ea	CA	4.94	236.00	159.00	---	395.00	**509.00**	308.00	387.0
46" x 46"	Inst	Ea	CA	5.80	361.00	186.00	---	547.00	**695.00**	437.00	541.0
Operable skylight, Velux model VS											
21-9/16" x 27-1/2"	Inst	Ea	CA	5.71	316.00	183.00	---	499.00	**638.00**	394.00	492.0
21-9/16" x 38-1/2"	Inst	Ea	CA	5.71	351.00	183.00	---	534.00	**679.00**	426.00	528.0
21-9/16" x 46-3/8"	Inst	Ea	CA	5.71	381.00	183.00	---	564.00	**713.00**	453.00	559.0
21-9/16" x 55"	Inst	Ea	CA	5.71	404.00	183.00	---	587.00	**739.00**	474.00	583.0
30-5/8" x 38-1/2"	Inst	Ea	CA	5.71	401.00	183.00	---	584.00	**736.00**	471.00	580.0
30-5/8" x 55"	Inst	Ea	CA	5.93	482.00	190.00	---	672.00	**840.00**	548.00	670.0
44-3/4" x 27-1/2"	Inst	Ea	CA	5.93	423.00	190.00	---	613.00	**772.00**	495.00	609.0
44-3/4" x 46-1/2"	Inst	Ea	CA	5.93	535.00	190.00	---	725.00	**901.00**	596.00	725.0
Fixed skylight, Velux model FS											
21-9/16" x 27-1/2"	Inst	Ea	CA	5.71	229.00	183.00	---	412.00	**538.00**	316.00	402.0
21-9/16" x 38-1/2"	Inst	Ea	CA	5.71	259.00	183.00	---	442.00	**573.00**	343.00	433.0
21-9/16" x 46-3/8"	Inst	Ea	CA	5.71	287.00	183.00	---	470.00	**605.00**	368.00	462.0
21-9/16" x 55"	Inst	Ea	CA	5.71	323.00	183.00	---	506.00	**646.00**	401.00	499.0
21-9/16" x 70-7/8"	Inst	Ea	CA	5.71	378.00	183.00	---	561.00	**710.00**	450.00	556.0
30-5/8" x 38-1/2"	Inst	Ea	CA	5.71	289.00	183.00	---	472.00	**607.00**	370.00	464.0
30-5/8" x 55"	Inst	Ea	CA	5.93	359.00	190.00	---	549.00	**698.00**	437.00	543.0
44-3/4" x 27-1/2"	Inst	Ea	CA	5.93	338.00	190.00	---	528.00	**674.00**	418.00	521.0
44-3/4" x 46-1/2"	Inst	Ea	CA	5.93	401.00	190.00	---	591.00	**747.00**	475.00	586.0
Fixed ventilation with removable filter on ventilation flap, Velux model FSF											
21-9/16" x 27-1/2"	Inst	Ea	CA	5.23	239.00	168.00	---	407.00	**526.00**	316.00	398.0
21-9/16" x 38-1/2"	Inst	Ea	CA	5.23	274.00	168.00	---	442.00	**566.00**	347.00	434.0
21-9/16" x 46-3/8"	Inst	Ea	CA	5.23	289.00	168.00	---	457.00	**584.00**	361.00	450.0
21-9/16" x 55"	Inst	Ea	CA	5.71	395.00	183.00	---	578.00	**729.00**	466.00	574.0
21-9/16" x 70-7/8"	Inst	Ea	CA	5.63	320.00	181.00	---	501.00	**638.00**	397.00	494.0
30-5/8" x 38-1/2"	Inst	Ea	CA	5.23	319.00	168.00	---	487.00	**618.00**	388.00	481.0
30-5/8" x 55"	Inst	Ea	CA	5.63	370.00	181.00	---	551.00	**696.00**	442.00	546.0
44-3/4" x 27-1/2"	Inst	Ea	CA	5.59	339.00	179.00	---	518.00	**659.00**	413.00	512.0
44-3/4" x 46-1/2"	Inst	Ea	CA	5.63	410.00	181.00	---	591.00	**742.00**	478.00	587.0

Description	Oper	Unit	Crew Size	Man-Hours Per Unit	Avg Mat'l Unit Cost	Avg Labor Unit Cost	Avg Equip Unit Cost	Avg Total Unit Cost	Avg Price Incl O&P	Avg Total Unit Cost	Avg Price Incl O&P
					Costs Based On Small Volume					**Large Volume**	
Low profile, acrylic double domes with molded edge.											
Preassembled units include plywood curb-liner, 16 oz. copper											
flashing for pitched or flat roof, screen, and all hardware											
Ventarama											
Ventilating type											
30" x 72"	Inst	Ea	CA	4.30	445.00	138.00	---	583.00	719.00	483.80	585.00
45-1/2" x 45-1/2"	Inst	Ea	CA	4.30	400.00	138.00	---	538.00	667.00	442.80	538.00
45-1/2" x 30"	Inst	Ea	CA	4.30	330.00	138.00	---	468.00	587.00	379.80	466.00
30" x 45-1/2"	Inst	Ea	CA	4.30	330.00	138.00	---	468.00	587.00	379.80	466.00
30" x 30"	Inst	Ea	CA	3.45	270.00	111.00	---	381.00	477.00	309.40	379.00
30" x 22"	Inst	Ea	CA	3.45	245.00	111.00	---	356.00	448.00	287.40	353.00
22" x 30"	Inst	Ea	CA	3.45	245.00	111.00	---	356.00	448.00	287.40	353.00
22" x 45-1/2"	Inst	Ea	CA	3.45	295.00	111.00	---	406.00	505.00	332.40	405.00
Note: The following percentage adjustment for Small Volume also applies to Large											
Add for insulated glass	Inst	%	---	---	30.0	---	---	---	---	---	---
Add for bronze tinted outer dome	Inst	Ea	---	---	35.00	---	---	---	40.30	---	36.20
Add for white inner dome	Inst	Ea	---	---	17.00	---	---	---	19.60	---	17.60
Add for roll shade											
22" width	Inst	Ea	---	---	100.00	---	---	---	115.00	---	104.00
30" width	Inst	Ea	---	---	108.00	---	---	---	124.00	---	112.00
45-1/2" width	Inst	Ea	---	---	115.00	---	---	---	132.00	---	119.00
Add for storm panel with weatherstripping											
30" x 30"	Inst	Ea	---	---	50.00	---	---	---	57.50	---	51.80
45-1/2" x 45-1/2"	Inst	Ea	---	---	69.00	---	---	---	79.40	---	71.40
Add for motorization	Inst	Ea	CA	5.37	200.00	172.00	---	372.00	489.00	284.00	362.00
Fixed type											
30" x 72"	Inst	Ea	CA	4.30	350.00	138.00	---	488.00	610.00	397.80	486.00
45-1/2" x 45-1/2"	Inst	Ea	CA	4.30	300.00	138.00	---	438.00	552.00	352.80	435.00
45-1/2" x 30"	Inst	Ea	CA	4.30	265.00	138.00	---	403.00	512.00	321.80	398.00
30" x 45-1/2"	Inst	Ea	CA	4.30	265.00	138.00	---	403.00	512.00	321.80	398.00
30" x 30"	Inst	Ea	CA	3.45	215.00	111.00	---	326.00	413.00	260.40	322.00
30" x 22"	Inst	Ea	CA	3.45	190.00	111.00	---	301.00	385.00	237.40	296.00
22" x 30"	Inst	Ea	CA	3.45	190.00	111.00	---	301.00	385.00	237.40	296.00
22" x 45-1/2"	Inst	Ea	CA	3.45	240.00	111.00	---	351.00	442.00	282.40	348.00
Note: The following percentage adjustment for Small Volume also applies to Large											
Add for insulated laminated glass	Inst	%	---	---	30.0	---	---	---	---	---	---

Skywindows

Labor costs are for installation of skywindows only,

no carpentry or roofing work included

**Ultraseal self-flashing units. Clear insulated glass, flexible flange,
no mastic or step flashing required. Other glazings available**

Description	Oper	Unit	Crew Size	Man-Hours Per Unit	Avg Mat'l Unit Cost	Avg Labor Unit Cost	Avg Equip Unit Cost	Avg Total Unit Cost	Avg Price Incl O&P	Avg Total Unit Cost	Avg Price Incl O&P
Skywindow E-Class											
22-1/2" x 22-1/2" fixed	Inst	Ea	CA	1.33	216.00	42.70	---	258.70	312.00	219.70	262.00
22-1/2" x 38-1/2" fixed	Inst	Ea	CA	1.33	248.00	42.70	---	290.70	349.00	248.70	295.00
22-1/2" x 46-1/2" fixed	Inst	Ea	CA	1.33	292.00	42.70	---	334.70	400.00	288.70	341.00
22-1/2" x 70-1/2" fixed	Inst	Ea	CA	1.33	320.00	42.70	---	362.70	432.00	313.70	370.00
30-1/2" x 30-1/2" fixed	Inst	Ea	CA	1.33	292.00	42.70	---	334.70	400.00	288.70	341.00
30-1/2" x 46-1/2" fixed	Inst	Ea	CA	1.33	356.00	42.70	---	398.70	473.00	345.70	407.00
46-1/2" x 46-1/2" fixed	Inst	Ea	CA	1.33	400.00	42.70	---	442.70	524.00	385.70	453.00
22-1/2" x 22-1/2" venting	Inst	Ea	CA	1.33	409.00	42.70	---	451.70	534.00	393.70	462.00
22-1/2" x 30-1/2" venting	Inst	Ea	CA	1.33	443.00	42.70	---	485.70	573.00	424.70	497.00
22-1/2" x 46-1/2" venting	Inst	Ea	CA	1.33	501.00	42.70	---	543.70	640.00	476.70	557.00

Description	Oper	Unit	Crew Size	Man-Hours Per Unit	Avg Mat'l Unit Cost	Avg Labor Unit Cost	Avg Equip Unit Cost	Avg Total Unit Cost	Avg Price Incl O&P	Avg Total Unit Cost	Avg Price Incl O&P
								Costs Based On Small Volume		Large Volume	
30-1/2" x 30-1/2" venting	Inst	Ea	CA	1.33	474.00	42.70	---	516.70	609.00	452.70	529.0
30-1/2" x 46-1/2" venting	Inst	Ea	CA	1.33	544.00	42.70	---	586.70	690.00	515.70	602.0

Step flash-pan flashed units. Insulated clear tempered glass. Deck-mounted step flash unit. Other glazings available

Skywindow Excel-10

Description	Oper	Unit	Crew Size	Man-Hours Per Unit	Avg Mat'l Unit Cost	Avg Labor Unit Cost	Avg Equip Unit Cost	Avg Total Unit Cost	Avg Price Incl O&P	Avg Total Unit Cost	Avg Price Incl O&P
22-1/2" x 22-1/2" fixed	Inst	Ea	CA	4.30	200.00	138.00	---	338.00	437.00	262.80	331.0
22-1/2" x 38-1/2" fixed	Inst	Ea	CA	4.30	210.00	138.00	---	348.00	449.00	271.80	342.0
22-1/2" x 46-1/2" fixed	Inst	Ea	CA	4.30	221.00	138.00	---	359.00	461.00	281.80	353.0
22-1/2" x 70-1/2" fixed	Inst	Ea	CA	4.30	256.00	138.00	---	394.00	501.00	312.80	389.0
30-1/2" x 30-1/2" fixed	Inst	Ea	CA	4.30	237.00	138.00	---	375.00	480.00	295.80	370.0
30-1/2" x 46-1/2" fixed	Inst	Ea	CA	4.30	286.00	138.00	---	424.00	536.00	339.80	420.0
22-1/2" x 22-1/2" vented	Inst	Ea	CA	4.30	335.00	138.00	---	473.00	592.00	384.80	471.0
22-1/2" x 30-1/2" vented	Inst	Ea	CA	4.30	361.00	138.00	---	499.00	622.00	407.80	498.0
22-1/2" x 46-1/2" vented	Inst	Ea	CA	4.30	431.00	138.00	---	569.00	703.00	470.80	570.0
30-1/2" x 30-1/2" vented	Inst	Ea	CA	4.30	415.00	138.00	---	553.00	684.00	456.80	554.0
30-1/2" x 46-1/2" vented	Inst	Ea	CA	4.30	463.00	138.00	---	601.00	739.00	499.80	603.0
Add for flashing kit	Inst	Ea	---	---	60.00	---	---	---	60.00	---	60.0

Low-profile insulated glass skywindow. Deck mount, clear tempered glass, self-flashing models. Other glazings available

Skywindow Genra-1

Description	Oper	Unit	Crew Size	Man-Hours Per Unit	Avg Mat'l Unit Cost	Avg Labor Unit Cost	Avg Equip Unit Cost	Avg Total Unit Cost	Avg Price Incl O&P	Avg Total Unit Cost	Avg Price Incl O&P
22-1/2" x 22-1/2" fixed	Inst	Ea	CA	4.30	178.00	138.00	---	316.00	412.00	242.80	308.0
22-1/2" x 30-1/2" fixed	Inst	Ea	CA	4.30	210.00	138.00	---	348.00	449.00	271.80	342.0
22-1/2" x 46-1/2" fixed	Inst	Ea	CA	4.30	258.00	138.00	---	396.00	504.00	314.80	391.0
22-1/2" x 70-1/2" fixed	Inst	Ea	CA	4.30	340.00	138.00	---	478.00	598.00	388.80	476.0
30-1/2" x 30-1/2" fixed	Inst	Ea	CA	4.30	243.00	138.00	---	381.00	486.00	301.80	376.0
30-1/2" x 46-1/2" fixed	Inst	Ea	CA	4.30	296.00	138.00	---	434.00	547.00	348.80	431.0
46-1/2" x 46-1/2" fixed	Inst	Ea	CA	4.30	378.00	138.00	---	516.00	642.00	422.80	515.0
22-1/2" x 22-1/2" vented	Inst	Ea	CA	4.30	349.00	138.00	---	487.00	608.00	396.80	485.0
22-1/2" x 30-1/2" vented	Inst	Ea	CA	4.30	370.00	138.00	---	508.00	633.00	415.80	507.0
22-1/2" x 46-1/2" vented	Inst	Ea	CA	4.30	440.00	138.00	---	578.00	713.00	478.80	580.0
30-1/2" x 30-1/2" vented	Inst	Ea	CA	4.30	410.00	138.00	---	548.00	679.00	451.80	549.0
30-1/2" x 46-1/2" vented	Inst	Ea	CA	4.30	480.00	138.00	---	618.00	759.00	514.80	621.0
46-1/2" x 46-1/2" vented	Inst	Ea	CA	4.30	590.00	138.00	---	728.00	886.00	613.80	735.0

Roof windows

Includes prefabricated flashing, exterior awning, interior rollerblind and insect screen. Sash rotates 180 degrees. Labor costs are for installation on roofs with 10- to 85-degree slope and include installation of roof window, flashing and sun screen accessories. Labor costs do not include carpentry or roofing work other than curb. Add for interior trim. Listed by actual dimensions, top hung

Aluminum-clad wood frame, double-insulated tempered glass

Description	Oper	Unit	Crew Size	Man-Hours Per Unit	Avg Mat'l Unit Cost	Avg Labor Unit Cost	Avg Equip Unit Cost	Avg Total Unit Cost	Avg Price Incl O&P	Avg Total Unit Cost	Avg Price Incl O&P
30-1/2" x 46-1/2" fixed	Inst	Ea	CA	5.93	590.00	190.00	---	780.00	964.00	645.00	782.0
46-1/2" x 46-1/2" fixed	Inst	Ea	CA	5.93	660.00	190.00	---	850.00	1040.00	708.00	854.0

Spas

Spas, with good quality fittings and faucets
See also Bathtubs

Description	Oper	Unit	Crew Size	Man-Hours Per Unit	Avg Mat'l Unit Cost	Avg Labor Unit Cost	Avg Equip Unit Cost	Avg Total Unit Cost	Avg Price Incl O&P	Avg Total Unit Cost	Avg Price Incl O&P
								Costs Based On Small Volume		**Large Volume**	
Detach & reset operations											
Whirlpool spa	Reset	Ea	SB	7.6	46.80	239.00	---	285.80	**406.00**	206.50	**291.00**
Remove operations											
Whirlpool spa	Demo	Ea	SB	3.8	25.50	120.00	---	145.50	**205.00**	105.10	**148.00**
Install rough-in											
Whirlpool spa	Inst	Ea	SB	14.5	72.30	457.00	---	529.30	**754.00**	374.50	**531.00**
Replace operations											
Jacuzzi Products, whirlpools and baths											
with color matched trim on jets and suction											
Builder Series											
Nova model, oval tub, with ledge											
60" x 42" x 18-1/2" H											
White / stock color	Inst	Ea	SB	11.4	1730.00	359.00	---	2089.00	**2520.00**	1672.00	**2010.00**
Premium color	Inst	Ea	SB	11.4	1850.00	359.00	---	2209.00	**2650.00**	1772.00	**2120.00**
72" x 42" x 20-1/2" H											
White / stock color	Inst	Ea	SB	11.4	2010.00	359.00	---	2369.00	**2830.00**	1902.00	**2270.00**
Premium color	Inst	Ea	SB	11.4	2150.00	359.00	---	2509.00	**3000.00**	2022.00	**2400.00**
Riva model, oval tub											
62" x 43" x 18-1/2" H											
White / stock color	Inst	Ea	SB	11.4	1770.00	359.00	---	2129.00	**2560.00**	1712.00	**2050.00**
Premium color	Inst	Ea	SB	11.4	1910.00	359.00	---	2269.00	**2730.00**	1822.00	**2180.00**
72" x 42" x 20-1/2" H											
White / stock color	Inst	Ea	SB	11.4	1970.00	359.00	---	2329.00	**2790.00**	1872.00	**2240.00**
Premium color	Inst	Ea	SB	11.4	2010.00	359.00	---	2369.00	**2840.00**	1912.00	**2280.00**
Tara model, corner unit, angled tub											
60" x 60" x 20-3/4" H											
White / stock color	Inst	Ea	SB	11.4	1770.00	359.00	---	2129.00	**2560.00**	1712.00	**2050.00**
Premium color	Inst	Ea	SB	11.4	1910.00	359.00	---	2269.00	**2730.00**	1822.00	**2180.00**
Torino model, rectangle tub											
66" x 42" x 20-1/2" H											
White / stock color	Inst	Ea	SB	11.4	1770.00	359.00	---	2129.00	**2560.00**	1712.00	**2050.00**
Premium color	Inst	Ea	SB	11.4	1910.00	359.00	---	2269.00	**2730.00**	1822.00	**2180.00**

				Costs Based On Small Volume						Large Volume	
Description	Oper	Unit	Crew Size	Man-Hours Per Unit	Avg Mat'l Unit Cost	Avg Labor Unit Cost	Avg Equip Unit Cost	Avg Total Unit Cost	Avg Price Incl O&P	Avg Total Unit Cost	Avg Price Incl O&P
Designer Collection											
Allusion model, rectangle tub											
66" x 36" x 26" H											
White / stock color	Inst	Ea	SB	11.4	3520.00	359.00	---	3879.00	4570.00	3152.00	3700.0
Premium color	Inst	Ea	SB	11.4	3560.00	359.00	---	3919.00	4620.00	3182.00	3740.0
72" x 36" x 26" H											
White / stock color	Inst	Ea	SB	11.4	3710.00	359.00	---	4069.00	4790.00	3312.00	3880.0
Premium color	Inst	Ea	SB	11.4	3750.00	359.00	---	4109.00	4840.00	3342.00	3920.0
72" x 42" x 26" H											
White / stock color	Inst	Ea	SB	11.4	3890.00	359.00	---	4249.00	5010.00	3462.00	4060.0
Premium color	Inst	Ea	SB	11.4	3940.00	359.00	---	4299.00	5060.00	3492.00	4100.0
Aura model											
72" x 60" x 20-1/2"											
White / stock color	Inst	Ea	SB	11.4	6020.00	359.00	---	6379.00	7450.00	5202.00	6070.0
Premium color	Inst	Ea	SB	11.4	6060.00	359.00	---	6419.00	7500.00	5242.00	6110.0
Ciprea model, with removable skirt											
72" x 48" x 20-1/2"											
White / stock color	Inst	Ea	SB	11.4	6020.00	359.00	---	6379.00	7450.00	5212.00	6070.0
Premium color	Inst	Ea	SB	11.4	6060.00	359.00	---	6419.00	7500.00	5242.00	6110.0
Fiore model, with removable skirt											
66" x 66" x 27-1/4"											
White / stock color	Inst	Ea	SB	11.4	7820.00	359.00	---	8179.00	9520.00	6692.00	7780.0
Premium color	Inst	Ea	SB	11.4	7870.00	359.00	---	8229.00	9570.00	6732.00	7820.0
Fontana model											
72" x 54" x 28" H, angled back, with RapidHeat											
White / stock color	Inst	Ea	SB	11.4	6330.00	359.00	---	6689.00	7810.00	5472.00	6370.0
Premium color	Inst	Ea	SB	11.4	6380.00	359.00	---	6739.00	7860.00	5502.00	6410.0
Options for whirlpool baths and spas											
Timer kit, 30 min, wall mount											
Polished chrome	Inst	Ea	---	---	71.40	---	---	71.40	71.40	58.80	58.8
Gold	Inst	Ea	---	---	83.30	---	---	83.30	83.30	68.60	68.6
Trim kit											
Polished chrome	Inst	Ea	---	---	125.00	---	---	125.00	125.00	103.00	103.0
Bright brass	Inst	Ea	---	---	167.00	---	---	167.00	167.00	138.00	138.0
Trip lever drain with vent											
Polished chrome	Inst	Ea	---	---	115.00	---	---	115.00	115.00	94.50	94.5
Bright brass	Inst	Ea	---	---	155.00	---	---	155.00	155.00	127.00	127.0
Gold	Inst	Ea	---	---	170.00	---	---	170.00	170.00	140.00	140.0
Turn drain with vent											
Polished chrome	Inst	Ea	---	---	146.00	---	---	146.00	146.00	120.00	120.0
Bright brass	Inst	Ea	---	---	184.00	---	---	184.00	184.00	151.00	151.0
Gold	Inst	Ea	---	---	189.00	---	---	189.00	189.00	155.00	155.0

Description	Oper	Unit	Crew Size	Man-Hours Per Unit	Costs Based On Small Volume						Large Volume	
					Avg Mat'l Unit Cost	Avg Labor Unit Cost	Avg Equip Unit Cost	Avg Total Unit Cost	Avg Price Incl O&P		Avg Total Unit Cost	Avg Price Incl O&P

Stairs

Stair parts

Balusters, stock pine

1-1/16" x 1-1/16"	Inst	LF	CA	.067	2.68	2.15	---	4.83	**6.31**	3.72	**4.73**
1-5/8" x 1-5/8"	Inst	LF	CA	.083	2.48	2.66	---	5.14	**6.85**	3.86	**5.01**

Balusters, turned

30" high
Pine	Inst	Ea	CA	.556	10.10	17.80	---	27.90	**38.30**	19.88	**26.60**
Birch	Inst	Ea	CA	.556	11.80	17.80	---	29.60	**40.30**	21.40	**28.40**

42" high
Pine	Inst	Ea	CA	.667	11.80	21.40	---	33.20	**45.60**	23.50	**31.60**
Birch	Inst	Ea	CA	.667	15.20	21.40	---	36.60	**49.50**	26.60	**35.20**

Newels, 3-1/4" wide

Starting	Inst	Ea	CA	2.22	163.00	71.20	---	234.20	**295.00**	191.70	**235.00**
Landing	Inst	Ea	CA	3.33	231.00	107.00	---	338.00	**426.00**	275.20	**339.00**

Railings, built-up

Oak	Inst	LF	CA	.267	16.30	8.57	---	24.87	**31.60**	20.03	**24.80**

Railings, subrail

Oak	Inst	LF	CA	.133	21.80	4.27	---	26.07	**31.40**	22.37	**26.70**

Risers, 3/4" x 7-1/2" high

Beech	Inst	LF	CA	.222	7.07	7.12	---	14.19	**18.80**	10.72	**13.80**
Fir	Inst	LF	CA	.222	2.04	7.12	---	9.16	**13.00**	6.13	**8.54**
Oak	Inst	LF	CA	.222	6.46	7.12	---	13.58	**18.10**	10.16	**13.20**
Pine	Inst	LF	CA	.222	2.04	7.12	---	9.16	**13.00**	6.13	**8.54**

Skirt board, pine

1" x 10"	Inst	LF	CA	.267	2.24	8.57	---	10.81	**15.40**	7.18	**10.10**
1" x 12"	Inst	LF	CA	.296	2.72	9.50	---	12.22	**17.40**	8.19	**11.40**

Treads, oak

1-16" x 9-1/2" wide
3' long	Inst	Ea	CA	.833	29.90	26.70	---	56.60	**74.50**	43.40	**55.40**
4' long	Inst	Ea	CA	.889	39.40	28.50	---	67.90	**88.20**	53.10	**67.00**

1-1/16" x 11-1/2" wide
3' long	Inst	Ea	CA	.833	32.00	26.70	---	58.70	**76.90**	45.20	**57.60**
6' long	Inst	Ea	CA	1.11	76.20	35.60	---	111.80	**141.00**	90.80	**112.00**

Note: The following percentage adjustment for Small Volume also applies to Large

For beech treads, ADD	Inst	%	---	---	40.0	---	---	---	---	---	---
For mitered return nosings, ADD	Inst	LF	CA	.222	11.40	7.12	---	18.52	**23.80**	14.67	**18.40**

				Costs Based On Small Volume						Large Volume	
Description	Oper	Unit	Crew Size	Man-Hours Per Unit	Avg Mat'l Unit Cost	Avg Labor Unit Cost	Avg Equip Unit Cost	Avg Total Unit Cost	Avg Price Incl O&P	Avg Total Unit Cost	Avg Price Incl O&P

Stairs, shop fabricated, per flight

Basement stairs, soft wood, open risers

3' wide, 8' high	Inst	Flt	2C	6.67	510.00	214.00	---	724.00	908.00	593.00	727.0

Box stairs, no handrails, 3' wide

Oak treads

2' high	Inst	Flt	2C	5.33	203.00	171.00	---	374.00	490.00	288.00	367.0
4' high	Inst	Flt	2C	6.67	435.00	214.00	---	649.00	822.00	525.00	649.0
6' high	Inst	Flt	2C	7.62	707.00	245.00	---	952.00	1180.00	792.00	962.0
8' high	Inst	Flt	2C	8.89	884.00	285.00	---	1169.00	1440.00	977.00	1180.0

Pine treads for carpet

2' high	Inst	Flt	2C	5.33	160.00	171.00	---	331.00	441.00	249.00	322.0
4' high	Inst	Flt	2C	6.67	273.00	214.00	---	487.00	635.00	377.00	479.0
6' high	Inst	Flt	2C	7.62	415.00	245.00	---	660.00	844.00	525.00	655.0
8' high	Inst	Flt	2C	8.89	517.00	285.00	---	802.00	1020.00	642.00	798.0
Stair rail with balusters, 5 risers	Inst	Ea	2C	1.78	238.00	57.10	---	295.10	359.00	251.30	301.0

Note: The following percentage adjustment for Small Volume also applies to Large

For 4' wide stairs, ADD	Inst	%	2C	---	25.0	5.0	---	---	---	---	

Open stairs, prefinished, metal stringers, 3'-6" wide treads, no railings

3' high	Inst	Flt	2C	5.33	592.00	171.00	---	763.00	937.00	642.00	774.0
4' high	Inst	Flt	2C	6.67	741.00	214.00	---	955.00	1170.00	804.00	970.0
6' high	Inst	Flt	2C	7.62	1290.00	245.00	---	1535.00	1850.00	1327.00	1570.0
8' high	Inst	Flt	2C	8.89	2070.00	285.00	---	2355.00	2810.00	2061.00	2430.0

Adjustments

For 3-piece wood railings and balusters, ADD

3' high	Inst	Ea	2C	1.78	203.00	57.10	---	260.10	319.00	219.30	264.0
4' high	Inst	Ea	2C	1.90	260.00	61.00	---	321.00	390.00	273.60	327.0
6' high	Inst	Ea	2C	2.05	400.00	65.80	---	465.80	559.00	404.50	478.0
8' high	Inst	Ea	2C	2.22	496.00	71.20	---	567.20	678.00	495.70	585.0
For 3'-6" x 3'-6" platform, ADD	Inst	Ea	2C	6.67	231.00	214.00	---	445.00	587.00	339.00	435.0

Curved stairways, oak, unfinished, curved balustrade

Open one side

9' high	Inst	Flt	2C	38.1	8670.00	1220.00	---	9890.00	11800.00	8645.00	10200.0
10' high	Inst	Flt	2C	38.1	9790.00	1220.00	---	11010.00	13100.00	9665.00	11400.0

Open both sides

9' high	Inst	Flt	2C	53.3	13600.00	1710.00	---	15310.00	18200.00	13430.00	15800.0
10' high	Inst	Flt	2C	53.3	14700.00	1710.00	---	16410.00	19500.00	14430.00	16900.0

Steel, reinforcing. See Concrete, page 69
Stucco. See Plaster, page 214
Studs. See Framing, page 133
Subflooring. See Framing, page 139

Suspended Ceilings

Suspended ceilings consist of a grid of small metal hangers which are hung from the ceiling framing with wire or strap, and drop-in panels sized to fit the grid system. The main advantage to this type of ceiling finish is that it can be adjusted to any height desired. It can cover a lot of flaws and obstructions (uneven plaster, pipes, wiring, and ductwork). With a high ceiling, the suspended ceiling reduces the sound transfer from the floor above and increases the ceiling's insulating value. One great advantage of suspended ceilings is that the area directly above the tiles is readily accessible for repair and maintenance work; the tiles merely have to be lifted out from the grid. This system also eliminates the need for any other ceiling finish.

1. **Installation Procedure**

   a. Decide on ceiling height. It must clear the lowest obstacles but remain above windows and doors. Snap a chalk line around walls at the new height. Check it with a carpenter's level.

   b. Install wall molding/angle at the marked line; this supports the tiles around the room's perimeter. Check with a level. Use nails (6d) or screws to fasten molding to studs (16" or 24" oc); on masonry, use concrete nails 24" oc. The molding should be flush along the chalk line with the bottom leg of the L-angle facing into the room.

   c. Fit wall molding at inside corners by straight cutting (90 degrees) and overlapping; for outside corners, miter cut corner (45 degrees) and butt together. Cut the molding to required lengths with tin snips or a hacksaw.

   d. Mark center line of room above the wall molding. Decide where the first main runner should be to get the desired border width. Position main runners by running taut string lines across room. Repeat for the first row of cross tees closest to one wall. Then attach hanger wires with screw eyes to the joists at 4' intervals.

   e. Trim main runners at wall to line up the slot with cross tee string. Rest trimmed end on wall molding; insert wire through holes; recheck with level; twist wire several times.

   f. Insert cross tee and tabs into main runner's slots; push down to lock. Check with level. Repeat.

   g. Drop 2' x 4' panels into place.

2. **Estimating Materials**

   a. Measure the ceiling and plot it on graph paper. Mark the direction of ceiling joists. On the ceiling itself, snap chalk lines along the joist lines.

   b. Draw ceiling layout for grid system onto graph paper. Plan the ceiling layout, figuring full panels across the main ceiling and evenly trimmed partial panels at the edges. To calculate the width of border panels in each direction, determine the width of the gap left after full panels are placed all across the dimension; then divide by 2.

   c. For purposes of ordering material, determine the room size in even number of feet. If the room length or width is not divisible by 2 feet, increase the dimension to the next larger unit divisible by 2. For example, a room that measured 11'-6" x 14'-4" would be considered a 12' x 16' room. This allows for waste and/or breakage. In this example, you would order 192 SF of material.

   d. Main runners and tees must run perpendicular to ceiling joists. For a 2' x 2' grid, cross and main tees are 2' apart. For a 2' x 4' grid, main tees are 4' apart and 4' cross tees connect the main tees; then 2' cross tees can be used to connect the 4' cross tees. The long panels of the grid are set parallel to the ceiling joists, so the T-shaped main runner is attached perpendicular to joists at 4' oc.

   e. Wall molding/angle is manufactured in 10' lengths. Main runners and tees are manufactured in 12' lengths. Cross tees are either 2' long or 4' long. Hanger wire is 12 gauge, and attached to joists with either screw eyes or hook-and-nail. Drop-in panels are either 8 SF (2' x 4') or 4 SF (2' x 2').

   f. To find the number of drop-in panels required, divide the nominal room size in square feet (e.g., 192 SF in above example) by square feet of panel to be used.

   g. The quantity of 12 gauge wire depends on the "drop" distance of the ceiling. Figure one suspension wire and screw eye for each 4' of main runner; if any main run is longer than 12' then splice plates are needed, with a wire and screw eye on each side of the splice. The length of each wire should be at least 2" longer than the "drop" distance, to allow for a twist after passing through runner or tee.

				Costs Based On Small Volume						Large Volume	
Description	Oper	Unit	Crew Size	Man-Hours Per Unit	Avg Mat'l Unit Cost	Avg Labor Unit Cost	Avg Equip Unit Cost	Avg Total Unit Cost	Avg Price Incl O&P	Avg Total Unit Cost	Avg Price Incl O&P

Suspended ceiling systems

Suspended ceiling panels and grid system

	Demo	SF	LB	.013	---	.35	---	.35	.52	.24	.3

Complete 2' x 4' slide lock grid system (baked enamel)
Labor includes installing grid and laying in panels

Luminous panels
Prismatic panels

Acrylic	Inst	SF	CA	.030	2.14	.96	---	3.10	3.91	2.62	3.2
Polystyrene	Inst	SF	CA	.030	1.32	.96	---	2.28	2.96	1.87	2.3

Ribbed panels

Acrylic	Inst	SF	CA	.030	2.27	.96	---	3.23	4.05	2.75	3.4
Polystyrene	Inst	SF	CA	.030	1.45	.96	---	2.41	3.11	2.00	2.5

Fiberglass panels
1/2" T

Plain white	Inst	SF	CA	.028	1.18	.90	---	2.08	2.70	1.72	2.2
Sculptured white	Inst	SF	CA	.028	1.32	.90	---	2.22	2.87	1.84	2.3

5/8" T

Fissured	Inst	SF	CA	.028	1.38	.90	---	2.28	2.93	1.90	2.4
Fissured, fire rated	Inst	SF	CA	.028	1.52	.90	---	2.42	3.10	2.03	2.5

Mineral fiber panels

1/2" T	Inst	SF	CA	.028	.97	.90	---	1.87	2.46	1.53	1.9
9/16" T	Inst	SF	CA	.028	1.56	.90	---	2.46	3.14	2.07	2.6

Note: The following percentage adjustment for Small Volume also applies to Large

For wood grained grid components

ADD	Inst	%		---	---	7.0	---	---	---	---	---

Panels only (labor to lay panels in grid system)

Luminous panels
Prismatic panels

Acrylic	Inst	SF	CA	.011	1.78	.35	---	2.13	2.58	1.89	2.2
Polystyrene	Inst	SF	CA	.011	.96	.35	---	1.31	1.63	1.14	1.4

Ribbed panels

Acrylic	Inst	SF	CA	.011	1.92	.35	---	2.27	2.74	2.01	2.4
Polystyrene	Inst	SF	CA	.011	1.10	.35	---	1.45	1.79	1.26	1.5

Fiberglass panels
1/2" T

Plain white	Inst	SF	CA	.009	.82	.29	---	1.11	1.38	.97	1.2
Sculptured white	Inst	SF	CA	.009	.96	.29	---	1.25	1.54	1.10	1.3

5/8" T

Fissured	Inst	SF	CA	.009	1.03	.29	---	1.32	1.62	1.16	1.4
Fissured, fire rated	Inst	SF	CA	.009	1.16	.29	---	1.45	1.77	1.28	1.5

Mineral fiber panels

5/8" T	Inst	SF	CA	.009	.62	.29	---	.91	1.15	.78	.9
3/4" T	Inst	SF	CA	.009	1.21	.29	---	1.50	1.82	1.32	1.6

Description	Oper	Unit	Costs Based On Small Volume							Large Volume	
			Crew Size	Man-Hours Per Unit	Avg Mat'l Unit Cost	Avg Labor Unit Cost	Avg Equip Unit Cost	Avg Total Unit Cost	Avg Price Incl O&P	Avg Total Unit Cost	Avg Price Incl O&P
Grid system components											
Main runner, 12' L pieces											
Enamel	Inst	Ea	---	---	6.10	---	---	6.10	6.10	5.56	5.56
Wood grain	Inst	Ea	---	---	6.99	---	---	6.99	6.99	6.38	6.38
Cross tees											
2' L pieces											
Enamel	Inst	Ea	---	---	1.10	---	---	1.10	1.10	1.00	1.00
Wood grain	Inst	Ea	---	---	1.23	---	---	1.23	1.23	1.13	1.13
4' L pieces											
Enamel	Inst	Ea	---	---	2.06	---	---	2.06	2.06	1.88	1.88
Wood grain	Inst	Ea	---	---	2.33	---	---	2.33	2.33	2.13	2.13
Wall molding, 10' L pieces											
Enamel	Inst	Ea	---	---	3.90	---	---	3.90	3.90	3.56	3.56
Wood grain	Inst	Ea	---	---	5.14	---	---	5.14	5.14	4.69	4.69
Main runner hold-down clips											
1/2" T panels (1000/carton)	Inst	Ctn	---	---	11.00	---	---	11.00	11.00	10.00	10.00
5/8" T panels (1000/carton)	Inst	Ctn	---	---	15.10	---	---	15.10	15.10	13.80	13.80

Telephone prewiring. See Electrical, page 107

Television antenna. See Electrical, page 107

Thermostat. See Electrical, page 107

Tile, ceiling. See Acoustical treatment, page 22

Tile, ceramic. See Ceramic tile, page 60

Tile, quarry. See Masonry, page 185

Tile, vinyl or asphalt. See Resilient flooring, page 218

Toilets, bidets, urinals

Toilets (water closets), vitreous china, 1.6 GPF,
includes seat, shut-off valve, connectors, flanges,
water supply valve

Description	Oper	Unit	Crew Size	Man-Hours Per Unit	Avg Mat'l Unit Cost	Avg Labor Unit Cost	Avg Equip Unit Cost	Avg Total Unit Cost	Avg Price Incl O&P	Avg Total Unit Cost	Avg Price Incl O&P
									(Small Volume)	(Large Volume)	
Frequently encountered applications											
Detach & reset operations											
Toilet	Reset	Ea	SA	1.78	34.00	64.70	---	98.70	**134.00**	76.40	103.00
Bidet	Reset	Ea	SA	1.78	34.00	64.70	---	98.70	**134.00**	76.40	103.00
Urinal	Reset	Ea	SA	1.78	34.00	64.70	---	98.70	**134.00**	76.40	103.00
Remove operations											
Toilet	Demo	Ea	SA	1.33	10.20	48.40	---	58.60	**82.80**	44.80	63.10
Bidet	Demo	Ea	SA	1.33	10.20	48.40	---	58.60	**82.80**	44.80	63.10
Urinal	Demo	Ea	SA	1.33	10.20	48.40	---	58.60	**82.80**	44.80	63.10
Install rough-in											
Toilet	Inst	Ea	SA	10.0	149.00	364.00	---	513.00	**705.00**	414.00	568.00
Bidet	Inst	Ea	SA	10.0	128.00	364.00	---	492.00	**681.00**	396.00	548.00
Replace operations											
Toilet / water closet											
One piece, floor mounted	Inst	Ea	SA	2.67	412.00	97.10	---	509.10	**617.00**	412.70	497.00
Two piece, floor mounted	Inst	Ea	SA	2.67	362.00	97.10	---	459.10	**559.00**	370.70	450.00
Two piece, wall mounted	Inst	Ea	SA	2.67	740.00	97.10	---	837.10	**993.00**	681.70	807.00
Bidet											
Deck-mounted fittings	Inst	Ea	SA	2.11	751.00	76.70	---	827.70	**977.00**	677.20	797.00
Urinal											
Wall mounted	Inst	Ea	SA	3.48	460.00	127.00	---	587.00	**715.00**	476.10	578.00
One piece, floor mounted											
American Standard Products											
Cadet, round											
White	Inst	Ea	SA	2.67	319.00	97.10	---	416.10	**509.00**	335.70	409.00
Color	Inst	Ea	SA	2.67	377.00	97.10	---	474.10	**576.00**	382.70	463.00
Premium color	Inst	Ea	SA	2.67	446.00	97.10	---	543.10	**656.00**	440.70	529.00
Cadet, elongated											
White	Inst	Ea	SA	2.67	377.00	97.10	---	474.10	**576.00**	382.70	463.00
Color	Inst	Ea	SA	2.67	451.00	97.10	---	548.10	**661.00**	443.70	534.00
Premium color	Inst	Ea	SA	2.67	524.00	97.10	---	621.10	**745.00**	503.70	603.00
Hamilton, elongated, space saving											
White	Inst	Ea	SA	2.67	316.00	97.10	---	413.10	**506.00**	332.70	406.00
Color	Inst	Ea	SA	2.67	412.00	97.10	---	509.10	**617.00**	412.70	497.00
Premium color	Inst	Ea	SA	2.67	481.00	97.10	---	578.10	**696.00**	468.70	562.00
Savona, elongated											
White	Inst	Ea	SA	2.67	575.00	97.10	---	672.10	**804.00**	546.70	652.00
Color	Inst	Ea	SA	2.67	722.00	97.10	---	819.10	**973.00**	666.70	790.00
Premium color	Inst	Ea	SA	2.67	808.00	97.10	---	905.10	**1070.00**	737.70	872.00

Description	Oper	Unit	Crew Size	Man-Hours Per Unit	Avg Mat'l Unit Cost	Avg Labor Unit Cost	Avg Equip Unit Cost	Avg Total Unit Cost	Avg Price Incl O&P	Avg Total Unit Cost	Avg Price Incl O&P
Lexington, elongated, low profile model											
White	Inst	Ea	SA	2.67	643.00	97.10	---	740.10	882.00	601.70	715.00
Color	Inst	Ea	SA	2.67	802.00	97.10	---	899.10	1070.00	733.70	867.00
Premium color	Inst	Ea	SA	2.67	904.00	97.10	---	1001.10	1180.00	816.70	963.00
Fontaine, elongated, pressure-assisted model											
White	Inst	Ea	SA	2.67	827.00	97.10	---	924.10	1090.00	753.70	890.00
Color	Inst	Ea	SA	2.67	1040.00	97.10	---	1137.10	1330.00	925.70	1090.00
Premium color	Inst	Ea	SA	2.67	1160.00	97.10	---	1257.10	1480.00	1028.70	1210.00

Kohler Products

Description	Oper	Unit	Crew Size	Man-Hours Per Unit	Avg Mat'l Unit Cost	Avg Labor Unit Cost	Avg Equip Unit Cost	Avg Total Unit Cost	Avg Price Incl O&P	Avg Total Unit Cost	Avg Price Incl O&P
Rialto, round											
White	Inst	Ea	SA	2.67	330.00	97.10	---	427.10	522.00	344.70	419.00
Color	Inst	Ea	SA	2.67	428.00	97.10	---	525.10	635.00	425.70	513.00
Premium color	Inst	Ea	SA	2.67	493.00	97.10	---	590.10	710.00	478.70	574.00
Rosario, round											
White	Inst	Ea	SA	2.67	444.00	97.10	---	541.10	653.00	437.70	527.00
Color	Inst	Ea	SA	2.67	577.00	97.10	---	674.10	806.00	547.70	653.00
Premium color	Inst	Ea	SA	2.67	663.00	97.10	---	760.10	906.00	618.70	735.00
San Martine, elongated											
White	Inst	Ea	SA	2.67	707.00	97.10	---	804.10	956.00	655.70	777.00
Color	Inst	Ea	SA	2.67	920.00	97.10	---	1017.10	1200.00	829.70	978.00
Premium color	Inst	Ea	SA	2.67	1060.00	97.10	---	1157.10	1360.00	943.70	1110.00
Revival, elongated, with lift knob											
White	Inst	Ea	SA	2.67	929.00	97.10	---	1026.10	1210.00	837.70	987.00
Color	Inst	Ea	SA	2.67	1210.00	97.10	---	1307.10	1540.00	1072.70	1260.00
Premium color	Inst	Ea	SA	2.67	1360.00	97.10	---	1457.10	1710.00	1192.70	1400.00

Two piece, floor mounted
American Standard Products

Description	Oper	Unit	Crew Size	Man-Hours Per Unit	Avg Mat'l Unit Cost	Avg Labor Unit Cost	Avg Equip Unit Cost	Avg Total Unit Cost	Avg Price Incl O&P	Avg Total Unit Cost	Avg Price Incl O&P
Cadet, elongated											
White	Inst	Ea	SA	2.67	289.00	97.10	---	386.10	475.00	310.70	381.00
Color	Inst	Ea	SA	2.67	362.00	97.10	---	459.10	559.00	370.70	450.00
Premium color	Inst	Ea	SA	2.67	401.00	97.10	---	498.10	604.00	402.70	487.00
Cadet, pressure assisted											
White	Inst	Ea	SA	2.67	460.00	97.10	---	557.10	671.00	451.70	542.00
Color	Inst	Ea	SA	2.67	564.00	97.10	---	661.10	792.00	537.70	641.00
Premium color	Inst	Ea	SA	2.67	633.00	97.10	---	730.10	871.00	594.70	707.00
Cadet, elongated, 16-1/2" H											
White	Inst	Ea	SA	2.67	315.00	97.10	---	412.10	504.00	331.70	405.00
Color	Inst	Ea	SA	2.67	388.00	97.10	---	485.10	588.00	391.70	474.00
Premium color	Inst	Ea	SA	2.67	427.00	97.10	---	524.10	633.00	423.70	511.00
Ravenna											
Round											
White	Inst	Ea	SA	2.67	203.00	97.10	---	300.10	376.00	239.70	299.00
Color	Inst	Ea	SA	2.67	251.00	97.10	---	348.10	431.00	279.70	344.00
Premium color	Inst	Ea	SA	2.67	286.00	97.10	---	383.10	471.00	307.70	377.00
Elongated											
White	Inst	Ea	SA	2.67	247.00	97.10	---	344.10	426.00	275.70	340.00
Color	Inst	Ea	SA	2.67	304.00	97.10	---	401.10	493.00	323.70	395.00
Premium color	Inst	Ea	SA	2.67	343.00	97.10	---	440.10	537.00	354.70	431.00

				Costs Based On Small Volume					Large Volume	
Description	Oper	Unit	Crew Size	Man-Hours Per Unit	Avg Mat'l Unit Cost	Avg Labor Unit Cost	Avg Equip Unit Cost	Avg Total Unit Cost	Avg Price Incl O&P	Avg Total Unit Cost	Avg Price Incl O&P
Town Square											
Round											
White	Inst	Ea	SA	2.67	328.00	97.10	---	425.10	520.00	342.70	418.00
Color	Inst	Ea	SA	2.67	426.00	97.10	---	523.10	632.00	423.70	510.00
Premium color	Inst	Ea	SA	2.67	492.00	97.10	---	589.10	709.00	477.70	573.00
Elongated											
White	Inst	Ea	SA	2.67	379.00	97.10	---	476.10	579.00	384.70	466.00
Color	Inst	Ea	SA	2.67	492.00	97.10	---	589.10	709.00	477.70	573.00
Premium color	Inst	Ea	SA	2.67	568.00	97.10	---	665.10	796.00	540.70	645.00
Yorkville, elongated, pressure-assisted model											
White	Inst	Ea	SA	2.67	756.00	97.10	---	853.10	1010.00	694.70	823.00
Color	Inst	Ea	SA	2.67	897.00	97.10	---	994.10	1170.00	811.70	956.00
Premium color	Inst	Ea	SA	2.67	989.00	97.10	---	1086.10	1280.00	887.70	1040.00

Kohler Products

Memoirs, round											
White	Inst	Ea	SA	2.67	229.00	97.10	---	326.10	406.00	261.70	324.00
Color	Inst	Ea	SA	2.67	297.00	97.10	---	394.10	485.00	317.70	389.00
Premium color	Inst	Ea	SA	2.67	342.00	97.10	---	439.10	536.00	354.70	431.00
Memoirs, elongated											
White	Inst	Ea	SA	2.67	355.00	97.10	---	452.10	551.00	364.70	443.00
Color	Inst	Ea	SA	2.67	461.00	97.10	---	558.10	673.00	452.70	544.00
Premium color	Inst	Ea	SA	2.67	531.00	97.10	---	628.10	753.00	509.70	609.00
Highline, elongated											
White	Inst	Ea	SA	2.67	207.00	97.10	---	304.10	380.00	242.70	303.00
Color	Inst	Ea	SA	2.67	265.00	97.10	---	362.10	447.00	290.70	358.00
Premium color	Inst	Ea	SA	2.67	305.00	97.10	---	402.10	493.00	323.70	396.00
Revival, elongated, with lift knob											
White	Inst	Ea	SA	2.67	530.00	97.10	---	627.10	752.00	509.70	609.00
Color	Inst	Ea	SA	2.67	688.00	97.10	---	785.10	934.00	639.70	758.00
Premium color	Inst	Ea	SA	2.67	794.00	97.10	---	891.10	1060.00	726.70	859.00

Two piece, wall mounted
American Standard Products

Glenwall, elongated											
White	Inst	Ea	SA	2.67	604.00	97.10	---	701.10	837.00	569.70	678.00
Color	Inst	Ea	SA	2.67	740.00	97.10	---	837.10	993.00	681.70	807.00
Premium color	Inst	Ea	SA	2.67	828.00	97.10	---	925.10	1090.00	754.70	891.00

Bidet, deck-mounted fitting
American Standard Products

Heritage											
White	Inst	Ea	SA	2.11	601.00	76.70	---	677.70	804.00	553.20	655.00
Color	Inst	Ea	SA	2.11	751.00	76.70	---	827.70	977.00	677.20	797.00
Savona											
White	Inst	Ea	SA	2.11	411.00	76.70	---	487.70	586.00	397.20	475.00
Color	Inst	Ea	SA	2.11	516.00	76.70	---	592.70	706.00	483.20	574.00
Pemium color	Inst	Ea	SA	2.11	573.00	76.70	---	649.70	772.00	530.20	628.00

Description	Oper	Unit	Crew Size	Man-Hours Per Unit	Avg Mat'l Unit Cost	Avg Labor Unit Cost	Avg Equip Unit Cost	Avg Total Unit Cost	Avg Price Incl O&P	Avg Total Unit Cost	Avg Price Incl O&P
						Costs Based On Small Volume					Large Volume	
Urinal, wall mounted, 1.0 GPF											
American Standard Products											
Lynnbrook											
White	Inst	Ea	SA	3.48	505.00	127.00	---	632.00	**767.00**	513.10	**621.00**
Color	Inst	Ea	SA	3.48	621.00	127.00	---	748.00	**900.00**	609.10	**731.00**
Innsbrook											
White	Inst	Ea	SA	3.48	373.00	127.00	---	500.00	**615.00**	404.10	**496.00**
Color	Inst	Ea	SA	3.48	460.00	127.00	---	587.00	**715.00**	476.10	**578.00**

				Costs Based On Small Volume					Large Volume	
Description	Oper	Unit	Crew Size	Man-Hours Per Unit	Avg Mat'l Unit Cost	Avg Labor Unit Cost	Avg Equip Unit Cost	Avg Total Unit Cost	Avg Price Incl O&P	Avg Total Unit Cost	Avg Price Incl O&P

Trash compactors

Includes wiring, connection, and installation in a pre-cutout area

Description	Oper	Unit	Crew Size	Man-Hours Per Unit	Avg Mat'l Unit Cost	Avg Labor Unit Cost	Avg Equip Unit Cost	Avg Total Unit Cost	Avg Price Incl O&P	Avg Total Unit Cost	Avg Price Incl O&P
White	Inst	Ea	EA	2.67	531.00	91.80	---	622.80	**745.00**	506.80	**604.00**
Colors	Inst	Ea	EA	2.67	574.00	91.80	---	665.80	**794.00**	541.80	**644.00**
To remove and reset	Reset	Ea	EA	.964	---	33.20	---	33.20	**48.40**	25.00	**36.50**
Material adjustments											
Stainless steel trim	R&R	LS	---	---	29.80	---	---	29.80	**34.20**	24.50	**28.20**
Stainless steel panel / trim	R&R	LS	---	---	63.80	---	---	63.80	**73.30**	52.50	**60.40**
"Black Glas" acrylic											
front panel / trim	R&R	LS	---	---	63.80	---	---	63.80	**73.30**	52.50	**60.40**
Hardwood top	R&R	LS	---	---	42.50	---	---	42.50	**48.90**	35.00	**40.30**

Trim. See Molding, page 186
Trusses. See Framing, page 139
Vanity units. See Cabinetry, page 53

Ventilation

Flue piping or vent chimney
Prefabricated metal, UL listed

Description	Oper	Unit	Crew Size	Man-Hours Per Unit	Avg Mat'l Unit Cost	Avg Labor Unit Cost	Avg Equip Unit Cost	Avg Total Unit Cost	Avg Price Incl O&P	Avg Total Unit Cost	Avg Price Incl O&P
Gas, double wall, galv. steel											
3" diameter	Inst	VLF	UB	.327	4.33	11.70	---	16.03	22.20	11.99	16.50
4" diameter	Inst	VLF	UB	.348	5.30	12.40	---	17.70	24.50	13.31	18.20
5" diameter	Inst	VLF	UB	.372	6.28	13.30	---	19.58	26.90	14.75	20.00
6" diameter	Inst	VLF	UB	.390	7.30	13.90	---	21.20	29.00	16.28	22.00
7" diameter	Inst	VLF	UB	.421	10.80	15.00	---	25.80	34.60	20.09	26.50
8" diameter	Inst	VLF	UB	.457	12.00	16.30	---	28.30	37.90	22.00	29.10
10" diameter	Inst	VLF	UB	.500	25.30	17.80	---	43.10	55.50	34.70	44.00
12" diameter	Inst	VLF	UB	.552	33.90	19.70	---	53.60	68.10	43.60	54.60
16" diameter	Inst	VLF	UB	.593	76.80	21.20	---	98.00	120.00	82.80	100.00
20" diameter	Inst	VLF	UB	.667	118.00	23.80	---	141.80	171.00	120.80	144.00
24" diameter	Inst	VLF	UB	.762	183.00	27.20	---	210.20	251.00	181.00	214.00
Vent damper, bi-metal, 6" flue											
Gas, auto., electric	Inst	Ea	UB	4.00	166.00	143.00	---	309.00	402.00	242.20	310.00
Oil, auto., electric	Inst	Ea	UB	4.00	200.00	143.00	---	343.00	441.00	271.20	344.00
All fuel, double wall, stainless steel											
6" diameter	Inst	VLF	UB	.381	35.80	13.60	---	49.40	61.30	41.12	50.50
7" diameter	Inst	VLF	UB	.410	46.10	14.60	---	60.70	74.60	50.90	61.90
8" diameter	Inst	VLF	UB	.444	53.80	15.80	---	69.60	85.30	58.50	70.80
10" diameter	Inst	VLF	UB	.471	78.10	16.80	---	94.90	115.00	80.80	96.90
12" diameter	Inst	VLF	UB	.516	105.00	18.40	---	123.40	148.00	105.70	126.00
14" diameter	Inst	VLF	UB	.571	138.00	20.40	---	158.40	189.00	136.30	161.00
All fuel, double wall, stainless steel fittings											
Roof support											
6" diameter	Inst	Ea	UB	.762	91.50	27.20	---	118.70	145.00	99.80	121.00
7" diameter	Inst	Ea	UB	.800	103.00	28.50	---	131.50	161.00	111.40	135.00
8" diameter	Inst	Ea	UB	.889	112.00	31.70	---	143.70	176.00	120.80	146.00
10" diameter	Inst	Ea	UB	.941	147.00	33.60	---	180.60	219.00	153.80	185.00
12" diameter	Inst	Ea	UB	1.07	178.00	38.20	---	216.20	261.00	182.90	219.00
14" diameter	Inst	Ea	UB	1.14	225.00	40.70	---	265.70	319.00	227.50	271.00
Elbow, 15 degree											
6" diameter	Inst	Ea	UB	.762	80.00	27.20	---	107.20	132.00	89.60	109.00
7" diameter	Inst	Ea	UB	.800	89.60	28.50	---	118.10	145.00	99.50	121.00
8" diameter	Inst	Ea	UB	.889	102.00	31.70	---	133.70	165.00	112.30	136.00
10" diameter	Inst	Ea	UB	.941	133.00	33.60	---	166.60	203.00	141.80	170.00
12" diameter	Inst	Ea	UB	1.07	163.00	38.20	---	201.20	243.00	169.90	203.00
14" diameter	Inst	Ea	UB	1.14	195.00	40.70	---	235.70	284.00	200.50	240.00
Insulated tee with insulated tee cap											
6" diameter	Inst	Ea	UB	.762	151.00	27.20	---	178.20	214.00	152.00	181.00
7" diameter	Inst	Ea	UB	.800	197.00	28.50	---	225.50	269.00	194.40	230.00
8" diameter	Inst	Ea	UB	.889	223.00	31.70	---	254.70	303.00	218.90	259.00
10" diameter	Inst	Ea	UB	.941	314.00	33.60	---	347.60	410.00	300.80	354.00
12" diameter	Inst	Ea	UB	1.07	442.00	38.20	---	480.20	564.00	415.90	487.00
14" diameter	Inst	Ea	UB	1.14	576.00	40.70	---	616.70	723.00	537.50	627.00

Description	Oper	Unit	Crew Size	Man-Hours Per Unit	Costs Based On Small Volume						Large Volume	
					Avg Mat'l Unit Cost	Avg Labor Unit Cost	Avg Equip Unit Cost	Avg Total Unit Cost	Avg Price Incl O&P		Avg Total Unit Cost	Avg Price Incl O&P
Joist shield												
6" diameter	Inst	Ea	UB	.762	46.10	27.20	---	73.30	93.20		59.70	74.9◖
7" diameter	Inst	Ea	UB	.800	49.90	28.50	---	78.40	99.60		64.50	80.8◖
8" diameter	Inst	Ea	UB	.889	61.40	31.70	---	93.10	118.00		76.10	94.8◖
10" diameter	Inst	Ea	UB	.941	69.80	33.60	---	103.40	130.00		85.40	106.0◖
12" diameter	Inst	Ea	UB	1.07	103.00	38.20	---	141.20	175.00		116.90	143.0◖
14" diameter	Inst	Ea	UB	1.14	128.00	40.70	---	168.70	207.00		141.50	172.0◖
Round top												
6" diameter	Inst	Ea	UB	.762	51.20	27.20	---	78.40	99.10		64.20	80.1◖
7" diameter	Inst	Ea	UB	.800	69.80	28.50	---	98.30	122.00		82.00	101.0◖
8" diameter	Inst	Ea	UB	.889	93.40	31.70	---	125.10	154.00		104.40	127.0◖
10" diameter	Inst	Ea	UB	.941	168.00	33.60	---	201.60	243.00		171.80	205.0◖
12" diameter	Inst	Ea	UB	1.07	239.00	38.20	---	277.20	332.00		236.90	281.0◖
14" diameter	Inst	Ea	UB	1.14	315.00	40.70	---	355.70	422.00		306.50	362.0◖
Adjustable roof flashing												
6" diameter	Inst	Ea	UB	.762	60.80	27.20	---	88.00	110.00		72.70	89.9◖
7" diameter	Inst	Ea	UB	.800	69.10	28.50	---	97.60	122.00		81.40	100.0◖
8" diameter	Inst	Ea	UB	.889	75.50	31.70	---	107.20	134.00		88.60	109.0◖
10" diameter	Inst	Ea	UB	.941	96.60	33.60	---	130.20	161.00		109.10	133.0◖
12" diameter	Inst	Ea	UB	1.07	125.00	38.20	---	163.20	200.00		135.90	165.0◖
14" diameter	Inst	Ea	UB	1.14	156.00	40.70	---	196.70	240.00		166.50	201.0◖
Minimum Job Charge	Inst	Job	UB	6.67	---	238.00	---	238.00	352.00		190.00	281.00

Wallpaper

Most wall coverings today are really not wallpapers in the technical sense. Wallpaper is a paper material (which may or not be coated with washable plastic). Most of today's products, though still called wallpaper, are actually composed of a vinyl on a fabric, not a paper, backing. Some vinyl/fabric coverings come prepasted.

Vinyl coverings with non-woven fabric, woven fabric or synthetic fiber backing are fire, mildew and fade resistant, and they can be stripped from plaster walls. Coverings with paper backing must be steamed or scraped from the walls. Vinyl coverings may also be stripped from gypsum wallboard, but unless the vinyl covering has a synthetic fiber backing, it's likely to damage the wallboard paper surface.

One gallon of paste (approximately two-thirds pound of dry paste and water) should adequately cover 12 rolls with paper backing and 6 rolls with fabric backing. The rougher the texture of the surface to be pasted, the greater the quantity of wet paste needed.

1. **Dimensions**

   a. Single roll is 36 SF. The paper is 18" wide x 24'-0" long.

   b. Double roll is 72 SF. The paper is 18" wide x 48'-0" long.

2. **Installation.** New paper may be applied over existing paper if the existing paper has butt joints, a smooth surface, and is tight to the wall. When new paper is applied direct to plaster or drywall, the wall should receive a coat of glue size before the paper is applied.

3. **Estimating Technique.** Determine the gross area, deduct openings and other areas not to be papered, and add 20% to the net area for waste. To find the number of rolls needed, divide the net area plus 20% by the number of SF per roll.

The steps in wallpapering

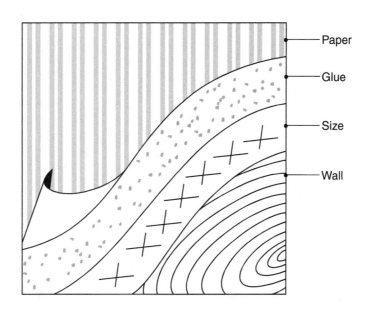

Paper — Glue — Size — Wall

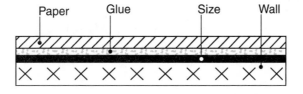

Paper   Glue   Size   Wall

Description	Oper	Unit	Crew Size	Man-Hours Per Unit	Avg Mat'l Unit Cost	Avg Labor Unit Cost	Avg Equip Unit Cost	Avg Total Unit Cost	Avg Price Incl O&P	Avg Total Unit Cost	Avg Price Incl O&P
					Costs Based On Small Volume					**Large Volume**	

Wallpaper

Butt joint

Includes application of glue sizing. Material prices are not given because they vary radically

Description	Oper	Unit	Crew Size	Man-Hours Per Unit	Avg Mat'l Unit Cost	Avg Labor Unit Cost	Avg Equip Unit Cost	Avg Total Unit Cost	Avg Price Incl O&P	Avg Total Unit Cost	Avg Price Incl O&P
Paper, single rolls, 36 SF/roll	Inst	Roll	QA	.549	---	18.70	6.36	25.06	**34.00**	16.24	**22.10**
Vinyl, with fabric backing (prepasted), single rolls, 36 SF/roll	Inst	Roll	QA	.615	---	20.90	7.13	28.03	**38.10**	18.23	**24.80**
Vinyl, with fabric backing (not prepasted), single rolls, 36 SF/roll	Inst	Roll	QA	.669	---	22.80	7.75	30.55	**41.40**	19.84	**26.90**

Labor adjustments

Note: The following percentage adjustments for Small Volume also apply to Large

ADD	Oper	Unit	Crew Size	Man-Hours Per Unit	Avg Mat'l Unit Cost	Avg Labor Unit Cost	Avg Equip Unit Cost	Avg Total Unit Cost	Avg Price Incl O&P	Avg Total Unit Cost	Avg Price Incl O&P
For ceiling 9'-0" high or more	Inst	%	QA	20.0	---	---	---	---	---	---	---
For kitchens, baths, and other rooms where floor area is less than 50.0 SF	Inst	%	QA	32.0	---	---	---	---	---	---	---

Water closets. See Toilets, page 274

Description	Oper	Unit	Costs Based On Small Volume						Large Volume	
			Crew Size	Man-Hours Per Unit	Avg Mat'l Unit Cost	Avg Labor Unit Cost	Avg Equip Unit Cost	Avg Total Unit Cost	Avg Price Incl O&P	Avg Total Unit Cost	Avg Price Incl O&P

Water heaters
(A.O. Smith Products)

Detach & reset operations

Water heater	Inst	Ea	SA	1.90	---	69.10	---	69.10	**102.00**	48.40	**71.10**

Remove operations

Water heater	Demo	Ea	SA	1.43	---	52.00	---	52.00	**76.40**	36.40	**53.40**

Replace operations
Electric
Standard model, tall
6-year Energy Saver, foam insulated

30 gal round	Inst	Ea	SA	2.86	522.00	104.00	---	626.00	**753.00**	502.70	**601.00**
40 gal round	Inst	Ea	SA	2.86	556.00	104.00	---	660.00	**792.00**	530.70	**633.00**
50 gal round	Inst	Ea	SA	2.86	624.00	104.00	---	728.00	**870.00**	586.70	**698.00**
55 gal round	Inst	Ea	SA	2.86	624.00	104.00	---	728.00	**870.00**	586.70	**698.00**
66 gal round	Inst	Ea	SA	3.81	845.00	138.00	---	983.00	**1180.00**	793.10	**943.00**
80 gal round	Inst	Ea	SA	3.81	940.00	138.00	---	1078.00	**1280.00**	871.10	**1030.00**
119 gal round	Inst	Ea	SA	3.81	1340.00	138.00	---	1478.00	**1740.00**	1197.10	**1410.00**

Standard model, short
6-year Energy Saver, foam insulated

30 gal round	Inst	Ea	SA	2.86	522.00	104.00	---	626.00	**753.00**	502.70	**601.00**
40 gal round	Inst	Ea	SA	2.86	556.00	104.00	---	660.00	**792.00**	530.70	**633.00**
50 gal round	Inst	Ea	SA	2.86	580.00	104.00	---	684.00	**819.00**	549.70	**656.00**

Standard model, low boy, top connection
6-year Energy Saver, foam insulated

6 gal round	Inst	Ea	SA	2.29	391.00	83.20	---	474.20	**572.00**	380.20	**456.00**
15 gal round	Inst	Ea	SA	2.29	417.00	83.20	---	500.20	**601.00**	401.20	**480.00**
19.9 gal round	Inst	Ea	SA	2.29	498.00	83.20	---	581.20	**695.00**	468.20	**557.00**
29 gal round	Inst	Ea	SA	2.86	546.00	104.00	---	650.00	**780.00**	521.70	**624.00**
40 gal round	Inst	Ea	SA	2.86	580.00	104.00	---	684.00	**819.00**	549.70	**656.00**
50 gal round	Inst	Ea	SA	2.86	644.00	104.00	---	748.00	**894.00**	603.70	**717.00**

Standard model, low boy, side connection
6-year Energy Saver, foam insulated

10 gal round	Inst	Ea	SA	2.29	388.00	83.20	---	471.20	**568.00**	377.20	**453.00**
15 gal round	Inst	Ea	SA	2.29	408.00	83.20	---	491.20	**592.00**	394.20	**472.00**
20 gal round	Inst	Ea	SA	2.29	442.00	83.20	---	525.20	**631.00**	422.20	**504.00**
29 gal round	Inst	Ea	SA	2.86	524.00	104.00	---	628.00	**755.00**	503.70	**603.00**
40 gal round	Inst	Ea	SA	2.86	556.00	104.00	---	660.00	**792.00**	530.70	**633.00**

Deluxe model, tall
10-year Energy Saver, foam insulated

30 gal round	Inst	Ea	SA	2.86	624.00	104.00	---	728.00	**870.00**	586.70	**698.00**
40 gal round	Inst	Ea	SA	2.86	658.00	104.00	---	762.00	**909.00**	614.70	**730.00**
50 gal round	Inst	Ea	SA	2.86	726.00	104.00	---	830.00	**988.00**	670.70	**794.00**
55 gal round	Inst	Ea	SA	2.86	726.00	104.00	---	830.00	**988.00**	670.70	**794.00**
66 gal round	Inst	Ea	SA	3.81	947.00	138.00	---	1085.00	**1290.00**	877.10	**1040.00**
80 gal round	Inst	Ea	SA	3.81	1040.00	138.00	---	1178.00	**1400.00**	955.10	**1130.00**
119 gal round	Inst	Ea	SA	3.81	1440.00	138.00	---	1578.00	**1860.00**	1287.10	**1510.00**

Description	Oper	Unit	Crew Size	Man-Hours Per Unit	Costs Based On Small Volume					Large Volume	
					Avg Mat'l Unit Cost	Avg Labor Unit Cost	Avg Equip Unit Cost	Avg Total Unit Cost	Avg Price Incl O&P	Avg Total Unit Cost	Avg Price Incl O&P
Deluxe model, short											
10-year Energy Saver, foam insulated											
30 gal round	Inst	Ea	SA	2.86	522.00	104.00	---	626.00	**753.00**	502.70	601.0♦
40 gal round	Inst	Ea	SA	2.86	556.00	104.00	---	660.00	**792.00**	530.70	633.0♦
Deluxe model, low boy, top connection											
10-year Energy Saver, foam insulated											
6 gal round	Inst	Ea	SA	2.29	493.00	83.20	---	576.20	**689.00**	464.20	552.0♦
15 gal round	Inst	Ea	SA	2.29	519.00	83.20	---	602.20	**719.00**	485.20	577.0♦
19.9 gal round	Inst	Ea	SA	2.29	600.00	83.20	---	683.20	**812.00**	552.20	654.0♦
29 gal round	Inst	Ea	SA	2.86	648.00	104.00	---	752.00	**898.00**	605.70	720.0♦
40 gal round	Inst	Ea	SA	2.86	682.00	104.00	---	786.00	**937.00**	633.70	752.0♦
50 gal round	Inst	Ea	SA	2.86	746.00	104.00	---	850.00	**1010.00**	687.70	814.0♦
Deluxe model, low boy, side connection											
10-year Energy Saver, foam insulated											
10 gal round	Inst	Ea	SA	2.29	430.00	83.20	---	513.20	**617.00**	412.20	493.0♦
15 gal round	Inst	Ea	SA	2.29	447.00	83.20	---	530.20	**637.00**	426.20	509.0♦
20 gal round	Inst	Ea	SA	2.29	476.00	83.20	---	559.20	**670.00**	450.20	536.0♦
29 gal round	Inst	Ea	SA	2.86	546.00	104.00	---	650.00	**780.00**	521.70	624.0♦
40 gal round	Inst	Ea	SA	2.86	573.00	104.00	---	677.00	**812.00**	544.70	649.0♦

Gas

Standard model, high recovery, tall

6-year Energy Saver

Description	Oper	Unit	Crew Size	Man-Hours Per Unit	Avg Mat'l Unit Cost	Avg Labor Unit Cost	Avg Equip Unit Cost	Avg Total Unit Cost	Avg Price Incl O&P	Avg Total Unit Cost	Avg Price Incl O&P
38 gal round, 50,000 btu	Inst	Ea	SA	6.15	780.00	224.00	---	1004.00	**1230.00**	804.00	976.0♦
48 gal round, 50,000 btu	Inst	Ea	SA	6.15	828.00	224.00	---	1052.00	**1280.00**	843.00	1020.0♦
50 gal round, 65,000 btu	Inst	Ea	SA	6.67	869.00	242.00	---	1111.00	**1360.00**	886.00	1070.0♦
65 gal round, 65,000 btu	Inst	Ea	SA	6.67	983.00	242.00	---	1225.00	**1490.00**	980.00	1180.0♦
74 gal round, 75,100 btu	Inst	Ea	SA	6.67	1150.00	242.00	---	1392.00	**1680.00**	1116.00	1340.0♦
98 gal round, 75,100 btu	Inst	Ea	SA	6.67	1830.00	242.00	---	2072.00	**2460.00**	1681.00	1990.0♦

Standard model, tall

6-year Energy Saver

Description	Oper	Unit	Crew Size	Man-Hours Per Unit	Avg Mat'l Unit Cost	Avg Labor Unit Cost	Avg Equip Unit Cost	Avg Total Unit Cost	Avg Price Incl O&P	Avg Total Unit Cost	Avg Price Incl O&P
28 gal round, 40,000 btu	Inst	Ea	SA	6.15	711.00	224.00	---	935.00	**1150.00**	746.00	910.0♦
38 gal round, 40,000 btu	Inst	Ea	SA	6.15	740.00	224.00	---	964.00	**1180.00**	770.00	938.0♦
50 gal round, 40,000 btu	Inst	Ea	SA	6.67	789.00	242.00	---	1031.00	**1260.00**	821.00	999.0♦

Standard model, short

6-year Energy Saver

Description	Oper	Unit	Crew Size	Man-Hours Per Unit	Avg Mat'l Unit Cost	Avg Labor Unit Cost	Avg Equip Unit Cost	Avg Total Unit Cost	Avg Price Incl O&P	Avg Total Unit Cost	Avg Price Incl O&P
30 gal round, 40,000 btu	Inst	Ea	SA	6.15	711.00	224.00	---	935.00	**1150.00**	746.00	910.0♦
40 gal round, 40,000 btu	Inst	Ea	SA	6.15	740.00	224.00	---	964.00	**1180.00**	770.00	938.0♦
50 gal round, 40,000 btu	Inst	Ea	SA	6.67	789.00	242.00	---	1031.00	**1260.00**	821.00	999.0♦

Deluxe model, high recovery, tall

10-year Energy Saver

Description	Oper	Unit	Crew Size	Man-Hours Per Unit	Avg Mat'l Unit Cost	Avg Labor Unit Cost	Avg Equip Unit Cost	Avg Total Unit Cost	Avg Price Incl O&P	Avg Total Unit Cost	Avg Price Incl O&P
38 gal round, 50,000 btu	Inst	Ea	SA	6.15	882.00	224.00	---	1106.00	**1340.00**	888.00	1070.0♦
48 gal round, 50,000 btu	Inst	Ea	SA	6.15	930.00	224.00	---	1154.00	**1400.00**	927.00	1120.0♦
50 gal round, 65,000 btu	Inst	Ea	SA	6.67	971.00	242.00	---	1213.00	**1470.00**	970.00	1170.0♦
65 gal round, 65,000 btu	Inst	Ea	SA	6.67	1000.00	242.00	---	1242.00	**1510.00**	994.00	1200.0♦
74 gal round, 75,100 btu	Inst	Ea	SA	6.67	1250.00	242.00	---	1492.00	**1790.00**	1201.00	1440.0♦
98 gal round, 75,100 btu	Inst	Ea	SA	6.67	1930.00	242.00	---	2172.00	**2580.00**	1761.00	2080.0♦

Description	Oper	Unit	Crew Size	Man-Hours Per Unit	Avg Mat'l Unit Cost	Avg Labor Unit Cost	Avg Equip Unit Cost	Avg Total Unit Cost	Avg Price Incl O&P	Avg Total Unit Cost	Avg Price Incl O&P
					Costs Based On Small Volume					Large Volume	

Deluxe model, tall
10-year Energy Saver
28 gal round, 40,000 btu	Inst	Ea	SA	6.15	813.00	224.00	---	1037.00	1260.00	830.00	1010.00
38 gal round, 40,000 btu	Inst	Ea	SA	6.15	842.00	224.00	---	1066.00	1300.00	854.00	1030.00
50 gal round, 40,000 btu	Inst	Ea	SA	6.67	891.00	242.00	---	1133.00	1380.00	905.00	1100.00

Deluxe model, short
10-year Energy Saver
30 gal round, 40,000 btu	Inst	Ea	SA	6.15	813.00	224.00	---	1037.00	1260.00	830.00	1010.00
40 gal round, 40,000 btu	Inst	Ea	SA	6.15	842.00	224.00	---	1066.00	1300.00	854.00	1030.00
50 gal round, 40,000 btu	Inst	Ea	SA	6.67	891.00	242.00	---	1133.00	1380.00	905.00	1100.00

Optional accessories
Water heater cabinets

24" x 24" x 72"											
heaters up to 40 gal	---	Ea	---	---	80.80	---	---	80.80	92.90	66.50	76.50
30" x 30" x 72"											
heaters up to 75 gal	---	Ea	---	---	89.30	---	---	89.30	103.00	73.50	84.50
36" x 36" x 83"											
heaters up to 100 gal	---	Ea	---	---	149.00	---	---	149.00	171.00	123.00	141.00
Water heater pans, by diameter											
20" pan (up to 18" unit)	---	Ea	---	---	17.70	---	---	17.70	20.30	14.60	16.70
24" pan (up to 22" unit)	---	Ea	---	---	21.10	---	---	21.10	24.20	17.40	20.00
26" pan (up to 24" unit)	---	Ea	---	---	25.70	---	---	25.70	29.50	21.10	24.30
28" pan (up to 26" unit)	---	Ea	---	---	29.80	---	---	29.80	34.20	24.50	28.20
Water heater stands											
22" x 22", heaters up to 52 gal	---	Ea	---	---	59.50	---	---	59.50	68.40	49.00	56.40
26" x 26", heaters up to 75 gal	---	Ea	---	---	83.30	---	---	83.30	95.80	68.60	78.90
Free-standing restraint system											
75 gal capacity	---	Ea	---	---	123.00	---	---	123.00	142.00	102.00	117.00
100 gal capacity	---	Ea	---	---	140.00	---	---	140.00	161.00	116.00	133.00
Wall mount platform restraint system											
75 gal capacity	---	Ea	---	---	115.00	---	---	115.00	132.00	94.50	109.00
100 gal capacity	---	Ea	---	---	132.00	---	---	132.00	152.00	109.00	125.00

Solar water heating systems

Complete closed loop solar system with solar electric water heater with exchanger, differential control, heating element, circulator, collector and tank, and panel sensors. The material costs given include the subcontractor's overhead and profit

82-gallon capacity collector
One deluxe collector	---	LS	---	---	3060.00	---	---	3060.00	---	2680.00	---
Two deluxe collectors											
Standard collectors	---	LS	---	---	3790.00	---	---	3790.00	---	3310.00	---
Deluxe collectors	---	LS	---	---	3900.00	---	---	3900.00	---	3410.00	---
Three collectors											
Economy collectors	---	LS	---	---	4450.00	---	---	4450.00	---	3890.00	---
Standard collectors	---	LS	---	---	4560.00	---	---	4560.00	---	3990.00	---
Four collectors											
Economy collectors	---	LS	---	---	4800.00	---	---	4800.00	---	4200.00	---
Standard collectors	---	LS	---	---	4920.00	---	---	4920.00	---	4310.00	---

				Costs Based On Small Volume					Large Volume	
			Man-Hours							
Description	Oper	Unit	Crew Size	Per Unit	Avg Mat'l Unit Cost	Avg Labor Unit Cost	Avg Equip Unit Cost	Avg Total Unit Cost	Avg Price Incl O&P	Avg Total Unit Cost	Avg Price Incl O&P

120-gallon capacity system

Three collectors

Description	Oper	Unit	Crew Size	Per Unit	Avg Mat'l	Avg Labor	Avg Equip	Avg Total	Avg Price	Avg Total	Avg Price
Economy collectors	---	LS	---	---	5510.00	---	---	5510.00	---	4820.00	--
Standard collectors	---	LS	---	---	5640.00	---	---	5640.00	---	4940.00	--
Four standard collectors	---	LS	---	---	6600.00	---	---	6600.00	---	5780.00	--
Five collectors											
Economy collectors	---	LS	---	---	7200.00	---	---	7200.00	---	6300.00	--
Standard collectors	---	LS	---	---	7320.00	---	---	7320.00	---	6410.00	--
Six collectors											
Economy collectors	---	LS	---	---	7800.00	---	---	7800.00	---	6830.00	--
Standard collectors	---	LS	---	---	7920.00	---	---	7920.00	---	6930.00	--

Material adjustments

Collector mounting kits, one required per collector panel

Description	Oper	Unit	Crew Size	Per Unit	Avg Mat'l	Avg Labor	Avg Equip	Avg Total	Avg Price	Avg Total	Avg Price
Adjustable position hinge	---	Ea	---	---	68.40	---	---	68.40	---	59.90	--
Integral flange	---	Ea	---	---	6.00	---	---	6.00	---	5.25	--
Additional panels											
Economy collectors	---	Ea	---	---	600.00	---	---	600.00	---	525.00	--
Standard collectors	---	Ea	---	---	720.00	---	---	720.00	---	630.00	--
Deluxe collectors	---	Ea	---	---	840.00	---	---	840.00	---	735.00	--

Individual components

Solar storage tanks, glass lined with fiberglass insulation

Description	Oper	Unit	Crew Size	Per Unit	Avg Mat'l	Avg Labor	Avg Equip	Avg Total	Avg Price	Avg Total	Avg Price
66 gallon	---	Ea	---	---	714.00	---	---	714.00	---	625.00	--
82 gallon	---	Ea	---	---	810.00	---	---	810.00	---	709.00	--
120 gallon	---	Ea	---	---	1230.00	---	---	1230.00	---	1080.00	--

Solar electric water heaters, glass lined with fiberglass insulation, heating element with thermostat

Description	Oper	Unit	Crew Size	Per Unit	Avg Mat'l	Avg Labor	Avg Equip	Avg Total	Avg Price	Avg Total	Avg Price
66 gallon, 4.5 KW	---	Ea	---	---	713.00	---	---	713.00	---	624.00	--
82 gallon, 4.5 KW	---	Ea	---	---	820.00	---	---	820.00	---	717.00	--
120 gallon, 4.5 KW	---	Ea	---	---	977.00	---	---	977.00	---	855.00	--
Additional element	---	Ea	---	---	39.60	---	---	39.60	---	34.70	--

Solar electric water heaters with heat exchangers, glass lined with fiberglass insulation, two copper exchangers, powered circulator, differential control, adjustable thermostat

Description	Oper	Unit	Crew Size	Per Unit	Avg Mat'l	Avg Labor	Avg Equip	Avg Total	Avg Price	Avg Total	Avg Price
82 gallon, 4.5 KW, 1/20 HP	---	Ea	---	---	1070.00	---	---	1070.00	---	932.00	--
120 gallon, 4.5 KW, 1/20 HP	---	Ea	---	---	1400.00	---	---	1400.00	---	1220.00	--

Water softeners

(Bruner/Calgon)

Automatic water softeners, complete with yoke with 3/4" I.P.S. supply

Description	Oper	Unit	Costs Based On Small Volume						Large Volume	
			Crew Size	Man-Hours Per Unit	Avg Mat'l Unit Cost	Avg Labor Unit Cost	Avg Equip Unit Cost	Avg Total Unit Cost	Avg Price Incl O&P	Avg Total Unit Cost	Avg Price Incl O&P

Single-tank units

Description	Oper	Unit	Crew Size	Man-Hrs Per Unit	Avg Mat'l Unit Cost	Avg Labor Unit Cost	Avg Equip Unit Cost	Avg Total Unit Cost	Avg Price Incl O&P	Avg Total Unit Cost	Avg Price Incl O&P
8,000 grain exchange capacity, 160 lbs salt storage, 6 gpm	---	Ea	---	---	509.00	---	---	509.00	**585.00**	396.00	**455.00**
15,000 grain exchange capacity, 200 lbs salt storage, 8.8 gpm	---	Ea	---	---	559.00	---	---	559.00	**643.00**	436.00	**501.00**
25,000 grain exchange capacity, 175 lbs salt storage, 11.3 gpm	---	Ea	---	---	780.00	---	---	780.00	**897.00**	607.00	**698.00**

Two-tank units, side-by-side

Description	Oper	Unit	Crew Size	Man-Hrs Per Unit	Avg Mat'l Unit Cost	Avg Labor Unit Cost	Avg Equip Unit Cost	Avg Total Unit Cost	Avg Price Incl O&P	Avg Total Unit Cost	Avg Price Incl O&P
30,000 grain exchange capacity, 1.0 CF	---	Ea	---	---	893.00	---	---	893.00	**1030.00**	695.00	**799.00**
45,000 grain exchange capacity, 1.5 CF	---	Ea	---	---	1010.00	---	---	1010.00	**1160.00**	783.00	**901.00**
60,000 grain exchange capacity, 2.0 CF	---	Ea	---	---	1460.00	---	---	1460.00	**1680.00**	1140.00	**1310.00**
90,000 grain exchange capacity, 3.0 CF	---	Ea	---	---	1570.00	---	---	1570.00	**1810.00**	1220.00	**1410.00**

Automatic water filters, complete with yoke and media with 3/4" I.P.S. supply

Two-tank units, side-by-side

Description	Oper	Unit	Crew Size	Man-Hrs Per Unit	Avg Mat'l Unit Cost	Avg Labor Unit Cost	Avg Equip Unit Cost	Avg Total Unit Cost	Avg Price Incl O&P	Avg Total Unit Cost	Avg Price Incl O&P
Eliminate rotten egg odor and rust	---	Ea	---	---	618.00	---	---	618.00	**711.00**	481.00	**554.00**
Eliminate chlorine taste	---	Ea	---	---	520.00	---	---	520.00	**598.00**	405.00	**466.00**
Clear up water discoloration	---	Ea	---	---	466.00	---	---	466.00	**535.00**	363.00	**417.00**
Clear up corroding pipes	---	Ea	---	---	484.00	---	---	484.00	**556.00**	377.00	**433.00**

Manual water filters, complete fiberglass tank with 3/4" pipe supplies with mineral packs

Description	Oper	Unit	Crew Size	Man-Hrs Per Unit	Avg Mat'l Unit Cost	Avg Labor Unit Cost	Avg Equip Unit Cost	Avg Total Unit Cost	Avg Price Incl O&P	Avg Total Unit Cost	Avg Price Incl O&P
Eliminate taste and odor	---	Ea	---	---	461.00	---	---	461.00	**530.00**	359.00	**413.00**
Eliminate acid water	---	Ea	---	---	392.00	---	---	392.00	**451.00**	305.00	**351.00**
Eliminate iron in solution	---	Ea	---	---	475.00	---	---	475.00	**546.00**	370.00	**425.00**
Eliminate sediment	---	Ea	---	---	392.00	---	---	392.00	**451.00**	305.00	**351.00**

Chemical feed pumps

Description	Oper	Unit	Crew Size	Man-Hrs Per Unit	Avg Mat'l Unit Cost	Avg Labor Unit Cost	Avg Equip Unit Cost	Avg Total Unit Cost	Avg Price Incl O&P	Avg Total Unit Cost	Avg Price Incl O&P
115 volt, 9 gal/day	---	Ea	---	---	273.00	---	---	273.00	**314.00**	213.00	**245.00**
230 volt, 9 gal/day	---	Ea	---	---	313.00	---	---	313.00	**360.00**	244.00	**280.00**

Weathervanes. See Cupolas, page 76

Windows

Description	Oper	Unit	Crew Size	Man-Hours Per Unit	Avg Mat'l Unit Cost	Avg Labor Unit Cost	Avg Equip Unit Cost	Avg Total Unit Cost	Avg Price Incl O&P	Avg Total Unit Cost	Avg Price Incl O&P
										Large Volume	

Windows, with related trim and frame

To 12 SF

Description	Oper	Unit	Crew	MH	Mat'l	Labor	Equip	Total	Price O&P	LV Total	LV Price
Aluminum	Demo Ea	LB		1.17	---	31.20	---	31.20	**46.40**	20.30	30.20
Wood	Demo Ea	LB		1.54	---	41.00	---	41.00	**61.10**	26.60	39.70
13 SF to 50 SF											
Aluminum	Demo Ea	LB		1.88	---	50.10	---	50.10	**74.60**	32.80	48.80
Wood	Demo Ea	LB		2.46	---	65.50	---	65.50	**97.60**	42.60	63.50

Aluminum

Vertical slide, satin anodized finish, includes screen and hardware

Single glazed

Description	Oper	Unit	Crew	MH	Mat'l	Labor	Equip	Total	Price O&P	LV Total	LV Price
1'-6" x 3'-0" H	Inst Set	CA		1.74	89.10	55.80	---	144.90	**186.00**	119.10	150.00
2'-0" x 2'-0" H	Inst Set	CA		1.74	81.00	55.80	---	136.80	**177.00**	111.60	141.00
2'-0" x 2'-6" H	Inst Set	CA		1.74	91.80	55.80	---	147.60	**189.00**	121.60	153.00
2'-0" x 3'-0" H	Inst Set	CA		1.90	98.60	61.00	---	159.60	**205.00**	130.80	164.00
2'-0" x 3'-6" H	Inst Set	CA		1.90	105.00	61.00	---	166.00	**213.00**	137.00	171.00
2'-0" x 4'-0" H	Inst Set	CA		1.90	111.00	61.00	---	172.00	**219.00**	142.50	177.00
2'-0" x 4'-6" H	Inst Set	CA		1.90	119.00	61.00	---	180.00	**228.00**	149.50	186.00
2'-0" x 5'-0" H	Inst Set	CA		2.05	124.00	65.80	---	189.80	**242.00**	157.70	196.00
2'-0" x 6'-0" H	Inst Set	CA		2.05	136.00	65.80	---	201.80	**255.00**	168.70	209.00
2'-6" x 3'-0" H	Inst Set	CA		1.90	108.00	61.00	---	169.00	**216.00**	139.50	174.00
2'-6" x 3'-6" H	Inst Set	CA		1.90	115.00	61.00	---	176.00	**223.00**	145.50	181.00
2'-6" x 4'-0" H	Inst Set	CA		1.90	122.00	61.00	---	183.00	**231.00**	152.50	189.00
2'-6" x 4'-6" H	Inst Set	CA		1.90	131.00	61.00	---	192.00	**242.00**	160.50	199.00
2'-6" x 5'-0" H	Inst Set	CA		2.05	136.00	65.80	---	201.80	**255.00**	168.70	209.00
2'-6" x 6'-0" H	Inst Set	CA		2.05	150.00	65.80	---	215.80	**271.00**	181.70	224.00
3'-0" x 2'-0" H	Inst Set	CA		1.90	104.00	61.00	---	165.00	**211.00**	135.80	170.00
3'-0" x 3'-0" H	Inst Set	CA		1.90	116.00	61.00	---	177.00	**225.00**	147.50	183.00
3'-0" x 3'-6" H	Inst Set	CA		1.90	124.00	61.00	---	185.00	**234.00**	154.50	191.00
3'-0" x 4'-0" H	Inst Set	CA		1.90	132.00	61.00	---	193.00	**244.00**	162.50	200.00
3'-0" x 4'-6" H	Inst Set	CA		1.90	143.00	61.00	---	204.00	**256.00**	172.50	212.00
3'-0" x 5'-0" H	Inst Set	CJ		4.21	149.00	124.00	---	273.00	**356.00**	219.10	280.00
3'-0" x 6'-0" H	Inst Set	CJ		4.71	163.00	138.00	---	301.00	**395.00**	241.50	310.00
3'-6" x 3'-6" H	Inst Set	CA		1.90	134.00	61.00	---	195.00	**245.00**	163.50	202.00
3'-6" x 4'-0" H	Inst Set	CA		2.22	143.00	71.20	---	214.20	**271.00**	179.50	222.00
3'-6" x 4'-6" H	Inst Set	CA		2.76	157.00	88.60	---	245.60	**313.00**	202.10	252.00
3'-6" x 5'-0" H	Inst Set	CJ		4.71	161.00	138.00	---	299.00	**392.00**	239.50	307.00
4'-0" x 3'-0" H	Inst Set	CA		1.90	134.00	61.00	---	195.00	**245.00**	163.50	202.00
4'-0" x 3'-6" H	Inst Set	CA		2.22	143.00	71.20	---	214.20	**271.00**	179.50	222.00
4'-0" x 4'-0" H	Inst Set	CJ		4.21	153.00	124.00	---	277.00	**361.00**	222.10	284.00
4'-0" x 4'-6" H	Inst Set	CJ		4.21	162.00	124.00	---	286.00	**372.00**	231.10	294.00
4'-0" x 5'-0" H	Inst Set	CJ		4.71	173.00	138.00	---	311.00	**406.00**	250.50	320.00

Craftsman Book Company
6058 Corte del Cedro
P.O. Box 6500
Carlsbad, CA 92018

Download all of Craftsman's Cost Databases for one low price with the Craftsman Site License.
http://www.craftsmansitelicense.com

24 hour order line
1-800-829-8123 Fax (760) 438-0398
Order online: http://www.craftsman-book.com

Name _____

Company _____

Address _____

City/State/Zip _____

Total enclosed _____ (In California add 7.25% tax)

☐ Send check or money order and we pay postage, or use your
☐ Visa ☐ MasterCard ☐ Discover or ☐ Amex

Card# _____

Expiration Date _____ Initials _____

E-Mail Address _____
(for tracking and special offers)

10-Day Money Back Guarantee

☐ 36.50 Basic Engineering for Builders
☐ 38.00 Basic Lumber Engineering for Builders
☐ 35.00 Building Contractor's Exam Preparation Guide
☐ 78.50 CD Estimator
☐ 29.95 Concrete Countertops
☐ 51.50 Contractor's Guide to QuickBooks Pro '06
☐ 39.50 Construction Estimating Reference Data with FREE stand-alone *Windows* estimating program on a CD-ROM.
☐ 41.75 Construction Forms & Contracts with a CD-ROM for both *Windows* and *Mac* .
☐ 39.00 Contractor's Guide to 1997 Building Code Revised
☐ 20.00 Contractor's Index to the 1997 UBC (loose-leaf)
☐ 65.00 Craftsman's Constr. Installation Encyclopedia
☐ 34.95 Drafting House Plans
☐ 39.50 Estimating Excavation
☐ 17.00 Estimating Home Building Costs
☐ 42.00 Excavation & Grading Handbook Revised
☐ 22.50 Finish Carpenter's Manual
☐ 39.00 Getting Financing & Devloping Land
☐ 28.00 National Building Cost Manual

☐ 54.00 National Concrete & Masonry Estimator with FREE stand-alone *Windows* estimating program on CD ROM.
☐ 52.50 National Construction Estimator with FREE stand-alone *Windows* estimating program on CD ROM.
☐ 52.75 National Electrical Estimator with FREE stand-alone *Windows* estimating program on CD ROM.
☐ 52.25 National Framing & Finish Carpentry Estimator with FREE stand-alone *Windows* estimating program on CD ROM.
☐ 53.75 National Home Improvement Estimator with FREE stand-alone *Windows* estimating program on CD ROM.
☐ 54.50 National Renovation & Insurance Repair Estimator with FREE stand-alone *Windows* estimating program on CD ROM.
☐ 53.50 National Repair & Remodeling Estimator with FREE stand-alone *Windows* estimating program on CD ROM.
☐ 28.25 Plumbing & HVAC Manhour Estimates
☐ 38.00 Roofing Construction & Estimating
☐ 32.50 Roof Cutter's Secrets
☐ 26.50 Rough Framing Carpentry
☐ **FREE Full Color Catalog**

Receive Fresh Cost Data Every Year – Automatically

Join Craftsman's Standing Order Club and automatically receive special membership discounts on annual cost books!

Qty.____ National Renovation & Ins. Repair Est.
Standing Order price $46.33

Qty.____ CD Estimator
Standing Order price $66.73

Qty. National Build. Cost Manual	Qty. National Const. Estimator	Qty. National Elect. Estimator	Qty. National Framing & Fin. Carp. Estimator	Qty. National Home Improv. Estimator	Qty. National Plumb. & HVAC Estimator	Qty. National Painting Cost Estimator	Qty. National Concrete & Masonry Est.	Qty. National Repair & Remodeling Est.
Standing Order price $23.80	Standing Order price $44.63	Standing Order price $44.84	Standing Order price $44.37		Standing Order price $45.26	Standing Order price $45.26	Standing Order price $45.05	Standing Order price $45.90	Standing Order price $45.48

How many times have you missed one of Craftsman's pre-publication discounts, or procrastinated about buying the updated book and ended up bidding your jobs using obsolete cost data that you "updated" on your own? As a Standing Order Member you never have to worry about ordering every year. Instead, you will receive a confirmation of your standing order each September. If the order is correct, do nothing, and your order will be shipped and billed as listed on the confirmation card. If you wish to change your address, or your order, you can do so at that time. Your standing order for the books you selected will be shipped as soon as the publications are printed. You'll automatically receive any pre-publication discount price.

Bill to:
Name _____
Company _____
Address _____
City/State/Zip _____

Ship to:
Name _____
Company _____
Address _____
City/State/Zip _____

Purchase order# _____
Phone # (____) _____
Signature _____

Mail This Card Today for a Free Full Color Catalog

Over 100 construction references at your fingertips with information that can save you time and money. Here you'll find information on carpentry, contracting, estimating, remodeling, electrical work and plumbing.

All items come with an unconditional 10-day money-back guarantee. If they don't save you money, mail them back for a full refund.

Name _____

Company _____

Address _____

City/State/Zip _____

Craftsman Book Company
6058 Corte del Cedro
P.O. Box 6500 • Carlsbad, CA 92018

BUSINESS REPLY MAIL

FIRST CLASS MAIL PERMIT NO. 271 CARLSBAD, CA

POSTAGE WILL BE PAID BY ADDRESSEE

Craftsman Book Company

6058 Corte del Cedro
P.O. Box 6500
Carlsbad, CA 92018-9974

BUSINESS REPLY MAIL

FIRST CLASS MAIL PERMIT NO. 271 CARLSBAD, CA

POSTAGE WILL BE PAID BY ADDRESSEE

Craftsman Book Company

6058 Corte del Cedro
P.O. Box 6500
Carlsbad, CA 92018-9974

BUSINESS REPLY MAIL

FIRST CLASS MAIL PERMIT NO. 271 CARLSBAD, CA

POSTAGE WILL BE PAID BY ADDRESSEE

Craftsman Book Company

6058 Corte del Cedro
P.O. Box 6500
Carlsbad, CA 92018-9974